Flows of Reactive Fluids

FLUID MECHANICS AND ITS APPLICATIONS
Volume 94

Series Editor: R. MOREAU
MADYLAM
Ecole Nationale Supérieure d'Hydraulique de Grenoble
Boîte Postale 95
38402 Saint Martin d'Hères Cedex, France

Aims and Scope of the Series

The purpose of this series is to focus on subjects in which fluid mechanics plays a fundamental role.

As well as the more traditional applications of aeronautics, hydraulics, heat and mass transfer etc., books will be published dealing with topics which are currently in a state of rapid development, such as turbulence, suspensions and multiphase fluids, super and hypersonic flows and numerical modeling techniques.

It is a widely held view that it is the interdisciplinary subjects that will receive intense scientific attention, bringing them to the forefront of technological advancement. Fluids have the ability to transport matter and its properties as well as to transmit force, therefore fluid mechanics is a subject that is particularly open to cross fertilization with other sciences and disciplines of engineering. The subject of fluid mechanics will be highly relevant in domains such as chemical, metallurgical, biological and ecological engineering. This series is particularly open to such new multidisciplinary domains.

The median level of presentation is the first year graduate student. Some texts are monographs defining the current state of a field; others are accessible to final year undergraduates; but essentially the emphasis is on readability and clarity.

For other titles published in this series, go to
www.springer.com/series/5980

Roger Prud'homme

Flows of Reactive Fluids

Springer

Roger Prud'homme
Institut Jean Le Rond d'Alembert
Université Pierre et Marie Curie and CNRS
75252 Paris Cedex 05
France
roger.prud_homme@upmc.fr

ISSN 0926-5112
ISBN 978-0-8176-4518-2 e-ISBN 978-0-8176-4659-2
DOI 10.1007/978-0-8176-4659-2
Springer New York Dordrecht Heidelberg London

Library of Congress Control Number: 2010930226

Mathematics Subject Classification (2010): 76Fxx, 76G25, 76H05, 76J20, 76L05, 76Nxx, 76Txx, 76Vxx, 80Axx, 80A25, 80A32

Printed on acid-free paper

Springer is part of Springer Science+Business Media (www.springer.com)

To Hayat.

In memory of Marcel and Simone Barrère.

Preface

Reactive fluids are present in many situations of great importance, such as in combustion chambers or around spacecraft re-entering the atmosphere. Analyzing the flow properties of such fluids represents one of the most difficult challenges to current technology. Indeed, all of the most difficult aspects of fluid mechanics appear to be grouped together in this research field! Such fluids are complex mixtures with compositions that vary rapidly in time and space. They are not usually at thermodynamic equilibrium, since the reaction times of the chemical reactions involved may not be negligible in comparison with the transit time of the fluid. However, the author of this book limits its scope to typical phenomena that are not very far from local equilibrium but can nevertheless exhibit the most important types of irreversible processes. The production of entropy is highly dependent on the chemical reaction pathway, which is difficult to simplify. Also, most of the classical problems that characterize fluid mechanics—such as turbulence, the presence of thin boundary layers or shear layers, and the propagation of acoustic waves and shock waves—are also present, and are much more difficult to analyze and describe than they are for homogeneous fluids, because reactive mixtures interact with these phenomena. For example, density is highly dependent on the chemical pathway since it is determined by the local and instantaneous production of chemical species, and so its value affects many other quantities through the equation of state and the balances of mass, momentum, and energy.

This book is a remarkable and quite pedagogical synthesis. It presents all of these problems in a logical and systematic way, step by step in the different chapters, without going into the complexities of the very many particular classes of them. Indeed, the author pays more attention to general concepts and to guiding ideas (thus justifying the formulation of a very general theory) than to combining them for particular applications. To be able to follow the text, the reader should understand mathematics to the level required for most graduate courses in fluid mechanics. This involves a knowledge of classical techniques such as multiple-scale analysis and matched asymptotic expansions, but without the need to dwell on their mathematical justification.

Dimensional analysis is proposed as a systematic and powerful tool for reducing the general set of equations to the relevant formulation for a given class of phenomena, and a list of the most important nondimensional numbers is given.

It was a true pleasure to look through this text, noting its good organization and reading some of the chapters and paragraphs in depth. I am convinced that readers of the book—it is aimed at graduate students as well as experienced scientists or engineers—will find an abundance of useful resources, and some will definitely keep it on their desk.

René Moreau
Professeur Émérite à Grenoble INP

Acknowledgments

I am indebted to all the people who have, in several ways, contributed to the final realization of this book: Professors Marcel Barrère, Paul Germain and Luigi Napolitano, who greatly contributed to my training in continuum mechanics and reactive fluids; Professor Nicola Bellomo for having proposed this book for publication by Springer/Birkhaüser; Tom Grasso, Editor at Birkhaüser Boston, who managed efficiently the realization of this book; and Professor René Moreau, who wrote the preface for this book and suggested publication in the *Fluid Mechanics and its Applications* (FMIA) series.

Early versions of the manuscript were critically read by Jean-Sylvestre Darrozès, Renée Gatignol, Paul Kuentzmann, Gérard Maugin and Pierre Sagaut, who made pertinent suggestions. I am also indebted to colleagues from the Université Pierre et Marie Curie (UPMC), the National Center for Scientific Research (CNRS), and the French Aerospace Lab (ONERA), including Nicolas Bertier, Francis Dupoirieux, Guillaume Legros, Lionel Matuszewski, Yves Mauriot, and Angela Vincenti, who kindly agreed to read some of the chapters in order to improve the English scientific terminology, as well as Cédric Croizet, Daniel Fruman and Denis Sébart, who helped in other ways.

I extend my appreciation to people who have granted me permission to reprint illustrations and to Christian Peak, for his copyediting work and English revisions to the final edition. My thanks also go to the directors of the Institute Jean le Rond d'Alembert (CNRS/UPMC) and ONERA/DEFA for their encouragement.

Roger Prud'homme

List of Symbols

Latin Symbols

a, b: van der Waals coefficients
a_j: activity of species j
A: chemical affinity
$\mathcal{A}$: adiabatic transformation
$\mathcal{A}_l$: chemical element
B: Spalding parameter, or Arrhenius coefficient
B, C, C': virial coefficients
B_{kj}: virial coefficients of a mixture
Bo: Bond number
c: velocity of sound, or molecular velocity
C: total number of moles per unit volume
C_f: friction coefficient
C_j: molar concentration per unit volume
C_p, C_v: specific heat at constant pressure or volume, respectively
Cr: crispation number
d: droplet diameter, molecular diameter, or distance
D: diffusion coefficient
$\mathbf{D}$: strain rate tensor
Da: first Damköhler parameter
D_T: thermal diffusion coefficient
e: internal energy per unit mass
E: internal energy
$\mathcal{E}$: equilibrium state
E_a: activation energy
$E(k)$: energy spectrum of turbulence

$E(t_e)$: distribution of residence times

$\mathcal{E}_j$: chemical species

f: undefined quantity, Helmholtz free energy per unit mass, or Blasius function

$\mathbf{f}$: force acting on a unit mass

$\mathbf{f}_j$: force acting on a unit mass of species j

F: Helmholtz free energy, or generalized force

$\mathbf{F}$: force

Fr: Froude number

g: Gibbs free enthalpy per unit mass, or the common value of λ/c_{pf} and ρD used in the study of droplet combustion when the Lewis number Le is equal to one

$g(\xi, \zeta)$: function of space variables in the Emmons problem

$g(U)$: function of the reduced velocity in the study of the planar detonation wave structure

$\mathbf{g}$: shear-force vector

G: Gibbs free enthalpy, reduced gradient of stagnation temperature in a detonation wave, filter function for wavenumbers, or production rate of entities per unit volume

G_C: concentration gradient

Gr: Grashov number

G_T: temperature gradient

h: enthalpy per unit mass

H: enthalpy

Hi Hickman number

$\mathbf{1}$: unit tensor

J: generalized flux

$\mathbf{J}$: total mass flux

$\mathbf{J}_j$: flux of species j

$\mathcal{J}_{Dj}$: diffusion flux of species j

k: Boltzmann constant, bulk viscosity, wavenumber, or kinetic energy per unit mass

K: kinetic energy, compressibility, number of chemical reactions in a mixture, or wavenumber

k_c: cut-off wavenumber

$k(T)$: specific reaction rate

K_a, K_C, K_p: equilibrium constants for activity, concentration, partial pressure, respectively

K_F: coefficient of turbulent exchange for quantity F

l: latent heat per unit mass, or mean free path

ℓ: integral scale of turbulence

ℓ_f: laminar premixed flame thickness

ℓ_K: microscale of Kolmogorov

ℓ_δ: reaction thickness in a laminar premixed flame

L: length, liquid, molar latent heat, number of chemical elements in a mixture, or crystal size

Le: Lewis number, or phenomenological coefficients

Lp: Prandtl mixing length

m: total mass

M: molecular mass

$\mathcal{M}$: mean molar mass

$\mathbf{M}$: bending stress tensor

m_j: mass of species j

$\mathcal{M}_j$: molar mass of species j

$\dot{m}$: mass flow rate

Ma: Bénard–Marangoni number

n: total number of moles

n_j: number of moles of species j

N: number of species, or number of molecules per unit volume

$\mathbf{N}$: unit normal vector to an interface

Nu: Nusselt number

p: thermodynamic pressure, or probability density

$\mathbf{P}$: pressure tensor

$\mathcal{P}$: mechanical power

Pr: Prandtl number

Pe: Peclet number

q: parameter, or heat flux

$\mathbf{q}$: heat flux vector

$\mathbf{q}'$: heat flux vector due to temperature gradient

$\dot{q}$: volume flow rate

Q: heat quantity

$\mathcal{Q}$: caloric power

$(q_f^0)_j$: enthalpy of formation per unit mass of species j

$(Q_f^0)_j = (H_0^0)_j$: molar enthalpy of formation of species j

r: perfect gas constant per unit mass, radius, or caloric power received per unit volume

R: universal molar gas constant, radius, or number of independent species in a mixture

r_N: rate of nucleation
$r_{T,p}$: energy of a chemical reaction
Ra: Rayleigh number
Re: Reynolds number
Ri: Richardson number
R_j: mass of j produced by a chemical reaction
R_r: chemical reaction

s: entropy per unit mass, or Arrhenius exponent
S: entropy
$\mathcal{S}$: surface
Sc: Schmidt number
Sh: Sherwood number
Sr: Strouhal number
St: Stanton number

t: time
T: absolute temperature
t_e: residence time
Ta: Taylor number
$T(k)$: turbulent transfer function

u, v, w: components of the velocity $\mathbf{v}$ with respect to x, y, z
u, v: coefficients in the equation of state for a real gas
U, U_∞: reference velocity
$\mathbf{v}$: barycentric velocity vector $\sum_{j=1}^{N} \mathbf{v}_j$
$\mathbf{v}_j$: velocity vector of species j
V: velocity, or force potential
$\mathbf{V}$: undefined vector, or composite velocity vector defined for the interface $\mathbf{V} = \mathbf{V}_{//} + \mathbf{w}\mathbf{N}$
$\mathcal{V}$: volume
$(\mathcal{V})$: manifold of equilibrium states in the thermodynamic space
Vi: surface viscosity number
$\mathbf{V}_j$: diffusion velocity of species $= \mathbf{v}_j - \mathbf{v}$
v_r, v_θ, v_z: components of $\mathbf{v}$ in cylindrical coordinates

w: normal velocity of a surface, or velocity vector in phase space
W: work
$\mathbf{W}$: local velocity vector of a discontinuity, or velocity vector of a fictitious motion
We: Weber number
$\dot{W}_F$: rate of production of quantity F

x, y, z: Cartesian coordinates
$\mathbf{x}$: position vector

X, Y, Z: coordinates in a relative frame

X_j: molar fraction of species j

Y_j: mass fraction of species j

Z: transfer function

Greek Symbols

α: coefficient of heat exchange, coefficient of dilatation, universal exponent for C_v, or constant in the linearized theory of droplet vaporization

$-\beta$: temperature gradient in a fluid layer

β_j: reduced concentration

β_{lj}: number of atoms of l in the jth molecule

β_T: reduced temperature

χ: compressibility coefficient

δ: boundary layer thickness, infinitesimal difference, or universal exponent along critical isotherm

Δ: difference, or Laplacian

ϵ: small parameter, or turbulent dissipation rate

ϕ: velocity potential, or pre-exponential factor of ψ

$\boldsymbol{\phi}$: linear density of torques

$\boldsymbol{\varphi}$: surface density of torques

φ_j: partial molar quantity associated with the quantity φ

$\Delta\varphi_m$: mixture quantity associated with φ

γ: isentropic coefficient c_p/c_v, or universal exponent for isothermal compressibility

Γ: circulation of the velocity vector

η: partial bulk viscosity, or reduced coordinate

κ: thermal diffusivity $\lambda/\rho c_p$, or mean curvature of flame surface

λ: coefficient of thermal conductivity, or scale factor

Λ: coefficient of load loss, or heat exchange coefficient

μ: coefficient of shear viscosity, or Gibbs free energy per mole

ν: kinematic viscosity, or universal critical exponent for the correlation length

ν_c: collision frequency

ν_j: algebraic stoichiometric coefficient $\nu_j = \nu_j'' - \nu_j'$

ν_j': stoichiometric coefficient of the direct reaction

ν_j'': stoichiometric coefficient of the reverse reaction

π: average normal pressure

Π_i: dimensionless ratio of physical quantities

θ: temperature, or angular coordinate

ϑ: volume per unit mass (inverse of density)

Θ: reduced stagnation temperature in a detonation wave

ρ: density

ρ_F: density of property F

ρ_j: partial density

σ: surface tension

$\boldsymbol{\sigma}$: surface tension tensor, or membrane stress tensor

Σ: surface area

$\mathbf{\Sigma}$: stress tensor

σ_T: temperature derivative of surface tension

τ: characteristic time, reduced enthalpy of a chemical reaction, or residence time

τ_T: thermal diffusion time

τ_v: lifetime of a vaporizing droplet

ω: rotation velocity, or pulsation of an oscillatory wave

$\boldsymbol{\omega}$: rotation vector

Ω: rotation velocity

ξ: progress variable per unit mass, reduced coordinate, or correlation length

ψ: stream function, or entity number in phase space

ζ: progress variable per unit volume, or reduced variable

$\boldsymbol{\zeta}$: position vector in phase space

$\dot{\zeta}$: production rate for a chemical reaction

Subscripts, Superscripts, and Other Symbols

$_a$: interface

$_c$: concentration

$_C$: critical point

$_{chem}$: chemical

D: direct, dissociation, or diffusive

$_e$: equilibrium flow, external, or exit section

$_f$: frozen flow, or flame

G: gas

$_i$, $_j$: species

$_i$: internal, or irreversible

$_L$: local

$_{NL}$: nonlocal

$_m$: mixture, or mass

$_{mech}$: mechanic
$_{O}$: opening section
$_{p}$: at constant pressure
$_{\mathcal{P}}$: partition wall
$_{r}$: chemical reaction, reference, or chemical reactor
$_{R}$: reverse, or recombination
$_{s}$: steady state, or surface
$_{S}$: surface
$_{t}$: turbulent
$_{T}$: temperature, or at constant temperature
$_{th}$: thermal
$_{V}$: vapor
$_{//}$: parallel to a surface
$_{\perp}$: normal to a surface
^{g}: gas
^{l}: liquid
$_{S}$: interface
^{S}: scalar
^{V}: vector
^{T}: second-order tensor
^{0}: reference quantity
$^{\bullet}$: pure simple substance
$\dot{()}$: per unit time, or for a production rate
$\bar{()}$: thermodynamic quantity per unit mole, or average quantity
$()'$: perturbation from average quantity
$\widetilde{()}$: transposed matrix, or Favre average
$()''$: perturbation from Favre average
$\tilde{d}$: reversible infinitesimal transformation
$\hat{()}$: pre-exponential factor
$\tilde{()}^{0}_{T}$: standard thermodynamic function
$\times$: vector product
$\otimes$: tensor product
$\cdot$: scalar product (contracted tensor product)
$:$: dyadic product (doubly contracted tensor product)
$\wedge$: exterior product
$*$: sonic conditions, or reference state
$<>$: ensemble average
$\boldsymbol{\nabla}$: nabla (gradient operator)
$[]^{+}_{-}$: jump in a quantity across an interface

$d()$: differential

$\delta()$: small variation

d/dt: material derivative

$\partial/\partial t$: partial time derivative

$d_{\mathbf{W}}/dt$: material derivative associated with the velocity $\mathbf{W}$ of a fictitious motion

Contents

1

Introduction

Homogeneous, heterogeneous and reactive flows are important in the aeronautical, space and chemical industries, industrial furnaces, nuclear engineering, the production of gas and electricity, the automotive industry, and many other fields. Chemical reactions play a role in many processes. They may be inherent to the process of interest—as in chemical engineering or combustion, or in the case of air pollution mechanisms. They can also be a consequence of motion—for instance, they occur during the atmospheric re-entry of a spacecraft, which causes intense heating to occur around its exterior, or in supersonic combustion, downstream of intense shock waves.

Reactive flows exhibit heterogeneities and/or variations in concentration, temperature, speed, and types of transfer phenomena, such as thermal conduction, diffusion or viscosity. Since these phenomena occur together (as well as with convection) in such flows, they are difficult to study. The modeling of reactive flows progressed in conjunction with the conquest of space, which gave birth to a new science—aerothermochemistry, and with advances in chemical and process engineering. The methods employed, the phenomena investigated, and the aims of such modeling differ from one field to the next; however, in all cases, the results obtained considerably enrich our knowledge of reactive flows. Thus, in order to study a given phenomenon, it can sometimes be interesting and useful to combine the methods of chemical engineering, aerodynamics, and combustion.

Sixty years ago only specialists in combustion were concerned with chemically reactive flows, and the theoretical models they employed were simple. For example, in the case of deflagration (Chap. 10), for which speeds are on the order of meters per second, aerodynamic effects were neglected, so the burning velocity depended only on the chemical processes involved and diffusion. In contrast, in the case of detonation, where velocities can reach several thousands of meters per second, aerodynamics was the predominent phenomenon, and chemical effects contributed only through the energy of the combustion reaction. It is only in more recent years that specific research

R. Prud'homme, *Flows of Reactive Fluids*,
DOI 10.1007/978-0-8176-4659-2_1, © Springer Science+Business Media, LLC 2010

on chemically reactive flows utilizing both aerodynamics and combustion has been conducted.

The combination of three different areas of science—aerodynamics, thermodynamics, and chemistry—was inspired by the works of P. Duhem and F. Jouguet, and amplified by Maurice Roy, who coined the term "aerothermodynamics," as well as Theodore von Kármán, who—insisting on the importance of chemistry—invented the term "aerothermochemistry," which is still used today.

We should also mention the more recent contributions to this field of Hirschfelder [118], Prigogine [213], de Groot [108], Penner [199], and E. Brun [34], who established the fundamental equations used as the basis for current research.

Other significant contributions include those of Barrère [7], Williams [290], Truesdell [278], Rajagopal and Tao [227], Villermaux [283], Slattery [260], Marble, Crocco, Soo [263], and many others. These authors have (directly or indirectly) enriched the science of homogeneous and heterogeneous reactive fluids by improving our knowledge of the thermodynamics of pure and mixed continuous media, the mechanics of mixtures, chemical reactors, combustion, interfaces, and multiphase flows.

Fig. 1.1. Theodore von Kármán (1881–1963), CalTech JPL (image from Wikipedia, http://www.wikipedia.org)

How should we approach the analysis of such flows? Let us take the example of the coupling between chemical kinetics and convection in a fluid. One hypothesis that comes to mind is to assume a chemical composition at equilibrium at the local pressure and temperature encountered within the fluid. Another hypothesis is to assume that chemical reactions can be ignored and that, in the absence of diffusion, the composition is constant. These are the two borderline cases of flows with chemical reactions. Such situations are characterized by the first parameter of similarity of Damköhler, the ratio of the mechanical timescale to the chemical timescale. When this ratio is much larger than unity, the reaction speeds up and the equilibrium composition is rapidly approached; if the ratio is smaller than unity, the composition freezes. If the Damköhler parameter is on the order of unity, the chemical reaction and convection are coupled and we have a relaxing flow.

Fig. 1.2. Marcel Barrère (1920–1996), ONERA (image from P. Kuentzmann, private communication, 2009)

We do not need to know the reaction rate constants in order to calculate the equilibrium composition; it is sufficient to know the equilibrium constants, which are thermodynamical quantities that define the probabilities that specific species will form. In contrast, for a relaxing flow, the details of all of the reaction pathways and the associated reaction rates must be known. However, the data on the chemical kinetics are generally sparse and/or uncertain, and so it is difficult to guess these reaction rates. The expressions that yield the reaction rates are often complex. If the reaction is unique and reversible, this rate is the time derivative of a progress variable, an exponential function of the chemical affinity of the reaction. Because of this exponential term, the analytical calculations are rather difficult, and linearization is conducted by assuming that the reactions occur close to equilibrium. If multiple reactions occur simultaneously in the system, linearization is not always possible. The complexity of these expressions for the chemical kinetics is another of the difficulties associated with aerothermochemistry!

For an adiabatic flow, entropy production depends only on the reactive process, and in the case of a unique reaction it is equal to the affinity term multiplied by the reaction rate. The borderline cases of equilibrium and fixed composition correspond to isentropic evolutions. In relaxing flows, entropy production depends (via the reaction pathway) on the reaction rates, indicating that the thermodynamics of a reactive flow are those of an irreversible process.

The aim of this book is to introduce the reader to a number of basic concepts and methods that are necessary to study reactive flows and transfer phenomena. Most of the chapters are based on the notes associated with a course on the physics of liquids given at the Université Pierre et Marie Curie (UPMC, Paris VI) and thematic lectures given at the École Polytechnique. Although the flows considered here generally comply with the principle of the local state, which means that they are never very far from equilibrium, a great many cases involving irreversible processes can be considered using this approach. However, electromagnetic phenomena are excluded, and radiation problems are only evoked briefly, thus limiting the scope of the book. In spite of these limitations, the contents of this book should be of interest to students, researchers and engineers who wish to be introduced to reactive flows and transfer phenomena.

Chapter 2 provides the equations of state for single- and multicomponent fluids in which each component is at thermodynamic equilibrium. Both require the definition of the state variables, the thermodynamic functions, the specific properties of the mixtures with or without chemical reactions, the conditions of thermodynamic stability, and finally the effect of surface tension.

In Chap. 3, the complementary or constitutive laws that are needed to analyze irreversible processes are stated. The guiding principles of the thermodynamics and chemical kinetics of irreversible processes are presented, while the kinetic theory of gases allows the transfer coefficients to be evaluated as a function of the temperature.

The mass, momentum, and energy balance equations for moving fluid mixtures are established in Chap. 4 for discrete systems, continuous media, and continuous media with simple discontinuities (e.g., shock waves). The additional use of the second law of thermodynamics then leads to the entropy balance. Finally, the probabilistic population balance—which is very useful in chemical engineering situations—is given.

The principles of dimensional analysis are presented and illustrated through a few practical examples in Chap. 5. The most commonly encountered dimensionless numbers are summarized. Dimensional analysis is then applied to more complex situations, such as viscous, thermal and chemical boundary layers, turbulent onset and development, etc., in the following chapters.

Ideal and real chemical reactors are considered in Chap. 6. We begin by considering the steady and unsteady regimes of an ideal perfectly stirred reactor. The concepts of macro- and micromixtures, which are crucial to real reactors and require the population balances established in Chap. 4, are then presented.

Coupling and interaction effects between various processes in the laminar regime are the subject of Chap. 7. Examples relate to chemical relaxation in a flowing fluid, thermodiffusion, combustion and flames, thermal osmosis, natural thermal diffusion and gravity (Rayleigh–Bénard convection), and finally to surface tension and viscosity (the Marangoni effect and the Bénard–Marangoni instability).

Fig. 1.3. Ilia Prigogine (1917–2003), ULB (image from A. Sanfeld, private communication, 2009)

The basic concepts of turbulent flows are presented in Chap. 8, initially for pure incompressible fluids, then for compressible fluids, and finally for reactive mixtures. Classical theories—Prandtl's mixing length, Kolmogorov's spectral theory, averaged balance equations—are recalled, as well as the models developed to implement CFD codes: k–ϵ, probability distribution functions, and large eddy simulation. Particular attention is paid to the various possible interactions between combustion and turbulence.

Chapter 9 relates to confined or unconfined wall boundary layers and shear layers between two flowing streams. Steady or unsteady, compressible or incompressible, stable or unstable laminar flows, with or without wall blowing or aspiration, are presented for the case of a single-component fluid. Mass and energy diffusion phenomena for a moving or stationary flat plate and for a rotating disc are examined. The Emmons problem—ablative burning of a plate under a flow of oxidizing gas—is solved. Various cases involving turbulent boundary layers are studied using multiple-scale analysis in association with dimensional analysis.

Chapter 10 deals with reactive and nonreactive waves. Studying the linear propagation of a small perturbation in a reactive mixture leads to the computation of the speed of sound. Continuous and discontinuous waves in a pure barotropic fluid are treated. We then consider combustion waves and the structures of detonation waves and spherical waves for small movements and blast waves.

Chapter 11 tackles interface phenomena. The bulk equations of an interfacial layer in orthogonal curvilinear coordinates are presented, and the general form of the surface balance law is established. Various types of interface are studied: simplified interfaces, premixed flames, boundary layers, and (using the method of virtual power) interfaces that are resistant to crumpling.

Lastly, some elements relating to multiphase flows are given in Chap. 12. First, macroscopic balance equations are developed for a simplified two-phase fluid, and three examples of their application then are provided: vortex flow, sound propagation, and spray generation. Various other phenomena, including frictional force, heat exchange, vaporization, and the combustion of spherical particles, are then studied at the mesoscopic scale.

The author hopes that the contents of this book will provide the reader with the fundamental equations of reactive flow for the most general cases involving homogeneous and heterogeneous fluids, along with boundary conditions that specify the transfer of mass, heat, and momentum to the interfaces. The methods used to solve these equations for general and simplified cases enable most of the problems encountered in this field to be approached with a certain degree of confidence.

2

Equations of State

Studying the motions of complex fluid media involves determining how local quantities that characterize the behavior of single- or multicomponent, homogeneous or heterogeneous media evolve over time and space.

The state variables of a mixture are defined in Sect. 2.1. Initially for continuous media at rest and at equilibrium, we introduce extensive quantities (i.e., quantities that depend on the size of the system) such as the masses and volumes of the various constituents of gases and ideal liquid mixtures, as well as their strains for solids. Specific quantities, such as chemical progress variables for reacting mixtures, are extensive quantities too. Intensive (i.e., bulk) quantities such as temperature, pressure, and strains are also presented. Moreover, when the medium is moving, velocity vectors, fluxes, and production rate terms must also be considered.

Section 2.2 considers thermodynamic aspects, and the notion of entropy (an extensive quantity) is introduced. Temperature, pressure, and species chemical potential (all intensive variables) are the conjugates of entropy, volume, and species mass, respectively. For solids, stresses are the intensive conjugates of strains. The laws of state (LOS) that express the relationship between the state variables for one component in a continuous medium at rest and equilibrium will remain valid even when the local values of these variables change (this is known as the *local state postulate*).

The LOS for mixtures are discussed in Sect. 2.3. Section 2.4 focuses on reactive mixtures, particularly those at chemical equilibrium.

Temperature, pressure (or stresses), and chemical potentials are introduced as the first partial derivatives of the internal energy with respect to their extensive conjugates in Sects. 2.2 to 2.4, while the second partial derivatives of energy are considered in Sect. 2.5. This leads, together with the second law of thermodynamics and the concept of stability at thermodynamic equilibrium, to inequalities involving quantities such as specific heats and characteristic speeds.

Section 2.6 addresses equilibrium at the fluid interface and the concept of surface tension, and applies this to the behavior of a bubble in a liquid.

R. Prud'homme, *Flows of Reactive Fluids*,
DOI 10.1007/978-0-8176-4659-2_2,

It should be pointed out that the laws of state can only be applied to equilibrium evolution problems. Other laws are needed to investigate irreversible processes, and these are described in Chap. 3.

However, the balance equations presented in Chap. 4 are required to completely solve problems involving complex media in motion, while multiphase media are addressed in Chap. 12.

2.1 Defining the State Variables of a Mixture

2.1.1 Classical Parameters

The thermodynamical and chemical states of a homogeneous mixture can be characterized by a certain number of extensive and intensive (bulk) parameters. The more accessible parameters are generally the temperature T (the absolute temperature in Kelvins), the pressure p, the volume of the considered system $\mathcal{V}$, and the mass m_j or the number of moles $n_j = m_j/\mathcal{M}_j$ of species j ($\mathcal{M}_j$ is the molar mass of species j). The total mass is

$$m = \sum_{j=1}^{N} m_j, \tag{2.1}$$

where N is the number of species in the mixture. The total number of moles in the control volume $\mathcal{V}$ is

$$n = \sum_{j=1}^{N} n_j. \tag{2.2}$$

We also define specific quantities such as the partial density of species j:

$$\rho_j = m_j/\mathcal{V}, \tag{2.3}$$

the molar concentration per unit volume:

$$C_j = n_j/\mathcal{V}, \tag{2.4}$$

the mean density:

$$\rho = \sum_{j=1}^{N} \rho_j = m/\mathcal{V}, \tag{2.5}$$

and the total number of moles per unit volume:

$$C = \sum_{j=1}^{N} C_j. \tag{2.6}$$

The mean molar mass is equal to

$$\mathcal{M} = \sum_{j=1}^{N} n_j \mathcal{M}_j / n. \tag{2.7}$$

The molar and mass fractions of the species are (respectively):

$$X_j = n_j/n,\; Y_j = m_j/m, \tag{2.8}$$

and we have

$$Y_j = \rho_j/\rho = (\mathcal{M}_j C_j) / (\mathcal{M} C) = \mathcal{M}_j X_j / \mathcal{M}. \tag{2.9}$$

We can also write

$$Y_j = \rho_j/\rho = \mathcal{M}_j C_j / \sum_{j=1}^{N} \mathcal{M}_j C_j = \mathcal{M}_j X_j / \sum_{j=1}^{N} \mathcal{M}_j X_j. \tag{2.10}$$

The unit mass of a homogeneous fluid mixture will generally be completely characterized by the $(N+1)$ variables $T, p, Y_1, Y_2, \dots Y_{N-1}$. The mass fraction Y_N does not appear, since

$$\sum_{j=1}^{N} Y_j = 1. \tag{2.11}$$

2.1.2 Chemical Progress Variable

A given chemical reaction can be symbolically represented as

$$\sum_{j=1}^{N} \nu'_j \mathcal{E}_j \rightarrow \sum_{j=1}^{N} \nu"_j \mathcal{E}_j \tag{2.12}$$

In the absence of mass diffusion, any variations in concentration are caused by this chemical reaction. In the presence of diffusion, it is not possible to distinguish the changes in concentration due to diffusion from those due to the chemical reaction. In this case, the concentrations comply with the balance laws for chemical species (see Chap. 4).

The changes in concentration due to a chemical reaction are characterized by a progress variable ξ per unit mass, or ζ per unit volume with $\zeta = \rho\xi$, such that

$$\begin{cases} dm_j = \nu_j \mathcal{M}_j d\xi,\; m_j = m_j^0 + \nu_j \mathcal{M}_j \xi \\ dn_j = \nu_j d\xi,\; n_j = n_j^0 + \nu_j \xi \end{cases} \tag{2.13}$$

$$d\rho_j = \nu_j \mathcal{M}_j d\zeta, \; dC_j = \nu_j \, d\zeta, \; C = C^0 + \nu\, \zeta, \; \nu = \sum_{j=1}^{N} \nu_j. \tag{2.14}$$

The expressions for X_j and Y_j are then as follows:

$$\begin{cases} X_j = \left(n_j^0 + \nu_j \xi\right) / \left(n^0 + \nu \xi\right) \\ \\ Y_j = \left(m_j^0 + \nu_j \mathcal{M}_j \xi\right) / m^0, \end{cases} \tag{2.15}$$

taking into account the conservation of mass during the reaction $\sum_{j=1}^{N} \nu_j \mathcal{M}_j = 0$. X_j and Y_j can also be written as functions of ζ.

When there are K reactions in total, K progress variables can be used.

Example: The combustion reactions of a H_2–O_2 mixture are shown below:

$$\begin{cases} H_2 \rightarrow 2H & \text{chain initiation} \\ H + O_2 \rightarrow OH + O & \text{chain branching} \\ O + H_2 \rightarrow OH + H & \text{chain branching} \\ H + OH \rightarrow H_2O & \text{chain breaking} \end{cases}$$

$$\|\nu_{ir}\| = \underbrace{\begin{Vmatrix} 2 & -1 & 1 & 1 & -1 \\ 0 & 1 & -1 & 0 & 0 \\ -1 & 0 & -1 & -1 & 0 \\ 0 & -1 & 0 & 0 & 0 \\ 0 & 1 & 1 & -1 & -1 \\ 0 & 0 & 0 & 1 & 1 \\ R_1 & R_2 & R_3 & R_4 & R_5 \end{Vmatrix}}_{\text{reaction } j} \quad \left.\begin{matrix} H \\ O \\ H_2 \\ O_2 \\ OH \\ H_2O \end{matrix}\right\} \text{species } j$$

2.1.3 Definitions Required to Describe a Reactive Fluid

In contrast to the case for the homogeneous mixture considered in Sects. 2.1.1 and 2.1.2, the properties of a flow are seldom uniform, and so it is not sufficient to simply provide the values of the quantities T, p, X_j. These quantities vary over space and time. There are, however, local parameters (specific or intensive) that define the state of the system at any point, so we then have $T(\mathbf{x}, t)$, $p(\mathbf{x}, t)$; $\rho(\mathbf{x}, t)$, $C_j(\mathbf{x}, t)$ (for unit volume) and $Y_j(\mathbf{x}, t)$, $n_j(\mathbf{x}, t)$ (for unit mass). This is the local state postulate.[1] Moreover, each (infinitely small) particle of fluid is also characterized by mechanical parameters. We define N velocity vectors $\mathbf{v}_j(\mathbf{x}, t)$, $j = 1, \ldots N$ and also:

- The barycentric velocity: $\mathbf{v} = (1/\rho) \sum_{j=1}^{N} \rho_j \, \mathbf{v}_j$ (see (2.5) for the definition of ρ)

[1]Regarding time, the local state postulate stipulates that the current internal energy state of a homogeneous system at any point during its evolution does not depend on the rate of evolution, and that it can be characterized using the same state variables as those that characterize the equilibrium state.

- The diffusion velocities: $\mathbf{V}_j = \mathbf{v}_j - \mathbf{v}$
- The species fluxes: $\mathbf{J}_j = \rho_j \, \mathbf{v}_j$
- The total mass flow: $\mathbf{J} = \rho \, \mathbf{v} = \sum_{j=1}^{N} \mathbf{J}_j$
- The diffusion fluxes: $\mathcal{J}_{Dj} = \rho_j \, \mathbf{V}_j$.

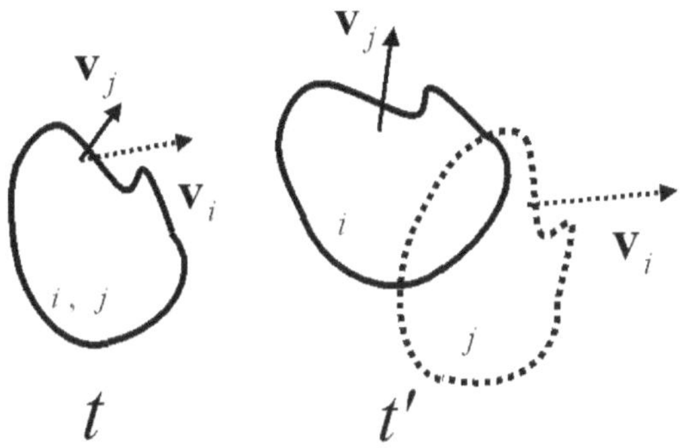

Fig. 2.1. Diffusion of species in a continuous medium

Due to the difference between the specific velocities $\mathbf{v}_j$ and $\mathbf{v}_i$, the domain common to both species i and j at time t is generally divided into two separate domains at time t'.

Each chemical species then exhibits its own motion (Fig. 2.1), although it is generally considered that these are not very different from the average motion. At time t, and per unit volume of the mixture, the mass of species j produced by the chemical reaction per unit time is

$$\dot{W}_j = \dot{W}_j(\mathbf{x}, t). \tag{2.16}$$

In the absence of diffusion ($\mathbf{V}_j = \mathbf{0}$), the following equations are obtained for only one reaction and for K reactions, respectively:

$$\begin{cases} \dot{W}_j = \nu_j \mathcal{M}_j \, \dot{\zeta}, \\ \dot{W}_j = \sum_{r=1}^{K} \nu_{jr} \mathcal{M}_j \, \dot{\zeta}_r. \end{cases} \tag{2.17}$$

2.1.4 The Case for Solids

The state of a solid depends on more variables than the state of a fluid. These additional variables allow us to characterize its strains and corresponding stresses, and they are most naturally presented in tensorial notation.

The thermodynamic state of a one-component solid does not depend only on the temperature and the specific volume, as is the case for a fluid, but also on the deformation variables. According to the classical theory for small deformations, the new variables form a symmetric 2D tensor $\boldsymbol{\epsilon}$. To define $\boldsymbol{\epsilon}$, we must first describe the continuous medium in Lagrange coordinates $(\mathbf{a}, t)$, where $\mathbf{a}$ is the material reference position at time $t = 0$. Any material position $\mathbf{x}$ at time t is defined as a function $\mathbf{x} = \boldsymbol{\Phi}(\mathbf{a}, t) = \mathbf{a} + \mathbf{X}(\mathbf{a}, t)$. For small deformations we can write $\mathbf{x} = \boldsymbol{\Phi}(\mathbf{a}, t) = \mathbf{a} + \eta\bar{\mathbf{x}}(\mathbf{a}, t)$, where η is a small parameter. The gradient $\mathbf{H}$ of $\mathbf{X}(\mathbf{a}, t)$, the components of which are H_{ij}, can be decomposed into its symmetric and antisymmetric parts $\boldsymbol{\epsilon}$ and $\mathbf{r}$:

$$\mathbf{H} = \boldsymbol{\epsilon} + \mathbf{r}, \ \boldsymbol{\epsilon} = \frac{1}{2}(\mathbf{H} + \widetilde{\mathbf{H}}), \ \mathbf{r} = \frac{1}{2}(\mathbf{H} - \widetilde{\mathbf{H}}).$$

The thermodynamic state of a deformed solid depends only on the symmetric part $\boldsymbol{\epsilon}$.

The following section briefly reviews the laws of thermodynamics and their application to one-component fluids and solids; subsequent sections will then consider composite fluid systems.

2.2 Thermodynamic Functions and Equation of State for Simple Fluids and Solids

2.2.1 Thermodynamics Reminder

The laws of thermodynamics govern the equilibrium states of closed (i.e., no mass exchange with the outside), simple, or composite systems [38, 97].

Among those that are possible for such a system, there are a subset of equilibrium states that form a continuous differential manifold $(\mathcal{M})$. This manifold $(\mathcal{M})$ can be visually interpreted as a surface or a family of surfaces. Points on this surface correspond to states of equilibrium, and so the region beyond this manifold corresponds to nonequilibrium states. Any trajectory through this space (broken line in Fig. 2.2) between equilibrium states $\mathcal{E}_1$ and $\mathcal{E}_2$ represents a thermodynamic transformation. If this trajectory occurs across the surface (continuous line in Fig. 2.2), any intermediate state between $\mathcal{E}_1$ and $\mathcal{E}_2$ is a state of equilibrium. Therefore, the continuous line in Fig. 2.2 represents *a reversible transformation.* The evolution of this line over time corresponds to the evolution of the equilibrium of this system. The broken line corresponds to an unspecified transformation and does not occur entirely within $(\mathcal{M})$.[2]

On the manifold $(\mathcal{M})$, we allow the existence of two differential forms $\widetilde{dQ}$ and $\widetilde{dW}$ that represent the elementary heat and work added to the system during the reversible transformation, respectively.

[2] A reversible transformation from $\mathcal{E}_2$ to $\mathcal{E}_1$ is also possible along the same line. This is not generally the case for an unspecified transformation. We will illustrate this when we discuss adiabatic transformations (second law of thermodynamics).

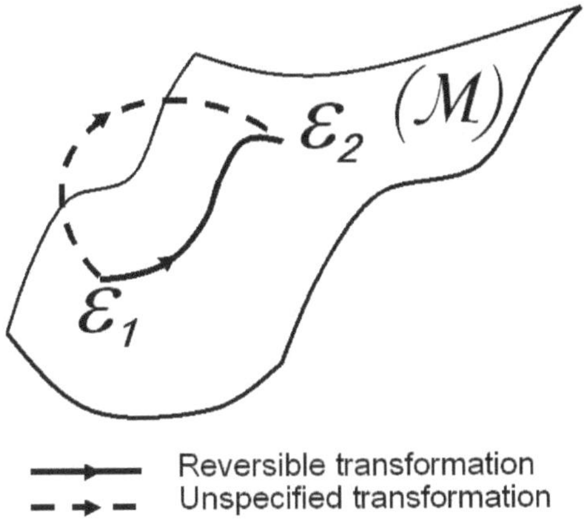

Fig. 2.2. Reversible and irreversible transformations

First Law of Thermodynamics

Q and W are the heat and work done by a closed system during an unspecified transformation between two states of equilibrium $\mathcal{E}_1$ and $\mathcal{E}_2$. Thus,

$$E_2 - E_1 = W + Q, \tag{2.18}$$

where E is a function called the *internal energy* that depends only on the equilibrium state of the system (E depends only on the variables of state and can be regarded as state variable itself). When $Q = 0$, the transformation is said to be "adiabatic."

For a reversible elementary transformation we have

$$dE = \tilde{d}W + \tilde{d}Q, \tag{2.19}$$

where $\tilde{d}W$ and $\tilde{d}Q$ are differential forms.

On the other hand, for an elementary (infinitesimal) unspecified transformation,

$$\delta E = \delta W + \delta Q \tag{2.20}$$

(in the general case, δW and δQ simply indicate small quantities rather than differential forms).

We can extend this to a system at equilibrium in Galilean reference frames with different velocities in states 1 and 2:

$$(E_2 + K_2) - (E_1 + K_1) = W + Q, \tag{2.21}$$

where K represents the kinetic energy of the system. For a reversible elementary transformation, we have

$$dE + dK = \widetilde{d}W + \widetilde{d}Q. \tag{2.22}$$

This relation corresponds to the principle of the conservation of energy (the first law of thermodynamics).

Second Law of Thermodynamics

The second law of thermodynamics is not elaborated in detail here. Only the Carathéodory statement [97] is recalled below.

Statement of the Principle of Carathéodory for a Closed System: Consider an equilibrium state for a system. In the vicinity of $\mathcal{E}_0$, there is a state $\mathcal{E}_1$ on the manifold $(\mathcal{M})$ that cannot be reached from $\mathcal{E}_0$ through an adiabatic transformation.

This statement implies the existence of a function of state S called the entropy, such that:

- $\widetilde{d}Q = T\,dS$ for an elementary reversible transformation (i.e., this relation is true on the manifold $(\mathcal{V})$).
- $S_2 - S_1 \geq 0$ for an unspecified adiabatic transformation. The equal sign applies if the transformation is reversible, and so in this case

$$dS = \widetilde{d}Q/T = 0,\ S_2 - S_1 = \int_{\mathcal{E}_1}^{\mathcal{E}_2} \widetilde{d}Q/T = 0.$$

Note: It can be shown that, for a simple system (dependent on two variables), there is always an integrating factor, and so the first part of the second law is always obeyed.

Extensions: For an unspecified infinitesimal transformation,

$$\delta Q = T(\delta S - \delta_i S),\ \delta_i S \geq 0. \tag{2.23}$$

For an unspecified transformation (not inevitably adiabatic),

$$S_2 - S_1 = \Delta_e S + \Delta_i S,\ \Delta_i S \geq 0. \tag{2.24}$$

In the general case, the main difficulty here involves determining the external variation in entropy $\Delta_e S$ (the entropy flux from the outside) and the internal variation $\Delta_i S$ (the entropy produced). These laws extend to transformations between nonequilibrium states if we can define (depending on the case) the functions E and S as well as the temperature T.

2.2.2 Properties of Simple Fluids at Equilibrium

When the fluid undergoes a reversible mechanical transformation, the elementary work done is equal to

$$\tilde{d}W = -p\,d\mathcal{V}. \tag{2.25}$$

Since it is also known that

$$\tilde{d}Q = T\,dS, \tag{2.26}$$

we can deduce that, for a closed fluid domain (no mass exchange with the outside),

$$dE = T\,dS - p\,d\mathcal{V}. \tag{2.27}$$

When the mass m of a one-component fluid is fixed, its internal energy is a function of S and $\mathcal{V}$. This function has partial derivatives with respect to S and $\mathcal{V}$ that are the temperature T and the negative of the pressure $(-p)$. The internal energy also depends on the mass or (equivalently) the number of moles of fluid considered (which is constant for a closed system):

$$E = E(S,\, \mathcal{V},\, m) \text{ or } E = E(S,\, \mathcal{V},\, n), \tag{2.28}$$

where n is the number of moles. Therefore,

$$T = \partial E/\partial S,\; -p = \partial E/\partial \mathcal{V}. \tag{2.29}$$

We then set

$$g = \partial E/\partial m \text{ or } \mu = \partial E/\partial n \tag{2.30}$$

where $\mu = \mathcal{M}g$, where $\mathcal{M}$ is the molar mass. This leads to the Gibbs equation

$$dE = T\,dS - p\,d\mathcal{V} + g\,dm \text{ or } dE = T\,dS - p\,d\mathcal{V} + \mu\,dn. \tag{2.31}$$

Establishing the Gibbs–Duhem Relation:

$$E(\lambda S,\, \lambda\mathcal{V},\, \lambda m) = \lambda E(S,\, \mathcal{V},\, m),\;\; \lambda \in \Re, \tag{2.32}$$

because $E, S, \mathcal{V}$, and m are extensive quantities. E is thus a homogeneous function of the first degree in $S, \mathcal{V}$, and m. Using Euler's theorem and the preceding definitions of T, p and g, we get

$$E(S,\, \mathcal{V},\, m) = T\,S - p\,\mathcal{V} + g\,m. \tag{2.33}$$

Demonstration:

$$E(\lambda S,\, \lambda\mathcal{V},\, \lambda m) = \lambda E(S,\, \mathcal{V},\, m)$$

$$\Big(\frac{\partial E}{\partial(\lambda S)}\,S + \frac{\partial E}{\partial(\lambda \mathcal{V})}\,\mathcal{V} + \frac{\partial E}{\partial(\lambda m)}\,m\Big)\,d\lambda = E(S,\, \mathcal{V},\, m)\,d\lambda$$

$$E(S,\, \mathcal{V},\, m) = \frac{\partial E}{\partial(\lambda S)}\,S + \frac{\partial E}{\partial(\lambda \mathcal{V})}\,\mathcal{V} + \frac{\partial E}{\partial(\lambda m)}\,m.$$

If $\lambda = 1$,

$$\Rightarrow\; E(S,\, \mathcal{V},\, m) = \frac{\partial E}{\partial S}\,S + \frac{\partial E}{\partial \mathcal{V}}\,\mathcal{V} + \frac{\partial E}{\partial m}\,m.$$

In the same way, we can obtain

$$E(S,\, \mathcal{V},\, n) = T\, S - p\, \mathcal{V} + \mu\, n. \tag{2.34}$$

Using this relation and the Gibbs equation, we can deduce the Gibbs–Duhem equation

$$0 = S\, dT - \mathcal{V}\, dp + m\, dg, \tag{2.35}$$

or

$$0 = S\, dT - \mathcal{V}\, dp + n\, d\mu. \tag{2.36}$$

Other Thermodynamic Functions

The usual thermodynamic functions that are derived from the internal energy E are the enthalpy H, the Helmholtz free energy F, and the Gibbs free enthalpy G. We have

$$H = E + p\mathcal{V},\ dH = T\, dS + \mathcal{V}\, dp + g\, dm. \tag{2.37}$$

The enthalpy is a function of the variables S, p and m. We can write

$$H = H(S,\, p,\, m),\ T = \partial H/\partial S,\ \mathcal{V} = \partial H/\partial p,\ g = \partial H/\partial m. \tag{2.38}$$

In the same manner, we obtain the following relations for the Helmholtz free energy:

$$\begin{cases} F = E - T\, S = -p\, \mathcal{V} + g\, m, \\ dF = -S\, dT - p\, d\mathcal{V} + g\, dm, \end{cases} \tag{2.39}$$

and for the Gibbs free enthalpy we obtain

$$\begin{cases} G = E - T\, S + p\, \mathcal{V} = g\, m, \\ dG = -S\, dT + \mathcal{V}\, dp + g\, dm. \end{cases} \tag{2.40}$$

The partial derivative g is thus the free enthalpy per unit mass, just as μ is the molar free enthalpy.
For unit mass and $\vartheta = 1/\rho$, we get

$$\begin{cases} e = T\, s - p\, \vartheta + g = e(s,\, \vartheta),\ de = T\, ds - p\, d\vartheta,\ 0 = s\, dT - \vartheta\, dp + dg, \\ h = e + p\vartheta = Ts + g = h(s,\, p),\ dh = T\, ds + \vartheta\, dp, \\ f = e - Ts = -p\vartheta + g = f(T,\, \vartheta),\ df = -s\, dT - p\, d\vartheta, \\ g = e - Ts + p\vartheta = g(T,\, p),\ dg = -s\, dT + \vartheta\, dp. \end{cases} \tag{2.41}$$

For one mole (indicated by bars above symbols),

$$\begin{cases} \overline{E} = T\,\overline{S} - p\overline{\mathcal{V}} + \mu = \overline{E}(\overline{S},\, \overline{\mathcal{V}}),\; d\overline{E} = T\,d\overline{S} - p\,d\overline{\mathcal{V}}, \\ 0 = \overline{S}\,dT - \overline{\mathcal{V}}\,dp + d\mu, \\ \overline{H} = \overline{E} + p\overline{\mathcal{V}} = \overline{H}(\overline{S},\, p),\; d\overline{H} = T\,d\overline{S} + \overline{\mathcal{V}}\,dp, \\ \overline{F} = \overline{E} - T\overline{S} = \overline{F}(T,\, \overline{\mathcal{V}}),\; d\overline{F} = -\overline{S}\,dT - p\,d\overline{\mathcal{V}}, \\ \overline{G} = \overline{E} - T\overline{S} + p\overline{\mathcal{V}} = \mu(T,\, p),\; d\mu = -\overline{S}\,dT + \overline{\mathcal{V}}\,dp. \end{cases} \tag{2.42}$$

For unit volume ($\rho_F = \rho\, f$),

$$\rho_E = T\,\rho_S - p + \mu\;\rho,\; d\rho_E = T\,d\rho_S + g\,d\rho,\; 0 = \rho_S\,dT - dp + \rho\,dg. \tag{2.43}$$

The equation $E = E(S,\, \mathcal{V},\, m)$, when $E(S,\, \mathcal{V},\, m)$ is a known function (or $E = me$, $e = e(s,\, \vartheta)$, where $e(s,\, \vartheta)$ is a known function), is the fundamental energy law for the fluid. If this law is known, all of the thermodynamic properties of the fluid are also known. This remark also applies to the other thermodynamic functions expressed in terms of their original variables. On the other hand, it is not sufficient if only one partial derivative is given; two of them are required, constituting the laws of state for the fluid.

2.2.3 Examples of Laws of State

Perfect Gases, Standard State

We have $p\bar{\mathcal{V}} = RT$, the law of state for one mole of a perfect gas [33, 109, 267]. Using the Gibbs equation, noting that $p/T = R/\overline{\mathcal{V}}$ and $\overline{S} = \overline{S}(\overline{E},\, \overline{\mathcal{V}})$, we get

$$\begin{cases} d\overline{S} = (1/T)d\overline{E} + (p/T)d\overline{\mathcal{V}}, \\ 1/T = \partial\overline{S}/\partial\overline{E},\; p/T = \partial\overline{S}/\partial\overline{\mathcal{V}} = R/\mathcal{V}. \end{cases} \tag{2.44}$$

Note that $\partial^2\overline{S}/\partial\overline{E}\,\partial\overline{\mathcal{V}} = \partial^2\overline{S}/\partial\overline{\mathcal{V}}\,\partial\overline{E}$ since $\overline{E}$ and $\overline{\mathcal{V}}$ are the natural independent variables of the problem of interest, or $\partial(1/T)/\partial\overline{\mathcal{V}} = \partial(p/T)/\partial\overline{E} = 0$ because $p/T = R/\overline{\mathcal{V}}$. Thus T depends only on $\overline{E}$, or conversely $\overline{E}$ is a function of T alone. The law of state for a perfect gas thus imposes a constraint, but does not provide the dependence of internal energy on temperature. We need the second law of state:

$$\overline{E} = \overline{E}_T^0. \tag{2.45}$$

Therefore, two laws of state are necessary to define the thermodynamic behavior of a simple fluid, or an energy law in canonical form (see Eqns. 2.42). The latter is, for the perfect gas,[3]

$$\mu = \mu_T^0 + RT\,ln(p/p^0). \tag{2.46}$$

The pressure p^0 is a standard pressure (generally $p^0 = 10^5 Pa$), and

[3]By deriving μ with respect to T and p, we can deduce the classical laws of state for a perfect gas: $\bar{S} = -\partial\mu/\partial T = -d\mu_T^0/dT - R\ln(p/p^0)$, $\overline{\mathcal{V}} = \partial\mu/\partial p = RT/p$.

$$\overline{E}_T^0 = \mu_T^0 - T d\mu_T^0/dT - RT. \tag{2.47}$$

In the same way,

$$\overline{H} = \overline{H}_T^0 = \overline{E}_T^0 + RT. \tag{2.48}$$

Heat capacities: For a reversible elementary transformation, we know that $\tilde{d}\overline{Q} = Td\overline{S}$. At constant volume, $d\overline{E} = \tilde{d}\overline{Q} = Td\overline{S} = \overline{C}_v dT$, where $\overline{C}_v$ is the molar heat capacity at constant volume. For a perfect gas, $\overline{C}_v$ is a function of T only, and we have $\overline{E} = \overline{E}_T^0 = \overline{E}_0^0 + \int_{T^0}^T \overline{C}_v(T)dT$, where $\overline{E}_0^0$ is a constant corresponding to the value of the molar internal energy at some standard temperature T^0.

At constant pressure, we have $d\overline{H} = \tilde{d}\overline{Q} = Td\overline{S} = \overline{C}_p dT$, where $\overline{C}_p$ is the molar heat capacity at constant pressure.

For a perfect gas, $\overline{C}_p$ is a function of T alone, $\overline{H} = \overline{H}_T^0 = \overline{H}_0^0 + \int_{T^0}^T \overline{C}_p(T)dT, \overline{C}_p = \overline{C}_v + R$, and $\overline{H}_0^0 = \overline{E}_0^0 + RT$.

The molar standard entropy of the perfect gas can be deduced: $\overline{S}_T^0 = \overline{S}_0^0 + \int_{T^0}^T (\overline{C}_p(T)/T)dT$, where $\overline{S}_0^0$ refers to the standard state defined by (T^0, p^0). For the molar entropy we have $\overline{S} = \overline{S}_T^0 - R\ln(p/p^0)$, and for the molar standard free enthalpy $\mu_T^0 = \overline{H}_T^0 - T\overline{S}_T^0$.

Use of a standard volume $\overline{\mathcal{V}}^0$***:*** Note that if we use the variables T, $\overline{\mathcal{V}}$ instead of T, p, a similar form is obtained for the entropy $\overline{S} = \breve{S}_T^0 + R\ln(\overline{\mathcal{V}}/\overline{\mathcal{V}}^0)$, but this time $\breve{S}_T^0 = \overline{S}_0^0 + \int_{T^0}^T (\overline{C}_v(T)/T)dT = \overline{S}_T^0 - R\ln(T/T^0)$. For the internal energy, we have $\overline{E} = \breve{E}_T^0$, where $\breve{E}_T^0 = \overline{E}_T^0$. For the molar Helmholtz free energy, $\overline{F} = \breve{F}_T^0 - RT\ln(\overline{\mathcal{V}}/\overline{\mathcal{V}}^0)$, where $\breve{F}_T^0 = \overline{E}_T^0 - T\overline{S}_T^0 + RT\ln(T/T_0)$.

The standard thermodynamic functions $\mu_T^0, \overline{H}_T^0, \ldots$ have been tabulated in other words. Note that the constants $\overline{E}_0^0$, $\overline{H}_0^0 = \overline{E}_0^0 + RT^0$ are considered arbitrary constants if the studied phenomenon involves just one pure gas. This is not the case if a reactive mixture is of interest, as we will show later.

Real Fluids

The state diagram for a pure substance is represented schematically in Fig. 2.3.

The perfect gas domain corresponds to small pressures and large volumes. We will discuss this point in more depth in the following with respect to van der Waals EOS (see Fig. 2.4 and the associated discussion).

Another important observation is the existence of the liquid–vapor critical point C (see Appendix A.5.1 for the thermodynamic properties of fluids near the critical point).

Since the law of state is generally given in the form $p = p(T, \overline{\mathcal{V}})$, where T is the temperature and $\overline{\mathcal{V}}$ is the molar volume, we will define a standard molar volume $\overline{\mathcal{V}}^0$ corresponding to the standard state of an ideal gas at the temperature T^0, where the standard pressure p^0 is given by $p^0\overline{\mathcal{V}}^0 = RT^0$. Such a standard state always exists.

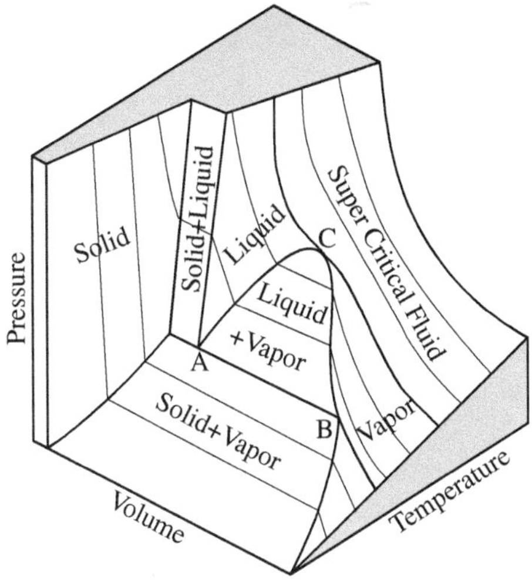

Fig. 2.3. State diagram for a one-component medium

Approximate method for determining thermodynamic functions: Starting from the equation of state, it is possible to determine the Helmholtz free energy as a function of T, $\overline{\mathcal{V}}$, and the other thermodynamic functions using an approximate method. To do this, we assume that there is a standard reference state defined by the molar volume $\overline{\mathcal{V}}^0$ such that, for $\overline{\mathcal{V}} > \overline{\mathcal{V}}^0$, the fluid can be considered a perfect gas; i.e., $p\overline{\mathcal{V}} = RT$. As $\partial\overline{F}/\partial\overline{\mathcal{V}} = -p(T, \overline{\mathcal{V}})$, we can write

$$\overline{F} = \breve{F}_T^0 - \int_{\overline{\mathcal{V}}^0}^{\overline{\mathcal{V}}} p d\overline{\mathcal{V}}.$$

Here $\breve{F}_T^0$ is the standard value of the molar free energy of the perfect gas at $\overline{\mathcal{V}} = \overline{\mathcal{V}}^0$. We write

$$\int_{\overline{\mathcal{V}}^0}^{\overline{\mathcal{V}}} p d\overline{\mathcal{V}} = \int_{\overline{\mathcal{V}}^0}^{\overline{\mathcal{V}}} (p - RT/\overline{\mathcal{V}}) d\overline{\mathcal{V}} + RTln(\overline{\mathcal{V}}/\overline{\mathcal{V}}^0).$$

This integral can be written as

$$\int_{\overline{\mathcal{V}}^0}^{\overline{\mathcal{V}}} (p - RT/\overline{\mathcal{V}}) d\overline{\mathcal{V}} = \int_{\overline{\mathcal{V}}^0}^{\infty} (p - RT/\overline{\mathcal{V}}) d\overline{\mathcal{V}} + \int_{\infty}^{\overline{\mathcal{V}}} (p - RT/\overline{\mathcal{V}}) d\overline{\mathcal{V}} \cong \int_{\infty}^{\overline{\mathcal{V}}} (p - RT/\overline{\mathcal{V}}) d\overline{\mathcal{V}},$$

since the fluid is considered to be a perfect gas if $\overline{\mathcal{V}} > \overline{\mathcal{V}}^0$. We then deduce that

$$\overline{F} \cong \breve{F}_T^0 - RT\ln(\overline{\mathcal{V}}/\overline{\mathcal{V}}^0) - \int_{\infty}^{\overline{\mathcal{V}}} (p - RT/\overline{\mathcal{V}}) d\overline{\mathcal{V}}.$$

van der Waals Fluids

The equation of state for a van der Waals fluid is

$$(\overline{V} - b)\,(p + a/\overline{V}^2) = RT. \tag{2.49}$$

This equation of state is the simplest one for real gases and liquids. It takes into account the minimum distance between the molecules and the interactions between them. For rarefied gases $a/\overline{V}^2 \ll p$, so $\overline{V} - b \cong (RT/p)(1 - a/p\overline{V}^2)$, and for a perfect gas (for $\overline{V} \gg b - aRT/p^2\overline{V}^2$) we have $\overline{V} = RT/p$. For liquids, $a/\overline{V}^2 \gg p$ ($a/\overline{V}^2$ is a few tens of thousands of atmospheres), so we can often ignore p, and thus $\overline{V} \cong \overline{V}_T^0$.

Proof: With the reduced variables $x = \overline{V}/\overline{V}_C$, $y = p/p_C$, $z = T/T_C$, the van der Waals state equation can be written as $y = 8z/(3x - 1) - 3/x^2$. Using these variables, the equation of state for a perfect gas becomes $y_{pft} = 8z/3x$. The fluid can be considered a perfect gas if $y - y_{pft} \ll 1$. The isobars $z = (3x-1)(y+3/x^2)/8$, $y = const.$ are shown in Fig. 2.4. It is clear that these curves tend to become straight lines when the relative volume increases. This proves that, for a van der Waals fluid, one can define a standard state on any isotherm at a given temperature T provided that the values of the parameters are chosen in the appropriate region. This result, which is related to the definition of a perfect gas, does not depend on the fluid considered. If ϵ is an arbitrarily small number, we can assume that the pressure in the van der Waals law is equal to that of a perfect gas with a relative error ϵ if

$$(y - y_{pft})/y_{pft} \leq \epsilon.$$

It is easy to see that if $x \geq 1/3\epsilon$, we also have $z \geq y/8\epsilon$.

We now directly determine the forms of the internal energy, the entropy, and the molar free enthalpy of a van der Waals fluid. We have

$$p = RT/(\overline{V}-b) - a/\overline{V}^2, d\overline{S} = d\overline{E}/T + (p/T)d\overline{V} = d\overline{E}/T + [R/(\overline{V}-b) - a/T\overline{V}^2]d\overline{V}$$

$$= (1/T)d(\overline{E} + a/\overline{V}) + R\,d\ln(\overline{V} - b),\ \overline{S} = \overline{S}(\overline{E} + a/\overline{V},\ \overline{V} - b).$$

Therefore, $\partial R/\partial(\overline{E} + a/\overline{V}) = 0 = \partial(1/T)/\partial(\ln(\overline{V} - b))$. It follows that $\overline{E} = \overline{E}_T^0 - a(1/\overline{V} - 1/\overline{V}^0)$, where $\overline{V}^0$ corresponds to a standard reference state, since T is not related to $(\overline{V} - b)$. We also have $\overline{S} = \overline{S}_T^0 + R\ \ln(\overline{V} - b)/(\overline{V}^0 - b)$, with $\overline{S}_T^0 = \overline{S}_0^0 + \int_{T^0}^{T} d\overline{E}_T^0/T$, since $d\overline{S} = d(\overline{E}_T^0/T) + R\,d\ \ln(\overline{V} - b)$.

Finally, $\overline{F} = \breve{F}_T^0 - a(1/\overline{V} - 1/\overline{V}^0) - RT\ \ln[(\overline{V} - b)/(\overline{V}^0 - b)]$, $\breve{F}_T^0 = \overline{E}_T^0 - T\,\breve{S}_T^0$, $\mu = \overline{E} - T\,\overline{S} + p\,\overline{V} = \breve{\mu}_T^0 - 2a(1/\overline{V} - 1/\overline{V}^0) - RT\ \ln[(\overline{V} - b)/(\overline{V}^0 - b)] + RT\left[\overline{V}/(\overline{V} - b) - \overline{V}^0/(\overline{V}^0 - b)\right]$.[4]

[4]If we apply the previous approximate method, we obtain the formula $\overline{F} = \breve{F}_T^0 - \ln(\overline{V}/\overline{V}^0) - RT\,ln\left[(\overline{V} - b)/\overline{V}\right] - a/\overline{V}$ for the Helmholtz free energy, which differs slightly from the exact one.

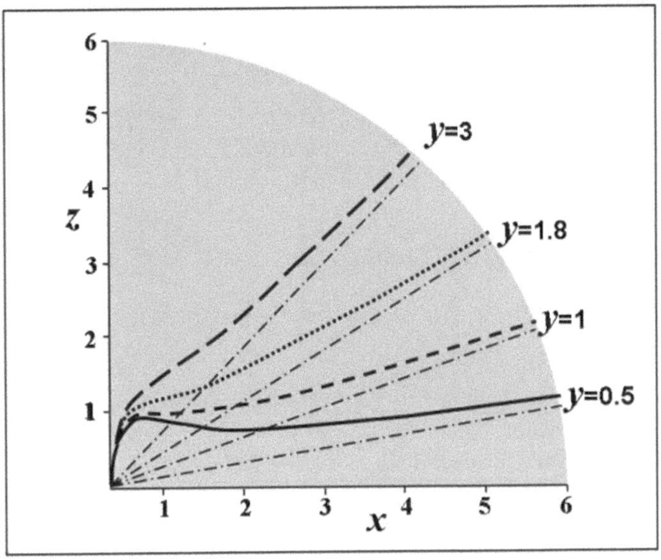

Fig. 2.4. Definition of a standard state: the plot shows isobars of $y = p/pc$ =const. in the $(x,\ z)$ plane

Other Laws for Dense Gases and Liquids

The following form is applicable to the van der Waals, Redlich–Kwong, and Peng–Robinson [44, 210, 233] equations of state:

$$p = \frac{RT}{\overline{V} - b} - \frac{a}{\overline{V}^2 + ub\overline{V} + wb^2}, \tag{2.50}$$

where u and w are coefficients.

For the Peng–Robinson equation,

$$u = 2,\ w = -1,\ b = \frac{0.07780\, RT_C}{p_C},\ a = \frac{0.45724\, R^2 T_C^2}{p_C}\left[1 + f\omega(1 - \sqrt{\frac{T}{T_C}})\right]^2,$$

$$f\omega = 0.37464 + 1.54226\omega - 0.26992\omega^2,$$

where the constant ω, called the acentric factor, has been tabulated for numerous species.[5]

[5]In thermodynamics, the acentric factor ω was originally used by Pitzer [204] as an expression in an equation for the compressibility factor. The term "acentric factor" was used to imply that the factor roughly charts the deviation of the intermolecular potential function from the corresponding function for simple fluids.

Thus, $\partial^2 p/\partial T^2 = -(\partial^2 a/\partial T^2)/(\overline{\mathcal{V}}^2 + 2b\overline{\mathcal{V}} - b^2)$.

Reid, Prausnitz, and Poling [233] provided the data needed to obtain the equation of state $p = p(T, \overline{\mathcal{V}})$ for some bodies as well as to determine the molar thermodynamic functions for perfect gases (heat capacities $\overline{C}_{p,j}$ J/mol K, standard enthalpy of formation $(\overline{Q}_f^0)_j$ at 298.2 K in J/mol, and free enthalpy of formation $(\mu_0^0)_j$ at 298.2K and 1 bar in J/mol.

Dilatable Incompressible Liquids

Liquids are generally not very compressible except when they are near the critical point.

- In other words, the compressibility coefficient (isothermal) as defined by $\chi = -\frac{1}{\overline{\mathcal{V}}}(\frac{\partial \overline{\mathcal{V}}}{\partial p})_T$ is generally small (on the order of 10^{-4} bar^{-1}). However, while it is small, neglecting compressibility implies that the speed of sound is infinite, whereas it is known to be finite (e.g., the speed of ultrasound is on the order of 1000 m/s). The resulting dilatation is characterized by the thermal dilatation coefficient (isobaric) $\alpha = \frac{1}{\overline{\mathcal{V}}}(\frac{\partial \overline{\mathcal{V}}}{\partial T})_p$.

- Let us start from $\overline{\mathcal{V}} = \overline{\mathcal{V}}_r\left[1 - \chi_L(p - p_r)\right]\left[1 + \alpha_L(T - T_r)\right]$ for the molar volume, where p_r and T_r are a reference pressure and temperature, and $\overline{\mathcal{V}}_r$ is a reference molar volume; χ_L and α_L are assumed to be constant. We deduce that $\chi = \chi_L/\left[1 - \chi_L(p - p_r)\right]$, $\alpha = \alpha_L/\left[1 + \alpha_L(T - T_r)\right]$, and so $\chi \cong \chi_L$, $\alpha \cong \alpha_L$ provided that $\chi_L\left|p - p_r\right|$ and $\alpha_L\left|T - T_r\right| \ll 1$. This latter formula corresponds to the linearization of a more general equation in the vicinity of p_r and T_r.

- The corresponding equation of state is

$$\rho = \rho_r/\left[(1 - \chi_L(p - p_r))(1 + \alpha_L(T - T_r))\right].$$

Let us consider the case of a liquid for which the molar volume is only a function of the temperature $\overline{\mathcal{V}} \cong \overline{\mathcal{V}}_T^0$. We will use the function $\overline{F}$, meaning that $\overline{F} = \overline{F}(T, \overline{\mathcal{V}})$ is a function of the temperature alone here; i.e., $\overline{F} = \overline{F}_T^0$. We deduce from this that

$$\mu = \overline{F}_T^0 + p\,\overline{\mathcal{V}}_T^0. \tag{2.51}$$

The term μ_T^0 is often used instead of $\overline{F}_T^0$. Then,

$$\left\{\begin{aligned} &\mu = \mu_T^0 + p\,\overline{\mathcal{V}}_T^0,\ \overline{F} = \mu_T^0,\\ &\overline{S} = -d\mu_T^0/dT - p\,d\overline{\mathcal{V}}_T^0/dT = \overline{S}_T^0 - p\,d\overline{\mathcal{V}}_T^0/dT,\\ &\overline{E} = \mu + T\overline{S} - p\overline{\mathcal{V}} = \mu_T^0(T) - T(d\mu_T^0/dT,\\ &+p\,d\overline{\mathcal{V}}_T^0/dT) = \overline{E}_T^0 - pT\,d\overline{\mathcal{V}}_T^0/dT,\\ &\overline{H} = \mu + T\overline{S} = \overline{E}_T^0 - p(T\,d\overline{\mathcal{V}}_T^0/dT - \overline{\mathcal{V}}_T^0). \end{aligned}\right. \tag{2.52}$$

Real Gases: The Virial Formulae

The virial formulae allow us to model real gases relatively accurately. They make it possible to express $\overline{\mathcal{V}}$ according to p and T or p as a function of $\overline{\mathcal{V}}$ and T using expansions. Thus, we have

$$\overline{\mathcal{V}} = RT/p + B + C'p. \tag{2.53}$$

Or, conversely,

$$p = \frac{RT}{\overline{\mathcal{V}}}(1 + \frac{B}{\overline{\mathcal{V}}} + \frac{C}{\overline{\mathcal{V}}^2} + \ldots). \tag{2.54}$$

Coefficients A,B,$C\ldots$ are functions of temperature[6].

By performing the expansions while assuming that the quantities p and $1/\overline{\mathcal{V}}$ are small, we find that $C = 2B + RTC'$. We then obtain the perfect gas law by canceling the coefficients of the virial B,$C\ldots$

The coefficient B characterizes the first-order molecular interactions. When $\overline{\mathcal{V}}$ is large (p is small), the van der Waals law becomes $p = \frac{RT}{\overline{\mathcal{V}}}(\frac{1}{1-b/\overline{\mathcal{V}}} - \frac{a}{RT\overline{\mathcal{V}}})$, and thus takes the form of a limited virial expansion $p = \frac{RT}{\overline{\mathcal{V}}}(1 + B/\overline{\mathcal{V}})$ or $\overline{\mathcal{V}} = RT/p + B$, where $B = b - a/RT$. By limiting the expansion of the virial to the coefficient B, we can easily calculate the thermodynamic functions by introducing functions of the temperature to be determined. Since $\overline{\mathcal{V}} = \partial\mu/\partial p = RT/p + B$, we find that

$$\begin{cases} \mu = \mu_T^0 + RT\,\ln(p/p^0) + Bp, \\ \overline{S} = -(\partial\mu/\partial T)_p = \overline{S}_T^0 - R\,\ln(p/p^0) - p\,dB/dT, \\ \overline{S}_T^0 = -d\mu_T^0/dT, \end{cases} \tag{2.55}$$

where p^0 is the standard pressure. Defining the fugacity f via $\mu = \mu_T^0 + RT\,\ln(f/p^0)$, we obtain

$$\ln f = \ln p + Bp/RT. \tag{2.56}$$

The molar free energy is a function of T and $\overline{\mathcal{V}}$. We have $p = RT(1+B/\overline{\mathcal{V}})/\overline{\mathcal{V}} = -\partial\overline{F}/\partial\overline{\mathcal{V}}$, which yields

$$\overline{F} = \overline{F}_T^0 - RT\,ln\overline{\mathcal{V}} + RTB/\overline{\mathcal{V}}. \tag{2.57}$$

Since $\mu = \overline{F} + p\overline{\mathcal{V}}$, then, starting from $\mu(T,p)$, we get

$$\mu = \mu_T^0 + RT\,\ln(RT/p^0\overline{\mathcal{V}}) + 2RTB/\overline{\mathcal{V}}, \tag{2.58}$$

[6]Other formulae can be found in the literature. So Stephenson [267] gives for Argon in homogeneous phase, an expansion with seven terms which is slightly different from the virial one.

which leads to

$$\overline{F} = \mu - p\overline{V} - \mu_T^0 + RT\,\ln(RT/p^0\overline{V}) + RTB/\overline{V} - RT \tag{2.59}$$

and gives the expression for $\overline{F}_T^0$ as a function of μ_T^0:

$$\overline{F}_T^0 = \mu_T^0 + RT\,\ln(RT/p^0) - RT. \tag{2.60}$$

The supercritical fluid case is presented in the Appendix (see Sect. A.5.1).

2.2.4 EOS for Solids

The internal energy of a solid is $E = E(S, \boldsymbol{\epsilon}, m)$, which depends on eight variables (or seven variables if we assume unit mass or consider one mole of the solid). The simplest equation of state is obtained for elastic solids. For a fluid, the conjugate of the volume is the negative of the pressure. For a solid, the conjugate of the strain tensor $\boldsymbol{\epsilon}$ is the stress tensor $\boldsymbol{\Sigma}$. In the framework of the small perturbation theory, the EOS for thermoelasticity is

$$\boldsymbol{\Sigma} = \lambda\,\mathrm{tr}(\boldsymbol{\epsilon})\mathbf{1} + 2\mu\boldsymbol{\epsilon} - 3K\alpha\theta\mathbf{1}, \tag{2.61}$$

where $\theta = T - T_0$ is the deviation from the reference temperature T_0, and $\lambda, \mu, K = (3\lambda + 2\mu)/3$ are coefficients calculated for isothermal elasticity, and α is the coefficient of dilatation.

We therefore have

$$\boldsymbol{\epsilon} = \frac{1+\nu}{E}\boldsymbol{\Sigma} - \frac{\nu}{E}\mathrm{tr}(\boldsymbol{\Sigma})\mathbf{1} + \alpha\,\theta\,\mathbf{1},$$

where E is Young's modulus and ν is the Poisson coefficient such that

$$E = \mu\frac{3\lambda + 2\mu}{\lambda + \mu}.$$

A second equation of state is needed to completely define the thermodynamic behavior. This equation gives the entropy:

$$\rho_0 s = 3K\alpha\,\mathrm{tr}(\boldsymbol{\epsilon}) + 2\beta\,\theta, \tag{2.62}$$

where $\beta = \rho_0 c_\epsilon / 2T_0$ and c_ϵ is the specific heat per unit mass of the solid at constant strain: $c_\epsilon = (\tilde{d}q/dT)_\epsilon = T(\partial s/\partial T)_\epsilon$. The specific heat per unit mass of the solid at constant stress can be defined in a similar manner to that for fluids; i.e., $c_{\boldsymbol{\Sigma}} = (\tilde{d}q/dT)_{\boldsymbol{\Sigma}} = T(\partial s/\partial T)_{\boldsymbol{\Sigma}}$. We then have $c_{\boldsymbol{\Sigma}} = c_\epsilon + 3K\alpha^2 T_0/\rho_0$. The free energy is

$$\rho_0 f = \frac{1}{2}(\lambda\,(\mathrm{tr}(\boldsymbol{\epsilon})^2 + 2\mu\boldsymbol{\epsilon}\cdot\boldsymbol{\epsilon}) - 3K\,\alpha\,\theta\,\mathrm{tr}\boldsymbol{\epsilon} - \beta\,\theta^2,$$

and the internal energy can be deduced from the relation $e = f + Ts$.

2.3 Properties of Mixtures

2.3.1 General Information

We will only treat gas mixtures and homogeneous liquid solutions here [71, 214, 109, 181]. The preceding arguments remain valid. In particular, for a closed system at equilibrium, we always have

$$dE = T\,dS - p\,d\mathcal{V} = \tilde{d}Q + \tilde{d}W. \tag{2.63}$$

If an unspecified component of the mixture is denoted by j, we have

$$E = E(S,\, \mathcal{V},\, n_j),\;\; j = 1, \ldots N. \tag{2.64}$$

In the same way, we always have

$$T = \partial E/dS,\;\; -p - \partial E/\partial \mathcal{V}. \tag{2.65}$$

We define the molar chemical potential μ_j of species j as

$$\mu_j = \partial E/\partial n_j, \tag{2.66}$$

so that

$$dE = T\,dS - p\,d\mathcal{V} + \sum_{j=1}^{N} \mu_j\,dn_j, \tag{2.67}$$

and so, according to the properties of homogeneity,

$$E = T\,S - p\,\mathcal{V} + \sum_{j=1}^{N} \mu_j\,n_j. \tag{2.68}$$

We extract from these two equations the Gibbs–Duhem relation

$$0 = S\,dT - \mathcal{V}\,dp + \sum_{j=1}^{N} n_j\,d\mu_j. \tag{2.69}$$

Note that, for a closed system in moving equilibrium (reversible transformation), it is necessarily the case that

$$\sum_{j=1}^{N} \mu_j\,dn_j = 0. \tag{2.70}$$

However, we will consider cases where this relation is not obeyed for closed systems. Each component will be regarded as being at equilibrium in the mixture, but there will not be a mutual equilibrium between the components. There are irreversible evolutions of chemical origin. In this case, we can always write $\delta E = \delta W + \delta Q$, but if δW is always equal to $-p\,\delta\mathcal{V}$, δQ will obey

$$\delta Q = T(\delta S - \delta_i S)$$

with $\delta_i S \geq 0$. Thus,

$$\sum_{j=1}^{N} \mu_j\,\delta n_j = -T\,\delta_i S \leq 0, \tag{2.71}$$

where the equals sign corresponds to reversible evolution. The symbol $\tilde{d}$ of the differential forms and the symbol d are obviously replaced by δ, which is characteristic of unspecified infinitesimal transformations.

Using the internal energy, we can also define the thermodynamic potentials

$$\begin{cases} H = E + p\mathcal{V} = TS + \sum_{j=1}^{N} \mu_j\, n_j = H(S,\, p,\, n_j), \\ dH = T\,dS + \mathcal{V}\,dp + \sum_{j=1}^{N} \mu_j\, dn_j, \\ F = E - TS = -p\mathcal{V} + \sum_{j=1}^{N} \mu_j\, n_j = F(T,\, \mathcal{V},\, n_j), \\ dF = -S\,dT - p\,d\mathcal{V} + \sum_{j=1}^{N} \mu_j\, dn_j, \\ G = E - TS + p\mathcal{V} = \sum_{j=1}^{N} \mu_j\, n_j = G(T,\, p,\, n_j), \\ dG = -S\,dT + \mathcal{V}\,dp + \sum_{j=1}^{N} \mu_j\, dn_j. \end{cases} \tag{2.72}$$

The thermodynamic functions for a given mixture cannot generally be deduced from the thermodynamic functions for the pure substances. Thus μ_j, the chemical potential of j in this mixture, is different from the thermodynamic potential $\mu_j^\bullet$ of the pure substance j (μ of the pure substance j).

2.3.2 Partial Molar Quantities

To express the laws of state in the absence of a known fundamental energy law $E(S,\, \mathcal{V},\, n_j)$ or $H(T,\, p,\, n_j)$, one often uses the partial molar quantities. If φ is an extensive thermodynamic function, the partial molar quantities of this extensive quantity will be

$$\overline{\varphi}_j = (\partial\varphi/\partial n_j)_{T,\, p,\, n_{i\neq j}}. \tag{2.73}$$

Thus,

$$\begin{cases} \overline{G}_j = \mu_j, \\ \overline{S}_j = (\partial S/\partial n_j)_{T,\,p,\,n_{i\neq j}} = -(\partial \mu_j/\partial T)_{p,\,n_i}, \\ \overline{\mathcal{V}}_j = (\partial \mathcal{V}/\partial n_j)_{T,\,p,\,n_{i\neq j}} = (\partial \mu_j/\partial p)_{T,\,n_i}. \end{cases} \tag{2.74}$$

Using the Gibbs equation, we find that

$$\begin{cases} \overline{E}_j = (\partial E/\partial n_j)_{T,\,p,\,n_{i\neq j}} = T\,\overline{S}_j - p\,\overline{\mathcal{V}}_j + \mu_j, \\ \overline{H}_j = T\,\overline{S}_j + \mu_j, \\ \overline{F}_j = -p\,\overline{\mathcal{V}}_j + \mu_j. \end{cases} \tag{2.75}$$

In the general case, we set

$$\varphi_j = n_j \overline{\varphi}_j. \tag{2.76}$$

The extensive quantity $\varphi(T,\, p,\, n_j)$ is homogeneous to the first degree with respect to n_j. It follows from this that

$$\varphi = \sum_{j=1}^{N} n_j \overline{\varphi}_j = \sum_{j=1}^{N} \varphi_j. \tag{2.77}$$

This result is valid for E,S,$\mathcal{V}$... and

$$\begin{cases} S = \sum_{j=1}^{N} n_j \overline{S}_j = \sum_{j=1}^{N} S_j, \\ \mathcal{V} = \sum_{j=1}^{N} n_j \overline{\mathcal{V}}_j = \sum_{j=1}^{N} \mathcal{V}_j, \\ E = \sum_{j=1}^{N} n_j \overline{E}_j = \sum_{j=1}^{N} E_j, \end{cases} \tag{2.78}$$

and we have[7]

$$\begin{cases} E_j = T S_j - p\mathcal{V}_j + \mu_j n_j, \\ H_j = T S_j + \mu_j n_j, \\ F_j = -p\mathcal{V}_j + \mu_j n_j. \end{cases} \tag{2.79}$$

[7] Initially, it is as if each species j constitutes an autonomous subsystem that is characterized by its own internal energy E_j and its own variables S_j,$\mathcal{V}_j$,n_j. If this was really the case, it would be necessary to obey the Gibbs relation for each subsystem; i.e., $dE_j = T\,dS_j - p\,d\mathcal{V}_j + \mu_j\,dn_j$. However, this is not the case in general. Indeed, $dE_j = T\,dS_j - p\,d\mathcal{V}_j + \mu_j\,dn_j + S_j\,dT - \mathcal{V}_j\,dp + n_j\,d\mu_j = T\,dS_j - p\,d\mathcal{V}_j + \mu_j\,dn_j + n_j(\overline{S}_j\,dT - \overline{\mathcal{V}}_j\,dp + d\mu_j)$; i.e., $dE_j = T\,dS_j - p\,d\mathcal{V}_j + \mu_j\,dn_j + n_j \sum_j (\partial \mu_j/\partial n_i)_{T,\,p,\,n_{k\neq i}}\,dn_i$. It would therefore be necessary, to ensure that the preceding assertion is obeyed, for $\sum_j (\partial \mu_j/\partial n_i)_{T,\,p,\,n_{k\neq i}}\,dn_i = 0$. This is not true for most cases. However, the results are simplified for ideal mixtures.

2.3.3 Ideal Mixture

If $\mu_j^\bullet$ is the molar thermodynamic potential of the pure substance j, we have $\mu_j^\bullet = \mu_j^\bullet(T, p)$. We know that $X_j = n_j/n$ is the molar fraction of j, where $n = \sum_j n_j$. A mixture is ideal if

$$\mu_j = \mu_j^\bullet + RT \ln X_j. \tag{2.80}$$

We then find that

$$\overline{S}_j = -\partial\mu_j/\partial T = -\partial\mu_j^\bullet/\partial T - R \ln X_j, \tag{2.81}$$

or

$$\overline{S}_j = \overline{S}_j^\bullet - R \ln X_j. \tag{2.82}$$

In the same way, $\overline{\mathcal{V}}_j = \partial\mu_j/\partial p = \partial\mu_j^\bullet/\partial p$; that is to say

$$\overline{\mathcal{V}}_j = \overline{\mathcal{V}}_j^\bullet. \tag{2.83}$$

Thus, partial molar volumes of the species j are unchanged compared to the components in a pure state. In the same way,

$$\begin{cases} \overline{E}_j = T\overline{S}_j - p\overline{\mathcal{V}}_j + \mu_j = T\overline{S}_j^\bullet - p\overline{\mathcal{V}}_j^\bullet + \mu_j^\bullet = \overline{E}_j^\bullet, \\ \overline{H}_j = T\overline{S}_j + \mu_j = T\overline{S}_j^\bullet + \mu_j^\bullet = \overline{H}_j^\bullet, \\ \overline{F}_j = -p\overline{\mathcal{V}}_j + \mu_j n_j = \overline{E}_j - T\overline{S}_j = \overline{E}_j^\bullet - T\overline{S}_j^\bullet + RT \ln X_j, \\ = \overline{F}_j^\bullet + RT \ln X_j. \end{cases} \tag{2.84}$$

2.3.4 Mixture Quantities

The mixture quantity associated with the parameter φ is the quantity

$$\Delta\varphi_m = \varphi - \sum_{j=1}^{N} n_j \overline{\varphi}_j^\bullet. \tag{2.85}$$

For an ideal mixture,

$$\begin{cases} \Delta G_m = RT \sum_j n_j \ln X_j, \ \Delta S_m = -R \sum_j n_j \ln X_j, \\ \Delta\mathcal{V}_m = 0, \ \Delta E_m = 0, \ \Delta H_m = 0. \end{cases} \tag{2.86}$$

2.3.5 Activity

The activity[8] of species j in a mixture is denoted a_j, where

$$\mu_j = \mu_j^\bullet + RT \ln a_j. \tag{2.87}$$

For an ideal mixture, $a_j = X_j$. In the general case,

$$\Delta G_m = RT \sum_j n_j \ln a_j. \tag{2.88}$$

2.3.6 Ideal Mixture of Perfect Gases

For a perfect gas, $\mu = \mu_T^0 + RT \ln(p/p^0)$; therefore,

$$\begin{cases} \mu_j^\bullet = (\mu_T^0)_j + RT \ln(p/p^0), \\ \mu_j = \mu_j^\bullet + RT \ln X_j. \end{cases} \tag{2.89}$$

The partial pressure is defined as follows:

$$p_j = pX_j. \tag{2.90}$$

It follows from this definition that

$$\begin{cases} p = \sum_j p_j, \\ \mu_j = (\mu_T^0)_j + RT \ln(pX_j/p^0) = \mu_j(T, p_j) \end{cases} \tag{2.91}$$

and

$$\begin{cases} G_j = n_j \mu_j = n_j \left[(\mu_T^0)_j + RT \ln(pX_j/p^0)\right], \\ dG_j = n_j[d(\mu_T^0)_j/dT + R \ln(pX_j/p^0)]dT + n_j RT/p_j \, dp_j + \mu_j \, dn_j \end{cases} \tag{2.92}$$

or

$$dG_j = -S_j \, dT + \mathcal{V} \, dp_j. \tag{2.93}$$

Recall that

$$G_j = G_j(T, p_j, n_j) = \mu_j n_j. \tag{2.94}$$

We can therefore see that the formalism used for simple systems will also be valid for component j, with the proviso that this component occupies the

[8]The chemical activity of a species is the active concentration of this species. In a solution, electrostatic interactions between different species reduce their potential reactivities. Therefore, the concentration term is corrected by a factor γ (which is less than unity) called the coefficient of activity. This occurs in the definition of chemical potential.

total volume of the mixture with a partial pressure of p_j at the temperature T. The classical relations are thus valid in this case; in particular the Euler, Gibbs, and Gibbs–Duhem relations:

$$\begin{cases} E_j = T\,S_j - p_j\,\mathcal{V} + \mu_j\,n_j = E_j(S_j,\,\mathcal{V},\,n_j), \\ dE_j = T\,dS_j - p_j\,d\mathcal{V} + \mu_j\,dn_j, \\ 0 = S_j\,dT - \mathcal{V}\,dp_j + n_j\,d\mu_j. \end{cases} \tag{2.95}$$

Here, the total system (mixture) can be regarded as the sum of N subsystems whose extensive properties are additive:

$$G = \sum_j G_j,\ E = \sum_j E_j,\ H = \sum_j H_j,\ S = \sum_j S_j,\ F = \sum_j F_j.$$

Therefore,

$$\begin{cases} E = \sum_j n_j(\overline{E}^0_T)_j = \sum_j n_j[(\overline{E}^0_0)_j + \int_{T^0}^T \overline{C}_{v,\,j}(T)\,dT], \\ H = \sum_j n_j(\overline{H}^0_T)_j = \sum_j n_j[(\overline{H}^0_0)_j + \int_{T^0}^T \overline{C}_{p,\,j}(T)\,dT], \\ (\overline{H}^0_0)_j = (\overline{E}^0_0)_j + RT^0,\ \overline{C}_{p,\,j} = \overline{C}_{v,\,j} + R, \\ S = \sum_j n_j[(\overline{S}^0_T)_j - R\,\ln(pX_j/p^0)], \\ (\overline{S}^0_T)_j = (\overline{S}^0_0)_j + \int_{T^0}^T (\overline{C}_{p,\,j}/T)\,dT, \\ F = \sum_j n_j[(\overline{F}^0_T)_j + RT\,\ln(pX_j/p^0)], \\ (\overline{F}^0_T)_j = (\overline{F}^0_0)_j + \int_{T^0}^T [\overline{C}_{p,\,j}(1 - T/T') - R - (\overline{S}^0_0)_j]\,dT', \\ (\overline{F}^0_0)_j = (\overline{E}^0_0)_j - T\,(\overline{S}^0_0)_j, \\ G = \sum_j n_j[(\mu^0_T)_j + RT\,\ln(pX_j/p^0)], \\ (\mu^0_T)_j = (\mu^0_0)_j + \int_{T^0}^T [\overline{C}_{p,\,j}(1 - T/T') - R - (\overline{S}^0_0)_j]\,dT', \\ (\mu^0_0)_j = (\overline{H}^0_0)_j - T\,(\overline{S}^0_0)_j. \end{cases} \tag{2.96}$$

It is important to note here that, as we mentioned in Sect. 2.2.3, the tabulated standard thermodynamical functions are not independent of each other.[9]

A number (equal to the number of elements in the periodic table) of basic species are defined from which all chemical species can be formed through chemical reactions. These basic species are generally stable at atmospheric temperatures, and their enthalpies are zero under standard conditions. However, the enthalpies of the other species are not zero under standard conditions, but are instead equal to their heats of formation from basic species (see for example [10]).

[9] If we use the variables T, $\mathcal{V}$, X_j instead of T, p, X_j, the Helmholtz free energy is then $F = \sum_j n_j[(\breve{F}^0_T)_j + RT\,ln(\mathcal{V}^0\,X_j/\mathcal{V})]$, $(\breve{F}^0_T)_j = (\breve{F}^0_0)_j + RT\,\ln(T/T^0)$.

2.3.7 A Mixture of Real Gases That Obeys the Virial Relation

For one mole of a real gas that obeys the virial relation (limited to coefficient B), we have

$$\begin{cases} \mu = \mu_T^0 + RT\,\ln(p/p^0) + Bp, \\ \overline{S} = -d\mu_T^0/dT - R\,\ln(p/p^0) - p\,dB/dT, \\ \overline{\mathcal{V}} = RT/p + B. \end{cases} \tag{2.97}$$

Starting from statistical thermodynamics, and considering first-order molecular interactions, it can be shown that

$$G = \sum_j n_j[RT\,\ln(p/p^0) + p\sum_{k,\,l} B_{k,\,l}X_kX_l + (\mu_T^0)_j + RT\,\ln X_j] \tag{2.98}$$

in a mixture, which leads to the following expression for the chemical potential of a binary mixture:

$$\mu_j = \mu_j^\bullet + RT\,\ln X_j - p\,X_i^2(B_{11} - 2B_{12} + B_{22}),\ i \neq j,$$

$$\mu_j^\bullet = (\mu_T^0)_j + RT\,\ln(p/p^0) + B_{jj}p.$$

For N species,

$$\mu_j = (\mu_T^0)_j + RT\,\ln(pX_j/p^0) + p\,(2\sum_k B_{kj}X_k - \sum_{k,\,l} B_{kj}X_kX_l). \tag{2.99}$$

The B_{ij} coefficients relate to the interactions between molecules of type i and j, so $B_{ij} = B_{ji}$. Note that the mixture is ideal, in the binary case, for $B_{12} = (1/2)\,(B_{11} + B_{22})$. However, this case is not supported by any experimental or theoretical justification, so it is a completely arbitrary assumption.

2.3.8 Liquid Solution

Let us consider a mixture of two liquids in the presence of their vapors, constituting an ideal mixture of perfect gases at a given T:

$$\begin{cases} G = \mu_1^g\,n_1^g + \mu_2^g\,n_2^g + \mu_1^l\,n_1^l + \mu_2^l\,n_2^l, \\ n_1^g + n_1^l = n_1 = \text{const.},\ n_2^g + n_2^l = n_2 = \text{const.}, \\ dG = -S\,dT + \mathcal{V}\,dp + (\mu_1^l - \mu_1^g)dn_1^l + (\mu_2^l - \mu_2^g)dn_2^l. \end{cases} \tag{2.100}$$

Let us consider a reversible evolution with p and T remaining constant. We have two differential forms $\widetilde{dW} = -p\,d\mathcal{V}$, $\widetilde{dQ} = T\,dS$ (see Fig. 2.5), and $dE = T\,dS - p\,d\mathcal{V}$, which leads to $dG = -S\,dT + \mathcal{V}\,dp$.

Therefore, for any reversible evolution of the system, with T and p constant, we have $\mu_1^l = \mu_1^g$ or $\mu_1^l = (\mu_T^0)_1^g + RT\,\ln(p_1/p^0)$, and $\mu_2^l = \mu_2^g$ or

$\mu_2^l = (\mu_T^0)_2^g + RT\,\ln(p_2/p^0)$. The equilibrium of each pure liquid with its vapor at the same temperature is given by $(\mu_1^\bullet)^l = (\mu_1^\bullet)^g = (\mu_T^0)_1^g + RT\,\ln(p_1^\bullet/p^0)$ and $(\mu_2^\bullet)^l = (\mu_2^\bullet)^g = (\mu_T^0)_2^g + RT\,\ln(p_2^\bullet/p^0)$, where $p_j^\bullet$ is the saturated vapor pressure of species j.

By subtraction,

$$\mu_1^l - (\mu_1^\bullet)^l = RT\,\ln(p_1/p_1^\bullet),\ \mu_2^l - (\mu_2^\bullet)^l = RT\,\ln(p_2/p_2^\bullet), \qquad (2.101)$$

where the activities $a_1 = p_1/p_1^\bullet$ and $a_2 = p_2/p_2^\bullet$, as defined by (2.87), and their physical significances are thus clarified.

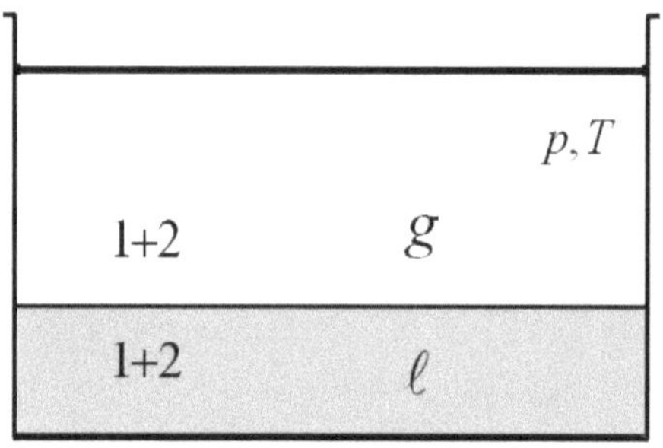

Fig. 2.5. Liquid solution in equilibrium with vapor

Our knowledge of the functions $a_i(T, p, n_i)$ as of $\mu_i^l(T)$ includes all of the equations of state for the solution, and we have $\mu_i = \mu_i^\bullet + RT\,\ln a_i$.

For an ideal liquid solution, we have $a_i = p_i/p_i^\bullet$ (Raoult's law), and so

$$\mu_i = \mu_i^\bullet + RT\,\ln X_i^l. \qquad (2.102)$$

Often, one of the components of a solution obeys Raoult's law, but not necessarily the other. How can we determine a_2, knowing that $a_1 = X_1$? We can write the Gibbs–Duhem equation for a given p and T: $n_1\,d\mu_1 + n_2\,d\mu_2 = 0$ or $X_1\,d\mu_1 + X_2\,d\mu_2 = 0$. This gives $X_1\,RT\,dX_1/X_1 + X_2\,RT\,da_2/a_2 = 0$, with $dX_1 + dX_2 = 0$. Then $da_2/a_2 = dX_2/X_2$, and

$$\ln(a_2) = \ln(X_2) + f(T, p). \qquad (2.103)$$

We set $a_2 = p_2/p_2^\bullet = b(T, p)\,X_2/p_2^\bullet$. If $b(T, p) = \text{const.}$, we obtain Henry's law.

2.3.9 Mixture of Real Fluids

Just as we did for real pure fluids in Sect. 2.2.3, we can model mixtures of real fluids that do not obey the virial relation. To do this, we first consider a standard reference state that is an ideal mixture of perfect gases corresponding to the low pressure limit (or a large molar volume) of a mixture of real fluids, and then consider the rules of mixture that govern the equation of state.

Rules of Mixture For a Real Mixture

The rules of mixture are semi-empirical. For van der Waals, Redlich–Kwong, or Peng–Robinson laws of state, we have [210, 233]

$$p = \frac{RT}{\mathcal{V} - b_m} - \frac{a_m}{\mathcal{V}^2 + ub_m + wb_m^2}. \tag{2.104}$$

$a_m = \sum_{i,\,j} X_i X_j \sqrt{a_i a_j}(1 - \bar{k}_{ij})$ and $b_m = \sum_i X_i b_i$ for instance. Knowledge of the correct rules of mixture and the correct coefficients k_{ij} (some data are available) is essential in order to deduce the laws of state.[10]

Several rules of mixture exist. If Q is a parameter, and Q_m is its value for a mixture, the general van der Waals expression is

$$Q_m = \sum_{i,\,j} X_i X_j Q_{i,\,j}. \tag{2.105}$$

$Q_{ii} = Q_i$ for the pure substance, and Q_{ij} where i and j are different corresponds to mixtures. Using this general expression, we can consider the following rules:

- $Q_{ij} = (Q_{ii} + Q_{jj})/2$, which gives $Q_m = \sum_i X_i Q_i$
- $Q_{ij} = \sqrt{Q_{ii} Q_{jj}}$, which gives $Q_m = (\sum_i X_i \sqrt{Q_i})^2$.

More sophisticated rules involve binary interaction parameters $k_{i,\,j}$ that are usually independent of temperature, pressure and concentration, such as:

- $Q_{ij} = k_{i,\,j}(Q_{ii} + Q_{jj})/2$, where for example $k_{ii} = 1$ and $k_{ij \neq i}$ is determined experimentally
- $Q_{ij} = k_{i,\,j}\sqrt{Q_{ii} Q_{jj}}$, with $k_{ii} = 1$.

Unlike a one-component fluid, there is no critical point for a mixture, but we can define a pseudo-critical state (see the Appendix, Sect. A.5.1).

[10]For example, the couple naphthalene–CO_2 yields $n^2 a_m = n_1^2 a_1 + 2(1 - k) n_1 n_2 \sqrt{n_1 n_2} + n_2^2 a_2$, $nb_m = n_1 b_1 + n_2 b_2$. The coefficients are: $a_1 = 0.39608(1 + 0.72782\sqrt{T/304.1})$, $b_1 = 2.66545 \times 10^{-5}$, $a_2 = 4.37138(1 + 0.81576\sqrt{T/748.4})$, $b_2 = 1.19533 \times 10^{-4}$.

Expression for the Free Energy of a Real Mixture

Euler's relation results from the first-degree homogeneous character of the thermodynamic potentials in the extensive quantities upon which they depend (and from the zero-degree homogeneous character of the thermodynamic potentials in the intensive quantities upon which they depend, if necessary). The free energy F of a mixture is expressed in canonical form as a function of its temperature T, its volume $\mathcal{V}$, and the molar amounts of the species n_j, $j = 1 \ldots, N$. The partial derivatives of F are:

$$(\frac{\partial F}{\partial T})_{\mathcal{V},n_j} = -S, \; (\frac{\partial F}{\partial \mathcal{V}})_{T,n_j} = -p, \; (\frac{\partial F}{\partial n_j})_{T,\mathcal{V},n_{i\neq j}} = \mu_j, \tag{2.106}$$

so the Euler and Gibbs relations are (respectively):

$$\begin{cases} F = -p\,\mathcal{V} + \sum_j \mu_j\, n_j, \\ dF = -S\,dT - p\,d\mathcal{V} + \sum_j \mu_j\, dn_j, \end{cases} \tag{2.107}$$

and the Gibbs–Duhem relation is found to be

$$0 = S\,dT - \mathcal{V}\,dp + \sum_j n_j\, d\mu_j. \tag{2.108}$$

We saw previously that if we know the rule of mixture it is then possible to express the pressure p of the mixture as a function of the temperature T, the volume $\mathcal{V}$, and the molar amounts of the species n_j, $j = 1 \ldots N$. These variables are the canonical variables of the free energy. We have

$$(\frac{\partial F}{\partial \mathcal{V}})_{T,\,n_j} = -p(T,\,\mathcal{V},\,n_j). \tag{2.109}$$

Integrating (2.109) with respect to $\mathcal{V}$ leads to

$$F(T,\,\mathcal{V},\,n_j) = F^0(T,\,\mathcal{V}^0,\,n_j) - \int_{\mathcal{V}^0}^{\mathcal{V}} p\,d\mathcal{V}, \tag{2.110}$$

where $\mathcal{V}^0$ is a standard volume. The standard state is thus characterized by the values T, $\mathcal{V}^0$, n_j. One can assume, as described in Sect. 2.2.3 for pure substances, that it is always possible to find a standard value $\mathcal{V}^0$ such that the mixture is an ideal mixture of perfect gases for values of $\mathcal{V}$ ranging between $\mathcal{V}^0$ and infinity. The preceding relation can then be written as

$$\begin{cases} F(T,\,\mathcal{V},\,n_j) = F^0(T,\,\mathcal{V}^0,\,n_j) - \int_{\mathcal{V}^0}^{\infty} p\,d\mathcal{V} - \int_{\infty}^{\mathcal{V}} p\,d\mathcal{V}, \\ F(T,\,\mathcal{V},\,n_j) = F^0(T,\,\mathcal{V}^0,\,n_j) - \int_{\infty}^{\mathcal{V}} (p - nRT/\mathcal{V})\,d\mathcal{V} - nRT\,\ln(\mathcal{V}/\mathcal{V}^0). \end{cases} \tag{2.111}$$

$F^0(T, \mathcal{V}^0, n_j)$ is given by the relation

$$\begin{cases} F^0 = \sum_j n_j[(\breve{F}_T^0)_j + RT\ \ln(\mathcal{V}^0 X_j/\mathcal{V})], \\ (\breve{F}_T^0)_j = (\overline{F}_0^0)_j + RT\ \ln(T/T^0). \end{cases} \tag{2.112}$$

Thus, knowledge of the rule of mixture that gives p as a function of T, $\mathcal{V}$ and n_j on the one hand, and the thermodynamic functions of the ideal mixture—i.e., mainly the quantities $(\overline{H}_0^0)_j$, $(\overline{S}_0^0)_j$, $\overline{C}_{p,\,j}(T)$—on the other hand, is sufficient to allow the determination of the free energy of the real fluid mixture.

2.4 Reactive Mixtures

2.4.1 Enthalpy of a Chemical Reaction

Consider a chemical reaction, such as that shown in (2.12) where ξ is the progress variable, in a homogeneous mixture. We define the energy $r_{T,p}$ released by this chemical reaction at a given T and p as

$$r_{T,p} = (\frac{\partial h}{\partial \xi})_{T,p}. \tag{2.113}$$

For a mixture of perfect gases we obtain (see [10])

$$r_{T,p} = \Delta H = -\sum_j \nu_j \mathcal{M}_j (q_f^0)_j, \tag{2.114}$$

where $(q_f^0)_j$ is the heat of formation per unit mass of the species under standard conditions, $(q_f^0)_j = (\overline{H}_0^0)_j/\mathcal{M}_j$, and ΔH is called the enthalpy of reaction.

Recall that, for a perfect gas, the standard enthalpy $(\overline{H}_T^0)_j$ per mole is

$$(\overline{H}_T^0)_j = (\overline{H}_0^0)_j + \int_{T^0}^{T} \overline{C}_{p,\,j}(T)\,dT.$$

The quantity ΔH is the energy released by the chemical reaction per mole for an ideal mixture of perfect gases at constant temperature and pressure.

For a combustion reaction, which is exothermic, ΔH is positive. The quantity

$$\Delta H\,\dot{\zeta} = -\sum_j \nu_j \mathcal{M}_j (q_f^0)_j\,\dot{\zeta} \tag{2.115}$$

is then the heat released during the combustion reaction. For a multireactive mixture (see Eq. 2.17), the heat released becomes[11]

[11] As noted in [206], the full term $\dot{W}'_T = -\sum_j \dot{W}_j h_j$, including the contributions from sensible heat enthalpy (i.e., the enthalpy due to an increase in temperature) terms, is also called the "heat released" by various authors.

$$\dot{W}_T = -\sum_j \dot{W}_j (q_f^0)_j = -\sum_{j,r} \nu_{jr} \mathcal{M}_j (q_f^0)_j \, \dot{\zeta}_r = \sum_{r=1}^{K} \Delta H_r \, \dot{\zeta}_r . \tag{2.116}$$

2.4.2 Entropy Production in a Homogeneous Reactive Mixture

In Sect. 2.3 we noted that the given laws of state did not imply that the system studied (the mixture) was necessarily at equilibrium. Each component j (which has a chemical potential μ_j that is a function of T, p, and n_j, and can also depend on n_i where $i \neq j$) of the system was considered a subsystem that was always at equilibrium; however, mutual equilibrium did not necessarily occur between such subsystems. The system was a homogeneous mixture with uniform thermodynamic properties, while the subsystem for each component was comparable to an open subsystem due to the potential for chemical reactions. It was apparent from (2.71) that, in all cases, the entropy produced during an infinitesimal transformation was

$$\delta_i S = -\frac{1}{T} \sum_j \mu_j \, \delta n_j . \tag{2.117}$$

For a single chemical reaction (as given by Eq. 2.12, with stoichiometric coefficients ν_j), in the absence of diffusion (which is the case for a homogeneous closed system with uniform properties), we have

$$dn_j = \nu_j \, d\xi , \tag{2.118}$$

where ξ is the progress variable for the reaction. The production of entropy becomes

$$\delta_i S = -\frac{\delta \xi}{T} \sum_j \nu_j \mu_j . \tag{2.119}$$

The chemical affinity A is defined as

$$A = -\sum_j \nu_j \mu_j . \tag{2.120}$$

In this case,

$$\delta_i S = \frac{A}{T} \, \delta \xi \geq 0 , \tag{2.121}$$

where an equals sign would correspond to the equilibrium of the mixture. Two equilibrium cases are possible:

- $\delta\xi = 0$, so the mixture is frozen; i.e., the concentrations of its components do not change, meaning that the chemical reaction does not take place or that it is completely inhibited
- The component concentrations can evolve during the reversible transformation, yielding $A = 0$; i.e., chemical equilibrium.

Aside from these two cases, equilibrium is not possible—the transformation is chemically irreversible and

$$\delta_i S = \frac{A}{T}\,\delta\xi > 0. \tag{2.122}$$

The Gibbs relation then becomes

$$\begin{cases} dE = T\,dS - p\,d\mathcal{V} + \sum_j \mu_j dn_j, \\ dE = T\,dS - p\,d\mathcal{V} - A\,d\xi. \end{cases} \tag{2.123}$$

The mixture is at equilibrium, or is undergoing a reversible transformation, if

$$A\,d\xi = 0. \tag{2.124}$$

We then derive the expression for the second law for reversible transformations:

$$dE = \tilde{d}Q + \tilde{d}W = T\,dS - p\,d\mathcal{V}.$$

For a chemically irreversible transformation,

$$\delta E = \delta Q + \delta W = T(\delta S - \delta_i S) - p\,d\mathcal{V}, \tag{2.125}$$

where the term $T\,\delta_i S$ is equal to $A\,d\xi = -\sum_j \mu_j dn_j$.

2.4.3 Chemical Reaction at Equilibrium for a Mixture of Perfect Gases

We know that, for species j in a mixture of perfect gases,

$$\mu_j = (\mu_T^0)_j + RT\,\ln(p_j/p^0), \tag{2.126}$$

where $(\mu_T^0)_j$ is a function of the temperature alone. The chemical affinity of a given reaction will be

$$A = -\sum_j \nu_j[\mu_T^0)_j + RT\,\ln(p_j/p^0)], \tag{2.127}$$

or, since $p_j = C_j RT$,

$$A = -\sum_j \nu_j[(\mu_T^0)_j + RT\,\ln(RT/p^0) + RT\,\ln C_j]. \tag{2.128}$$

By setting

$$\ln K_C = -(1/RT)\sum_j \nu_j[(\mu_T^0)_j + RT\,\ln(RT/p^0)], \tag{2.129}$$

we therefore obtain

$$\prod_{j=1}^{N}(C_j)^{\nu_j} = K_C\,\exp(-A/RT). \tag{2.130}$$

Note that K_C is a function of T alone. When the reaction is at chemical equilibrium,

$$\prod_{j=1}^{N}(C_j)^{\nu_j} = K_C(T). \tag{2.131}$$

The function $K_C(T)$ is termed the equilibrium constant of concentrations.[12]

2.4.4 Mixture of Perfect Gases in Chemical Equilibrium

Equilibrium Composition

The results in the previous section were obtained for a given chemical reaction within the mixture. Let us consider now a multireactive mixture. Suppose that all of the species in the mixture can be derived from a certain number of basic species present in the mixture:

$$\mathcal{E}_i = \sum_{l=1}^{L}\beta_{li}\,\mathcal{A}_l,\ i = (L+1),\ldots N. \tag{2.132}$$

The conservation equations for the basic species are:

$$n_l + \sum_{i=l+1}^{N}\beta_{li}\,n_i = n_l^0,\ l = 1,\ldots L, \tag{2.133}$$

meaning that the n_i, $i = (L+1),\ldots N$, form a system of progress variables, and the affinities of formation for the reactions are then

$$\begin{cases} -A_i = \mu_i - \sum_{l=1}^{L}\beta_{li}\mu_l,\ i = (L+1),\ldots N, \\ -A_i = (\mu_T^0)_i - \sum_{l=1}^{L}\beta_{li}(\mu_T^0)_l \\ -RT(\sum_{l=1}^{L}\beta_{li} - 1)\ln(RT/p^0) + RT\,\ln(\prod_l(C_l/C_i)). \end{cases} \tag{2.134}$$

[12] Note that there is also an equilibrium constant of partial pressures $K_p(T)$, defined as $\ln K_p = -(1/RT)\sum_j \nu_j(\mu_T^0)_j$. The chemical equilibrium relation then becomes $\prod_{j=1}^{N}(p_j/p^0)^{\nu_j} = K_p(T)$.

By setting

$$\ln K_{C,i} = [\sum_{l=1}^{L} \beta_{li}(\mu_T^0)_l - (\mu_T^0)_i + RT(\sum_{l=1}^{L} \beta_{li} - 1)\ln(RT/p^0)]/RT, \quad (2.135)$$

we obtain

$$\prod_{l=1}^{L}(C_l/C_i) = K_{C,i}\exp(-A_i/RT),\ i = (L+1)\dots N. \quad (2.136)$$

At equilibrium, A_i vanishes and we get

$$\prod_{l=1}^{L}(C_l/C_i) = K_{C,i},\ i = (L+1)\dots N. \quad (2.137)$$

We can solve the system of equations obtained by noting the relation $\rho n_i = C_i$, where n_i is taken for the unit mass of mixture ($n_i = Y_i/\mathcal{M}_i$).

Specific Heats and Other Coefficients at Chemical Equilibrium

In order to analyze a fluid mixture flow while assuming chemical equilibrium, we need to:

1. Calculate the mixture composition at a given pressure and temperature from the previous equilibrium relations
2. Determine the partial derivatives of the thermodynamic functions.

We will not discuss the numerical methods that can be used to calculate the mixture composition in detail here (for more on this, see for example [10]).

At constant composition, for a mass m of an ideal mixture of perfect gases (called a "frozen mixture" or a "mixture with a frozen composition;" m is a given mass that may be unit mass), we simply have

$$C_{pf} = (\frac{\tilde{d}Q}{dT})_{p,n_j} = \frac{1}{T}(\frac{\partial S}{\partial T})_{p,n_j} = (\frac{\partial H}{\partial T})_{p,n_j} = \sum_{j=1}^{N} n_j \bar{C}p,j, \quad (2.138)$$

or, for a mixture with unit mass,

$$c_{pf} = (\frac{\tilde{d}q}{dT})_{p,n_j} = \frac{1}{T}(\frac{\partial s}{\partial T})_{p,n_j} = (\frac{\partial h}{\partial T})_{p,n_j} = \sum_{j=1}^{N} Y_j \bar{C}p,j. \quad (2.139)$$

For the isentropic exponent γ_f and the characteristic speed a_f, we have (respectively):

$$\begin{cases} \gamma_f = (\partial \ln p/\partial \ln \rho)_{s,n_j} = c_{pf}/c_{vf} = c_{pf}/(c_{pf} - nR), \\ a_f^2 = c_{pf}\, nRT/(c_{pf} - nR). \end{cases} \tag{2.140}$$

This type of result makes it possible to determine the variables as functions of one of them provided the coefficients obtained do not vary too much in the vicinity of a given state. We then get laws of the form

$$p\,T^{-\gamma_f/(\gamma_f - 1)} = \text{const.}$$

At chemical equilibrium, the problem is more complex. The chemical composition (the values of $n_{je}, j = 1 \ldots N$) is obtained by noting $A_i = 0,\ i = L+1, \ldots N$. In particular, we must also determine partial derivatives that characterize the variation in composition:

$$\begin{cases} D_{T,j} = (\partial \ln n_j/\partial \ln T)_p,\ D_{p,j} = (\partial \ln n_j/\partial \ln p)_T, \\ D_T = (\partial \ln n/\partial \ln T)_p,\ D_p = (\partial \ln n/\partial \ln p)_T, \end{cases} \tag{2.141}$$

where $n(p,\, T) = \sum_{j=1}^{N} n_j = p/\rho RT$ is the total number of moles of a unit mass of the mixture (these coefficients D_T, D_p vanish for a frozen mixture).[13]

We then obtain

$$\begin{cases} c_{pe} = (\partial h/\partial T)_{p,n_{je}} = c_{pf} + \sum_{j=1}^{N} n_j\, \bar{H}_j\, D_{T,j}, \\ \gamma_e = (\partial \ln p/\partial \ln \rho)_{s,n_{je}} = c_{pe}/[c_{pe}(1 - D_p) - nR(1 + D_T)^2]. \end{cases} \tag{2.142}$$

Note that it is not sufficient to replace c_{pf} with c_{pe} in order to calculate equilibrium quantities such as γ_e, a_e, or $(\partial \ln T/\partial \ln p)_e$. For instance, we have $(\partial \ln T/\partial \ln p)_{sn_{je}} = (1 + D_T)\, nR/c_{pe}$, even when $(\partial \ln T/\partial \ln p)_{sn_j} = nR/c_{pf}$. Also note that we have not discussed the signs of the different coefficients that we introduced in this section. Such a discussion needs to take thermodynamic stability into account; this is addressed in Sect. 2.5.

2.4.5 Unspecified Multireactive Mixtures

Let us assume that the number of reactions K is fixed and that the progress variables are $\xi_1, \xi_2, \ldots \xi_K$. There are then K chemical affinities that vanish at equilibrium. These reactions are not inevitably independent; i.e., the rank of the matrix for ν_{jr} can be lower than K. If this is the case, we must choose R independent relations among the K present and thus define a number R of "reduced" progress variables $\overline{\xi_1}, \overline{\xi_2}, \ldots \overline{\xi_R}$.

---

[13]The coefficients can be determined by the Jacobian method, as performed by Barrère and Prud'homme [10].

2.4.6 Reactive Solutions

This time, consider a reactive liquid solution in equilibrium with its vapor at a temperature T and pressure p. For each component we have

$$\mu_i = \mu_i^\bullet + RT \ln a_i. \tag{2.143}$$

The entropy production due to chemical irreversibility is

$$\delta_i S = -\frac{1}{T}\sum_j \mu_j \,\delta n_j. \tag{2.144}$$

In the case of only one reaction, we set

$$A = -\sum_j \nu_j \mu_j, \quad \delta n_j = \nu_j \,\delta\xi, \tag{2.145}$$

and we find that

$$\delta_i S = (A/T)\,\delta\xi. \tag{2.146}$$

This time,

$$A = -\sum_j \nu_j \mu_j^\bullet - RT \ln(a_j)^{\nu_j}. \tag{2.147}$$

Setting

$$\ln K_a = -\sum_j \nu_j \mu_j^\bullet, \tag{2.148}$$

we get

$$\prod_{j=1}^{N} (a_j)^{\nu_j} = K_a(T, p) \exp(-A/RT), \tag{2.149}$$

and, at equilibrium,

$$\prod_{j=1}^{N} (a_j)^{\nu_j} = K_a(T, p). \tag{2.150}$$

2.4.7 Extension to Nonequilibrium Mixtures

The results obtained above also allow us to consider cases where the chemical affinities do not cancel. For instance, the rate of entropy production for a one-reaction mixture will be

$$\delta_i S = (A/T)\,\dot{\xi}. \tag{2.151}$$

Chemical kinetics (see Sect. 3.4) allows us to obtain the expression for the reaction rate $\dot{\xi}$ as a function of the state variables and then to determine the time evolution of the mixture.

2.5 Thermodynamic Stability

Sections 2.1 to 2.3 discussed the significance of the first derivatives of the internal energy with respect to the extensive quantities on which the internal energy depends. For a mixture of N chemical species [181], we have

$$E = E(S, \mathcal{V}, n_1, n_2, \dots n_N), \tag{2.152}$$

and (2.65) and (2.66) lead us to deduce that

$$T = \partial E/dS, \; -p = \partial E/\partial \mathcal{V}, \; \mu_j = \partial E/\partial n_j. \tag{2.153}$$

The chemical potential g_j per unit mass was also introduced, where $g_j = \mu_j/\mathcal{M}_j$, and this leads to results similar to those shown above:

$$\begin{cases} E = E(S, \mathcal{V}, m_1, m_2, \dots m_N), \\ T = \partial E/dS, \; -p = \partial E/\partial \mathcal{V}, \; g_j = \partial E/\partial m_j. \end{cases} \tag{2.154}$$

We see here that the second law of thermodynamics leads, in the case of stable equilibrium, to conditions on the partial second derivatives with respect to the extensive variables of the internal energy, and this has consequences for the usual quantities and relations. This is thermodynamic stability.

2.5.1 The Stability Matrix

We now consider a mixture of N species distributed on both sides of a non-deformable, adiabatic and impermeable surface inside a container. The walls of the container are isolated from the outside (see Fig. 2.6a).

Each subsystem is at thermodynamic equilibrium, with fixed concentrations. If we remove the wall (we assume that this requires a negligible amount of work and that no chemical reaction takes place during the transformation), the mixture then occupies the entire container and will naturally reach a new state of equilibrium. This is a characteristic of a medium with thermodynamic stability (see Fig. 2.6b).

We then obtain the following relations:

$$\begin{cases} E + E' = E_f, \; S + S' \leq S_f, \\ \mathcal{V} + \mathcal{V}' = \mathcal{V}_f, \; n_j + n'_j = n_{jf}. \end{cases} \tag{2.155}$$

Then

$$E(S, \mathcal{V}, n_j) + E(S', \mathcal{V}', n'_j) = E(S_f, \mathcal{V} + \mathcal{V}', n_j + n'_j). \tag{2.156}$$

Since $T = \partial E/dS \geq 0$,

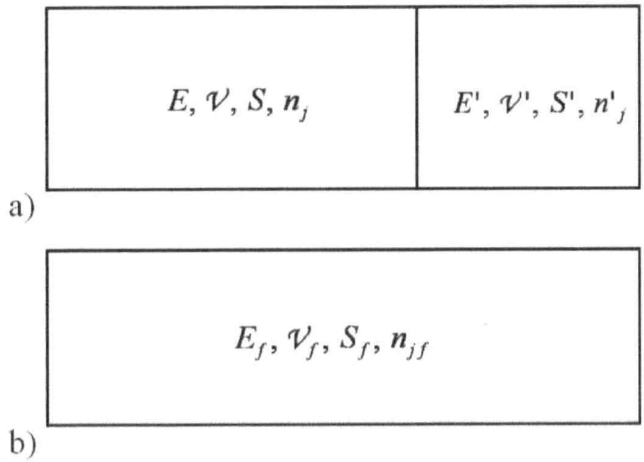

Fig. 2.6. A chemical mixture: **a** in a container with two sections; **b** after removing the inner surface

$$E(S_f, \ldots) \geq E(S + S', \ldots). \tag{2.157}$$

It follows that

$$E(S,\, \mathcal{V},\, n_j) + E(S',\, \mathcal{V}',\, n_j') \geq E(S + S',\, \mathcal{V} + \mathcal{V}',\, n_j + n_j'). \tag{2.158}$$

The first-degree homogeneous character of the function $E = E(S,\, \mathcal{V},\, n_j)$ enables us to write

$$\frac{1}{2}[E(S,\, \mathcal{V},\, n_j) + E(S',\, \mathcal{V}',\, n_j')] \geq E(\frac{S + S'}{2},\, \frac{\mathcal{V} + \mathcal{V}'}{2},\, \frac{n_j + n_j'}{2}). \tag{2.159}$$

This leads us to conclude that $E = E(S,\, \mathcal{V},\, n_j)$ *is a convex function*. It is easy to show that this result implies a positive definite stability matrix:

$$\left\| \begin{array}{ccccc} \partial^2 E/\partial S^2 & \partial^2 E/\partial S \partial \mathcal{V} & \partial^2 E/\partial S \partial n_1 & \ldots & \partial^2 E/\partial S \partial n_N \\ \partial^2 E/\partial \mathcal{V} \partial S & \partial^2 E/\partial \mathcal{V}^2 & \partial^2 E/\partial \mathcal{V} \partial n_1 & \ldots & \partial^2 E/\partial \mathcal{V} \partial n_N \\ \partial^2 E/\partial n_1 \partial S & \partial^2 E/\partial n_1 \partial \mathcal{V} & \partial^2 E/\partial n_1^2 & \ldots & \partial^2 E/\partial n_1 \partial n_N \\ \ldots & \ldots & \ldots & \ldots & \ldots \\ \ldots & \ldots & \ldots & \ldots & \ldots \\ \partial^2 E/\partial n_N \partial S & \partial^2 E/\partial n_N \partial \mathcal{V} & \partial^2 E/\partial n_N \partial n_1 & \ldots & \partial^2 E/\partial n_N^2 \end{array} \right\| . \tag{2.160}$$

In this case of a symmetrical matrix, the necessary and sufficient condition for this is that all of the principal minors should be positive. Thus,[14]

[14]Similar results are obtained if the molar amounts of the species n_j are replaced with the masses of the species m_j.

$$\begin{cases} \partial^2 E/\partial S^2 = \partial T/\partial S > 0, \\ \partial^2 E/\partial \mathcal{V}^2 = -\partial p/\partial \mathcal{V} > 0, \\ (\partial^2 E/\partial S^2)(\partial^2 E/\partial \mathcal{V}^2) - (\partial^2 E/\partial S \partial \mathcal{V})^2 > 0, \\ \partial^2 E/\partial n_j^2 = \partial \mu_j/\partial n_j > 0, \ \ldots \end{cases} \tag{2.161}$$

What are the consequences of this for a mixture of unit mass? We need to consider the relation

$$\sum_{j=1}^{N} \mathcal{M}_j \, n_j = 1. \tag{2.162}$$

This relation does not modify the convexity of the function E, which then becomes e and depends on $N+1$ independent variables instead of $N+2$.

This result can be generalized if an additional condition on concentration variations is given. For a certain number R of independent progress variables, we have

$$n_j = n_j^0 + \sum_{r=1}^{R} \nu_{jr} \xi_r, \tag{2.163}$$

with

$$\sum_{j=1}^{N} \mathcal{M}_j \, n_j^0 = 1 \tag{2.164}$$

and

$$\sum_{j=1}^{N} \mathcal{M}_j \, \nu_{jr} = 0. \tag{2.165}$$

The corresponding chemical affinities A_r are

$$A_r = -\sum_{j=1}^{N} \nu_{jr} \, \mu_j, \tag{2.166}$$

and we have

$$\begin{cases} e = e(s, \, \vartheta, \, \xi_1, \, \xi_2, \, \ldots \xi_R), \\ de = T \, ds - p \, d\vartheta - \sum_{r=1}^{R} A_r \, d\xi_r. \end{cases} \tag{2.167}$$

Let us now set, according to convention,

$$\partial^2 e/\partial x \, \partial y = e_{xy}. \tag{2.168}$$

The convexity of the internal energy function results in positive definite character for the matrix

$$\left\| \begin{array}{ccccc} e_{ss} & e_{s\vartheta} & e_{s\xi_1} & \cdots & e_{s\xi_R} \\ e_{vs} & e_{\vartheta\vartheta} & e_{\vartheta\xi_1} & \cdots & e_{\vartheta\xi_R} \\ e_{\xi_1 s} & e_{\xi_1\vartheta} & e_{\xi_1\xi_1} & \cdots & e_{\xi_1\xi_R} \\ \cdots & \cdots & \cdots & \cdots & \cdots \\ \cdots & \cdots & \cdots & \cdots & \cdots \\ e_{\xi_R s} & e_{\xi_R\vartheta} & e_{\xi_R\xi_1} & \cdots & e_{\xi_R\xi_R} \end{array} \right\| . \tag{2.169}$$

We now examine the consequences of this result.

2.5.2 Case of a Simple Fluid

If the fluid comprises only one component, we can write

$$e_{ss} > 0, \; e_{\vartheta\vartheta} > 0, \; e_{ss}e_{\vartheta\vartheta} - e_{s\vartheta}^2 > 0. \tag{2.170}$$

Specific Heats

$$\begin{cases} c_v = (\tilde{d}q/dT)_\vartheta = T(\partial s/\partial T)_\vartheta = T/e_{ss} > 0, \\ c_p = (\tilde{d}q/dT)_p = T(\partial s/\partial T)_p, \end{cases} \tag{2.171}$$

but if $p = cte$, we have

$$\begin{cases} -dp = 0 = e_{\vartheta s}\, ds + e_{\vartheta\vartheta}\, d\vartheta, \\ dT = e_{ss}\, ds + e_{s\vartheta}\, d\vartheta = (e_{ss} - e_{s\vartheta}^2/e_{\vartheta\vartheta})\, ds. \end{cases} \tag{2.172}$$

The specific heat at constant pressure can therefore be written as

$$c_p = T(e_{ss} - e_{s\vartheta}^2/e_{\vartheta\vartheta})^{-1} > 0. \tag{2.173}$$

Moreover,

$$c_p - c_v = (T/e_{ss})(e_{ss}e_{\vartheta\vartheta}/e_{s\vartheta}^2 - 1)^{-1} > 0. \tag{2.174}$$

Characteristic Speed

$$(\partial p/\partial \rho)_s = -\vartheta^2(\partial p/\partial \vartheta)_s = v^2\, e_{\vartheta\vartheta} > 0. \tag{2.175}$$

We therefore set

$$c^2 = (\partial p/\partial \rho)_s. \tag{2.176}$$

2.5.3 Case of a Mixture with One Degree of Chemical Freedom

If the mixture has a frozen chemical composition, the results are similar to those shown above:

$$\begin{cases} c_{vf} > 0, \; c_{pf} > 0, \; c_{pf} - c_{vf} > 0, \\ c_f^2 = (\partial p/\partial \rho)_{s\xi}. \end{cases} \tag{2.177}$$

If the mixture is at chemical equilibrium, we have

$$\begin{cases} A = 0, \\ c_{ve} = T(\partial s/\partial T)_{\vartheta A} = T(e_{ss} - e_{s\xi}^2/e_{\xi\xi})^{-1} > 0, \\ c_{pe} > 0, \; c_{pe} - c_{ve} > 0, \; c_e^2 = (\partial p/\partial \rho)_{sA}, \\ c_e^2 = \vartheta^2(e_{\vartheta\vartheta} - e_{\vartheta\xi}^2/e_{\xi\xi}) \le c_f^2. \end{cases} \tag{2.178}$$

On the other hand, we cannot say anything about the signs of quantities such as

$$a = (\partial p/\partial \xi)_{s\rho} = -e_{\vartheta\xi} \tag{2.179}$$

or

$$b = (\partial p/\partial A)_{s\rho} = e_{\vartheta\xi}/e_{\xi\xi}. \tag{2.180}$$

The b/a ratio, which is equal to $(-e_{\xi\xi})$, is always negative.

2.5.4 Case of a Mixture with Several Degrees of Chemical Freedom

The inequalities obtained in this case are the same as those obtained for only one degree of freedom.

We can also consider thermodynamic states in which some of the ξ_r are constant and the A_r corresponding to the other ξ_r are equal to zero. We can then define new inequalities, specific heats, and characteristic speeds, such as

$$c_K^2 = (\partial p/\partial \rho)_{s\,\xi_1 \ldots \xi_K\, A_{K+1} \ldots A_R}. \tag{2.181}$$

2.6 Surface Tension

To a first approximation, the fluid–fluid interface can be considered an autonomous two-dimensional medium that is characterized by thermodynamic properties, in the same manner as for three-dimensional media (a more precise theory is presented in Chap. 11). We will first present thermostatic relations for the one-component case, and then address multicomponent fluids.

2.6.1 One-Component Fluid–Fluid Interfaces

For a one-component interface with *one mole* of surface area $\overline{\Sigma}$, a surface tension σ, and at a temperature T, we can write

$$\overline{E}_a = T\overline{S}_a + \sigma\overline{\Sigma} + \mu_a = \overline{E}_a(\overline{S}_a, \overline{\Sigma}). \tag{2.182}$$

We then obtain the following Gibbs and Gibbs–Duhem relations:[15]

$$\begin{cases} d\overline{E}_a = Td\overline{S}_a + \sigma\, d\overline{\Sigma}, \\ 0 = \overline{S}_a\, dT + \overline{\Sigma}\, d\sigma + d\mu_a. \end{cases} \tag{2.183}$$

For the liquid–vapor interface of a pure substance, it is generally assumed that the surface tension is a function of temperature alone. We then have $\sigma = \sigma(T)$. The usual thermodynamic properties are easily deduced using the interfacial free energy F_a. For one mole,

$$\begin{cases} \overline{F}_a = \overline{E}_a - T\,\overline{S}_a = \sigma\,\overline{\Sigma} + \mu_a = \overline{F}_a(T, \overline{\Sigma}), \\ d\,\overline{F}_a = -\overline{S}_a\, dT + \sigma\, d\,\overline{\Sigma}. \end{cases} \tag{2.184}$$

Therefore, $-\partial\overline{S}_a/\partial\overline{\Sigma} = d\sigma/dT$ is a function of T only. $d\sigma/dT$ is usually negative. We deduce from this relation that

$$\overline{S}_a = -(d\sigma/dT)\overline{\Sigma} + (\overline{S}^0_T)_a. \tag{2.185}$$

Then

$$\begin{cases} d\,\overline{F}_a = [(d\sigma/dT)\overline{\Sigma} - (\overline{S}^0_T)_a]\, dT + \sigma\, d\overline{\Sigma}, \\ \overline{F}_a = (\overline{F}^0_0)_a + \sigma\,\overline{\Sigma} - \int_{T^0}^{T} (\overline{S}^0_T)_a\, dT, \\ (\overline{F}^0_0)_a = \text{const}. \end{cases} \tag{2.186}$$

Or, setting $(\overline{F}^0_T)_a = (\overline{F}^0_0)_a - \int_{T^0}^{T} (\overline{S}^0_T)_a\, dT$,

$$\overline{F}_a = (\overline{F}^0_T)_a + \sigma\,\overline{\Sigma}. \tag{2.187}$$

Due to the evident relation

$$\mu_a = \overline{F}_a - \sigma\,\overline{\Sigma}, \tag{2.188}$$

it follows that

$$\mu_a = (\overline{F}^0_T)_a, \tag{2.189}$$

---

[15]It is more common to consider *unit area*, in which case we get $\overline{e}_a = T\,\overline{s}_a + \sigma + \mu_a\overline{n}_a$, and then the relations can be deduced from (2.183) and $\overline{n}_a\overline{\Sigma} = 1$. However, we will retain the present formalism, which is also valid.

where $(\overline{F}^0_T)_a$ is a function of T alone. The internal energy is obtained by writing $\overline{E}_a = \overline{F}_a + T\,\overline{S}_a$ or $\overline{E}_a = (\overline{F}^0_T)_a + \sigma\,\overline{\Sigma} + T\,[(\overline{S}^0_T)_a - (d\sigma/dT)\overline{\Sigma}]$; moreover,

$$\overline{E}_a = (\overline{E}^0_T)_a + (\sigma - T\,d\sigma/dT)\overline{\Sigma}, \tag{2.190}$$

with

$$(\overline{E}^0_T)_a = (\overline{F}^0_0)_a + T\,(\overline{S}^0_T)_a - \int_{T^0}^{T} (\overline{S}^0_T)_a dT. \tag{2.191}$$

Example: A Vapor Bubble Inside a Liquid

Let us consider, as an example, a vapor bubble inside a liquid under zero-gravity conditions (see Fig. 2.7). The system is at a uniform constant temperature T and is subjected to a constant pressure p through the use of a piston.

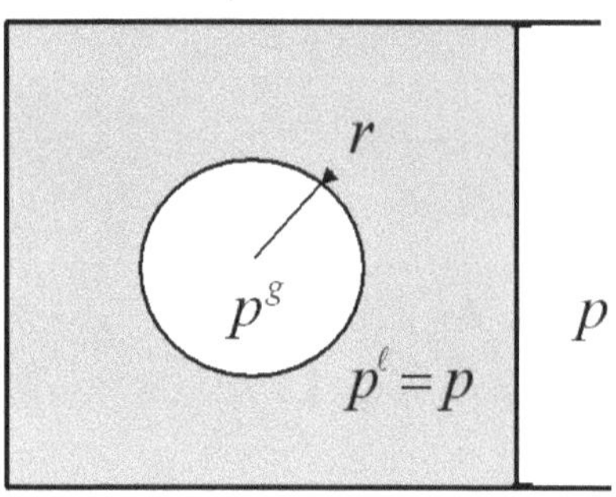

Fig. 2.7. A vapor bubble inside a liquid under zero-gravity conditions

For each of the three subsystems we have

$$\begin{cases} dE^g = T\,dS^g - p^g\,d\mathcal{V}^g + \mu^g\,dn^g, \\ dE^l = T\,dS^l - p^l\,d\mathcal{V}^l + \mu^l\,dn^l, \\ dE_a = T\,dS_a - \sigma\,d\Sigma + \mu_a\,dn_a. \end{cases} \tag{2.192}$$

For the whole system we have

$$\begin{cases} E = E^g + E^l + E_a, \; S = S^g + S^l + S_a, \\ \mathcal{V} = \mathcal{V}^g + \mathcal{V}^l, \; n = n^g + n^l + n_a, \\ dE = T\,dS - p\,d\mathcal{V} + (p^l - p^g)d\mathcal{V}^g + \sigma\,d\Sigma, \\ +(\mu^g - \mu^l)dn^g + (\mu_a - \mu^l)dn_a. \end{cases} \tag{2.193}$$

For a reversible transformation we always have $dE = \tilde{d}W + \tilde{d}Q$, $\tilde{d}W = -p\,d\mathcal{V}$, $\tilde{d}Q = T\,dS$. Therefore, for an unspecified infinitesimal transformation, $\delta E = T\,\delta S - p\,\delta\mathcal{V} - T\,\delta_i S$, assuming that the piston displacement is reversible. It follows that

$$-T\,\delta_i S = (p^l - p^g)\delta\mathcal{V}^g + (\mu^g - \mu^l)\delta n^g + (\mu_a - \mu^l)\delta n_a + \sigma\,\delta\Sigma \leq 0. \tag{2.194}$$

Because

$$\begin{cases} \mathcal{V}^g = 4/3\,\pi\,r^3, \; \delta\mathcal{V}^g = 4\,\pi\,r^2\,\delta r, \\ \Sigma = 4\,\pi\,r^2, \; \delta\,\Sigma = 8\,\pi\,r\,\delta r = 2\,\delta\mathcal{V}^g/r, \end{cases} \tag{2.195}$$

and n^g and n_a may be considered independent variables, we then obtain

$$-T\,\delta_i S = (p^l - p^g + 2\,\sigma/r)\delta\mathcal{V}^g + (\mu^g - \mu^l)\delta n^g + (\mu_a - \mu^l)\delta n_a \leq 0. \tag{2.196}$$

At equilibrium (i.e., during a reversible transformation),

$$p^g = p^l + 2\,\sigma/r, \; \mu^g = \mu^l = \mu_a. \tag{2.197}$$

Exercise:

Calculate the saturated vapor pressure of a vapor bubble in a liquid at a pressure p and temperature T, and then compare it with that obtained for a liquid with a planar surface ($p^\bullet$).

Solution: For the bubble, $p^g = p^l + 2\,\sigma/r$, $\mu^g - \mu^l = 0$, $(\mu_T^0)^g + RT\,\ln(p^g/p^0) = (\mu_T^0)^l + p^l\,\overline{\mathcal{V}}^l$. For the planar surface,

$$(\mu_T^0)^g + RT\,\ln(p^\bullet/p^0) = (\mu_T^0)^l + p^\bullet\,\overline{\mathcal{V}}^l,$$

$$RT\,\ln(p^g/p^\bullet) = (p^l - p^\bullet)\overline{\mathcal{V}}^l.$$

However,

$$p^g = p^l + 2\,\sigma/r,$$

$$RT\,\ln(p^g/p^\bullet) = (p^l - p^\bullet - 2\,\sigma/r)\overline{\mathcal{V}}^l.$$

For small values of r, we get $\ln(p^g/p^\bullet) = -(2\,\sigma/r)(\overline{\mathcal{V}}^l/RT) = -B/r$.

Example of Water at 300K: $\ln(p^g/p^\bullet) \cong -10^{-9}/r$, where r is in meters. The vapor pressure decreases rapidly with the bubble radius, and vanishes when r vanishes. However, we must take into account the fact that, for a radius that is of the same order of magnitude as the intermolecular distances within the liquid, the surface tension is dependent on r. The liquid pressure $p^l = p$ is smaller than the vapor pressure by the quantity $2\,\sigma/r$. At equilibrium, p can reach very small values for very small radii. Therefore, for a pure confined liquid, it will be necessary to apply significant force to the liquid in order to create a vapor bubble.

Remark: We now invert the above problem by considering a liquid drop suspended in its vapor [33]. Again using water as an example, we have $ln(p^g/p^\bullet) \cong 10^{-9}/r$ for small values of r, and $p^l - p^g = 2\,\sigma/r$. For small droplets, the vapor pressure is very high compared to the saturated vapor pressure of the planar (or weakly curved) liquid surface. This phenomenon is unstable. There is an equilibrium radius r^* for a given vapor pressure $p^g > p$, and any droplet that is smaller than r^* tends to evaporate, while any droplet that is larger than r^* tends to increase.

Finally, we should mention that capillarity influences chemical reactions in bubbles, as demonstrated by A. Sanfeld et al. This is easily understood by noting that the concentrations in a gas mixture at equilibrium are sensitive to the pressure, and that higher pressures occur in small bubbles, as we have seen. More on this issue can be found in [245].

Cases with a Nonuniform Surface Temperature

When there is a nonuniform surface temperature, we cannot assume that the system is in complete thermodynamic equilibrium. Examples of motions induced by nonuniform surface tension due to surface temperature fluctuations are given in Chap. 7 (Sects. 7.9.1 and 7.9.2). Of course, local equilibrium is always present, and a simplified equation of state can be obtained by assuming a constant derivative $d\sigma/dT$ for the surface tension. However, more detailed empirical EOS formulae are also available in this case.

2.6.2 Multicomponent Fluid–Fluid Interfaces

For the multicomponent case, we can write an expression for the internal energy of a given mass of interface as a function of temperature, specific area, and specific molar amounts of species at equilibrium. We then have

$$\begin{cases} E_a = T\,S_a + \sigma\,\Sigma + \sum_{j=1}^{N} \mu_j\, n_{ja}, \\ dE_a = T\,dS_a + \sigma\,d\Sigma + \sum_{j=1}^{N} \mu_j\, dn_{ja}, \\ 0 = S_a\,dT + \Sigma\,d\sigma + \sum_{j=1}^{N} n_{ja}\, d\mu_j. \end{cases} \tag{2.198}$$

Numerous expressions for the multicomponent interface EOS can be found in the literature, but it is difficult to find a general formula. In addition, surface tension is very sensitive to the presence of surfactants. Fortunately, in some common cases, we can assume that the derivatives $\partial\sigma/\partial n_{ja}$ are constant.

3

Transfer Phenomena and Chemical Kinetics

Fundamental physical quantities such as mass, momentum and energy can be exchanged between portions of a system, or created during transformations through different processes such as convection, diffusion and chemical reactions. Convection is a type of mass exchange that is dependent upon the mean material velocity. Momentum and energy transport are generally associated with mass transport. Aside from mean material motions, momentum and energy can also be exchanged through molecular interactions. Indeed, molecular quantities are transferred by collisions between molecules, as explained in Sect. 3.3.1. Molecular collisions lead to momentum transfer due to viscosity and to internal energy transfer through thermal conduction. Self-diffusion is the diffusion of molecules in a one-component medium. Chemical mixtures involve the diffusion of species, and the mass diffusion flux has already been defined in Sect. 2.1.3. All of these diffusive fluxes have associated transfer coefficients.

All of the above is true of laminar flows. However, in turbulent flows transfers are often much more efficient than in laminar flows because convection is enhanced by random processes at the macroscopic scale. It is convenient to treat this phenomenon in a similar manner to molecular collisions and thus to define turbulent fluxes and turbulent transfer coefficients. Chapter 8 describes the advantages and limitations of such an approach. The important difference is that molecular transfer coefficients depend only on local quantities such as temperature, pressure and concentration, which are in fact molecular averages, while turbulent transfer coefficients are much more complex.

In mixtures that undergo chemical reactions, chemical species are produced and removed, leading to positive (source) or negative (sink) production terms in the balance equations. These terms obey the laws of chemical kinetics and involve specific reaction rates.

We will limit ourselves in this chapter to molecular transfer and laminar flows.

Section 3.1 considers two examples, each of which involves only one elementary irreversible process: near-equilibrium chemical relaxation and heat

R. Prud'homme, *Flows of Reactive Fluids*,
DOI 10.1007/978-0-8176-4659-2_3, © Springer Science+Business Media, LLC 2010

transfer between discrete systems. In these two cases it is possible to express the entropy production rate as a product of two terms, one of which can be interpreted as the cause of the process and the other as the effect. The simplest possible relation between cause and effect is proportionality, and the corresponding proportionality coefficient is called the phenomenological coefficient.

In Sect. 3.2, we then generalize the previous method to tackle phenomena with multiple transfer processes, and entropy production[1] allows us to introduce generalized forces and fluxes that are connected by linear laws with matrices of phenomenological coefficients.

A simplified kinetic theory of gases is presented in Sect. 3.3.1. This allows us to obtain the transfer coefficients from a molecular-scale analysis and to determine their dependence on state variables. This analysis particularly concerns one-component systems.

More discussion of transfer coefficient determination is provided in Sect. A.6 and, in particular, transfer coefficient formulae that are valid for chemical mixtures are detailed in Sect. A.6.3.

Finally, Sect. 3.4 is devoted to chemical kinetics.

3.1 General Information on Irreversible Phenomena

The irreversible character of a phenomenon is not associated with the changes in entropy between the initial and final states of the considered system, but with the entropy produced. Indeed, we have [213]

$$\varDelta S = \varDelta_e S + \varDelta_i S. \tag{3.1}$$

The transformation of interest will be irreversible if $\varDelta_i S > 0$. From this point of view, the simplest systems to study are discrete systems. These systems consist of subsystems that are each individually at equilibrium, but in mutual disequilibrium with each other. Thus, the overall system is not in equilibrium, except in particular situations. It is possible to apply the knowledge gathered for simple systems at equilibrium to this type of system.

We will also see that similar methods can be applied to systems with continuous spatial variations due to the local state postulate (see Sect. 2.1.3). We will initially examine two types of discrete system.

3.1.1 A Chemical Reaction Near Equilibrium

In a monoreactive homogeneous mixture [9, 181, 213], we have

---

[1]Note that transfer phenomena—just like chemical production—are necessarily associated with irreversibility, and they result in increased entropy, as we show in Chap. 4.

$$\begin{cases} n_j = n_j^0 + \nu_j \xi, \\ dE = T\,dS - p\,d\mathcal{V} - A\,d\xi, \\ A = -\sum_j \nu_j \mu_j. \end{cases} \tag{3.2}$$

The internal energy of the closed system with uniform properties of interest is thus a function of the three variables S, $\mathcal{V}$, and ξ:

$$E = E(S,\, \mathcal{V},\, \xi). \tag{3.3}$$

We have

$$T = \partial E/\partial S,\ -p = \partial E/\partial \mathcal{V},\ -A = \partial E/\partial \xi. \tag{3.4}$$

The entropy production rate is

$$\dot{\delta_i S} = (A/T)\dot{\xi} \geq 0. \tag{3.5}$$

Two limit cases then appear:

1. A frozen mixture corresponding to $\dot{\xi} = 0$
2. A mixture at chemical equilibrium characterized by $A = 0$.

In the three-dimensional state space, the manifolds $(\mathcal{M})$ are thus the surfaces ξ =const. and $A = 0$ (see Fig. 3.1).

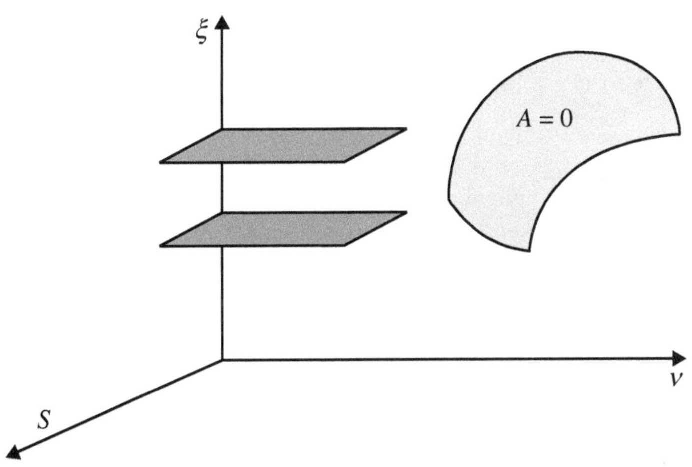

Fig. 3.1. Equilibrium surfaces for a mixture

Any displacement of a point on one of these surfaces involves a reversible transformation (evolving equilibrium). This is the case for any evolution with ξ =const. As soon as ξ varies, the evolution is reversible only if the representative point is on and remains on surface $A = 0$.

Any point in this space thus appears to be an equilibrium point a priori. However, the surfaces represented here are not simply loci of equilibrium points. The surfaces ξ =const. lead to horizontal stratification. If an evolution takes place on one of these surfaces, the chemical reaction does not take place; the mixture is inert and behaves like a simple system at equilibrium. Only the $A = 0$ surface corresponds to true chemical evolution. Generally, any point located outside this surface will attempt to approach it through chemical reaction because of the stability represented by equilibrium (see Sect. 2.5), and the corresponding phenomenon will be irreversible. However, at the same time external constraints can tend to move away from equilibrium.

If the $A = 0$ surface does exist, both the reaction and its opposite need to exist, so we can write

$$\sum_{j=1}^{N} \nu'_j \mathcal{E}_j \rightleftharpoons \sum_{j=1}^{N} \nu''_j \mathcal{E}_j. \tag{3.6}$$

The ratio A/T tends to make the system evolve towards the chemical equilibrium characterized by $A/T = 0$. The evolution is characterized by a production rate $\dot{\xi}$ that will be positive if $A/T > 0$ and negative in the opposite case. At complete equilibrium, $\dot{\xi}$ is equal to zero and the system does not evolve for constant values of T and p (or S and $\mathcal{V}$). It can be said that A/T is a generalized force and that $\dot{\xi}$ is a generalized flux. We can express the causality relation by writing that $\dot{\xi}$ is related to A/T. Close to equilibrium (in the vicinity of the surface $A = 0$), we can carry out a Taylor series expansion of this function. If we limit this to the first order, we obtain:

$$\dot{\xi} = L\, A/T, \tag{3.7}$$

where L is called the " phenomenological coefficient." This coefficient depends on S, $\mathcal{V}$ and ξ (or T, p and ξ), and is positive according to the second law.

If near-equilibrium evolution occurs while maintaining two parameters constant, for example E and $\mathcal{V}$ (isolated system), we can write

$$A/T = (\partial(A/T)/\partial\xi)^{e}_{E,\mathcal{V}}\,(\xi - \xi^{e}), \tag{3.8}$$

where ξ^e is the equilibrium value corresponding to the given E and $\mathcal{V}$. We have

$$\dot{\xi} = L(\partial(A/T)/\partial\xi)^{e}_{E,\mathcal{V}}\,(\xi - \xi^{e}). \tag{3.9}$$

The quantity A/T occurs in the Gibbs relation written in the form

$$dS = \frac{1}{T}dE + \frac{p}{T}d\mathcal{V} + \frac{A}{T}d\xi, \tag{3.10}$$

and corresponds to the derivative

$$\frac{A}{T} = (\frac{\partial S}{\partial \xi})_{E,\mathcal{V}}. \tag{3.11}$$

Therefore,

$$(\frac{\partial A/T}{\partial \xi})^e_{E,\mathcal{V}} = (\frac{\partial^2 S}{\partial \xi^2})_{E,\mathcal{V}}. \tag{3.12}$$

It can be shown (due to the stability associated with chemical equilibrium) that $(\partial^2 S/\partial \xi^2)_{E,\mathcal{V}}$ is always negative; inversely, $(\partial^2 E/\partial \xi^2)_{S,\mathcal{V}}$ is always positive (Sect. 2.5). This means that

$$-L(\partial(A/T)/\partial\xi)^e_{E,\mathcal{V}} = 1/\tau. \tag{3.13}$$

The parameter τ is called the chemical time when E and $\mathcal{V}$ are constant. The evolution equation then becomes

$$\frac{d\xi}{dt} + \frac{\xi - \xi^e}{\tau} = 0 \tag{3.14}$$

with τ =const. and ξ^e =const.

Let us now consider a more complex case without two constant parameters (i.e., constant E and $\mathcal{V}$) but for which it is possible to prove that the system is close to equilibrium at any time. The type of equilibrium we are referring to is an evolving equilibrium. The target quantities ξ^e and τ will be functions of time on this occasion.

Exercise: Let us simplify matters by assuming that τ is constant. Suppose that $\overline{\xi} = \xi/\xi_r$ and $\overline{t} = t/t_r$, where ξ_r and t_r are reference values, and that $\overline{\xi^e}(\overline{t}) = -\overline{t}$. We wish to determine the solutions for near-equilibrium evolution. We have

$$\frac{d\overline{\xi}}{d\overline{t}} + \frac{\overline{\xi} + \overline{\overline{t}}}{\overline{\tau}} = 0.$$

The solution is then (see Fig. 3.2)[2]

$$\overline{\xi} = \exp -\overline{t}/\overline{\tau} - \overline{t} + \overline{\tau}.$$

[2] t_r can be thought of as a mechanical time τ_m that is characteristic of the evolutions of the external (i.e., flow rather than chemical) parameters, $\tau_m = |dt/dln\mathcal{V}|^e$ for example. Close to equilibrium, the chemical time τ is very small compared to τ_m. Note that the evolution is always out of equilibrium and tends to proceed at a constant distance $\overline{\xi} - \overline{\xi}^e = \overline{\tau}$ from the reference evolving equilibrium. There is relaxation. The chemical reaction tends to bring chemical equilibrium closer, but the evolutions of the external parameters ($\mathcal{V}(t)$ for example) tend to move the system away from equilibrium. Similar phenomena are found when the local state postulate is applied to flows in propulsion nozzles, where the expansion of gas leads to chemical freezing.

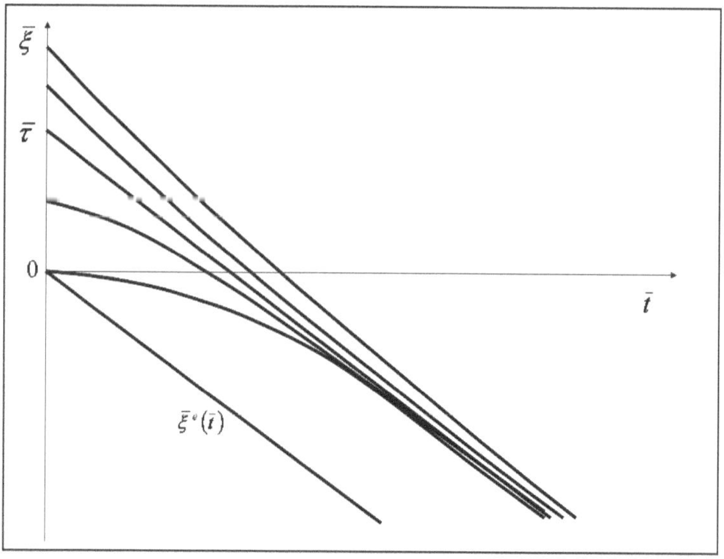

Fig. 3.2. A simple example of chemical relaxation at constant E and $\mathcal{V}$, with linear equilibrium evolution relaxation curves for several initial values of the progress variable ξ

In this case, the distance from equilibrium increases because the reactions slow down due to thermal effects, resulting in an increase in the chemical time τ. The chemical time τ is generally compared with the mechanical time τ_m (used as the reference time) in order to evaluate variations from chemical equilibrium. The ratio $D_a = \tau_m/\tau$ is the first Damköhler parameter. If $D_a \gg 1$, the evolution takes place near equilibrium. If D_a is much smaller than unity, the chemical reactions are very slow compared to the mechanical process and we have frozen chemical evolution (see Sect. 7.2 for more information). An evolution of this type is shown in Fig. 3.3, where the surface $A = 0$ corresponds to chemical equilibrium and the surfaces ξ =const. to frozen flows.

3.1.2 Thermal Exchange

In certain situations, heat exchange occurs between two media, such as when the temperature of each medium is uniform but is a function of time. The heat transfer occurs in a zone of space between the two media that is comparable to a nonadiabatic wall $\mathcal{P}$. It is generally assumed that the quantities present in subsystem $\mathcal{P}$, such as internal energy, mass, entropy, etc., are infinitely small. However, $\mathcal{P}$ is the center of heat transfer and entropy production. Suppose that two media 1 and 2 are evolving and that the wall $\mathcal{P}$ is fixed and nondeformable (see Fig. 3.4). The laws of thermodynamics give us

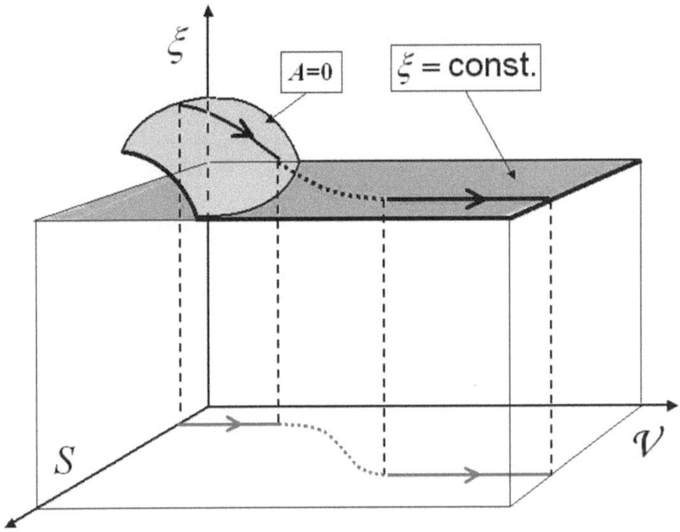

Fig. 3.3. Chemical relaxation in a nozzle: representation of the thermodynamic path in the (volume, entropy, progress variable) coordinate system

$$\delta E_1 = \delta Q_1 + \delta W_1 = \delta Q_{e1} + \delta Q_{21} + \delta W_{e1}, \tag{3.15}$$

where δQ_{e1} and δW_{e1} are the elementary heat and work, which we assume are reversibly received by subsystem 1 from outside, while $\delta Q_{21} = -\delta Q$ is the heat received from subsystem 2 through the wall. Then

$$\begin{cases} \delta E_2 = \delta Q_{e2} + \delta Q_{12} + \delta W_{e2}, \\ \delta E_{\mathcal{P}} = 0 = \delta Q_{12} + \delta Q_{21} \to \delta Q_{21} = -\delta Q_{12}, \end{cases} \tag{3.16}$$

and for the whole system

$$\delta E = \delta E_1 + \delta E_2 + \delta E_{\mathcal{P}} = \delta Q_e + \delta W_e, \tag{3.17}$$

where $\delta Q_e = \delta Q_{e1} + \delta Q_{e2}$, $\delta W_e = \delta W_{e1} + \delta W_{e2}$. The second law yields

$$\begin{cases} \delta S_1 = (1/T_1)(\delta Q_{e1} + \delta Q_{21}), \\ \delta S_2 = (1/T_2)(\delta Q_{e2} + \delta Q_{12}), \\ \delta S_{\mathcal{P}} = 0 = \delta Q_{21}/T_2 + \delta Q_{12}/T_1 + \delta_i S_{\mathcal{P}}. \end{cases} \tag{3.18}$$

Here, the first two relations correspond to reversible transformations, and the last to an unspecified transformation. Therefore, the entropy production of the wall is

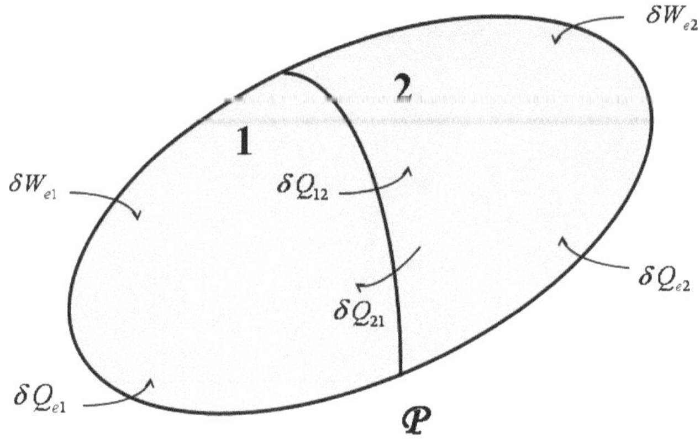

Fig. 3.4. Thermal exchange between two parts (1 and 2) of a system through a wall $\mathcal{P}$

$$\delta_i S_{\mathcal{P}} = \delta Q(\frac{1}{T_2} - \frac{1}{T_1}). \tag{3.19}$$

For the overall system,

$$\delta S = \frac{\delta Q_{e1}}{T_1} + \frac{\delta Q_{e2}}{T_2} + \delta_i S, \tag{3.20}$$

but we also have

$$\delta S = \delta S_1 + \delta S_2 + \delta S_{\mathcal{P}} = \frac{\delta Q_{e1} + \delta Q_{21}}{T_1} + \frac{\delta Q_{e2} + \delta Q_{12}}{T_2}. \tag{3.21}$$

It then follows that

$$\begin{cases} \delta_i S = \delta Q_{21}/T_1 + \delta Q_{12}/T_2 = \delta Q(1/T_2 - 1/T_1), \\ = \delta_i S_{\mathcal{P}} \geq 0, \ \delta Q = \delta Q_{12}, \end{cases} \tag{3.22}$$

which shows that the wall is indeed the center of the irreversibility. We of course find that δQ is positive if $T_1 > T_2$.

Equilibrium is reached in two cases: if $\delta Q = 0$ (adiabatic wall) or if $T_2 = T_1$ (mutual thermal equilibrium).

Ignoring these two cases, we introduce an exchange coefficient $h(T_1, T_2)$ such that the heat flux $\dot{Q}$ from subsystem 1 to subsystem 2 is

$$\dot{Q} = \delta Q/\delta t = -h(T_2 - T_1), \tag{3.23}$$

and so

$$\delta_i \dot{S} = h(T_2 - T_1)^2/T_1 T_2. \tag{3.24}$$

The coefficient h is obviously positive.

Cases of irreversible evolution where one of the subsystems cannot exchange heat with the outside are easily represented graphically. Let us take the simple example where media 1 and 2 are nondeformable and where subsystem 2 is separated from the outside by an adiabatic wall (Fig. 3.5). Suppose that the heat capacities C_1 and C_2 are constant. We have

$$E_1 = C_1 T_1,\ E_2 = C_2 T_2,\ \delta Q = \delta Q_{12} = \delta E_2. \tag{3.25}$$

We then write $dS = dE_1/T_1 + dE_2/T_2$ or

$$dS = \frac{dE}{T_1} + (\frac{1}{T_2} - \frac{1}{T_1})dE_2,\ S - S(E,\, E_2).$$

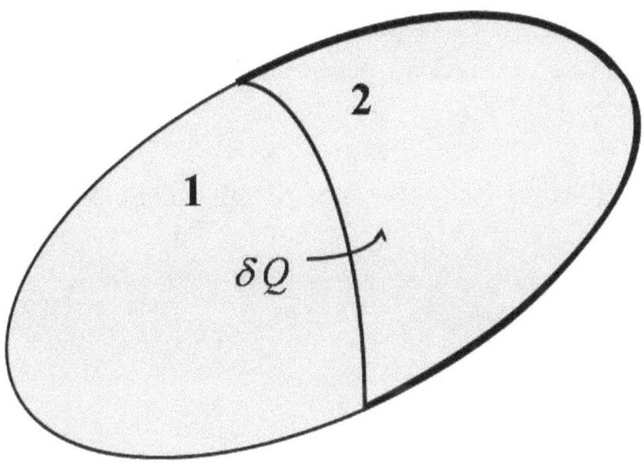

Fig. 3.5. Simplified case corresponding to that of Fig. 3.4 where media 1 and 2 are nondeformable and where subsystem 2 is separated from the outside by an adiabatic wall

E_2 is similar to a progress variable for heat exchange between subsystems 1 and 2. The two equilibrium cases are then as follows:

1. $\delta Q = 0$ or $E_2 =$const. (frozen exchange)
2. $T_2 = T_1$ or $E_2/C_2 = E/(C_1 + C_2)$ (thermal equilibrium).

Figure 3.6 highlights the various situations.

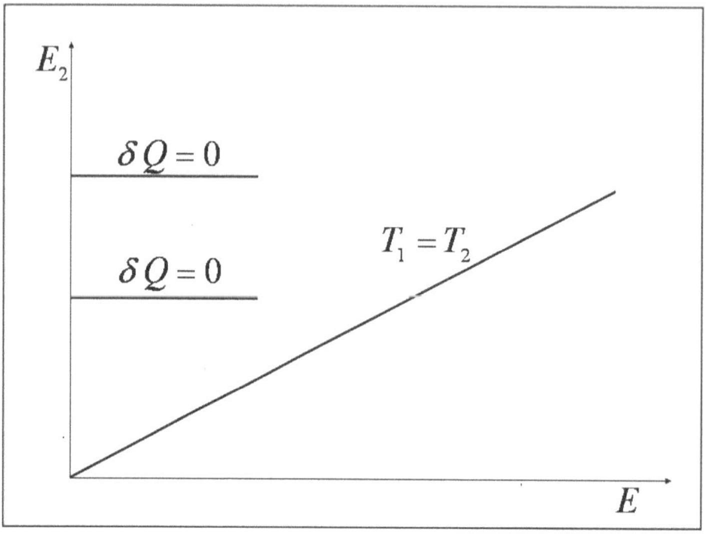

Fig. 3.6. Equilibrium situations for the simplified case corresponding to Fig. 3.5

Out of equilibrium, the existence of an exchange coefficient h makes it possible to write

$$\frac{dE_2}{dt} + h(\frac{1}{C_1} + \frac{1}{C_2})(E_2 - \frac{C_2}{C_1 + C_2}E) = 0, \tag{3.26}$$

and the coefficient

$$\tau^{-1} = h(\frac{1}{C_1} + \frac{1}{C_2}) \tag{3.27}$$

is like the inverse of a heat exchange relaxation time. Note that the situation with the chemical reaction is similar, although simpler. At thermal equilibrium, we have $E_2^e = (C_2/(C_1 + C_2))E$, so that

$$\frac{dE_2}{dt} + \frac{E_2 - E_2^e}{\tau} = 0. \tag{3.28}$$

E_2 can be fixed or variable as a function of time, depending on the case in question. It is worth noting here that irreversible phenomena do not always arise in such a simple manner; consider, for example, friction phenomena (solid friction in particular) and plasticity. However, this treatment will be applicable to most of the situations evoked in this chapter.

3.2 Presenting the Coefficients of Transfer via the Thermodynamics of Irreversible Processes

3.2.1 Basic Relations

In this section we generalize the preceding results to the case where there are several simultaneous irreversible phenomena and to the case where there are continuous media [108]. In all of the cases in which this theory can be applied, we can write the entropy production rate per unit volume in the form

$$\dot{W}_S = J^S F^S + J^V \cdot F^V + J^T \colon F^T, \tag{3.29}$$

where the F parameters are generalized forces (like the ratio A/T or the jump of $1/T$ when crossing an interface, or a gradient of $1/T$, ...) and the J parameters are generalized fluxes (chemical production, heat flux, ...). The exponents S, V, and T denote scalar, vectorial, and second-order tensor here. These quantities F or J must form a system of independent variables. Table (3.1) describes the various types of phenomena.

	S	S	V	V	T
F	Ratio A/T	Volume expansion $\nabla \cdot \mathbf{v}$	Temperature gradient or jump	Chemical potential gradient or jump	Strain rate deviator tensor
J	Chemical production $\dot{\xi}$	Bulk viscosity	Conductive heat flux	Mass diffusion flux	Tensor of viscous stress

Table 3.1. Tensorial order of generalized forces and fluxes

Paradoxically, this theory is seldomly used for chemical irreversibilities. Chemical kinetics generally provides more precise and nonlinear expressions for the production rates $\dot{\xi}$. During reversible evolution, $\dot{W}_S$, the fluxes, and the forces all vanish. This case aside, the generalized fluxes can be regarded as functions of the generalized forces. These functions depend on the state parameters of the system. When the irreversibility is not too strict, we can define the following linearized relations between fluxes and forces, termed "phenomenological relations:"

$$\begin{cases} J^S = L^{SS}F^S + L^{SV}F^V + L^{ST}F^T, \\ J^V = L^{VS}F^S + L^{VV}F^V + L^{VT}F^T, \\ J^T = L^{TS}F^S + L^{TV}F^V + L^{TT}F^T. \end{cases} \tag{3.30}$$

We need to consider the influence of spatial symmetry on these phenomenological equations. It has been shown that the nondiagonal matrix coefficients are null. There are only relations between generalized fluxes and generalized forces of same tensorial order [108]. In addition, there are relations between the components of the matrix coefficients. These relations are known as Onsager reciprocal relations.[3] In particular, if electromagnetic phenomena do not occur, the matrix is symmetric.[4]

3.2.2 Species Diffusion, Heat Conduction, and Viscosity

For multicomponent fluid flows, there are three types of flux that play a role in transport phenomena:

- The diffusion flux of species j, $\mathcal{J}_{Dj} = \rho_j \mathbf{V}_j$
- The heat flux, $\mathcal{J}_E = \mathbf{q}$
- The viscous momentum flux, $\mathcal{J}_{\mathbf{Mv}} = \mathbf{P} - p\mathbf{1}$.

It is apparent that coupling is possible between the diffusion flux and the heat flux, which share the same tensorial order. No intrinsic coupling will take place between these fluxes and the momentum flux. The simplest phenomenological relations, for the case of no coupling between transfer phenomena, are the following:

$$\begin{cases} \mathcal{J}_{\mathbf{Dj}} = -\rho D \boldsymbol{\nabla} \mathbf{Y_j} \text{ (Fick's law)} \\ \mathbf{q} = -\lambda \boldsymbol{\nabla} \mathbf{T} \text{ (Fourier's law)} \\ \mathbf{P} - p\mathbf{1} = -\mu(\boldsymbol{\nabla} \otimes \mathbf{v} + \widetilde{\boldsymbol{\nabla} \otimes \mathbf{v}}) - k \boldsymbol{\nabla} \cdot \mathbf{v}\, \mathbf{1} \text{ (Newton's law).} \end{cases} \tag{3.31}$$

D is the coefficient of diffusion, λ is the thermal conductivity, μ is the shear viscosity coefficient, $k = \eta - 2\mu/3$ where η is the bulk viscosity, and $\mathbf{P} = -\boldsymbol{\Sigma}$ where $\boldsymbol{\Sigma}$ is the stress tensor.

It will be shown in Chap. 4 that, in the case of a fluid with a thermodynamic pressure p and a pressure tensor $\mathbf{P}$, the entropy production rate includes the dissipative (irreversible) term $-(1/T)(\mathbf{P} - p\mathbf{1})\mathbf{:}\boldsymbol{\nabla} \otimes \mathbf{v}$. Since the

[3] Starting from the microscopic property of the "time reversal invariance" of the equations of motion for individual particles, which implies that the particles retrace their former paths if all velocities are reversed, it is possible to derive a macroscopic theorem, as done by Onsager.

[4] In the presence of an external magnetic field, the property of "time reversal invariance" implies that the particles retrace their former paths only if the particle velocities *and* the magnetic field are reversed. As a consequence, the Onsager relations must be modified.

P tensor is symmetrical, we can replace $\nabla \otimes \mathbf{v}$ with its symmetrical part $(1/2)(\nabla \otimes \mathbf{v} + \widetilde{\nabla \otimes \mathbf{v}})$. To highlight the scalar parts of these tensors, they should be written as follows:

$$\mathbf{D} = \frac{1}{2}(\nabla \otimes \mathbf{v} + \widetilde{\nabla \otimes \mathbf{v}} - \frac{2}{3}\nabla \cdot \mathbf{v}\,\mathbf{1}) + \frac{1}{3}\nabla \cdot \mathbf{v}\,\mathbf{1}, \tag{3.32}$$

where the trace of the first tensor, $(1/2)(\nabla \otimes \mathbf{v} + \widetilde{\nabla \otimes \mathbf{v}} - (2/3)\nabla \cdot \mathbf{v}\,\mathbf{1})$, is equal to zero.

In the same way,

$$\mathbf{P} - p\mathbf{1} = \mathbf{P} - \pi\mathbf{1} + (\pi - p)\mathbf{1}, \tag{3.33}$$

where

$$\pi = \frac{1}{3}\mathrm{tr}(\mathbf{P}), \tag{3.34}$$

so that $\mathrm{tr}(\mathbf{P} - \pi\mathbf{1}) = 0$. The quantity π is the mean normal pressure. Therefore, the viscosity term for the entropy production becomes

$$-\frac{1}{2T}(\mathbf{P} - \pi\mathbf{1}):(\nabla \otimes \mathbf{v} + \widetilde{\nabla \otimes \mathbf{v}} - \frac{2}{3}\nabla \cdot \mathbf{v}\mathbf{1}) - \frac{1}{T}(\pi - p)\nabla \cdot \mathbf{v}, \tag{3.35}$$

The two phenomenological relations are then

$$\begin{cases} \mathbf{P} - \pi\mathbf{1} = -\mu[\nabla \otimes \mathbf{v} + \widetilde{\nabla \otimes \mathbf{v}} - (2/3)\nabla \cdot \mathbf{v}\,\mathbf{1}], \\ \pi - p = -k\nabla \cdot \mathbf{v}. \end{cases} \tag{3.36}$$

For diffusion and heat transfer, there is a coupled effect known as thermal diffusion (see Sects. 7.3 and A.6.3). In addition, the fluxes $\mathcal{J}_{\mathrm{Dj}}$ are not independent since their sum is zero. In fact, the corresponding entropy production term utilizes not ∇Y_j but $\nabla(g_j/T)$, which also depends on the variation in the temperature and the pressure gradient. The heat flux $\mathbf{q}$ includes the conductive flux and a duffision-related flux. These issues will be addressed again in Chap. 7. The phenomenological relations are valid for both continuous systems and discrete systems. In the latter case, the Fourier analysis reduces to the heat exchange law shown in Sect. 3.1.

Finally, note that the linearized theory of irreversible processes has its limitations. There are alternative theories, such as that of Jou et al. [130], as mentioned in Sect. 5.3.

3.3 Other Ways of Presenting the Transfer Coefficients

3.3.1 Presenting the Transfer Coefficients via the Simplified Kinetic Theory of Gases

The elementary *kinetic theory of gases* offers a first approximation of the transfer coefficients [46, 118]. Each molecule of diameter d has a sphere of influence of radius d. The volume swept by this molecule, which has an average velocity of $\bar{c}$ (average quadratic velocity for all species), is $\pi d^2 \bar{c}$ per unit time (see Fig. 3.7). The total number of molecules per unit volume is N, and so the number of collisions per molecule per unit time is thus

$$\nu_c = \pi d^2 \bar{c} N, \tag{3.37}$$

and the mean free path l is equal to

$$l = \bar{c}/\nu_c = 1/\pi d^2 N. \tag{3.38}$$

The collision frequency and mean free path were calculated by assuming that the molecules in the swept volume are motionless. If we account for the movements of the target molecules, we obtain

$$l = 1/(\sqrt{2}\pi d^2 N). \tag{3.39}$$

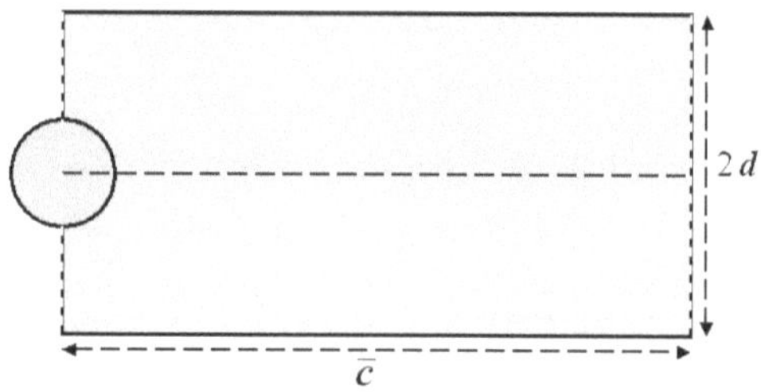

Fig. 3.7. Volume swept by a molecule

Now let us consider the unit flux density of a property F (f per unit mass) through a surface at coordinate x_0. To simplify matters, we assume that the molecular movements (agitations) occur parallel to the x axis (see Fig. 3.8). The molecules coming from $x < x_0$ transport $f(x_0 - l/2)$ of the property per unit mass on average. Only a collision can change f. The molecules coming from $x > x_0$ transport $f(x_0 + l/2)$ of the property per unit mass. The modulus of the molecular flow rate, from the left and from the right, is proportional to $N\bar{c}$.

Thus, in the positive x direction, the flux crossing the surface per unit area will be proportional to

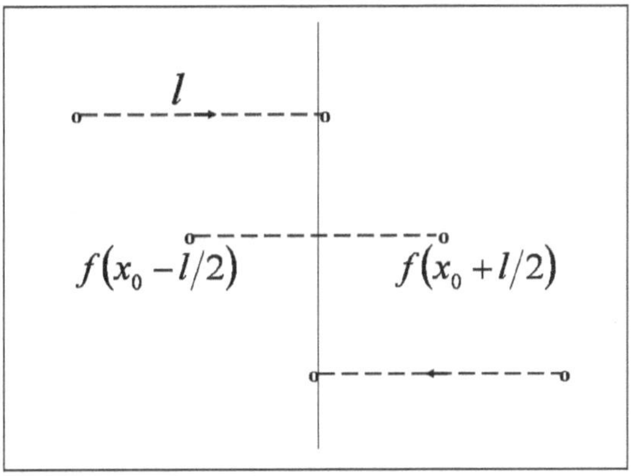

Fig. 3.8. Molecular transfer

$$\mathcal{J}_F \propto MN\bar{c}[f(x_0 - l/2) - f(x_0 + l/2)]. \tag{3.40}$$

This result is valid for a fluid at rest and a surface at rest. However, it is also appropriate for a fluid moving under the assumption that each component is in local equilibrium, and the x_0 surface then follows the motion with a mean velocity of $\mathbf{v}$. We then have

$$\mathcal{J}_F = a_F\, M\, N\, \bar{c}\, [f(x_0 - l/2) - f(x_0 + l/2)]. \tag{3.41}$$

By expanding the right hand side according to the powers of l to the first order, we get

$$\mathcal{J}_F = -a_F MN\bar{c}l(\partial f/\partial x)_{x_0}, \tag{3.42}$$

or, using the expression for l,

$$\mathcal{J}_F = -a_F(M\bar{c}\partial f/\partial x)/(\sqrt{2}\pi d^2). \tag{3.43}$$

The mean molecular velocity at translational equilibrium (in the case of monatomic molecules) is $M\bar{c}^2/2 = 3kT/2$, where k is the Boltzmann constant ($r = R/\mathcal{M} = k/M$). Therefore, $\bar{c}^2 = 3rT$. From these considerations, we can deduce that

$$\mathcal{J}_F = -a_F \frac{M}{\pi d^2}\sqrt{\frac{3rT}{2}}\frac{\partial f}{\partial x}. \tag{3.44}$$

Let us apply these relations successively to the mass of species j, the internal energy, and the momentum; i.e.,

$$f = M_j N_j / MN = Y_j,\ f = c_v T,\ f = \mathbf{v}. \tag{3.45}$$

We obtain the following relations that are valid for simple cases:

$$\begin{cases} \mathcal{J}_{\mathbf{Dj}} = -a_D MN\bar{c}l\boldsymbol{\nabla}\mathbf{Y_j}, \\ \mathbf{q} = --a_T MN\bar{c}lc_v\boldsymbol{\nabla}\mathbf{T}, \\ \mathbf{P} - p\mathbf{1} = -a_\mu MN\bar{c}l\boldsymbol{\nabla}\otimes\mathbf{v}. \end{cases} \tag{3.46}$$

Therefore, the transfer coefficients defined in Sect. 3.2 for the diffusion, thermal conduction and viscosity are equal to

$$D = a_D\bar{c}l,\ \lambda = a_T\rho\bar{c}lc_v,\ \mu = a_\mu\rho\bar{c}l. \tag{3.47}$$

Thus ρD, λ and μ are proportional to $\sqrt{T}$. Since $\rho = p/rT$, the coefficient of diffusion D at a given pressure is proportional to $T^{2/3}$. For a polyatomic gas, $\bar{c}$ can still be considered—to a first approximation—to be proportional to $\sqrt{T}$.

Three dimensionless numbers are usually used to compare the preceding effects:

- The Prandtl number $Pr = \mu c_p/\lambda = \gamma a_T/a_\mu$ with $\gamma = c_p/c_v$. The ratio a_T/a_μ is taken to be equal to 2/5, giving $Pr = 2\gamma/5$ for monatomic gases. In the case of polyatomic gases, the Eucken relation gives us $\lambda/\mu = c_v + 9R/4\mathcal{M}$, which leads to the expression $Pr = 4\gamma/(9\gamma - 5)$. This latter relation also applies to monatomic gases, for which $\gamma = 5/3$, giving the previous result $Pr = 2\gamma/5 = 2/3$.
- The Schmidt number $Sc = \mu/\rho D = a_\mu/a_D$. This number is sensitive to the variation in the molar mass $Sc = 0.145\mathcal{M}^{0.556}$. It is an empirical relation.
- The Lewis number $Le = Sc/Pr = a_T/\gamma a_D$; i.e.,

$$Le \cong 0.03625\frac{9\gamma - 5}{\gamma}\mathcal{M}^{0.556}.$$

Note that for numerous gases, these dimensionless numbers are on the order of 1.

3.3.2 More Precise Estimation of the Transfer Coefficients

A more precise estimate of the transfer coefficients requires an analysis of the dynamics of the collisions [10, 46, 118]. The theory, which will not be described here in detail, takes into account the deviations in the molecular trajectories that are caused by interaction potentials. It implies the experimental determination of the collision cross-sections and collision integrals that occur in the Boltzmann gas equation. These considerations are examined further in the Appendix (see Sect. A.6).

In Sects. 3.2 and 3.3.1, our discussion of transport coefficients was mainly limited to single species. While mass diffusion coefficients were introduced, these coefficients were termed "self-diffusion coefficients" in (3.31), and the transfer coefficients did not depend on the species concentrations. Again, this topic is investigated further in the Appendix (see Sect. A.6).

3.3.3 Liquids and Dense Gases

The case for a fluid close to its critical point is presented in the Appendix (see Sect. A.5.1).

The classical kinetic theory of gases is not valid for liquids because the intermolecular distances are of the same order of magnitude as the dimensions of the molecules. Experimental measurements show substantial differences from the gas case. Figures 3.9 and 3.10, extracted from the work *Molecular Theory of Gases and Liquids*, by Hirschfelder, Curtiss and Bird [118], show how the pressure and the temperature (divided by their critical values) influence the viscosity μ and the thermal conductivity λ of carbon dioxide and argon.

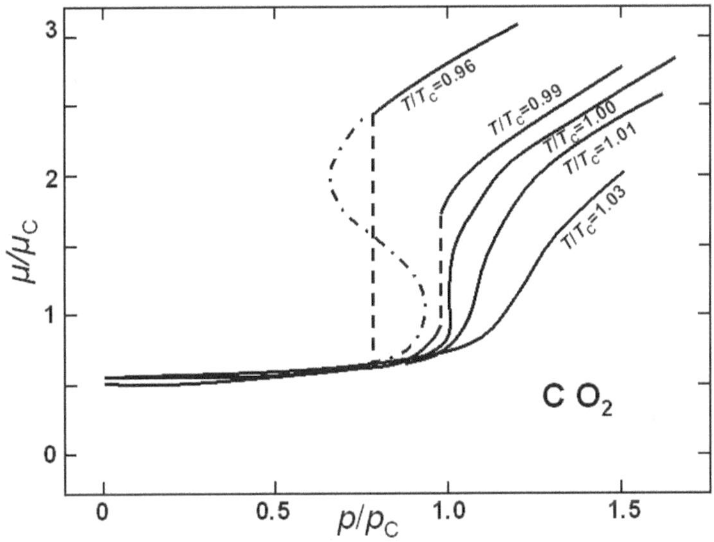

Fig. 3.9. Viscosity of CO_2 (after [118]; reprinted with the permission of Wiley)

Several theories have been used to determine the transfer coefficients of liquids and dense gases, including:

- The principle of the corresponding states which gives results only for viscosity

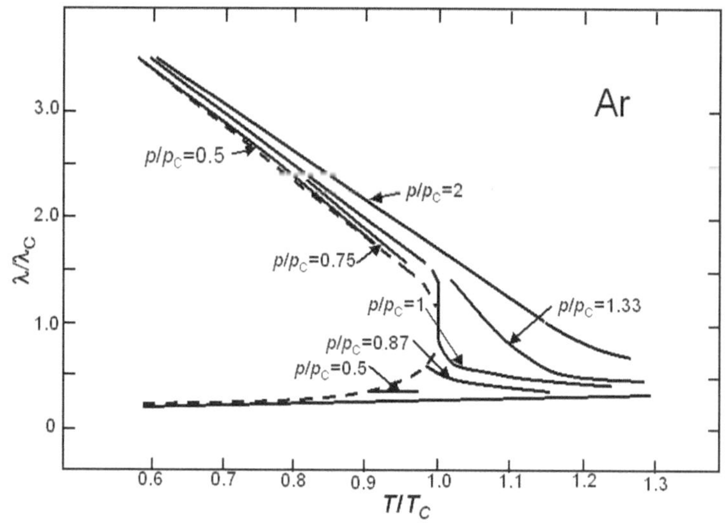

Fig. 3.10. Thermal conductivity of Ar (after [118]; reprinted with the permission of Wiley)

- The theory of the activated complex (Eyring) [105], which is appropriate for plastic flows and liquids, but not for gases
- The theory of Enskog, which is appropriate for real gases
- The rigorous theory, based on Liouville's equation, which is valid for dense gases but is very complex, especially when dealing with mixtures.

We might hope that the methods used to describe the liquid state and the chemical reactivity in a liquid phase, such as those based on Langevin's equation, will produce interesting results for reactive mixtures in the presence of convection.

The similarity parameters are significantly affected by the strong molecular interactions that occur in liquids. The Prandtl number, for example, can vary considerably, as shown in Fig. 3.11.

3.4 Elements of Chemical Kinetics

In this section we consider the chemical kinetics of cases where the chemical process can be divided into a set of elementary reactions [10, 105, 118]. The chemical production rates are assumed to obey polynomial expressions for the concentrations.

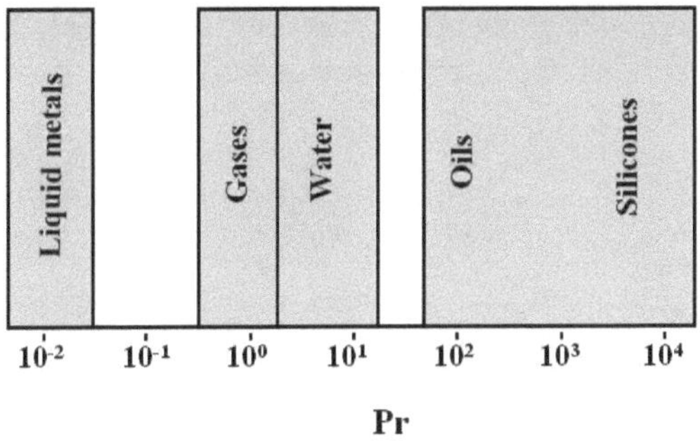

Fig. 3.11. Prandtl numbers of various fluids

Consider a mixture that includes N chemical species, each of which is represented by a different value of the index j. In this case, a chemical reaction can be represented as

$$\sum_{j=1}^{N} \nu'_j \mathcal{E}_j \overset{k(T)}{\rightarrow} \sum_{j=1}^{N} \nu''_j \mathcal{E}_j, \tag{3.48}$$

where ν_j is the algebraic stoichiometric coefficient of species j, which was introduced in Sect. 2.4.2, and

$$\nu_j = \nu''_j - \nu'_j. \tag{3.49}$$

The production rate of the species, in mass per unit volume and unit time is

$$\dot{W}_j = \nu_j \mathcal{M}_j \dot{\zeta}, \tag{3.50}$$

where

$$\dot{\zeta} = k(T) \prod_{i=1}^{N} C_i^{\nu'_i}. \tag{3.51}$$

The specific reaction rate $k(T)$ generally takes the form

$$k(T) = BT^s \exp(-E_a/RT), \tag{3.52}$$

where B and s are constants and E_a is the activation energy of the reaction. Such a formula can be established using various theories: the theory of the activated complex (Eyring), collision theory, etc. The values of B, s, and E_a are tabulated for many elementary chemical reactions (see, for example, [222]). Such a simple formula is only valid for elementary reactions. Also, determining the reactive schemes and thus elucidating all of the elementary reactions is a not inconsiderable task.

If the opposite of an elementary reaction also takes place, we can write the direct (D) and reverse (R) reactions as follows:

$$\sum_{j=1}^{N} \nu'_j \mathcal{E}_j \rightleftharpoons \sum_{j=1}^{N} \nu''_j \mathcal{E}_j. \tag{3.53}$$

The production rates are then

$$\begin{cases} \dot{W}_j = \nu_j \mathcal{M}_j (\dot{\zeta}_D - \dot{\zeta}_R), \\ \dot{\zeta}_D = k_D(T) \prod_{i=1}^{N} C_i^{\nu'_i}, \\ \dot{\zeta}_R = k_R(T) \prod_{i=1}^{N} C_i^{\nu''_i}. \end{cases} \tag{3.54}$$

We then have

$$\begin{cases} \dot{W}_j = \nu_j \mathcal{M}_j \dot{\zeta}, \\ \dot{\zeta} = \dot{\zeta}_D - \dot{\zeta}_R = \dot{\zeta}_R \left(\dot{\zeta}_D / \dot{\zeta}_R - 1 \right), \end{cases} \tag{3.55}$$

or

$$\dot{\zeta} = \dot{\zeta}_R \left(\frac{k_D}{k_R} \prod_i C_i^{-\nu_i} - 1 \right). \tag{3.56}$$

We established in Sect. 2.4.3 that, for an ideal gas mixture,

$$\prod_{i=1}^{N} (C_i)_i^{\nu} = K_C \exp(-A/RT), \tag{3.57}$$

so that

$$\dot{\zeta} = \dot{\zeta}_R \left(\frac{k_D}{k_R K_C} \exp(A/RT) - 1 \right). \tag{3.58}$$

At chemical equilibrium, we must have $\dot{\zeta}_D \cong \dot{\zeta}_R$ as well as $A \cong 0$. It follows that

$$\frac{k_D}{k_R} = K_C(T). \tag{3.59}$$

The specific reaction rates of the direct and reverse reactions are thus connected by a simple thermodynamic equation. We can deduce that

$$\dot{\zeta} = \dot{\zeta}_R (\exp(A/RT) - 1). \tag{3.60}$$

Thus, chemical kinetics makes it possible to express $\dot{\zeta}$ in terms of the state parameters, which completes the results of Sect. 2.4.7.

Close to equilibrium, the preceding relation can be linearized according to A. We obtain

$$\dot{\zeta} = \dot{\zeta}_R(A/RT). \tag{3.61}$$

The entropy production rate

$$\delta_i\dot{S} = \dot{\zeta}(A/T) = \dot{\zeta}_R\left[\exp(A/RT) - 1\right] A/T \tag{3.62}$$

becomes

$$\delta_i\dot{S} = \dot{\zeta}_R(A^2/RT^2), \tag{3.63}$$

and is of course always positive.

This result is in agreement with the classical theory of linearized irreversible processes. Chemical affinity can be regarded as a generalized force, and the rate of reaction $\dot{\zeta}$ can be thought of as the generalized flux that results from its action. This flux and force are scalars. We can then write

$$F^S = A/T, \; J^S = \dot{\zeta} \tag{3.64}$$

and

$$\begin{cases} \delta_i\dot{S} = \dot{W}_S = J^S\,F^S, \\ J^S = L\,F^S, \end{cases} \tag{3.65}$$

with

$$L = \dot{\zeta}_R/R. \tag{3.66}$$

Note that, close to equilibrium, A/T and $\dot{\zeta}$ are first order and infinitely small, whereas $\dot{W}_S$ is second order and infinitely small and can be regarded as a negligible quantity in many cases. We can see that it could be interesting to formulate certain reactive flow problems using the entropy function, which can be regarded as being constant in cases close to equilibrium. In the next chapter we find that the same remark remains valid for any transfer phenomenon close to equilibrium.

Also note that the kinetics of heterogeneous systems are considered in Sect. 11.4.1.

4

Balance Equations for Reactive Flows

The thermodynamic properties of one-component systems and multicomponent fluid mixtures were presented in Chap. 2, although the thermodynamic systems considered were mainly discrete systems. Flux and production expressions that are important for fluid flows were then given in Chap. 3. Therefore, we now need to derive the balance equations for these fluid flows.

In order to better understand the derivation of the relevant bulk balance law equations, in Sect. 4.1 we consider the jump from discrete to continuous systems for the particular case of one-dimensional thermal transfer in a motionless medium.

Then, in Sect. 4.2, we review the important concepts of the material derivative and strain in a one-component three-dimensional medium, which are subsequently applied to multicomponent fluids.

The mass balance for species j in a composite medium is then established in Sect. 4.3, and the total mass balance is deduced.

In Sect. 4.4 we derive the general balance equation of any property F,[1] initially based on the mean material motion $\mathbf{v}(\mathbf{x}, t)$, and then based on an arbitrary continuous nonspecific motion $\mathbf{W}(\mathbf{x}, t)$ (this approach will be useful when addressing discontinuities such as shock waves).[2]

This general balance equation is then applied to the momentum balance in Sect. 4.5 and to the energy balance in Sect. 4.6.

The flux and entropy production are determined for a discrete system in Sect. 4.7, before the general balance equation is applied to the entropy in a continuous medium in Sect. 4.8, and the balance laws of discontinuities in a continuous medium are considered in Sect. 4.9.

[1]The variation in a given property F in a control volume $\mathcal{V}$ over time is usually assumed to result from the fluxes through its boundary Σ and internal production. The local balance equations for the continuous fluid mechanics of mixtures are then deduced via the fundamental lemma for the mechanics of continuous media.

[2]Discontinuities with internal properties that have their own balance equations are treated in Chap. 11.

R. Prud'homme, *Flows of Reactive Fluids*,
DOI 10.1007/978-0-8176-4659-2_4, © Springer Science+Business Media, LLC 2010

Finally, other methodologies for deriving the balance laws are discussed in Sect. 4.10 with an eye on chemical reactors, and the total deterministic and probabilistic population balances are then written.

Tensor notation is explained in Sect. A.1.

At this point we have the necessary basic equations needed to study continuous fluid flows with transfer phenomena and chemical reactions; our task then is to solve the system of equations obtained. We will focus on this task in the following chapters.

4.1 Passage to the Continuum: Example of Thermal Transfer in a Continuous Medium at Rest

In this section we consider the first example of a continuous system that has a nonhomogeneous thermodynamic state and does not comprise a discrete sum of subsystems. We assume a one-dimensional process in order to simplify matters (Fig. 4.1). The energy balance equation is[3]

$$\rho\frac{\partial e}{\partial t}\Sigma\delta x = [q - (q + \frac{\partial q}{\partial x}\delta x)]\Sigma \tag{4.1}$$

if we take an average value over the section thickness δx for the internal energy per unit mass $e(x, t)$. The density ρ is assumed to be constant. The specific heat c obeys $de = c\,dT$, and is taken to be constant.

Therefore, the energy balance becomes

$$\rho\frac{\partial e}{\partial t} + \frac{\partial q}{\partial x} = 0 \text{ or } \rho c\frac{\partial T}{\partial t} + \frac{\partial q}{\partial x} = 0. \tag{4.2}$$

The entropy balance equation is determined in a similar way (see Fig. 4.2):

$$(\rho\frac{\partial s}{\partial t} - \dot{W}_S)\Sigma\delta x = [\frac{q}{T} - (\frac{q}{T} + \frac{\partial(q/T)}{\partial x}\delta x)]\Sigma, \tag{4.3}$$

where the entropy production rate $\delta_i\dot{S}$ per unit volume is denoted $\dot{W}_S$. We then obtain

$$\rho\frac{\partial s}{\partial t} + \frac{\partial(q/T)}{\partial x} = \dot{W}_S. \tag{4.4}$$

Each section thickness, as small as required, is regarded as a subsystem in equilibrium. We therefore obtain in this case

$$de = T\,ds \quad \text{or} \quad \frac{\partial e}{\partial t} = T\frac{\partial s}{\partial t} \text{ (Gibbs)}. \tag{4.5}$$

This relation makes it possible to eliminate s and e between the energy and entropy balances. We get

---

[3]We ignore energy source terms due to, for example, radiative transfer here.

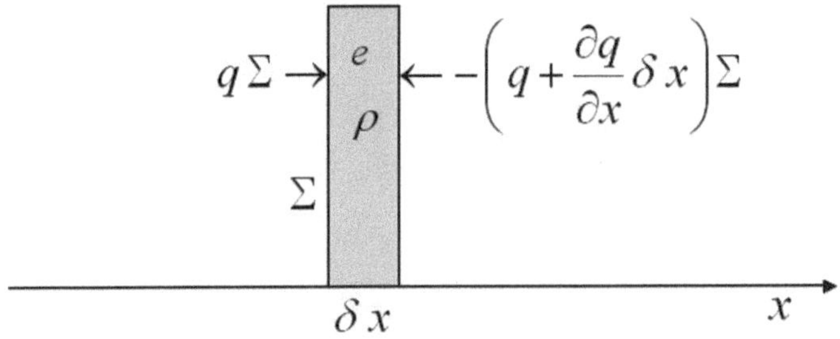

Fig. 4.1. One-dimensional energy balance

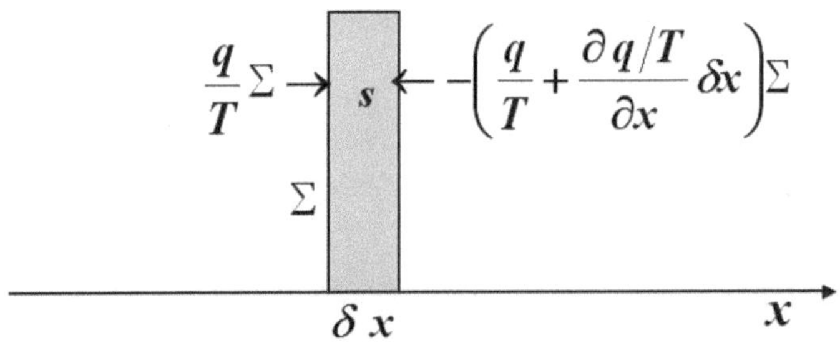

Fig. 4.2. One-dimensional entropy balance

$$\dot{W}_S = q\frac{\partial(1/T)}{\partial x}. \tag{4.6}$$

This relation reveals the generalized flux and force: $J = q$ and $F = \partial(1/T)/\partial x$, respectively. Here, the principles of irreversible thermodynamics (Sect. 3.2) give

$$J = LF \text{ or } q = L\frac{\partial(1/T)}{\partial x}. \tag{4.7}$$

By setting $\lambda = L/T^2$, we obtain the well-known Fourier law

$$q = -\lambda\frac{\partial T}{\partial x}. \tag{4.8}$$

The energy balance becomes

$$\rho\frac{\partial e}{\partial t} - \frac{\partial}{\partial x}(\lambda\frac{\partial T}{\partial x}) = 0. \tag{4.9}$$

Based on the selected assumptions, and assuming that λ is constant, we obtain the Fourier equation

$$\frac{\partial T}{\partial t} - \frac{\lambda}{\rho c}\frac{\partial^2 T}{\partial x^2} = 0. \tag{4.10}$$

We generally set $\lambda/\rho c = \kappa$, the thermal diffusivity, so that

$$\frac{\partial T}{\partial t} - \kappa\frac{\partial^2 T}{\partial x^2} = 0. \tag{4.11}$$

Finally, the entropy production rate becomes

$$\dot{W}_S = L(\frac{\partial(1/T)}{\partial x})^2 \tag{4.12}$$

or

$$\dot{W}_S = \frac{\lambda}{T^2}(\frac{\partial T}{\partial x})^2. \tag{4.13}$$

If $(\partial T/\partial x)$ is an infinitely small first-order quantity, then $\dot{W}_S$ is second order and infinitely small as long as (λ/T^2) is a finite quantity. We will see, particularly in Sect. 4.6, that these results can be generalized to three-dimensional processes.

4.2 Reminder of the Concepts of the Material Derivative and Strain in a Simple Medium

The material derivative and strain are essential concepts for analyzing deformable moving media such as fluids [96, 99]. Consider a property $f(\mathbf{x}, t)$

per unit mass of a material particle in a simple continuous medium (i.e., a medium with only one component). If we follow the motion of this particle, the time derivative is

$$\frac{df}{dt} = \frac{\partial f}{\partial t} + \mathbf{v} \cdot \nabla f, \tag{4.14}$$

where $\mathbf{v}$ is the velocity vector of the particle. This quantity[4] is called the material derivative.

Let us now consider an elementary material vector $\delta \mathbf{x}$ that is as short as required. The ends of the vector are located at $\mathbf{x}$ and $\mathbf{x} + \delta \mathbf{x}$ at time t. These ends, and thus the vector itself, are transported by the motion. We have

$$\frac{d(\delta x)}{dt} = \mathbf{K} \cdot \delta x, \tag{4.15}$$

with

$$\mathbf{K} = \nabla \otimes \mathbf{v}. \tag{4.16}$$

We then find that the material volume element $\delta \mathcal{V}$ obeys

$$\frac{d(\delta \mathcal{V})}{dt} = tr(\mathbf{K})\, \delta \mathcal{V} = \nabla \cdot \mathbf{v}\, \delta \mathcal{V}. \tag{4.17}$$

$\nabla \cdot \mathbf{v}$ is the rate of volume dilatation (the bulk dilatation rate). The tensor $\nabla \otimes \mathbf{v}$ can be split into a symmetrical $\mathbf{D} = \frac{1}{2}(\mathbf{K} + \tilde{\mathbf{K}})$ part tensor of the strain rate and an antisymmetric $\boldsymbol{\Omega} = \frac{1}{2}(\mathbf{K} - \tilde{\mathbf{K}})$ part tensor of the rate of rotation.

Note About the Velocity Gradient

The preceding formulae are easily proven as follows. Consider the motions $\mathbf{x}$ and $\mathbf{x}'$ of two points (material particles) in a continuous medium that are close to each other between the times t and $t + dt$, where dt is small (Fig. 4.3).

By limiting the Taylor series expansions to the first order, we get $\mathbf{x}(t + dt) = \mathbf{x}(t) + (d\mathbf{x}/dt)dt = \mathbf{x}(t) + \mathbf{v}dt$, $\mathbf{x}'(t+dt) = \mathbf{x}'(t) + \mathbf{v}'dt$, where $\mathbf{x}' = \mathbf{x} + \delta\mathbf{x}$, $\mathbf{v}' = \mathbf{v} + \delta\mathbf{v}$ $\Rightarrow \delta\mathbf{x}(t + dt) = \delta\mathbf{x} + \delta\mathbf{v}dt$; i.e., $d(\delta x)/dt = \delta\mathbf{v} = \nabla \otimes \mathbf{v} \cdot \delta\mathbf{x}$.

---

[4]This concept of the material derivative is easily introduced if one considers a shift from Lagrange coordinates to Euler coordinates. Let $(\mathbf{x}, t)$ be the Euler coordinates of a particle, and $(\mathbf{a}, \tau)$ be its Lagrange coordinates ($\mathbf{a}$; for example, the initial position of the particle). We then write $\mathbf{x} = \phi(\mathbf{a}, \tau)$, with $\mathbf{a} = \phi(\mathbf{a}, 0)$. Let us now formulate the change in coordinates for the elementary quantities: $d\mathbf{x} = \partial\phi/\partial\mathbf{a} \cdot d\mathbf{a} + \partial\phi/\partial\tau\, d\tau$. Material velocity $\mathbf{v}$ is easily defined in Lagrange coordinates; if $\mathbf{a}$ is fixed, $d\mathbf{x} = (\partial\phi/\partial\tau)\, d\tau = \mathbf{v}d\tau$; i.e., $d\mathbf{x}/dt = (\partial\phi/\partial t)_{\mathbf{a}=\text{const.}} = \mathbf{v}$.

We define the material derivative of a quantity $F(\mathbf{a}, \tau)$ in a similar way: $dF = (\partial F/\partial\mathbf{a}) \cdot d\mathbf{a} + (\partial F/\partial\tau)\, d\tau$. In Lagrange coordinates, the material derivative of the quantity considered is $\partial F/\partial\tau$. We now have $F(\mathbf{a}, \tau) = f(\mathbf{x}, t) = f(\phi(\mathbf{a}, \tau, t))$. Moreover: $df = (\partial f/\partial\phi) \cdot d\phi + (\partial f/\partial t)dt$, $df/dt = (\partial f/\partial x) \cdot d\mathbf{x}/dt + \partial f/\partial t = \nabla f \cdot \mathbf{v} + \partial f/\partial t$. Finally, in Euler coordinates, the material derivative is $df/dt = \partial f/\partial t + \mathbf{v} \cdot \nabla f$ and is equal to $(\partial F/\partial t)_{\mathbf{a}=\text{const.}}$. In the case of steady (permanent) motion, $\partial f/\partial t = 0$, so $\mathbf{v} = \mathbf{v}(\mathbf{x})$ in Euler variables.

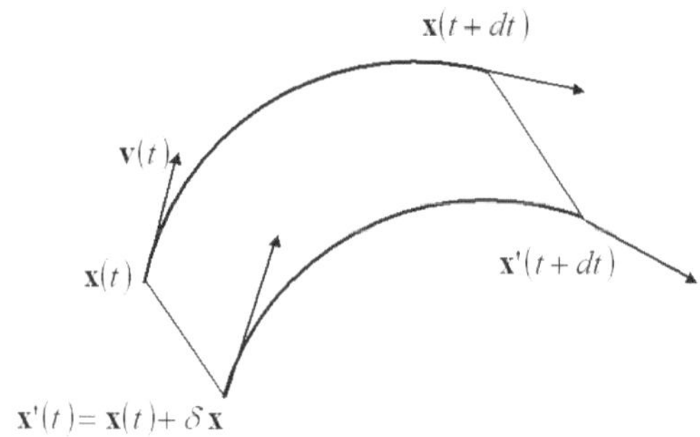

Fig. 4.3. Motion in a continuous medium

Rate of dilatation: $\delta\mathcal{V} = (\delta\mathbf{x}, \delta\mathbf{y}, \delta\mathbf{z})$ where $\delta\mathbf{x} = \delta x\,\mathbf{i}$, $\delta\mathbf{y} = \delta y\,\mathbf{j}$, $\delta\mathbf{z} = \delta z\,\mathbf{k}$. It can be verified that $d(\delta\mathcal{V})/dt = \boldsymbol{\nabla}\cdot\mathbf{v}\,\delta\mathcal{V}$, which is the rate of volume dilatation (the bulk strain rate). $\delta\mathcal{V}$ is an elementary material volume that follows the motion.

Demonstration: $\delta\mathcal{V} = \delta x\,\delta y\,\delta z = \delta\mathbf{x}\cdot(\delta\mathbf{y}\times\delta\mathbf{z})$, $d(\delta\mathcal{V})/dt = (d(\delta\mathbf{x})/dt)\cdot(\delta\mathbf{y}\times\delta\mathbf{z}) + \ldots = (\boldsymbol{\nabla}\otimes\mathbf{v}\cdot\delta\mathbf{x})\cdot(\delta\mathbf{y}\times\delta\mathbf{z}) + \ldots$ For $\delta\mathbf{x} = \delta x\,\mathbf{i}$, $\delta\mathbf{y} = \delta y\,\mathbf{j}$, $\delta\mathbf{z} = \delta z\,\mathbf{k}$, we get $d(\delta\mathcal{V})/dt = (u_{,x}\mathbf{i}+v_{,x}\mathbf{j}+w_{,x}\mathbf{k})\cdot\mathbf{i}\,\delta y\,\delta z+\ldots = u_{,x}\,\delta x\,\delta y\,\delta z+\ldots = (u_{,x}+v_{,y}+w_{,z})\,\delta\mathcal{V} = (\boldsymbol{\nabla}\cdot\mathbf{v})\,\delta\mathcal{V}$.

Mass conservation: The mass present in $\delta\mathcal{V}$ is preserved during the motion. If ρ is the local density, $\delta m = \rho\delta\mathcal{V} = \text{const.}$ during the motion. Thus, $d(\delta m)/dt = d\rho/dt\,\delta\mathcal{V} + \rho\,d(\delta\mathcal{V})/dt$; in other words,

$$\frac{1}{\delta\mathcal{V}}\frac{d(\delta\mathcal{V})}{dt} = -\frac{1}{\rho}\frac{d\rho}{dt} = \boldsymbol{\nabla}\cdot\mathbf{v}. \tag{4.18}$$

4.3 Mass Balance of Species j and Total Mass Balance in a Composite Medium

The fluid considered here is a chemical mixture of species. It is assumed to consist of only one phase; multiphase media will be investigated in this context in Chap. 12. Each species in the mixture can be regarded as having the properties of a continuous fluid, and can exchange quantities with the other species present in the mixture. The most rigorous method of deriving the mass balance of the given species j is to consider its motion. A control surface (Σ_j), which has an arbitrary initial position and bounds a volume $(\mathcal{V}_j)$, is assumed to follow the motion $\mathbf{v}_j$ of species j. The material derivative associated with this motion is denoted d_j/dt. Only chemical reactions are likely to modify the mass of species j in volume $(\mathcal{V}_j)$. This yields the following balance equation written in integral form [10, 34, 85, 108, 290]:

$$\frac{d_j}{dt}\int_{\mathcal{V}_j} \rho_j \, d\mathcal{V} = \int_{\mathcal{V}_j} \dot{W}_j \, d\mathcal{V}. \tag{4.19}$$

The derivative of the integral implies the evaluation of

$$\frac{d_j(\rho_j \, d\mathcal{V})}{dt} = \frac{d_j \rho_j}{dt} \, d\mathcal{V} + \rho_j \frac{d_j(d\mathcal{V})}{dt}. \tag{4.20}$$

According to the results of Sect. 4.2,

$$\begin{cases} d_j \rho_j / dt = \partial \rho_j / \partial t + \mathbf{v}_j \cdot \boldsymbol{\nabla} \rho_j, \\ d_j(d\mathcal{V})/dt = \boldsymbol{\nabla} \cdot \mathbf{v}_j \, d\mathcal{V}. \end{cases} \tag{4.21}$$

The balance equation then becomes

$$\int_{\mathcal{V}_j} \left(\frac{\partial \rho_j}{\partial t} + \mathbf{v}_j \cdot \boldsymbol{\nabla} \rho_j + \rho_j \boldsymbol{\nabla} \cdot \mathbf{v}_j - \dot{W}_j\right) d\mathcal{V} = 0 \tag{4.22}$$

or

$$\int_{\mathcal{V}_j} \left(\frac{\partial \rho_j}{\partial t} + \boldsymbol{\nabla} \cdot (\rho_j \mathbf{v}_j) - \dot{W}_j\right) d\mathcal{V} = 0. \tag{4.23}$$

The *fundamental lemma of the mechanics of continuous media* [99] makes it possible to use the local balance equation. According to this lemma, if the preceding equation is valid whatever the initial control volume of the considered flow, and if the properties and their first derivatives are continuous, the quantity under the integral sign is equal to zero. We assume these conditions hold, and so

$$\frac{\partial \rho_j}{\partial t} + \boldsymbol{\nabla} \cdot (\rho_j \mathbf{v}_j) - \dot{W}_j = 0. \tag{4.24}$$

Inversely, if Σ delimits a fixed control volume $\mathcal{V}$, the integration of (4.24) leads to

$$\frac{\partial}{\partial t}\int_{\mathcal{V}} \rho_j \, d\mathcal{V} + \int_{\Sigma} \rho_j \mathbf{v}_j \cdot \mathbf{n} \, d\Sigma = \int_{\mathcal{V}} \dot{W}_j d\mathcal{V}. \tag{4.25}$$

The various terms present in the mass balance of species j inside the motionless volume $(\mathcal{V})$ (which is a thermodynamically open system) are clearly apparent here: the time derivative of the mass of species j in volume $(\mathcal{V})$; the mass flow $J_j = \rho_j \mathbf{v}_j$ through a unit area of surface (Σ); and the chemical production rate in volume $(\mathcal{V})$. If we carry out a summation of the terms in the local balance (4.24) over the index j, $j = 1, 2 \ldots N$, we get

$$\frac{\partial \rho}{\partial t} + \boldsymbol{\nabla} \cdot (\rho \mathbf{v}) = 0, \tag{4.26}$$

where (see Sect. 2.1.3)

$$\rho = \sum_{j=1}^{N} \rho_j, \ \rho \mathbf{v} = \sum_{j=1}^{N} \rho_j \mathbf{v}_j. \tag{4.27}$$

The total mass produced is obviously equal to zero, so that

$$\sum_{j=1}^{N} \dot{W}_j = 0. \tag{4.28}$$

It is easily verified that this result is compatible with the laws of the chemical kinetics given in Chap. 3. Indeed,

$$\sum_j \dot{W}_j = \sum_{j,r} \nu_{jr} \mathcal{M}_j \dot{\zeta}_r = \sum_r \dot{\zeta}_r \sum_j \nu_{jr} \mathcal{M}_j. \tag{4.29}$$

However,

$$\sum_j \nu_{jr} \mathcal{M}_j = 0 \tag{4.30}$$

simply expresses the conservation of mass during reaction r.
The diffusion flux is defined via

$$\mathcal{J}_{Dj} = \rho_j \mathbf{V}_j = \rho_j(\mathbf{v}_j - \mathbf{v}), \tag{4.31}$$

so that the local balance equation for species j becomes

$$\frac{\partial \rho_j}{\partial t} + \boldsymbol{\nabla} \cdot \mathcal{J}_{Dj} + \boldsymbol{\nabla} \cdot (\rho_j \mathbf{v}) = \dot{W}_j. \tag{4.32}$$

By setting $\rho_j/\rho = Y_j$, the mass fraction of species j, and by using the conservation equation for the overall mass balance, we deduce that

$$\rho \frac{dY_j}{dt} + \boldsymbol{\nabla} \cdot \mathcal{J}_{Dj} = \dot{W}_j, \tag{4.33}$$

where the material derivative d/dt is that of the barycentric motion

$$\frac{dY_j}{dt} = \frac{\partial Y_j}{\partial t} + \mathbf{v} \cdot \boldsymbol{\nabla} Y_j. \tag{4.34}$$

4.4 General Balance Equation for a Property *F*

4.4.1 Balance Equation Based on the Mean Material Motion v(x, *t*)

The result (4.33) obtained for a mass fraction Y_j—which contains three local terms: the material derivative (obtained by following the motion of the barycentric velocity $\mathbf{v}$; see Eq. 4.27), the divergence in the unit flux through a surface due to this motion, and the production rate—formally extends to any physical property F [10, 99].

If F is an extensive physical property (mass, momentum, energy, ...) of the fluid (more generally, a continuous medium), and f is its value per unit mass, then, if we follow a volume ($\mathcal{V}$) during its motion, we get

$$\frac{d}{dt}\int_{\mathcal{V}} \rho\, f\, d\mathcal{V} + \int_{\partial\mathcal{V}} \boldsymbol{\mathcal{J}}_F \cdot \mathbf{n}\, dS = \int_{\mathcal{V}} \dot{W}_F\, d\mathcal{V}, \tag{4.35}$$

where $\mathbf{n}$ is the external unit vector normal to the closed surface $\partial\mathcal{V}$ that surrounds (and is presumed to be connected to) the volume $\mathcal{V}$. We have (see Fig. 4.4)

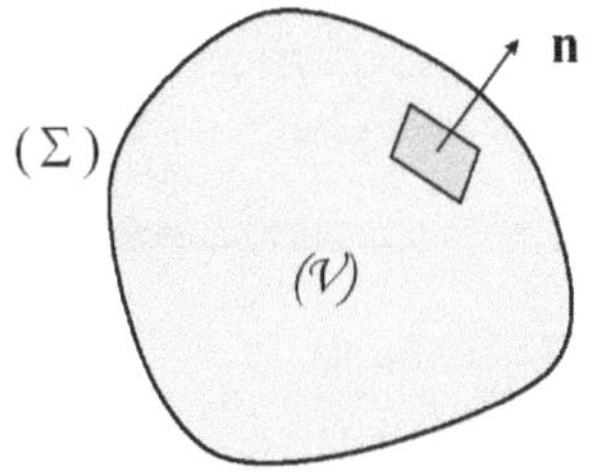

Fig. 4.4. The control volume

$$\frac{d}{dt}\int_{\mathcal{V}} \rho\, f\, d\mathcal{V} = \int_{\mathcal{V}} \frac{d(\rho\, f\, d\mathcal{V})}{dt} = \int_{\mathcal{V}} \frac{d(\rho\, f)}{dt}\, d\mathcal{V} + \int_{\mathcal{V}} \rho\, f\, \frac{d(d\mathcal{V})}{dt}. \tag{4.36}$$

Taking into account the strain equation (4.18, i.e., $d(d\mathcal{V})/dt = \boldsymbol{\nabla}\cdot\mathbf{v}\, d\mathcal{V}$), we deduce that

$$\frac{d}{dt}\int_{\mathcal{V}} \rho\, f\, d\mathcal{V} = \int_{\mathcal{V}} (\frac{d(\rho\, f)}{dt}\, d\mathcal{V} + \rho\, f\, \nabla\cdot\mathbf{v})\, d\mathcal{V}. \tag{4.37}$$

By applying the *Stokes–Ostrogradski theorem*, we also get

$$\int_{\partial\mathcal{V}} \boldsymbol{\mathcal{J}}_F \cdot \mathbf{n}\, dS = \int_{\mathcal{V}} \boldsymbol{\nabla}\cdot\boldsymbol{\mathcal{J}}_F\, d\mathcal{V}. \tag{4.38}$$

The general balance equation thus becomes

$$\int_{\mathcal{V}} (\frac{d(\rho\, f)}{dt} + \boldsymbol{\nabla}\cdot\boldsymbol{\mathcal{J}}_F + \rho\, f\, \nabla\cdot\mathbf{v} - \dot{W}_F)\, d\mathcal{V} = 0. \tag{4.39}$$

When we apply the fundamental lemma as in Sect. 4.3, where there must not be any discontinuity within the volume ($\mathcal{V}$) of interest in the flow, this quantity is zero. The local balance equation of F is thus written

$$\frac{d(\rho f)}{dt} + \rho f \nabla \cdot \mathbf{v} + \boldsymbol{\nabla} \cdot \boldsymbol{\mathcal{J}}_F = \dot{W}_F. \tag{4.40}$$

In this equation, we can identify the flux term that characterizes the transfer of F through the unit area on the moving surface (see Chap. 3), as well as the production rate of F per unit volume. We modify the local balance equation (4.40) by applying the mass conservation equation (4.26), which then becomes

$$\rho \frac{df}{dt} + \boldsymbol{\nabla} \cdot \boldsymbol{\mathcal{J}}_F = \dot{W}_F. \tag{4.41}$$

Another form of the local balance equation is obtained by introducing the unit flux through a fixed surface

$$\mathbf{J}_F = \boldsymbol{\mathcal{J}}_F + \rho f \mathbf{v}, \tag{4.42}$$

which gives

$$\frac{\partial(\rho f)}{\partial t} + \boldsymbol{\nabla} \cdot \mathbf{J}_F = \dot{W}_F. \tag{4.43}$$

By applying one or the other of these equations to the mass of species j, we of course derive the results we obtained previously. The forms (4.41) and (4.43) of the local balance equation for a property F are particular cases of a more general formula, as we will show.

4.4.2 Balance Equation for a Property F Based on an Arbitrary Continuous Motion $\mathbf{W}(\mathbf{x},\, t)$

By analogy to the real motion $\mathbf{v}$, but where $\mathbf{v}$ is replaced with $\mathbf{W}$, we obtain the following integral equation:

$$\frac{d_{\mathbf{W}}}{dt} \int_{\mathcal{V}} \rho f \, d\mathcal{V} + \int_{\partial \mathcal{V}} \mathbf{J}_{\mathbf{W}F} \cdot \mathbf{n} \, dS = \int_{\mathcal{V}} \dot{W}_F \, d\mathcal{V}.$$

The production $\int_{\mathcal{V}} \dot{W}_F \, d\mathcal{V}$ is independent of the field $\mathbf{W}(\mathbf{x},\, \mathbf{t})$. In order to determine the flux expression $\mathbf{J}_{\mathbf{W}F}$, we write

$$\begin{cases} d_{\mathbf{W}}(\rho f \, d\mathcal{V})/dt + \boldsymbol{\nabla} \cdot \mathbf{J}_{\mathbf{W}F} \, d\mathcal{V} = \dot{W}_F \, d\mathcal{V}, \\ d_{\mathbf{W}}(\rho f)/dt + \rho f \, \boldsymbol{\nabla} \cdot \mathbf{W} + \boldsymbol{\nabla} \cdot \mathbf{J}_{\mathbf{W}F} = \dot{W}_F, \\ \partial \rho f/\partial t + \boldsymbol{\nabla} \cdot (\mathbf{J}_{\mathbf{W}F} + \rho f \mathbf{W}) = \dot{W}_F, \end{cases} \tag{4.44}$$

where $d_{\mathbf{W}}/dt = \partial/\partial t + \mathbf{W} \cdot \boldsymbol{\nabla}$.

The flux $\mathbf{J}_{\mathbf{W}F} + \rho f \, \mathbf{W}$ is independent of $\mathbf{W}$. We therefore have

$$\mathbf{J}_{\mathbf{W}F} + \rho f \, \mathbf{W} = \mathbf{J}_F = \mathcal{J}_F + \rho f \, \mathbf{v}. \tag{4.45}$$

Here, $\mathbf{J}_{\mathbf{W}F}$ is the flux of F relative to velocity $\mathbf{W}$, $\mathbf{J}_F = \mathbf{J}_{\mathbf{O}F}$ is the flux relative to any motionless surface (i.e., surface at rest), and $\mathcal{J}_F = \mathbf{J}_{\mathbf{v}F}$ is the flux relative to the real velocity $\mathbf{v}$. This latter flux is the only "physical" quantity.

Finally, the following expression for the flux vector and the corresponding balance law are obtained:

$$\begin{cases} \mathbf{J}_{\mathbf{W}\,F} = \mathcal{J}_F + \rho \, f \, (\mathbf{v} - \mathbf{W}), \\ d_{\mathbf{W}}(\rho f)/dt + \rho f \, \boldsymbol{\nabla} \cdot \mathbf{W} + \boldsymbol{\nabla} \cdot \mathbf{J}_{\mathbf{W}F} = \dot{W}_F. \end{cases} \tag{4.46}$$

This result is particularly useful for establishing the balance laws at discontinuities.

4.5 Momentum Balance

We now associate a velocity $\mathbf{v}_j$, a pressure tensor $\mathbf{P}_j$, and an external force per unit mass $\mathbf{f}_j$ with each species j. Only the result obtained by summing the quantities over the index j is shown below:

$$\rho \frac{d\mathbf{v}}{dt} + \boldsymbol{\nabla} \cdot \mathbf{P} = \sum_j \rho_j \mathbf{f}_j. \tag{4.47}$$

Regarding the quantities introduced in Sect. 4.4.1, we have $F = m\mathbf{v}$; $f = \rho\mathbf{v}$ is the momentum per unit volume; $\mathcal{J}_F = \mathbf{P} = -\boldsymbol{\Sigma}$ is the total pressure tensor, which is assumed to be symmetric;[5] and $\dot{W}_F = \sum_j \rho_j \mathbf{f}_j$ is the bulk force per unit volume (applied far from the volume).

4.6 Energy Balance

When extended to moving objects, the first law of thermodynamics (per unit time) is expressed as follows [99]:

$$\frac{dE}{dt} + \frac{dK}{dt} = \mathcal{P} + \mathcal{Q}. \tag{4.48}$$

dE/dt and dK/dt are the time derivatives of the internal energy and the kinetic energy of the fluid present in a volume ($\mathcal{V}$) moving at the barycentric

[5] As mentioned by de Groot and Mazur [108], this assumption is usually made in hydrodynamics, but strictly speaking it is only justifiable for systems that consist of spherical molecules or very low density systems. The pressure tensor may contain an antisymmetric part for other systems.

velocity $\mathbf{v}$ of the mixture. $\mathcal{P}$ and $\mathcal{Q}$ are respectively the mechanical power and calorific power supplied to this volume. According to (4.37), the quantity F present in the volume $(\mathcal{V})$ verifies

$$\frac{dF}{dt} = \frac{d}{dt}\int_{\mathcal{V}} \rho f \, d\mathcal{V} = \int_{\mathcal{V}} (\frac{\partial(\rho f)}{\partial t} + \rho f \boldsymbol{\nabla}\cdot\mathbf{v})\, d\mathcal{V}, \tag{4.49}$$

and, taking into account that the total mass is conserved (4.26), we obtain

$$\frac{dF}{dt} = \int_{\mathcal{V}} \rho \frac{df}{dt}\, d\mathcal{V}. \tag{4.50}$$

Applied to E and K, this result gives

$$\begin{cases} dE/dt = \int_{\mathcal{V}} \rho \, de/dt \, d\mathcal{V}, \\ dK/dt = (d/dt)\int_{\mathcal{V}} \sum_j \rho_j \, (v_j^2/2)\, d\mathcal{V} \cong \int_{\mathcal{V}} d(v^2/2)/dt \, d\mathcal{V}, \end{cases} \tag{4.51}$$

since, ignoring the squared diffusion velocity terms and noting that $\sum_j \rho_j \mathbf{V}_j = \sum_j \mathcal{J}_{Dj} = 0$,

$$\sum_j \rho_j v_j^2 = \sum_j \rho_j (\mathbf{v} + \mathbf{V}_j)^2 \cong \rho v^2. \tag{4.52}$$

Scalar multiplication of the two sides of the momentum equation (4.47)by $\mathbf{v}$ yields

$$\rho \mathbf{v}\cdot\frac{d\mathbf{v}}{dt} = -\mathbf{v}\cdot\boldsymbol{\nabla}\cdot\mathbf{P} + \mathbf{v}\cdot\sum_j \rho_j \mathbf{f}_j. \tag{4.53}$$

Then, taking into account the notation $\boldsymbol{\nabla}\otimes\mathbf{v} : \mathbf{P} = \frac{\partial v_\alpha}{\partial x_\beta} P_{\alpha\beta}$ (see Sect. A.1),

$$\frac{1}{2}\rho\frac{dv^2}{dt} = -\boldsymbol{\nabla}\cdot(\mathbf{v}\cdot\mathbf{P}) + \boldsymbol{\nabla}\otimes\mathbf{v} : \mathbf{P} + \sum_j \rho_j \mathbf{v}_j \cdot \mathbf{f}_j - \sum_j \mathcal{J}_{Dj}\cdot\mathbf{f}_j. \tag{4.54}$$

It follows that

$$\frac{dK}{dt} = -\int_{\partial\mathcal{V}} (\mathbf{v}\cdot\mathbf{P})\cdot\mathbf{n} dS + \int_{\mathcal{V}} \sum_j \rho_j \mathbf{v}_j\cdot\mathbf{f}_j d\mathcal{V} + \int_{\mathcal{V}} \boldsymbol{\nabla}\otimes\mathbf{v} : \mathbf{P} d\mathcal{V} - \int_{\mathcal{V}} \sum_j \mathcal{J}_{Dj}\cdot\mathbf{f}_j d\mathcal{V}. \tag{4.55}$$

The mechanical power supplied to the system is the power from external forces:

$$\mathcal{P} = -\int_{\partial\mathcal{V}} (\mathbf{v}\cdot\mathbf{P})\cdot\mathbf{n} dS + \int_{\mathcal{V}} \sum_j \rho_j \mathbf{v}_j \cdot \mathbf{f}_j d\mathcal{V}. \tag{4.56}$$

We define the power from internal stresses as follows:

$$\mathcal{P}_i = \int_{\mathcal{V}} \boldsymbol{\nabla} \otimes \mathbf{v} : \mathbf{P} d\mathcal{V} - \int_{\mathcal{V}} \sum_j \mathcal{J}_{Dj} \cdot \mathbf{f}_j d\mathcal{V}. \tag{4.57}$$

We then obtain the kinetic energy theorem

$$\frac{dK}{dt} = \mathcal{P} + \mathcal{P}_i, \tag{4.58}$$

so the first law of thermodynamics (4.48) becomes

$$\frac{dE}{dt} = \mathcal{Q} - \mathcal{P}_i. \tag{4.59}$$

The calorific power supplied includes a bulk term corresponding to an energy contribution per unit volume and a surface term:

$$\mathcal{Q} = \int_{\mathcal{V}} r \, d\mathcal{V} - \int_{\partial\mathcal{V}} \mathbf{q} \cdot \mathbf{n} dS. \tag{4.60}$$

Therefore, the internal energy balance becomes

$$\int_{\mathcal{V}} \rho \frac{de}{dt} \, d\mathcal{V} = \int_{\mathcal{V}} r \, d\mathcal{V} - \int_{\partial\mathcal{V}} \mathbf{q} \cdot \mathbf{n} dS - \int_{\mathcal{V}} \boldsymbol{\nabla} \otimes \mathbf{v} : \mathbf{P} d\mathcal{V} + \int_{\mathcal{V}} \sum_j \mathcal{J}_{Dj} \cdot \mathbf{f}_j d\mathcal{V}. \tag{4.61}$$

The resulting local internal energy balance is

$$\rho \frac{de}{dt} + \boldsymbol{\nabla} \cdot \mathbf{q} = r - \boldsymbol{\nabla} \otimes \mathbf{v} : \mathbf{P} + \sum_j \mathcal{J}_{Dj} \cdot \mathbf{f}_j. \tag{4.62}$$

By introducing the strain tensor $\mathbf{D}$, the symmetrical part of the gradient velocity tensor

$$\mathbf{D} = \frac{1}{2}(\boldsymbol{\nabla} \otimes \mathbf{v} + \widetilde{\boldsymbol{\nabla} \otimes \mathbf{v}}), \tag{4.63}$$

and, given that the pressure tensor $\mathbf{P}$ (the opposite of the stress tensor $\boldsymbol{\Sigma}$) is a symmetrical tensor, we finally obtain

$$\rho \frac{de}{dt} + \boldsymbol{\nabla} \cdot \mathbf{q} = r - \mathbf{D} : \mathbf{P} + \sum_j \mathcal{J}_{Dj} \cdot \mathbf{f}_j. \tag{4.64}$$

4.7 Flux and Entropy Production in a Discrete System

In this section we consider the following discrete system: a nondeformable container of volume $\mathcal{V}$ that is delimited by an impermeable and adiabatic wall and separated into two parts by a fixed and porous wall [111, 213]. Initially, although both compartments contain the same fluid, the fluid in each compartment is at its own equilibrium. We will study a small transformation of this system following fluid exchange through the porous wall (see Fig. 4.5).

Let us suppose that a small mass δm moves from compartment 2 to compartment 1 during a small time increment δt. Consider the system at the time when the two subsystems have masses and internal energies of $(m_1 + \delta m)$ and $(m_2 - \delta m)$, $(e_1 + \delta e_1)$ and $(e_2 + \delta e_2)$, respectively. For each part, the first law of thermodynamics gives

$$\begin{cases} (m_1 + \delta m)\,(e_1 + \delta e_1) - (m_1\,e_1 + \delta m\,e_2) \cong (p_2/\rho_2)\delta m, \\ (m_2 - \delta m)\,(e_2 + \delta e_2) - (m_2 - \delta m)\,e_2 \cong -(p_2/\rho_2)\delta m, \end{cases} \tag{4.65}$$

or

$$\begin{cases} m_1\,\delta e_1 + \delta m\,(e_1 - e_2) \cong (p_2/\rho_2)\delta m, \\ m_2\,\delta e_2 \cong -(p_2/\rho_2)\delta m. \end{cases} \tag{4.66}$$

By summation,

$$\delta(m_1 e_1) + \delta(m_2 e_2) = 0. \tag{4.67}$$

Relation 4.67 confirms that variations in the internal energy of one subsystem are compensated for by those of the other subsystem, which we would expect considering that the overall system is isolated.

Let us now apply the second law of thermodynamics. The entropy variations for the subsystems are:

$$\begin{cases} m_1\,\delta s_1 + \delta m\,(s_1 - s_2), \\ m_2\,\delta s_2. \end{cases} \tag{4.68}$$

The entropy variation for the total system is thus

$$\delta_i S = m_1\,\delta s_1 + m_2\,\delta s_2 + \delta m\,(s_1 - s_2) \geq 0. \tag{4.69}$$

The Gibbs relation (see Eq. 2.31)

$$\delta s = \frac{1}{T}\delta e + \frac{p}{T}\delta\vartheta \tag{4.70}$$

and the expression for s given by the Euler relation (2.33)

$$s = \frac{1}{T}e + \frac{p}{T}\vartheta - \frac{g}{T} \tag{4.71}$$

lead to the relation

$$\delta_i S = \frac{\delta(m_1\,e_1)}{T_1} + \frac{\delta(m_2\,e_2)}{T_2} + \delta m(\frac{g_2}{T_2} - \frac{g_1}{T_1}). \tag{4.72}$$

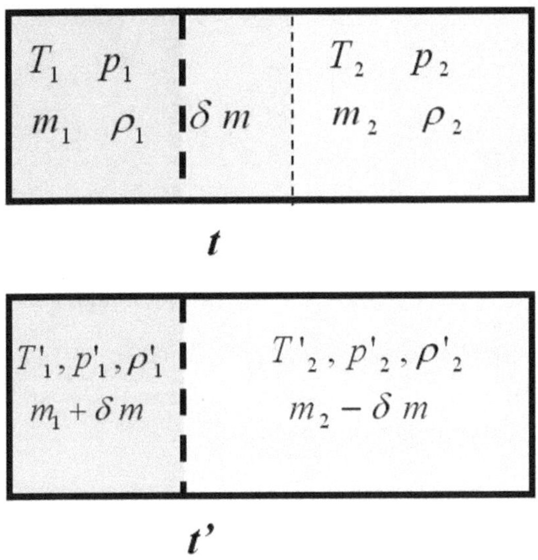

Fig. 4.5. Exchange through a porous wall: $T_i' = T_i + \delta T_i$, $p_i' = p_i + \delta p_i$, $\rho_i' = \rho_i + \delta\rho_i$

By noting that the volume of each compartment is constant, we get

$$\begin{cases} \delta(m_1\,\vartheta_1) = m_1\,\delta\vartheta_1 + \vartheta_1\delta m = 0, \\ \delta(m_2\,\vartheta_2) = m_2\,\delta\vartheta_2 - \vartheta_2\delta m = 0. \end{cases} \tag{4.73}$$

Then, from relations (4.67) and (4.72), we deduce that

$$\delta_i S = \delta(m_1\,e_1)\left(\frac{1}{T_1} - \frac{1}{T_2}\right) + \delta m\left(\frac{g_2}{T_2} - \frac{g_1}{T_1}\right). \tag{4.74}$$

Note that, since these equations are additive and the variations in the internal energy and entropy of the total system are the sums of the variations of its subsystems, we can derive the same result directly by writing

$$\begin{cases} \delta(m_1 e_1) + \delta(m_2 e_2) = 0, \\ \delta_i S = \delta(m_1\,s_1) + \delta(m_2\,s_2) \geq 0. \end{cases} \tag{4.75}$$

The internal energy flux received by subsystem i per unit time is thus

$$q_i = \overset{\bullet}{\overline{\delta(m_i e_i)}}, \tag{4.76}$$

and the corresponding entropy flux is

$$\overset{\bullet}{\overline{\delta_e S_i}} = \overset{\bullet}{\overline{\delta(m_i s_i)}}\,. \tag{4.77}$$

The volume flux is

$$\overline{\dot{\delta(m_i\vartheta_i)}} = 0. \tag{4.78}$$

In addition,

$$\delta(m_i s_i) = m_i\left(\frac{\delta e_i}{T_i} + \frac{p_i}{T_i}\delta\vartheta_i\right) + \delta m_i\left(\frac{e_i}{T_i} + \frac{p_i}{T_i}\vartheta_i - \frac{g_i}{T_i}\right). \tag{4.79}$$

Based on this, we find that

$$\overline{\dot{\delta_e S_i}} = \frac{1}{T_i}\overline{\dot{\delta(m_i e_i)}} - \frac{g_i}{T_i}\overline{\dot{\delta(m_i)}}, \tag{4.80}$$

or, for subsystem 1 (removing the index 1),

$$\overline{\dot{\delta_e S}} = \frac{1}{T}(q - g\dot{m}), \tag{4.81}$$

where $\dot{m}$ is the mass flow and q is the received heat flow, as defined by (4.76).

In the presence of several species, we obtain

$$\overline{\dot{\delta_e S}} = \frac{1}{T}(q - \sum_j g_j \dot{m}_j), \tag{4.82}$$

where $\dot{m}_j$ is the mass flux of species j through the porous wall. The entropy production rate will be

$$\overline{\dot{\delta_i S}} = -q\Delta\left(\frac{1}{T}\right) + \sum_j \dot{m}_j \Delta\left(\frac{g_j}{T}\right), \tag{4.83}$$

where

$$\Delta\left(\frac{1}{T}\right) = \frac{1}{T_2} - \frac{1}{T_1}, \quad \Delta\left(\frac{g_j}{T}\right) = \frac{g_{j2}}{T_2} - \frac{g_{j1}}{T_1}. \tag{4.84}$$

Note here that the heat flux (internal energy flux)

$$q = \overline{\dot{\delta(me)}} \tag{4.85}$$

includes a conductive part and a part resulting from mass diffusion through the wall. Also note that the fluxes q and $\dot{m}_j$ are entering fluxes for the subsystem of interest. The conduction and the convection of internal energy will be separated out in Chap. 7 (see Sect. 7.6), where the entropy production rate will be written in a form that clearly shows the terms resulting from the jump in temperature (some are present in $\Delta(g/T)$) and those resulting from the jump in concentration.

4.8 Entropy Balance in a Continuous Medium

Each element of the surface $(d\Sigma)$, which has a normal $\mathbf{n}$ and follows the barycentric motion velocity $\mathbf{v}$, is traversed by a quantity of heat $\mathbf{q} \cdot \mathbf{n}\, d\Sigma$ and by a mass of species j diffused per unit time $\mathcal{J}_{Dj} \cdot \mathbf{n}\, d\Sigma$. The normal $\mathbf{n}$ is a unit vector, and its orientation is selected such that the energy or mass contribution is positive in its direction. By extending the formula (4.82), the corresponding entropy flux vector is thus found to be [10, 85, 108, 111]

$$\mathcal{J}_S = \frac{1}{T}(\mathbf{q} - \sum_j g_j \mathcal{J}_{Dj}), \tag{4.86}$$

and its scalar product by $\mathbf{n}\, d\Sigma$ is the quantity of entropy crossing $(d\Sigma)$.

Note: By applying the Gibbs relation

$$dS = \frac{1}{T}dE + \frac{p}{T}d\mathcal{V} - \frac{1}{T}\sum_j g_j dm_j \tag{4.87}$$

to fluxes, we can also formally write (as done by Napolitano [181])

$$\mathcal{J}_S = \frac{1}{T}\mathcal{J}_E + \frac{p}{T}\mathcal{J}_\mathcal{V} - \frac{1}{T}\sum_j g_j \mathcal{J}_{Dj}. \tag{4.88}$$

Since the volume flux $\mathcal{J}_\mathcal{V}$ is zero, and $\mathcal{J}_E$ is equal to $\mathbf{q}$, we obtain the preceding formula (4.86).

We have also assumed a bulk energy contribution r, so the external entropy contribution for the volume $(\mathcal{V})$ is

$$\dot{S}_e = \int_\mathcal{V} \frac{r}{T} d\mathcal{V} - \int_{\partial\mathcal{V}} \frac{1}{T}(\mathbf{q} - \sum_j g_j \mathcal{J}_{Dj}) \cdot \mathbf{n} d\Sigma. \tag{4.89}$$

According to the second law, we have

$$\dot{S} = \dot{S}_e + \dot{S}_i, \;\; \dot{S}_i \geq 0, \tag{4.90}$$

where

$$\begin{cases} \dot{S} = (d/dt)\int_\mathcal{V} \rho s d\mathcal{V} = \int_\mathcal{V} \rho\, ds/dt\, d\mathcal{V}, \\ \dot{S}_i = \int_\mathcal{V} \dot{W}_S\, d\mathcal{V}. \end{cases} \tag{4.91}$$

The integral equation for the entropy balance is then

$$\int_\mathcal{V} \rho \frac{ds}{dt} d\mathcal{V} - \int_\mathcal{V} \frac{r}{T} d\mathcal{V} + \int_{\partial\mathcal{V}} \frac{1}{T}(\mathbf{q} - \sum_j g_j \mathcal{J}_{Dj}) \cdot \mathbf{n} d\Sigma = \int_\mathcal{V} \dot{W}_S d\mathcal{V}. \tag{4.92}$$

Using the Stokes–Ostrogradski theorem and the fundamental lemma, we then get the local equation

$$\rho \frac{ds}{dt} + \nabla \cdot \left(\frac{\mathbf{q} - \sum_j g_j \mathcal{J}_{Dj}}{T}\right) = \frac{r}{T} + \dot{W}_S. \tag{4.93}$$

The Gibbs relation applies to the material derivative; in other words,

$$\frac{ds}{dt} = \frac{1}{T}\frac{de}{dt} + \frac{p}{T}\frac{d\vartheta}{dt} - \sum_j \frac{g_j}{T}\frac{dY_j}{dT}. \tag{4.94}$$

By combining these two last equations and taking into account the expressions obtained in the preceding paragraphs for de/dt, $d\vartheta/dt = d(1/\rho)/dt$, and dY_j/dT, we find that

$$\begin{aligned} \dot{W}_S = &- \textstyle\sum_j (g_j/T)\dot{W}_j - [(\pi - p)/T]\nabla \cdot \mathbf{v} + \mathbf{q} \cdot \nabla(1/T) \\ &- \textstyle\sum_j \mathcal{J}_{Dj} \cdot [\nabla(g_j/T) - \mathbf{f_j}/T] \\ &-(1/2T)(\nabla \otimes \mathbf{v} + \widetilde{\nabla \otimes} \mathbf{v} - (2/3)\nabla \cdot \mathbf{v}\mathbf{1}) : (\mathbf{P} - \pi\mathbf{1}). \end{aligned} \tag{4.95}$$

In order to correctly apply the laws of thermodynamics for irreversible processes, we need to separate out the independent generalized forces in this expression. This is done in (4.95) for the viscous terms, and will be performed later for the other terms (see Chap. 7).

4.9 Balance Laws for Discontinuities in Continuous Media

Various discontinuities can arise in fluid flows: solid walls, surfaces that separate phases, shock waves or combustion waves, contact surfaces, etc. The preceding balance equations were derived for the continuous parts of flows, and so they do not apply to discontinuities. To address discontinuities, we need to establish the boundary conditions at the surface of the discontinuity and connect the conditions present on both sides of this surface. This is the objective when deriving balance equations for discontinuities. The surface considered can be motionless or moving, and deformable or nondeformable. It may or may not be the focus of production processes or internal fluxes (chemical reactions at the surface, surface tension, etc). The balance equations are easy to establish for quantities with vanishing interfacial production, which therefore have no associated internal fluxes (momentum flux, such as surface tension) or accumulation terms. The fluxes of mass, energy and momentum relative to the normal of the surface are then conserved as the surface is traversed. It is not the same for the entropy, but we can generally deduce the entropy production from the other equations, just as we can in the absence of a discontinuity. We will limit ourselves here to the case of separation interfaces between fluid zones, although the method is also applicable to surfaces that separate solids from fluids. We are therefore interested in the quantities that exhibit conservative fluxes; i.e., in those whose flux components normal to the interface with the discontinuity (Fig. 4.6) are preserved upon traversing it: volume, momentum, and total energy.

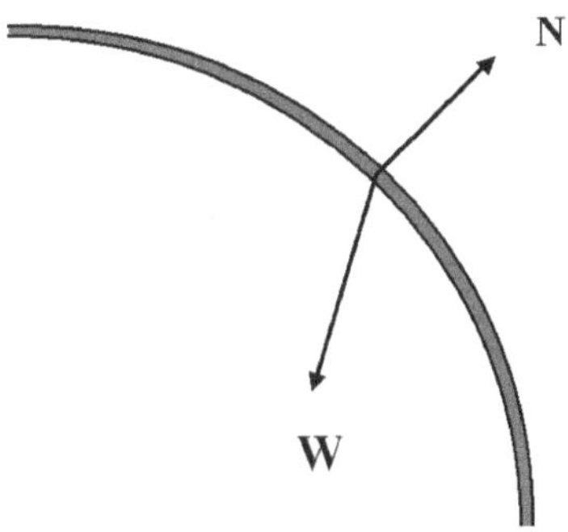

Fig. 4.6. Representation of a fluid discontinuity

The balance equation that expresses the conservation of the flux of the quantity F normal to the interface with the (moving) discontinuity upon crossing this interface is as follows:

$$[\mathcal{J}_F + \rho f(\mathbf{v} - \mathbf{W})]_-^+ \cdot \mathbf{N} = 0, \tag{4.96}$$

where $\mathbf{W}$ is the local velocity of the discontinuity, $\mathbf{N}$ is the normal unit vector orientated from the $(-)$ side towards the $(+)$ side, and $[\varphi]_-^+$ is the difference $\varphi^+ - \varphi^-$.[6]

We now apply this equation to successive cases. For the overall mass $F = m$,

$$[\rho(\mathbf{v} - \mathbf{W})]_-^+ \cdot \mathbf{N} = 0. \tag{4.97}$$

For the species $F = m_j = mY_j$,

$$[\mathcal{J}_{Dj} + \rho Y_j(\mathbf{v} - \mathbf{W})]_-^+ \cdot \mathbf{N} = 0. \tag{4.98}$$

For the momentum $F = m\mathbf{v}$,

$$[\mathbf{P} + \rho\mathbf{v} \otimes (\mathbf{v} - \mathbf{W})]_-^+ \cdot \mathbf{N} = 0. \tag{4.99}$$

[6]When there are also internal processes, it is necessary to add an accumulation term (for unsteady phenomenona and convection) and a tangential flow term (see Chap. 11) to the left hand side of the general balance equation. For the right hand side, we also need a term for the production of the quantity F per unit area, $\dot{W}_{Fa}$, as generally applied for the entropy in Eq. 4.101.

For the total energy $F = E + K$,

$$[\mathbf{q} + \mathbf{v} \cdot \mathbf{P} + \rho(e + k)(\mathbf{v} - \mathbf{W})]_-^+ \cdot \mathbf{N} = 0. \tag{4.100}$$

For the entropy, which is not a conservative quantity, we get

$$[\mathcal{J}_S + \rho s(\mathbf{v} - \mathbf{W})]_-^+ \cdot \mathbf{N} = \dot{W}_{Sa} \geq 0. \tag{4.101}$$

4.10 Other Methodologies for Balance Laws

The local forms of the balance equations have a broad field of application. However, they are inevitably not suitable for solving all fluid flow problems. In addition to their complexity, they are difficult to solve. They are too precise in certain cases and insufficient in others. In chemical engineering for example, the deterministic total balance is a better option since it allows certain chemical reactors (known as "ideal reactors;" see Chap. 6) to be studied. On the other hand, turbulent reactive flows and "real" chemical reactors may require probabilistic balance equations.

4.10.1 Total Deterministic Balance

In theory, in this case, we need only write the preceding balance equations in integral form. We generally need to take into account the fact that the entry and the exit points for a chemical reactor consist of pipes along which the properties of the fluids are relatively uniform. In addition, the reactor, or the particular reactor element of interest, is assumed to have nondeformable walls [9]. The weight balance for species j will then be

$$\frac{dm_j}{dt} = \dot{m}_{jo} - \dot{m}_{je} + R_j, \; j = 1, 2, \ldots N, \tag{4.102}$$

where m_j is the mass of species j present in the reactor, $\dot{m}_{jo}$ is the rate of mass flow into the reactor, $\dot{m}_{je}$ is the rate of mass flow out of the reactor, and R_j the mass produced by chemical reactions.

Let us apply the first law of thermodynamics to the open reactor shown in Fig. 4.7 between the times t and $t + \delta t$. We have

$$(E + \delta E + \sum_j \dot{m}_{je} e_{je} \delta t) - (E + \sum_j \dot{m}_{jo} e_{jo} \delta t) = \frac{p_o}{\rho_o} \dot{m}_o \delta t - \frac{p_e}{\rho_e} \dot{m}_e \delta t - \dot{Q}(T) \delta t, \tag{4.103}$$

where $\dot{Q}(T)$ indicates the heat released per unit time through the walls. We then deduce that

$$\frac{dE}{dt} = \sum_j \dot{m}_{jo} h_{jo} - \sum_j \dot{m}_{je} h_{je} - \dot{Q}(T). \tag{4.104}$$

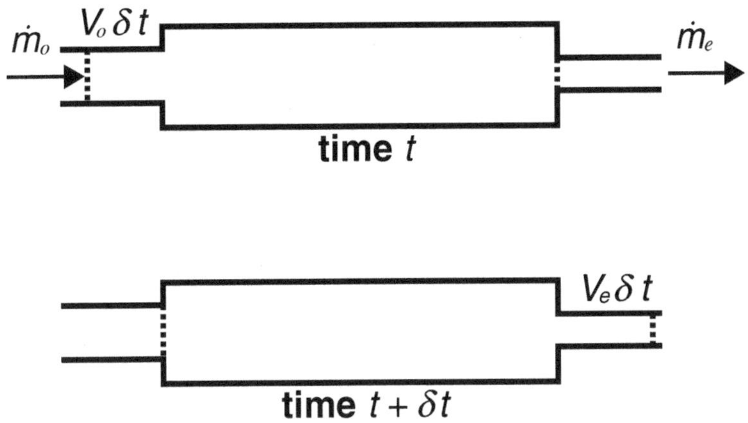

Fig. 4.7. Mass balance in an open reactor

These $N+1$ equations make it possible to solve many problems. Thus, for an ideal perfectly stirred reactor, we assume that the specific or intensive quantities are identical in the exit section of the reactor and at any point inside the reactor. We assume a very turbulent fluid mixture, which leads to a simple formula (see Chap. 6).

4.10.2 Probabilistic Population Balance

The probabilistic method for the population balance (Danckwerts, 1951) is applicable to chemical reactors [283]. It is assumed that an entity of fluid at the position $\mathbf{x}$ at time t is characterized by a certain number of quantities $\zeta_1, \zeta_2, \ldots$, combined into a vector $\boldsymbol{\zeta}$. At time t, the number of entities present in an element of volume $d\mathbf{x}d\boldsymbol{\zeta}$ in this phase space is then, by definition,

$$\psi(\mathbf{x}, \boldsymbol{\zeta}, t)d\mathbf{x}d\boldsymbol{\zeta}, \tag{4.105}$$

and the total number of entities present in a control volume $(\mathcal{V})$ in the phase space is then

$$N = \int_{\mathcal{V}} \psi(\mathbf{x}, \boldsymbol{\zeta}, t)d\mathbf{x}d\boldsymbol{\zeta}. \tag{4.106}$$

The probability that an entity at position $\mathbf{x}$ is in the state defined by $\boldsymbol{\zeta}$ at time t is

$$f = \psi/N. \tag{4.107}$$

The population balance is

$$\frac{\partial}{\partial t}(\int_{\mathcal{V}} \psi d\mathcal{V}) + \int_{\partial\mathcal{V}} \psi \mathbf{V} \cdot d\mathbf{\Sigma} = \int_{\mathcal{V}} G d\mathcal{V}, \tag{4.108}$$

where $(\partial\mathcal{V})$ is the motionless closed surface that delimits the volume, and $d\mathbf{\Sigma}$ is the element of area that is orientated normally to $(\partial\mathcal{V})$. The vector $\mathbf{V}$ is the velocity vector of the phase space. The production rate of entities per unit of volume in this space is G by definition. By setting $\mathbf{w} = d\boldsymbol{\zeta}/dt$, the local equation becomes

$$\frac{\partial\psi}{\partial t} + \boldsymbol{\nabla} \cdot (\psi \mathbf{v}) + \sum_i \frac{\partial(\psi w_i)}{\partial \zeta_i} = G. \tag{4.109}$$

By integrating this equation over the physical volume $(\mathcal{V}_r)$ of an open reactor with a inlet volume flow rate of $\dot{q}_o$ and an exit volume flow rate of $\dot{q}_e$, we obtain

$$\frac{1}{\mathcal{V}_r}\frac{\partial(\bar{\psi}\mathcal{V}_r)}{\partial t} + \frac{\dot{q}_e\bar{\psi}_e - \dot{q}_o\bar{\psi}_o}{\mathcal{V}_r} + \sum_i \frac{\partial(\bar{\psi} w_i)}{\partial \zeta_i} = \bar{G}, \tag{4.110}$$

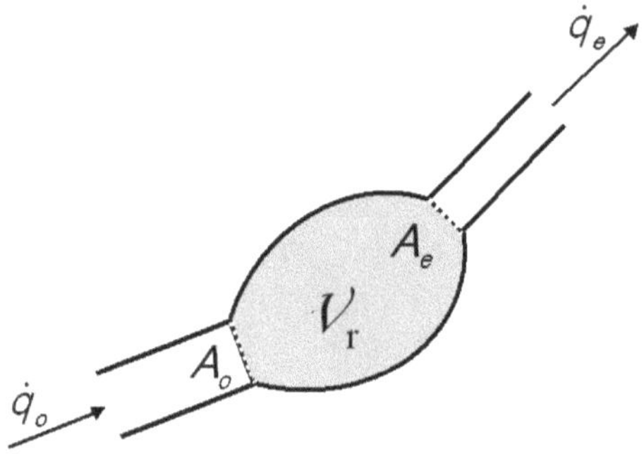

Fig. 4.8. Crystallizer

where the terms for the inlet are indicated by a subscript o and those for the exit are indicated by a subscript e. This relation also contains the average quantities per unit volume

$$\bar{\psi} = \frac{1}{\mathcal{V}_r}\int_{\mathcal{V}_r} \psi d\mathcal{V}, \; \bar{G} = \frac{1}{\mathcal{V}_r}\int_{\mathcal{V}_r} G d\mathcal{V} \tag{4.111}$$

and the average volume flow rates

$$\bar{\psi}_o = \frac{1}{\dot{q}_o}\int_{A_o} \psi \mathbf{v} \cdot d\mathbf{A}, \; \bar{\psi}_e = \frac{1}{\dot{q}_e}\int_{A_e} \psi \mathbf{v} \cdot d\mathbf{A}. \tag{4.112}$$

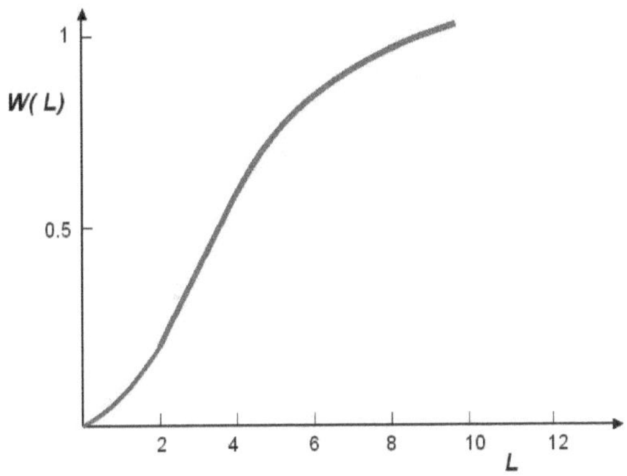

Fig. 4.9. Results for a crystallizer (J. Villermaux, private communication, 1986; see also [283])

This balance equation can be applied for example to a crystallizer (see Fig. 4.8). The aqueous solution that enters the reactor is homogeneous, and the rate of crystal nucleation in the reactor is r_N. The crystal size L increases at a certain speed $W = dL/dt = r_0 L^b$, where r_0 and b are constants. The reactor is assumed to be perfectly stirred. In the steady regime and for $b = 0$, we obtain a law for $W(L)$, which is the proportion of the total crystallized mass m_T that is smaller than L (see Fig. 4.9), and this law agrees remarkably well with that observed experimentally. The derivative $W'(L)$ is

$$W'(L) = \frac{m'(L)}{m_T} = \frac{1}{6L_0}(\frac{L}{L_0})^3 \exp{(-L/L_0)}, \tag{4.113}$$

where $L_0 = \vartheta r_0/\dot{q}$, and $\dot{q} = \dot{q}_o = \dot{q}_e$ are volume flow rates.

5

Dimensionless Numbers and Similarity

This chapter presents the principles of dimensional analysis and the similarity method. Section 5.1 introduces basic aspects of dimensional analysis such as the definition of Π_i ratios and Vashi–Buckingham's theorem. The practical utility of dimensional analysis is then emphasized using the determination of head loss in a cylindrical pipe as an example. Indeed, dimensional analysis is a necessary step in the analysis of any physical problem [63].

The similarity method is defined in Sect. 5.2 and applied to the case of a flexible balloon that is subjected to a wind moving at given speed.

The concepts of dimensional analysis can also help us to solve equations by identifying interesting groups of variables, as illustrated in Sect. 5.3, where the analytical solutions for a heat transfer problem (self-similar solution) are obtained.

The dimensionless numbers defined in Sect. 5.4 are the analytically determined products of various quantities raised to certain powers. These quantities are specific parameters of the case of interest: a transfer coefficient, the size of the chemical reactor, the average heat gradient, the characteristic time of a phenomenon, etc. They must be expressed in a coherent system of units. These dimensionless numbers are very useful for evaluating the most influential phenomena in a flow during a preliminary analysis.

However, applications of dimensional analysis extend further than this. Changes in stability are often characterized by curves or surfaces of a space that are defined based on such numbers.

It is sometimes possible to study phenomena without the need to resolve the whole system of equations that govern them, and, based on some assumptions, to draw very useful conclusions through the use of dimensional analysis and dimensionless numbers in conjunction with experimental observations.

A table of some of the dimensionless numbers that are currently used is provided at the end of this chapter.

R. Prud'homme, *Flows of Reactive Fluids*,
DOI 10.1007/978-0-8176-4659-2_5,

5.1 Elements of Dimensional Analysis: Π_i Ratios

5.1.1 Basic Considerations

The solution to a given problem identifies relations between the characteristic physical quantities $F_1, F_2, \ldots, F_n$ of the medium of interest. These quantities are measured using a coherent system of units. The basic idea behind dimensional analysis is that the physical relations are independent of the units used to measure these quantities [2, 34].

Let us consider the final relation of a given problem,

$$f(F_1, F_2, \ldots F_n) = 0. \tag{5.1}$$

We know that the form of this relation will not depend on the system units used to measure these quantities.

If $F_i = x_i\, e_i$ in a given system of units e_i, then, for another system of units e'_i, we have

$$x'_i\, e'_i = x_i\, e_i. \tag{5.2}$$

Any quantity can be expressed, from a dimensional point of view, using three primary quantities: length L, mass M, and time T. In other words,

$$[F_i] = [L^{\alpha_i}\, M^{\beta_i}\, T^{\gamma_i}]. \tag{5.3}$$

For example, pressure can be expressed as $[p] = [L^{-1}MT^{-2}]$.[1]

We can also consider quantities other than the primary quantities. If

$$\begin{cases} [X] = [L^{a_{11}}M^{a_{12}}T^{a_{13}}], \\ [Y] = [L^{a_{21}}M^{a_{22}}T^{a_{23}}], \\ [Z] = [L^{a_{31}}M^{a_{32}}T^{a_{33}}], \end{cases} \tag{5.4}$$

the dimensional equation for the quantity F_i will be

$$[F_i] = [L^{\alpha_i}\, M^{\beta_i}\, T^{\gamma_i}] \text{ or } [F_i] = [X^{x_i}\, Y^{y_i}\, Z^{z_i}], \tag{5.5}$$

provided that the matrix of a_{kl} is regular. Since these two expressions for $[F_i]$ are equivalent,

$$\begin{cases} a_{11}x_i + a_{21}y_i + a_{31}z_i = \alpha_i, \\ a_{12}x_i + a_{22}y_i + a_{32}z_i = \beta_i, \\ a_{13}x_i + a_{23}y_i + a_{33}z_i = \gamma_i, \end{cases} \tag{5.6}$$

which makes it possible to calculate the coefficients x_i, y_i, and z_i.

---

[1]Certain quantities are defined based on primary quantities and universal constants. The dimensions of these constants are neglected. As an example, for the temperature θ, we have $[\theta] = [L^2T^{-2}]$ (note that we use θ to denote temperature in this chapter for obvious reasons).

Let us assume that the table of the coefficients α_i, β_i, γ_i (the exponents of L, M, T) is of rank $p\,(p \leq 3)$. There are then $n - p$ dimensionless ratios (starting from the fundamental p) among n quantities, which are classified in such a way that

$$X = F_1,\; Y = F_2,\; \ldots \tag{5.7}$$

The products are thus of the form

$$\Pi_i = \frac{F_i}{F_1^{x_i}\, F_2^{y_i}\, \ldots},\; i = p + 1, \ldots, n. \tag{5.8}$$

5.1.2 Vashi–Buckingham or Π Theorem

A final relation, $f(F_1,\, F_2, \ldots F_n) = 0$ between n quantities, can be written in the form of an equation between $n - p$ ratios, where p is the maximum number of dimensionally independent quantities in the considered equation. This is the *Vashi–Buckingham theorem* (developed in 1890). We therefore have

$$\varphi(\Pi_{p+1}, \ldots \Pi_n) = 0, \tag{5.9}$$

or similarly

$$\Pi_{p+1} = \psi(\Pi_{p+2}, \ldots \Pi_n). \tag{5.10}$$

It is important to note that dimensional analysis does not provide the functions φ and ψ here. The forms of these functions can only be obtained experimentally (or through complete mathematical analysis of the problem).

5.1.3 Practical Utility of Dimensional Analysis

Dimensional analysis can be used to

- Reduce the number of variables (this sometimes requires additional assumptions)
- Search for similar solutions
- Present the results in a simplified form
- Search for experimental or numerical solutions.

Procedure: The usual procedure employed for dimensional analysis is as follows:

1. Initially write down all of the quantities (such as independent variables and physical constants) that are likely to appear in f, and then eliminate those that are considered to play a minor role.
2. Choose basic quantities $X, Y \ldots$ that are known to vary (i.e., are not constant).

3. Form the Π_i ratios.
4. Produce the law $\varphi(\Pi_{p+1}, \dots \Pi_n) = 0$ (or the law in ψ). This law can sometimes be written in the form $\Pi_{p+1}^{\alpha} \Pi_{p+2}^{\beta} \dots =$const.

5.1.4 Example: Head Loss in a Cylindrical Pipe

The surface roughness is characterized by an average thickness e (see Fig. 5.1).

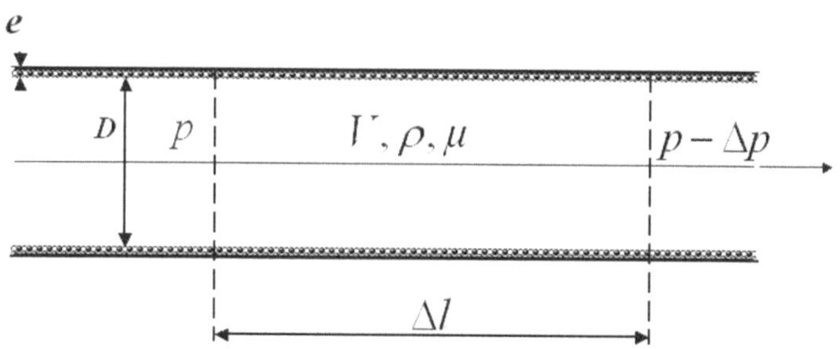

Fig. 5.1. Head loss in a cylindrical pipe

The dimensions of the active quantities are given in Table 5.1.

F_i	D	Δl	e	ρ	V	Δp	μ
L	1	1	1	−3	1	−1	−1
M	0	0	0	1	0	1	1
T	0	0	0	0	−1	−2	−1

Table 5.1. Dimensions of the active quantities

We now follow the previously mentioned steps. First, we choose the basic parameters: $\{X, Y, Z\} = \{D, \rho, V\}$ (the base change matrix is regular). $[\Delta l] = [D]$, $[e] = [D]$, $[\mu] = [L^{-1}MT^{-1}] = [X^{x_\mu}Y^{y_\mu}Z^{z_\mu}]$. We find that $[\mu] = [D\rho V]$. In the same way, $[\Delta p] = [\rho V^2]$.

The Π_i ratios are therefore

$$\Pi_{\Delta l} = \Delta l/D,\ \Pi_e = e/D = \epsilon,\ \Pi_{\Delta p} = \Delta p/\rho V^2,\ \Pi_\mu = \mu/\rho V D = 1/Re,$$

where ϵ is the relative roughness and Re is the Reynolds number.

We now apply the Vashi–Buckingham theorem:

$$\Delta p/\rho V^2 = \psi_1(\Delta l/D,\ \epsilon,\ Re). \tag{5.11}$$

Experimental result:

$$\Delta p/\rho V^2 = (\Delta l/D)\, \psi_2(\epsilon,\ Re). \tag{5.12}$$

One significant quantity is the head-loss coefficient, $\Lambda = (2D\Delta p)/(\rho V^2 \Delta l)$. Therefore, we have $\Lambda = \psi(\epsilon, Re)$. This result has been verified experimentally for laminar flow and turbulent flow. In the particular case of laminar flow,[2]

$$\Lambda = 64/Re. \tag{5.13}$$

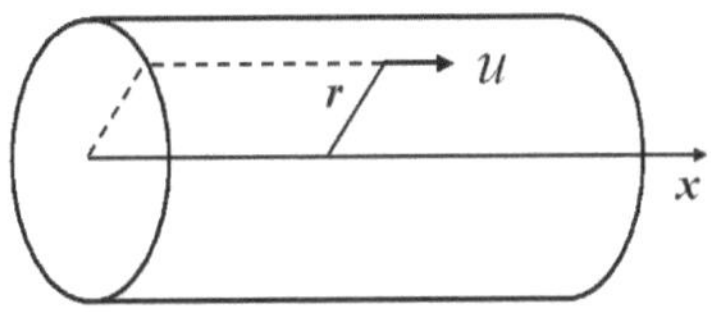

Fig. 5.2. Cylindrical symmetry

5.2 Similarity

5.2.1 Definition

In order to simplify the process of analyzing a real phenomenon (relating to, for example, hydraulic engineering, aeronautics, or engines), we often first attempt to analyze a reduced-scale model that exhibits a similar phenomenon, and then extend the results of this analysis to the real phenomenon.

However, before the solution for the model can be extended to the real phenomenon, there are a number of *similarity rules* that must be verified.

The principles of dimensional analysis are applicable here because *changing the units of measurement for the quantities F_i is equivalent to considering two different scales S and S' for the same physical problem*, where the homologous parameters are x_i and x'_i (both of which use the same units). When all of the similarity conditions are verified, there is said to be perfect similarity between the model and real systems, although this is often difficult to achieve in practice. A scale factor λ_i is associated with each parameter. For perfect similarity,

---

[2]**Reminder of Poiseuille flow in the laminar regime.** We suppose that $u = w = 0$, ρ =const., and the regime is steady (see Fig. 5.2). $\nabla \cdot \mathbf{v} = \mathbf{0}$ gives $\partial u/\partial x = 0$ or $u = u(r)$. Therefore, in turn, we have:

- Momentum equations: $\partial p/\partial x = \mu \Delta u(r)$ (Laplacian), $\partial p/\partial r = 0$, from which we get $p = p(x)$
- Solution of $(\mu/r)(d/dr)(r\, du/dr) = dp/dx$: dp/dx =const. $= -\Delta p/\Delta l$, $u = (\Delta p/4\mu\Delta l)(D^2/4 - r^2)$
- If U is the mean velocity of the flow, $\pi D^2 U/4 = \int urdrd\theta$, yielding $\Delta p/\Delta l = 32\mu U/D^2$, $\Lambda = 2D\Delta p/\rho U^2 \Delta l = 64\mu/\rho UD = 64/Re$.

we must have both $f(F_1, F_2, \dots F_n) = 0$ and $f(\lambda_1 F_1, \lambda_2 F_2, \dots \lambda_n F_n) = 0$. The Π_i ratios must take the form

$$\Pi_i = F_i / F_1^{x^i} F_2^{y^i} \dots = \lambda_i F_i / \lambda_1^{x^i} F_1^{x^i} \lambda_2^{y^i} F_2^{y^i} \dots, \; i = p+1, \dots, n$$

to obey the Vashi–Buckingham theorem and ensure the invariance of the relations $\varphi(\Pi_{p+1}, \dots \Pi_n) = 0$. We will then have

$$\lambda_i / \lambda_1^{x^i} \lambda_2^{y^i} = 1, \; i = p+1, \dots, n \tag{5.14}$$

Moreover, it will be necessary to make sure that usual simplifications are still valid for the model (for example, the viscosity may be negligible in one case but not in the other).

5.2.2 Example: Similarity of a Flexible Balloon Subjected to a Wind

The following problem was handed to students to solve as part of the course given by J. Bouttes at École Polytechnique [29].

We would like to produce a reduced-scale model of a large balloon; the model should possess the same skin tension as the large balloon (see Fig. 5.3a).

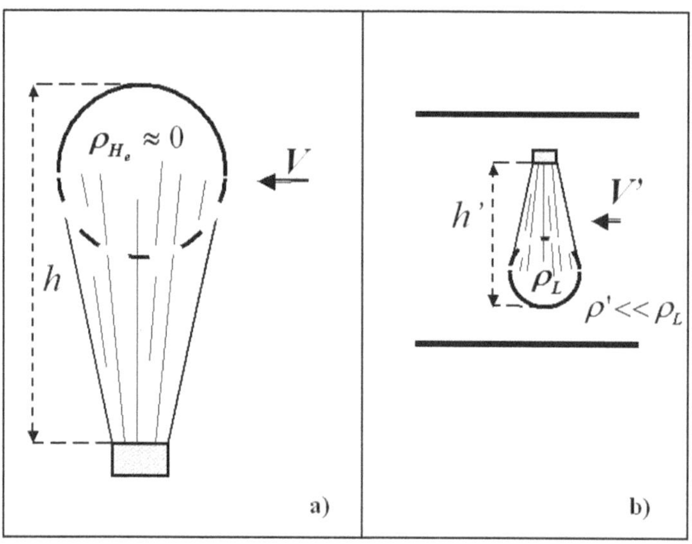

Fig. 5.3. **a** Large balloon; **b** model in a wind tunnel

Various phenomena are involved here. First, we consider the surface tension due to the hydrostatic pressure. This pressure $\Delta p \propto \rho g h$, so the surface tension $\sigma \propto R\Delta p$ (where R is the radius of the balloon). To obtain the same tension in both the skin of the balloon and the skin of the model, we must have $R\Delta p = R'\Delta p'$ (where R' is the radius of the model) or $R\rho h = R'\rho' h'$; i.e., $\rho'/\rho = Rh/R'h' \cong (R/R')^2 = \lambda^2$.

Since the model is much smaller than the balloon, ρ' must be much larger than ρ. However, this means that the model exhibits different behavior to that of the balloon in a wind tunnel.

We can solve this problem by filling the balloon with a liquid of density ρ_L. The preceding results remain valid if the model is positioned upside down in the tunnel (see Fig. 5.3b): $\rho_L/\rho = (R/R')^2 = \lambda^2$. If the liquid used is water, $\lambda^2 = 1000$ and so $R/R' = \lambda \approx 30$, which is a reasonable value.

Now, if ρ' is the density of the air around the model, the aerodynamic pressure for an incompressible flow (Mach number $M < 0.25$, which we assume has been verified) is $\Delta p \propto \rho V^2$. Since the induced skin tensions need to be equal, we have $\rho V^2 R = \rho' V'^2 R'$, and so $R/R' = \rho' V'^2/\rho V^2$.

Assuming that the air has the same density in the atmosphere and the wind tunnel, we get $V'/V = \sqrt{\lambda}$. If we use water in the model, this implies that the wind speed needs to be 5–6 times larger in the tunnel.

We now consider the viscous stress. The Reynolds number Re is used for the viscous stress:

$$\frac{Re'}{Re} = \frac{\rho' V' R'}{\mu'} \frac{\mu}{\rho V R},$$

where $R'/R = 1/\lambda$.

If the temperature and pressure in the wind tunnel are the same as those in the atmosphere, ρ and μ are also the same for both the atmosphere and the wind tunnel, and so $Re'/Re = V'R'/VR = 1/\sqrt{\lambda}$. In this case, the Reynolds numbers for the atmosphere and the wind tunnel are very different.

In order to obtain similarity, we could increase the pressure at a constant temperature (so that $\rho \propto p$):

$$\frac{Re'}{Re} = 1 = \frac{\rho' V' R'}{\rho V R} = \sqrt{\frac{p'}{p}} \sqrt{\frac{\rho' V'^2}{\rho V^2} \frac{R'}{R}} = \sqrt{\frac{p'}{p} \frac{1}{\sqrt{\lambda}}}.$$

We therefore require $p'/p = \lambda$ (for water, $p' \approx 30p$).

We could also decrease the temperature (i.e., use a cryogenic wind tunnel):

$$\frac{Re'}{Re} = 1 = \frac{\sqrt{\rho'}}{\sqrt{\rho}} \sqrt{\frac{\rho' V'^2}{\rho V^2}} \sqrt{\frac{R'}{R}} \sqrt{\frac{T}{T'}},$$

because $\rho \propto p/T$ and $\mu \propto \sqrt{T}$. If we impose $p = p'$, we find that $T'/T = 1/\sqrt{\lambda}$. For water, which must be heated so that it does not turn into ice, we get $T' \approx T/5$.

5.3 Analytical Search for the Solutions to a Heat Transfer Problem (Self-Similar Solution)

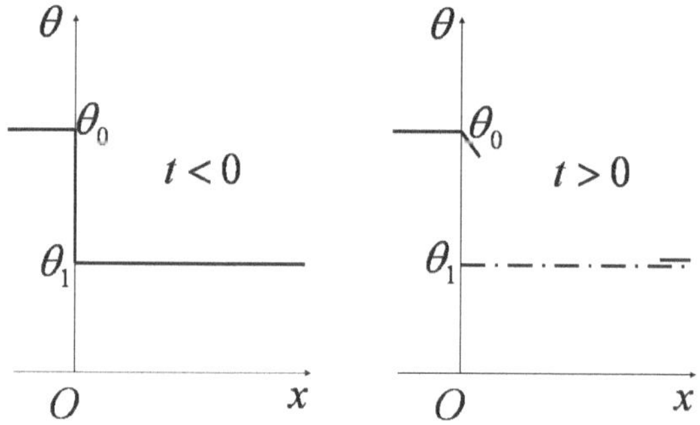

Fig. 5.4. Heat transfer in a semi-infinite medium

The equation for heat transfer in a continuous quiescent medium was established in Sect. 4.1. For evolution along a plane section, we have (see Eq. 4.10)

$$\frac{\partial \theta}{\partial t} - \kappa \frac{\partial^2 \theta}{\partial x^2} = 0. \tag{5.15}$$

The solution to a given problem takes the form $\theta = f(x, t, \kappa)$. We look for the solutions that, if λ, μ, ν need to be determined, give $f(x, t, \kappa) = f(\lambda x, \mu t, \nu \kappa), \forall \{x, t, \kappa\}$; i.e., the self-similar solutions. In other words, the solution $f(x, t, \kappa)$ must be invariant upon scaling. We know that $\theta = f(x, t, \kappa)$ can be written in the form of a dimensionless ratio. Indeed, Table 5.2 gives us $\Pi_\kappa = \kappa t / x^2$, so $\theta = \varphi(x^2/\kappa t)$.

Since $x^2/\kappa t = \lambda^2/\mu\nu \, x^2/\kappa t$, then $\lambda^2 = \mu\nu$.

$\mathbf{F_i}$	x	t	κ
L	1	0	2
M	0	0	0
T	0	1	−1

Table 5.2. Dimensions

We therefore look for solutions of the preceding form. These do exist under some circumstances (i.e., when the initial data and the boundary conditions allow it).

Let us now consider the exact problem of heat transfer in a semi-infinite medium (see Fig. 5.4) [147]. For $t < 0$, we have $\theta = \theta_0$ for $x < 0$ and $\theta = \theta_1$ for $x > 0$, (adiabatic wall at $x = 0$ for $t < 0$). For $t > 0$, we always have $\theta = \theta_0$ for $x < 0$, as well as $\theta = \theta_0$ at $x = 0$. For infinite t, we have $\theta = \theta_1$, $\forall x$. Setting $\eta = x/\sqrt{\kappa t}$, the equation to solve becomes

$$\frac{d^2\theta}{d\eta^2} + \frac{\eta}{2}\frac{d\theta}{d\eta} = 0.$$

The solution takes the form $\theta = A\,\mathrm{erf}(\eta/2) + B$ for $t > 0$, $x \geq 0$, where $\mathrm{erf}(s) = \frac{2}{\sqrt{\pi}}\int_0^s e^{-u^2}du$, $\mathrm{erf}(\infty) = 1$. The initial condition $t \to 0$ gives us $\eta \to \infty$, so $A+B = \theta_1$. The boundary condition $x = 0$ gives $\eta = 0$, $\theta = \theta_0$ for $t > 0$, and so $B = \theta_0$. Thus,

$$\theta = \theta_0 + (\theta_1 - \theta_0)\,\mathrm{erf}(x/\sqrt{4\kappa t}). \tag{5.16}$$

This solution always leads, when $x \to \infty$ ($\eta \to \infty$), to $\theta = \theta_1$. When $t \to \infty$, we obtain $\theta = \theta_0$ for finite x (see Fig. 5.5). Note that, as soon as $t > 0$, the disturbance arising from the contact temperature at $t = 0$ exists throughout the half-plane $x > 0$. The propagation is thus instantaneous. Actually, instantaneous propagation does not occur, but this model of linear irreversible thermodynamics does not make it possible to take this into account. Thus, to analyze propagation phenomena, it is important to consider nonlinear irreversible thermodynamics, as done for example by Jou et al. [130].

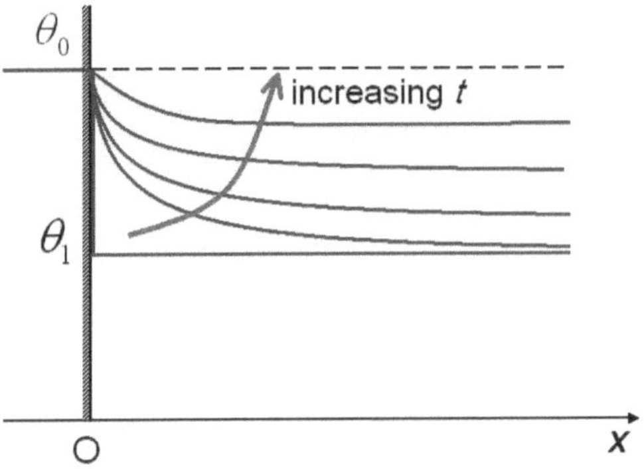

Fig. 5.5. Temperature evolution in a semi-infinite medium

5.4 A Few Dimensionless Numbers

In fluid mechanics, similarity reflects the relationship between several types of forces that arise during flows [2, 110]. In most flows, inertial forces (which are required to change the motions of the fluid particles) have the greatest magnitude. We successively compare them with the other types of forces—gravity, viscosity, surface tension, and compressibility—which allows us to derive the Froude, Reynolds, Weber and Mach numbers, respectively.

The value of the Froude number reflects the relative importance of inertial forces compared to buoyancy forces. Froude similarity reflects the relationship between inertial forces and gravitational forces; it is important in cases where the positions and the motions of the fluid particles can be influenced by these two types of forces, especially in the case of flow in the presence of a free surface.

The Reynolds number characterizes the relative importance of inertial forces and viscous forces. The Reynolds similarity respects the ratio of these forces and is most important in cases of laminar flow and turbulent flow. It is important to note the general result that, in any problem that involves both viscous forces and gravity, a study based on similarity—which is often the only kind possible—will require the use of both the Froude number and the Reynolds number. In such cases, we must often assume approximate similarity by preserving only one of these two numbers.

The Grashof number is presented in the form of the square of a particular Reynolds number associated with the speed induced by the buoyancy forces: $V_G = \sqrt{Lg\Delta\rho/\rho}$.

The Weber number characterizes the relative importance of inertial forces and forces related to surface tension. Weber similarity need only be respected if there is a free surface where surface tension phenomena can occur.

The Mach number characterizes the relative importance of the inertial forces and the elastic compressive forces of the fluid ($c = \sqrt{(\partial p/\partial \rho)_s}$). Mach similarity should only be respected in problems where there is wave propagation in a compressible medium.

Most dimensionless numbers can be defined as ratios of characteristic times. We have already seen that the Reynolds number is the ratio of the momentum diffusion time divided by the viscosity L^2/ν to the momentum convection time L/V. The Peclet number is the ratio of the thermal diffusion time L^2/κ to the convection time L/V. Similarly, the Prandtl number Pr is the ratio of the thermal diffusion time to the viscous diffusion time, the Schmidt number Sc is the ratio of the molecular diffusion time for the species to the viscous diffusion time, the Lewis number is the ratio of the molecular diffusion time of the species to the thermal diffusion time, while the Mach number M is the ratio of the acoustic time L/c to the convection time. The first Damköhler parameter Da is the ratio of the mechanical time to the chemical time. We can also define the Marangoni number Ma as the ratio of the thermal diffusion time to a capillary time corresponding to the velocity induced by spatial variations in the surface tension $V_\sigma = \Delta\sigma/\mu$.

Some dimensionless numbers can be defined as products of other dimensionless numbers, and others as ratios of fluxes (as in the case of the Nusselt number Nu). The Hickman number appears in the case of a flat, horizontal evaporation surface for rapidly evaporating liquids (for example those in a vacuum). The liquid is then unstable to local variations in the vaporization rate; local surface depressions are produced by the force exerted on the surface (vapor recoil) by the rapidly departing vapor, while sustained liquid flows are driven by the shear exerted on the liquid surface by the vapor [94].

Table 5.3 provides a list of some of the dimensionless numbers that are encountered in fluid mechanics. There are many dimensionless numbers that are not included in the table, such as the Rossby number, which makes it possible to compare the effect of inertial forces to the effect of Coriolis forces; the Stanton (or Margoulis) number St, which is the ratio of a convection coefficient to the product $\rho c_p V$; the Strouhal number Sr, which is used in Sect. 8.2.1 and is the convection time divided by a characteristic swirl emission time; and the Taylor number Ta, which occurs in Taylor–Couette instability theory (see Sect. 9.2.3).

Each dimensionless number has its own range of validity, and its relevance depends on the problem of interest. They allow complex problems to be solved without the need to solve all of the equations associated with each problem, they can provide self-similar solutions, and they allow experiments to be performed on models by applying the similarity principles.

Listing all known dimensionless numbers here would be unrealistic, as every type of problem has its own associated dimensionless numbers.

Name of the dimensionless number	**Associated phenomena**	**Expression for the dimensionless number**
Bond number	Superimposed fluids with surface tension	$Bo = \sigma/gL^2\Delta\rho$
Crispation number	Viscosity, thermal diffusion, and surface tension	$Cr = \mu\kappa/\sigma^*\delta$ (Sect. 11.4.1)
First Damköhler parameter	Chemical reaction in a flow	$Da = \tau_{mec}/\tau_{ch}$ (Sect. 7.2)
Euler number	Static and dynamic pressure	$Eu = p/\rho V^2$
Froude number	Inertia and gravity	$Fr = V^2/gL$
Grashof number	Thermal or chemical buoyancy and viscosity	$Gr = \Delta\rho g L^3/\rho\nu^2$
Hickman number	Vapor recoil over a flat liquid surface	$Hi = (\frac{d\dot{m}}{dT})^* \frac{\dot{m}^* \beta\delta^2\mu_V}{\rho_L\kappa_L\sigma*}$ $(1/\rho_V - 1/\rho_L)$ [94]
Lewis number	Heat diffusion and mass diffusion	$Le = Sc/Pr = \kappa/D$
Mach number	Convection and elastic compressive forces	$M = V/c$
Marangoni number (chemical)	Gradient of surface tension, species diffusion, and viscosity	$Ma_{ch} = \sigma_C G_C \delta^2/D\mu$
Marangoni number (thermal)	Gradient of surface tension, thermal diffusion, and viscosity	$Ma_{th} = \sigma_T G_T \delta^2/\kappa\mu$ (Sect. 7.9.2)
Nusselt number	Convection coefficient and thermal conduction	$Nu = \alpha L/\kappa$ (Sect. 12.4.2)
Peclet number	Convection and heat conduction	$Pe = VL/\kappa$
Prandtl number	Viscosity and heat diffusion	$Pr = \nu/\kappa$

Rayleigh number (chemical)	As for Gr_{ch} but with mass diffusion	$Ra_{ch} = Gr_{ch}Sc$ $= \Delta\rho_{ch}gL^3/\nu D$
Rayleigh number (thermal)	As for Gr_{th} but with thermal diffusion	$Ra_{th} = Gr_{th}Pr$ $= \Delta\rho_{th}gL^3/\nu\kappa$
Reynolds number	Convection and viscosity	$Re = \rho VL/\mu$
Richardson number	Superimposed flows and mixing layers	$Ri = (\rho_L - \rho_G)g\delta_G$ $/\rho_G(V_L - V_G)^2$ [94]
Schmidt number	Viscosity and species diffusion	$Sc = \nu/D$
Sherwood number (Nusselt for conc.)	Convection coefficient and species diffusion	$Sh = \alpha_C L/D$
Surface viscosity number	Surface viscosity and bulk shear viscosity	$Vi = (\kappa_a + \epsilon_a)/\mu D$ (Sect. 11.4.1)
Weber number	Inertia and surface tension	$We = \rho V^2 L/\sigma$

Table 5.3. Dimensionless numbers. Here, σ designates the surface tension, g is the acceleration due to gravity, L is a characteristic length, ρ is the density, $\Delta\rho = \rho^* - \rho$, μ is the viscosity, $\nu = \mu/\rho$ is the kinematic viscosity, κ is the thermal diffusivity, τ_{mech} is a mechanical time, τ_{ch} is the chemical time, p is the pressure, V is the velocity, $\dot{m}$ is the unit mass flow rate of vaporization, (*) denotes a reference state, ($_G$) denotes a gas, ($_V$) denotes a vapor, ($_L$) denotes a liquid, δ is the thickness of a layer or boundary layer, ($-\beta$) is the thermal gradient in a thermal boundary layer of thickness δ, D is the mass diffusivity, c is the speed of sound, $\sigma_C = d\sigma/dC$, $\sigma_T = d\sigma/dT$, G_T is the temperature gradient, G_C is the concentration gradient, α is a heat transfer coefficient, α_C is a mass transfer coefficient, ($_{ch}$) denotes a chemical effect, ($_{th}$) denotes a temperature effect, and κ_a and ϵ_a are surface viscosities

6

Chemical Reactors

Chemical engineering is a very important field with its own specific methods. Classical methods of analyzing and modeling aerodynamics and combustion are often not the most suitable methods for solving problems associated with chemical reactors, the subject of this chapter. For example, the deterministic and probabilistic population balance equations given in Sect. 4.10 are used instead of the general balance equation of Sect. 4.4. The latter is too refined for such problems, and it can be difficult to use to interpret experimental results.

The differences between ideal and real reactors are underlined in the first section of this chapter. Then, in Sect. 6.2, the behavior of one well-known type of ideal reactor—a homogeneous perfectly stirred reactor in either the steady or unsteady regime—is described. Paradoxically, there are some similarities between the stability properties of this type of chemical reactor and those of a premixed flame reaction zone, and we will make use of this observation in Chap. 8.

The residence time distribution is studied in Sect. 6.3. The basic residence time distribution equations are derived and then applied to ideal reactors—a homogeneous perfectly stirred reactor in the steady regime, a piston reactor, Poiseuille flow—and finally to real reactors.

6.1 Ideal and Real Reactors

In the broadest sense, a chemical reactor is a zone where chemical transformations can occur. This definition covers a very wide range of situations found in the chemical industry, aircraft engines, rockets, chemical engines, etc.

The configuration of a particular reactor depends on many factors related to the nature of the chemical transformation that takes place there: whether the reaction is homogeneous or heterogeneous, deviation from equilibrium, whether a catalyst is present, turbulence, thermodynamic state (pressure, temperature, etc.).

R. Prud'homme, *Flows of Reactive Fluids*,
DOI 10.1007/978-0-8176-4659-2_6, © Springer Science+Business Media, LLC 2010

The study of real reactors requires the development of experimental techniques and theoretical investigative tools. The residence time of the chemical species in a reactor is a significant concept, and studying the residence time distribution often makes it possible to comprehend the process in all of its complexity. Studying extreme cases—i.e., ideal reactors—is a very useful approach because they are relatively simple and are often a good approximation to real cases. It is often unrealistic to attempt to solve the local balance equations without using simplifying assumptions. Total balances that take the principal characteristics of the reactor into account are generally preferred. For example, in some cases we can ignore transfer phenomena (thermal conduction, diffusion, viscosity). Sometimes (in the case of an isothermal reactor), only a species balance equation is used [9]. For one type of ideal reactor known as a "perfectly stirred" reactor, we only need to distinguish between the inlet conditions and those of the whole reactor, which are also the exit conditions. This assumption is not valid for a "piston" reactor, where the flow is assumed to evolve through successive sections until it reaches the exit. Lastly, in most of the cases considered in this chapter, the effects of pressure variations and viscosity are neglected, which means that the momentum equation cannot be used.

We now consider various types of ideal reactors in turn (see Fig. 6.1). We study their steady regimes and sometimes their transient states, as well as the stabilities of their working regimes.

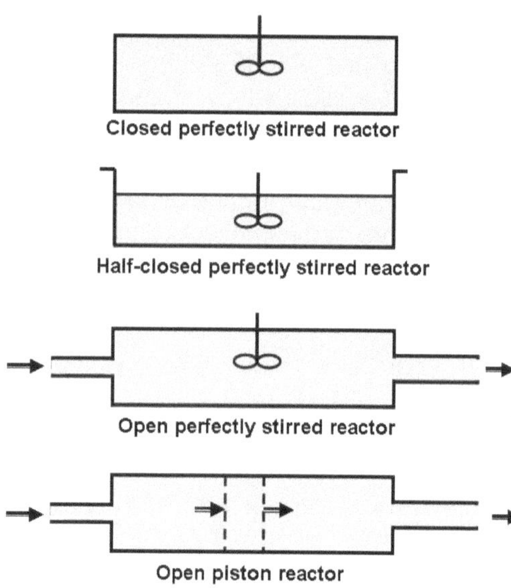

Fig. 6.1. Various types of ideal reactors

6.2 Homogeneous Perfectly Stirred Chemical Reactors

In this section, we will study perfectly stirred reactors. After presenting the basic equations for such reactors, we study their steady regimes and their stability and stabilization conditions.

Note that a similar study (not presented here) can be performed for tubular reactors[1] (see for instance [196]) assuming plug flow (see Sect. 6.3.3).

Simplified chemical reactor concepts are not relevant for combustion waves. They are studied in Chap. 10 by solving the general balance equations of reactive fluid flows involving transfer phenomena.

6.2.1 Basic Equations of Perfectly Stirred Reactors

We apply the total balance equations (see Sect. 4.10.1) to an open homogeneous perfectly stirred reactor [5, 9, 283]:

$$\frac{dm_j}{dt} = \dot{m}_{jo} - \dot{m}_{je} + R_j, \tag{6.1}$$

$$\frac{dE}{dt} = \sum_j \dot{m}_{jo} h_{jo} - \sum_j \dot{m}_{je} h_{je} - \dot{Q}(T). \tag{6.2}$$

"Perfectly stirred" means that the thermodynamic and chemical parameters are the same at any point in the reactor, once the inlet section has been traversed. In particular, $T_e = T$, $h_{je} = h_j$, $C_{je} = C_j$. In the case considered here, we assume that the pressure p is roughly constant, so that, if $\mathcal{V}_r$ is the volume of the reactor, $dE/dt = dH/dt$ where $H = \mathcal{V}_r \sum_j \rho_j h_j$. Finally the mass flow rates $\dot{m}_j$ are assumed to be proportional to the densities, so that, if $\dot{q}$ is the volume flow rate,

$$\dot{m}_j = \rho_j \dot{q}. \tag{6.3}$$

By setting the crossing time

$$\tau = \mathcal{V}_r / \dot{q} \tag{6.4}$$

equal to the average residence time, we have

$$\begin{cases} \tau\, d\rho_j/dt = \rho_{jo} - \rho_j + \tau \dot{W}_j, \\ \tau\, d(\sum_j \rho_j h_j)/dt = \sum_j (\rho_{jo} h_{jo} - \rho_j h_j) - \dot{Q}/\dot{q}, \end{cases} \tag{6.5}$$

given that $R_j = \mathcal{V}_r \dot{W}_j$.

[1] In a tubular reactor, the feed enters at one end of a cylindrical tube and the product stream leaves at the other end. The long tube and the lack of provision for stirring prevents complete mixing of the fluid in the tube, so the properties of the flowing stream will vary in both the radial and the axial directions.

When only one chemical reaction is involved,

$$\dot{W}_j = \nu_j \mathcal{M}_j \dot{\zeta}, \tag{6.6}$$

where ν_j is the stoichiometric coefficient and $\dot{\zeta}$ is the reaction rate in moles per unit time. We also have

$$d\rho_j = \nu_j \mathcal{M}_j d\zeta, \tag{6.7}$$

where ζ is the progress variable of the reaction such that

$$\rho_j = \rho_{jo} + \nu_j \mathcal{M}_j \zeta. \tag{6.8}$$

The species balance equation is therefore

$$\tau\, d\zeta/dt = -\zeta + \tau\, \dot{\zeta}. \tag{6.9}$$

For the energy equation, we note that

$$h_j - (q_f^0)_j + \int_{T^0}^{T} c_{p,\,j} dT, \tag{6.10}$$

where $(q_f^0)_j$ is the heat of formation and $c_{p,\,j}$ the specific heat at constant pressure per unit mass of species j; thus

$$d(\sum_j \rho_j h_j)/dt = \sum_j (h_j d\rho_j/dt + \rho_j c_{p,\,j} dT/dt). \tag{6.11}$$

We have

$$\begin{aligned} \textstyle\sum_j h_j d\rho_j/dt &= \textstyle\sum_j h_j[(\rho_{jo} - \rho_j)/\tau + \dot{W}_j] \\ &= \textstyle\sum_j h_j[(\rho_{jo} - \rho_j)/\tau + \nu_j \mathcal{M}_j \dot{\zeta}] \\ &= \textstyle\sum_j h_j(\rho_{jo} - \rho_j)/\tau + \dot{\zeta} \sum_j \nu_j \mathcal{M}_j h_j. \end{aligned} \tag{6.12}$$

If ΔH is the average molar enthalpy of reaction then

$$\Delta H = \sum_j \nu_j \mathcal{M}_j h_j \tag{6.13}$$

and by setting

$$\rho c_p = \sum_j \rho_j c_{p,\,j} \tag{6.14}$$

we obtain

$$\sum_j (\rho_{jo} - \rho_j) h_j - \tau \Delta H \dot{\zeta} + \rho c_p \tau dT/dt = \sum_j (\rho_{jo} h_{jo} - \rho_j h_j) - \dot{Q}/\dot{q}. \tag{6.15}$$

Simplifying and setting

$$\sum_j \rho_{je}(h_{jo} - h_j) = \sum_j \rho_{jo} \int_T^{T_e} c_{p,\,j} dT \cong \rho c_p (T_o - T) \tag{6.16}$$

yields

$$\tau dT/dt = T_o - T + \Delta^* \tau \dot{\zeta} - Q^* \tag{6.17}$$

with

$$\Delta^* = \Delta H/\rho c_p,\ Q^* = \dot{Q}/\rho c_p \dot{q}. \tag{6.18}$$

The system governing the behavior of the reactor is thus

$$\begin{cases} \tau\, d\zeta/dt = -\zeta + \tau\, \dot{\zeta}, \\ \tau dT/dt = T_o - T + \Delta^* \tau \dot{\zeta} - Q^*. \end{cases} \tag{6.19}$$

Let us now study the case for a first-order reaction:

$$A \overset{k(T)}{\rightarrow} B \tag{6.20}$$

where

$$k(T) = k_0\, e^{-T_a/T}. \tag{6.21}$$

We have

$$\dot{\zeta} = k_0\, e^{-T_a/T} C_A = k_0\, e^{-T_a/T} (C_{Ao} - \zeta). \tag{6.22}$$

By setting $X = \zeta / C_{Ao}$, we find that

$$\dot{\zeta} = C_{Ao}\, k_0\, e^{-T_a/T} (1 - X). \tag{6.23}$$

Let us assume that the law of heat exchange is

$$Q^* = K(T - T_0) \tag{6.24}$$

where T_0 is the outside temperature, and set

$$\begin{cases} f(X,\, T) = k_0\,(1 - X) e^{-T_a/T}, \\ T^* = \Delta^* C_{Ao}. \end{cases} \tag{6.25}$$

The balance equations (6.19) then take the form

$$\begin{cases} \tau\, dX/dt = -X + \tau\, f(X,\, T), \\ \tau dT/dt = T_o - T + \tau\, T^* f(X,\, T) - K(T - T_0). \end{cases} \tag{6.26}$$

6.2.2 Steady Regimes of Perfectly Stirred Reactors

The steady mode (index s) is characterized by the following two equations in the plane (T_s, X_s) (see Fig. 6.2):

$$X_s = \tau\, k_0\, e^{-T_a/T_s}/(1 + \tau\, k_0\, e^{-T_a/T_s}), \tag{6.27}$$

$$X_s = (K+1)T_s/T^* - (T_o + KT_0)/T^*. \tag{6.28}$$

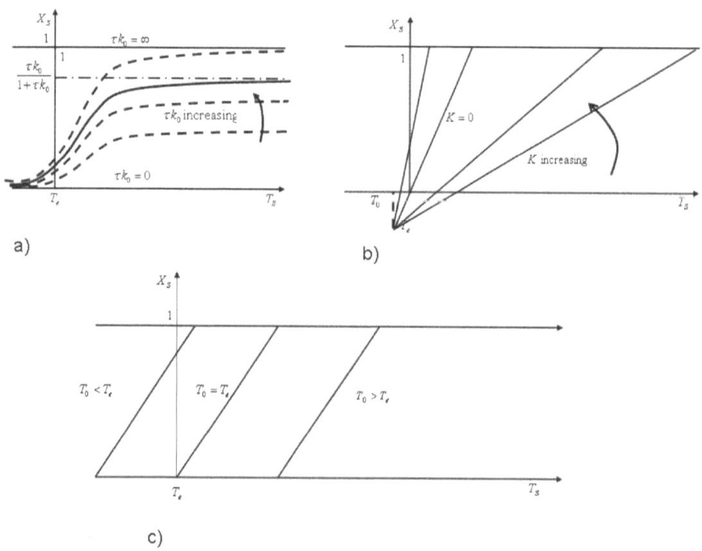

Fig. 6.2. Steady regime (Eqs. 6.27 and 6.28): **a** Influence of τk_0, which is similar to the first Damkölher parameter $D_a = \tau/\tau_{chim}$. For a given T_s, increasing τk_0 (a faster reaction or a slower flow) increases X_s. **b** Influence of the heat exchange coefficient K for a given T_0 and T_o (here $T_0 < T_o$). **c** Influence of the exchanger temperature for a given K

The solutions correspond to the intersections of the curves for Eqs. 6.27 and 6.28 (see Fig. 6.3).

We can also represent the solutions in the plane (X_s, f_s), where

$$f_s = f(X_s, T_s) = k_0\,(1 - X_s)e^{-T_a/T_s}. \tag{6.29}$$

The two following equations are then obtained:

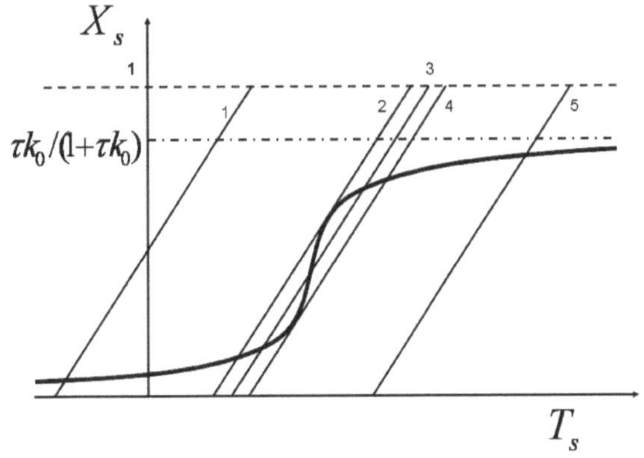

Fig. 6.3. Various solutions for T_a with various values of T_0 and for a given τk_0 and K

$$\begin{cases} f_s = X_s/\tau, \\ f_s = k_0\,(1 - X_s)e^{-T_a(1+K)/(T_o+KT_0+T^* X_s)}. \end{cases} \tag{6.30}$$

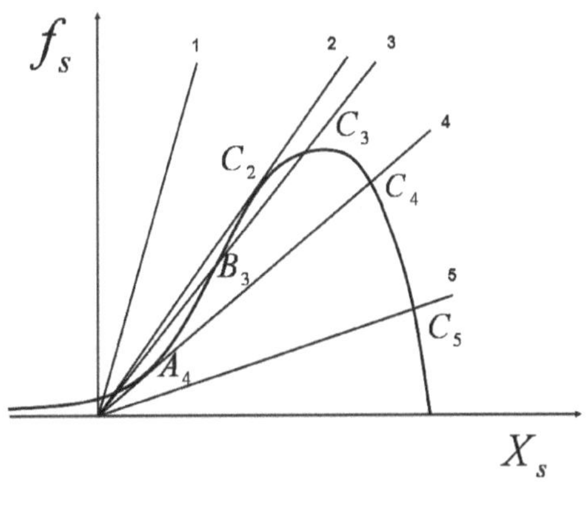

Fig. 6.4. Steady solutions in the (X_S, f_S) plane

Here, the influence of the crossing time defined by (6.4) is highlighted by the slope $1/\tau$ of the straight line that crosses the origin (see Fig. 6.4). We have one or three operating points, or exceptionally two, and the points A_j correspond to very weak reaction rates. It may be useful to make X_s as large as possible and the crossing time τ quite low. A runaway reaction can be obtained by decreasing the mass flow rate (and thus $1/\tau$) or by increasing the temperature of the exchanger.

6.2.3 Stability Analysis of a Perfectly Stirred Reactor

It is important to assess the operating point stability. To do this, we study the evolution over time after a deviation from the steady mode:

$$x_1 = (X - X_s)/(1 - X_s),\ x_2 = (T - T_s)/T_s. \tag{6.31}$$

We then obtain

$$\begin{cases} dx_1/d\theta = -x_1 - a_1[1 - (1 - x_1)\exp(a_2 x_2/(1 + x_2))], \\ dx_2/d\theta = -a_3 x_2 - a_4[1 - (1 - x_1)\exp(a_2 x_2/(1 + x_2))], \end{cases} \tag{6.32}$$

with

$$\begin{cases} \theta = t/\tau,\ a_1 = \tau k_0 e^{-T_a/T_s},\ a_2 = T_a/T_s, \\ a_3 = 1 + K,\ a_4 = T^*(1 - X_s)a_1/T_s. \end{cases} \tag{6.33}$$

After linearizing the right hand sides, we obtain

$$\begin{cases} dx_1/d\theta = -(1 + a_1)x_1 + a_1 a_2 x_2, \\ dx_2/d\theta = -a_4 x_1 + (a_2 a_4 - a_3)x_2. \end{cases} \tag{6.34}$$

The characteristic equation for the system is

$$\lambda^2 - S\lambda + P = 0, \tag{6.35}$$

with

$$\begin{cases} S = -(1 + a_1 + a_3 - a_2 a_4), \\ P = (1 + a_1)a_3 - a_2 a_4. \end{cases} \tag{6.36}$$

The roots are real and distinct if

$$\Delta = (1 + a_1 - a_3)^2 + 2a_2 a_4(1 - a_1 - a_3) + a_2^2 a_4^2 > 0. \tag{6.37}$$

Conjugate imaginary roots occur when $\Delta < 0$, and a double root occurs when $\Delta = 0$.

First case: real nonzero roots ($\Delta > 0$):

There is stability ($\lambda_1 < 0$, $\lambda_2 < 0$) provided that P is positive and that S is negative. Any low-amplitude variation around the steady state tends to be damped according to an exponential law. If only one of the roots is negative or if both of the roots are positive there is instability.

Second case: conjugate imaginary roots ($\Delta < 0$)

Here we have $P > 0$ in all cases. The real part of each root λ is equal to $S/2$. Stability will thus be assured if $S < 0$ and the oscillatory solution is then damped. If S is positive there is instability. The oscillatory solution is not damped for vanishing S.

Third case: double roots ($\Delta = 0$):

The solution is stable if $S < 0$.

Fourth case: one of the roots is equal to zero ($P = 0$):

There is asymptotic stability. The variation tends to stabilize towards a nonzero value when $\theta \to \infty$.

In brief, the stability of the linearized system is only ensured for $P > 0$ and $S < 0$. One of these conditions can be represented graphically. Indeed, in the plane (T_s, X_s), if p_1 is the slope of the curve $X_s = \tau k_0 e^{-T_a/T_s}/(1 + \tau k_0 e^{-T_a/T_s})$ and p_2 is the slope of the straight line $X_s = (K+1)T_s/T^* - (T_o + KT_0)/T^*$, differentiation yields

$$1 - p_1/p_2 = [(1 + a_1)a_3 - a_2 a_4]/a_3(1 + a_1) = P/a_3(1 + a_1). \qquad (6.38)$$

Thus, P will be positive if the slope of the straight line is higher than that of the curve, and P will be negative in the opposite case. Thus, there are sufficient conditions for instability when $p_2 < p_1$, which only occurs between the points A and B. The arc of curve AB is therefore a zone of unstable stationary modes with respect to small disturbances (Fig. 6.5).

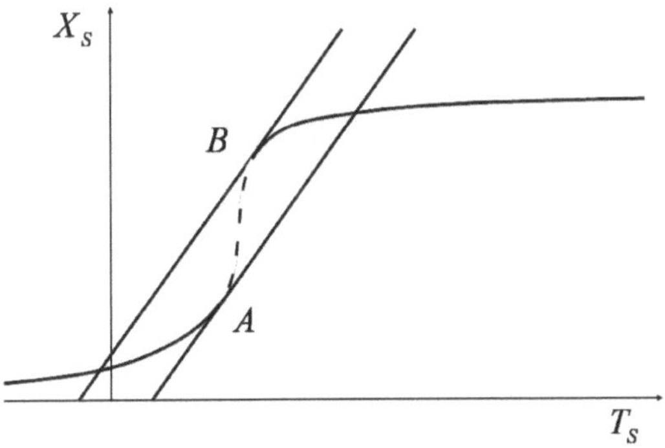

Fig. 6.5. Unstable steady modes

We can only completely describe the unsteady behavior of the reactor and elucidate the response to large disturbances by studying the nonlinear equations

$$\begin{cases} \tau\, dX/dt = F(X,\, T), \\ \tau\, dT/dt = G(X,\, T). \end{cases} \tag{6.39}$$

Following Admundson and Bilous [18], we can track the evolution by dividing the plane $(T,\, X)$ into areas according to the signs of $F(X,\, T)$ and $G(X,\, T)$ and carrying out a representation in this plane. Figure 6.6 qualitatively shows the result of doing this when there is only one steady state [219].

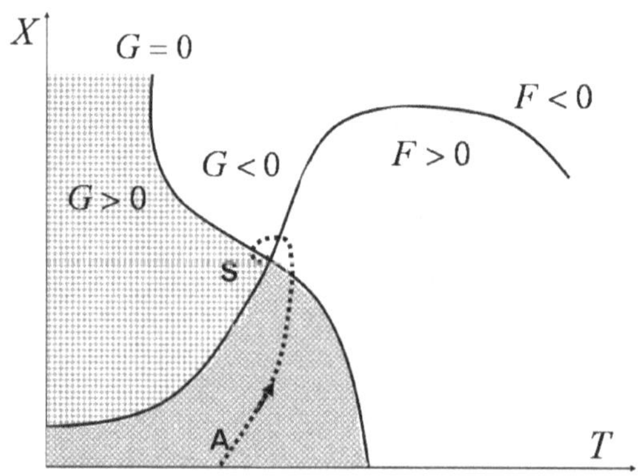

Fig. 6.6. Case for a single steady state. The path AS indicates a steady state with oscillating stability

For three steady states, we obtain a boundary line xy (see Fig. 6.7).

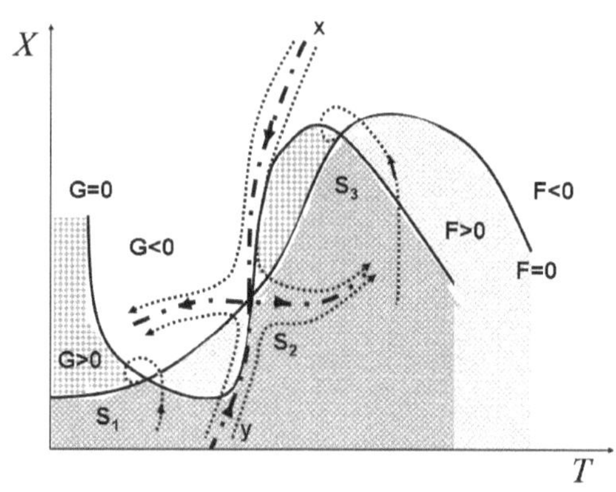

Fig. 6.7. Case for three steady states

These figures are obtained for a reversible reaction $A \rightleftharpoons B$; slightly different from the preceding scenario.

Aris and Amundson [5, 6] obtained the numerical solutions for an irreversible reaction $A \rightarrow B$ with three steady states (see Fig. 6.8, as well as [219]).

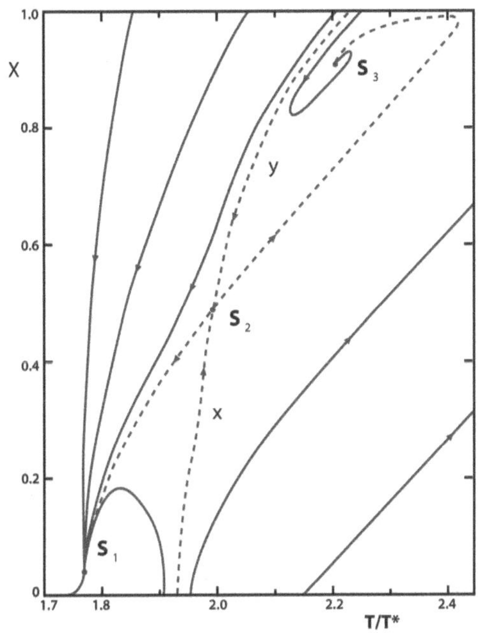

Fig. 6.8. Numerical solutions for three steady states (after [6])

In Fig. 6.8, note that all of the unsteady evolutions on the left side of the boundary line xy converge towards the steady state S_1. For example, if we wish to obtain the stable steady state S_3, which is the only interesting one, we will have to preheat the reactor (as mentioned above).

There are two problems with this type of reactor:

- The time needed to reach a stable steady state
- The existence of unstable steady states.

These problems can be solved by adjusting the conditions of the reactor appropriately, which makes it possible to accelerate towards the steady state or to stabilize a naturally unstable operating point.

We can, for example, adjust the flow rate of the external cooling fluid according to the exit temperature of the reactor. This makes it possible to modify the exchange coefficient K, which becomes $K + m$.

The stability conditions of the linearized system thus become

$$\begin{cases} 1 + a_1 + a_3 + m - a_2 a_4 > 0 \\ (1 + a_1)(a_3 + m) - a_2 a_4 > 0 \end{cases} \tag{6.40}$$

Note that if m is sufficiently large it will be always possible to stabilize the operating point.

However, such control is difficult to achieve in reality. We can also obtain stabilization, but only in certain situations, by using the concentration as the signal instead of the temperature of the reactor. In this case, the heat flux with the exchanger becomes a function $Q(T, X)$. Linearization then gives us

$$\begin{cases} dx_1/d\theta = -(1 + a_1)x_1 + a_1 a_2 x_2, \\ dx_2/d\theta = -(a_4 + \mu)x_1 + (a_2 a_4 - a_3)x_2, \end{cases} \tag{6.41}$$

where μ characterizes the influence of the concentration on Q, which is presumed to be linear. The stability conditions become

$$\begin{cases} 1 + a_1 + a_3 - a_2 a_4 > 0, \\ (1 + a_1)a_3 - a_2 a_4 + \mu a_1 a_2 > 0. \end{cases} \tag{6.42}$$

It is seen that adjusting the concentration only influences the second condition; i.e., P. We can only stabilize points where the first condition is obeyed.

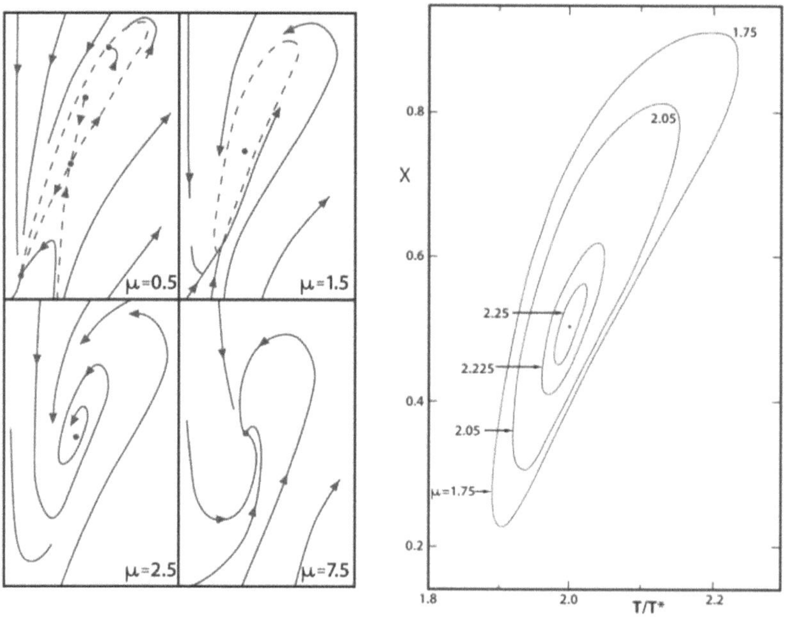

Fig. 6.9. Stabilization and limiting cycles (after [5])

Aris and Amundson (see [5, 6], as well as [219]) studied the nonlinear system numerically with increasing values of T for this reactor. They observed the stabilization of the steady state S_2 and, for intermediate values of μ, the onset of limiting cycles for increasing values of the control parameter μ (see Fig. 6.9).

6.3 Residence Time Distribution

6.3.1 Basic Equations

For real chemical reactors, we can make use of the probabilistic balance equation described in Chap. 4. For a reactor of constant volume $\mathcal{V}_r$, and with entry and exit volume flow rates of $\dot{q}_o$ and $\dot{q}_e$, this equation becomes [283]:

$$\frac{\partial \bar{\psi}}{\partial t} + \frac{\dot{q}_e \bar{\psi}_e - \dot{q}_o \bar{\psi}_o}{\mathcal{V}_r} + \sum_i \frac{\partial(\bar{\psi} w_i)}{\partial \zeta_i} = \bar{G}, \tag{6.43}$$

where ψ is the distribution function of the phase space

$$\psi = \psi(\mathbf{x}, \boldsymbol{\zeta}, t). \tag{6.44}$$

The averages $\bar{\psi}$, $\bar{\psi}_o$ and $\bar{\psi}_e$ are defined as

$$\bar{\psi} = \frac{1}{\mathcal{V}_r} \int_{\mathcal{V}_r} \psi \, d\mathcal{V}, \; \bar{\psi}_o = \frac{1}{\dot{q}_o} \int_{A_o} \psi \mathbf{v} \cdot d\mathbf{A}, \; \bar{\psi}_e = \frac{1}{\dot{q}_e} \int_{A_e} \psi \mathbf{v} \cdot d\mathbf{A}, \tag{6.45}$$

where $\mathbf{v}$ is the velocity vector and A_o and A_e are the entry and exit sections of the chemical reactor, respectively.

Various models have been elaborated by choosing the parameters ζ_j and the corresponding w_j as well as the average production $\bar{G}$.

At the experimental level, we can use the concepts of age and residence time of the molecules in the reactor. To do this, we neglect the effects of chemical reactions on the flow, which is considered to be an inert fluid. We define the age (α) distribution $I(\alpha)$ such that $I(\alpha)\, d\alpha$ is the fraction of the molecules in the reactor that have ages of between α and $\alpha + d\alpha$. The residence time (t_e) distribution $E(t_e)$ is defined such that $E(t_e)\, dt_e$ represents the fraction of the molecules in the exit flow that have been in the reactor for times ranging from t_e to $t_e + dt_e$. The residence time distribution can be obtained experimentally by measuring the concentration or flow rate of a tracer at the exit section. If a tracer impulse is introduced at time $t = 0$ (the time profile for the impulse is a Dirac delta function) at the inlet, and $C(x_e, t_e)$ is the concentration of tracer at the exit section, we have

$$E(t_e) = \frac{\int_{A_e} C(\mathbf{x}_e, t_e) \, \mathbf{v}(\mathbf{x}_e, t_e) \cdot d\mathbf{A}}{\int_{t_e=0}^{\infty} \int_{A_e} C(\mathbf{x}_e, t_e) \, \mathbf{v}(\mathbf{x}_e, t_e) \cdot d\mathbf{A} dt_e}. \tag{6.46}$$

The crossing time τ is equal to the average residence time; i.e.,

$$\tau = \bar{t}_e = \int_0^\infty t_e E(t_e)\, dt_e. \tag{6.47}$$

We now give three examples of well-defined reactors.

6.3.2 Homogeneous Well-Stirred Reactor in Steady Mode

In the absence of any chemical reactions in a well- (or perfectly) stirred reactor, we have

$$dC/dt = (C_o - C)/\tau \tag{6.48}$$

for $t > 0$, where τ is the crossing time (equal to $\mathcal{V}_r/\dot{q}$). Given that $C_o(t) = C_0\,\delta(t)$, where $\delta(t)$ is the Dirac delta function, the solution to this is

$$C = C_0\, e^{-t/\tau},\ t > 0. \tag{6.49}$$

Therefore,

$$\int_{A_e} C\,u\,dA = C_0 e^{-t/\tau} \int_{A_e} u\,dA = C_0\,\dot{q}\,e^{-t/\tau}, \tag{6.50}$$

where u is the velocity, so that

$$E(t_e) = e^{-t_e/\tau} / \int_0^\infty e^{-t_e/\tau}\, dt_e = e^{-t_e/\tau}/\tau. \tag{6.51}$$

6.3.3 Piston Reactor

A "piston" reactor (or "plug flow" reactor) is an ideal tubular reactor in which the flow is carried in planar sections where the parameters are uniform (complete mixing in the radial direction, a uniform velocity profile across the radius) and there is no mixing in the axial direction (i.e., the direction of flow, assuming plug flow). An impulse injection of dye at the entry section, in the absence of diffusion, will give the same signal at the exit section when the crossing time τ has elapsed, so that

$$E(t_e) = \delta(t_e - \tau). \tag{6.52}$$

6.3.4 Poiseuille Flow

In Poiseuille flow, the velocity profile is given by (see Sect. 5.1.4 for definitions)

$$u(r) = \frac{\Delta p}{4\mu \Delta l}(R^2 - r^2),\ R = D/2. \tag{6.53}$$

The residence time is thus related to r:

$$t_e = \Delta l/u(r) = \Delta l/a(R^2 - r^2),\ a = \Delta p/4\mu\Delta l. \tag{6.54}$$

The flow between t_e and $t_e + dt_e$ is

$$u(r)\,2\pi r\,dr = 2\pi ar(R^2 - r^2)\,dr, \tag{6.55}$$

and the total flow is given by integrating between 0 and R:

$$2\pi aR^4/4. \tag{6.56}$$

Thus,

$$E(t_e)\,dt_e = 4\pi u(r)\,dr/aR^4 = 4(R^2 - r^2)r\,dr/R^4. \tag{6.57}$$

As

$$dt_e = \frac{\Delta l}{a}\frac{2r\,dr}{(R^2 - r^2)^2},\ r \in [0,\ R], \tag{6.58}$$

we find that

$$\begin{cases} E(t_e) = \tau^2/2t_e^3,\ t_e \in [\tau/2,\ \infty[, \\ E(t_e) = 0,\ t_e \in [0,\ \tau/2[, \end{cases} \tag{6.59}$$

where

$$\tau = t_e = 2\Delta l/aR^2. \tag{6.60}$$

The solutions for these three examples are shown in Fig. 6.10, together with the shape of the residence time distribution for a real reactor.

6.3.5 Real Reactor

A correlation between the age distribution and the residence time distribution can be obtained using the population balance. The age α then characterizes a particle of fluid, so that $\zeta = \alpha$ and $w = d\alpha/dt = 1$ in steady flow. We then have

$$\dot{q}_e\bar{\psi}_e/\mathcal{V}_r + \partial\bar{\psi}/\partial\alpha = 0,\ \alpha > 0. \tag{6.61}$$

However, the age distribution is equal to $\bar{\psi}/N$, where $N = \int_0^\infty \bar{\psi}\,d\alpha$, and the residence time distribution is given by $\bar{\psi}_e/N$. Writing $\tau = \mathcal{V}_r/\dot{q}_e$, and assuming an incompressible fluid, we then obtain

$$dI(\alpha)/d\alpha + E(\alpha)/\tau = 0. \tag{6.62}$$

For example, in the perfectly stirred case, we find

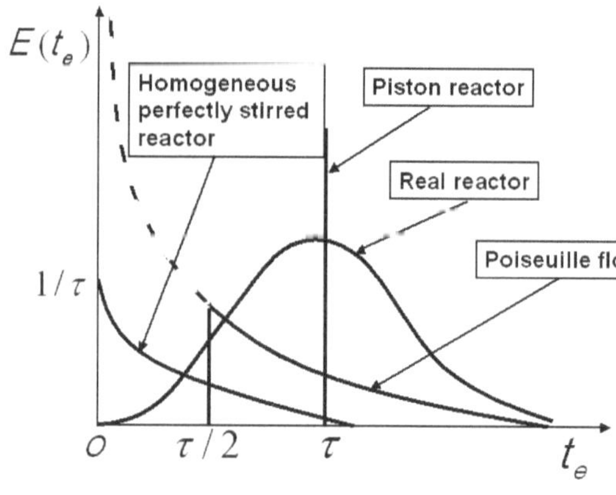

Fig. 6.10. Residence time distributions of chemical reactors

$$I(\alpha) = E(\alpha) = e^{-\alpha/\tau}/\tau. \tag{6.63}$$

In the general case, we have

$$\tau\, I(t) = 1 - \int_0^t E(t_e)\, dt_e. \tag{6.64}$$

This relation is easy to interpret if we imagine that dye is injected into an incompressible fluid according to the Heaviside function. Integration yields

$$\mathcal{V}_r \int_0^t I(\alpha)\, d\alpha = \int_0^t \dot{q} dt' \int_{t'}^{\infty} E(t_e)\, dt_e, \tag{6.65}$$

where $\int_{t'}^{\infty} E(t_e)\, dt_e$ represents the fraction of the exit flow that is older than t', and therefore colorless. We then write a balance relation at time $t > 0$: the quantity of colored fluid in the reactor at time t is equal to the quantity of colorless fluid which has exited the reactor since the beginning the injection $(t = 0)$.

The residence time distribution characterizes the macro mixture; i.e., the relative displacement of the fluid particles.

The micro mixture is related to the interactions and content exchanges that occur between the aggregates[2] and their environment (see Sect. 8.1). The micro mixture strongly depends on the small-scale turbulence present, and so it is difficult to study. Chemical engineers have used a different approach to model development compared to fluid mechanists.

---

[2]Every real fluid is partially segregated, since it consists of aggregates of molecules formed by shear effects arising from mechanical agitation and friction. These aggregates exchange matter with one another through diffusion [276].

In the IEA (interaction, exchange with the average) model, for example, the concentration C of a chemical aggregate conforms to the equation

$$dC/d\alpha = h(C - \bar{C}) + \dot{C}(C), \tag{6.66}$$

where h is the frequency of exchange and $\bar{C}$ is the average concentration

$$\bar{C} = \int_0^\infty I(\alpha)C(\alpha)\, d\alpha. \tag{6.67}$$

The average production rate $\bar{\dot{C}}$ is equal to

$$\bar{\dot{C}} = \int_0^\infty I(\alpha)\, \dot{C}(C(\alpha))\, d\alpha. \tag{6.68}$$

This leads to the integro-differential equation

$$dC/d\alpha = h(C - \int_0^\infty I(\alpha)C(\alpha)\, d\alpha) + \dot{C}(C). \tag{6.69}$$

To solve this, we must know the initial concentration probability density.

7

Coupled Phenomena

Basic equations for fluid flows with chemical reactions and transfer phenomena were described in Chaps. 2 to 4, and initial tools for the analysis of these flows were given in Chap. 5. In Chap. 6, the basic equations were applied to relatively simple problems involving no transport phenomena. This chapter presents and solves more complex examples, and particular attention is directed towards the concepts of coupling and interaction. The goal of this chapter is to examine problems of increasing complexity, in the knowledge that even more difficult cases will be adressed in subsequent chapters.

Section 7.1 deals with the various kinds of phenomena encountered in fluid flows; some simple examples are provided.

In Sect. 7.2, we study a flow without molecular transfer—more complex than the one described in Chap. 6—that involves coupling between a chemical reaction and a compressible flow. A monodimensional analysis of a subsonic/supersonic nozzle flow is presented, along with an analysis of the transonic throat zone that utilizes an asymptotic expansion method (the corresponding bidimensional analysis is performed in Sect. 10.3).

Section 7.3 deals with the interaction between heat and mass diffusion, which has the same tensorial order as in the Onsager theory. Particular attention is paid to defining convenient phenomenological coefficients.

When multiple transfer phenomena are coupled with chemical reactions—as is often the case in combustion—the problem is more complex, and we need approximations such as the classical Shvab–Zel'dovich approximation presented in Sect. 7.4.

Section 7.5 examines the case of a mixture with an interface at which vaporization occurs.

Thermal osmosis provides us with the opportunity, in Sect. 7.6, to introduce an interesting property that was discovered by I. Prigogine: coupled systems evolve so as to minimize their entropy production.

In Sect. 7.7, we consider laminar flames resulting from interactions between chemical kinetics and dissipative flow. Reference is also made to Chap. 11, where there is an explanation of stretched flames.

R. Prud'homme, *Flows of Reactive Fluids*,
DOI 10.1007/978-0-8176-4659-2_7, © Springer Science+Business Media, LLC 2010

Instabilities are often caused by coupled phenomena. In this chapter we present two types of instability that are of historical interest. One of these was first observed by Henry Bénard, and was then the subject of half a century of discussions. This type of instability occurs due to buoyancy forces when a fluid layer between two horizontal walls is heated from below. It is known as the Rayleigh–Bénard instability, and is presented in Sect. 7.8.

If the upper wall is removed, the buoyancy forces are then placed in competition with the forces generated by surface tension gradients resulting from local temperature fluctuations on the free surface. This leads to the so-called Bénard–Marangoni instability (Sect. 7.9.2). Note that the occurrence of Bénard–Marangoni instability, or any other type of instability (some of which are presented in Chaps. 8 and 9), can lead to the onset of turbulence (see Chap. 8).[1]

Also note that the elementary knowledge of surface tension provided in Chap. 2 is sufficient to be able to study the problems described in Sect. 7.8. However, we will treat more complex surface phenomena in Chap. 11.

7.1 General Information

7.1.1 Types of Coupled Phenomena

The parameters associated with a fluid flow (velocity, temperature, density, concentrations, etc.) conform to thermodynamic relations, such as laws of state and constitutive relations, and obey balance equations. The latter contain various kinds of terms.

Thus, the local balance equation of a property F (f per unit mass) is written

$$\rho \frac{df}{dt} + \boldsymbol{\nabla} \cdot \boldsymbol{\mathcal{J}}_F = \dot{W}_F. \tag{7.1}$$

Using the global form, and following the particle motion, leads to

$$\frac{d}{dt} \int_{\mathcal{V}} \rho f d\mathcal{V} + \int_{\partial \mathcal{V}} \boldsymbol{\mathcal{J}}_F \cdot \mathbf{n} dS = \int_{\mathcal{V}} \dot{W}_F d\mathcal{V}. \tag{7.2}$$

On the left hand side, the first term is the rate of variation of F if we follow the motion velocity $\mathbf{v}$, while the second term is the outgoing relative flux. The right hand side contains the internal production. The so-called material derivative df/dt (or Lagrangian derivative) is equal to $\partial f/\partial t + \mathbf{v} \cdot \boldsymbol{\nabla} f$ and comprises a nonstationary term and a convection term.

The nonstationary part and the convective flow both appear in the global balance for a volume limited by a fixed surface ($\partial \mathcal{V}$):

---

[1] We know that instability and turbulence evolve differently in confined systems at rest in the reference configuration and in open systems where there is a reference flow.

$$\frac{\partial}{\partial t}\int_{\mathcal{V}} \rho f d\mathcal{V} + \int_{\partial\mathcal{V}} \rho f \mathbf{v}\cdot\mathbf{n} dS + \int_{\partial\mathcal{V}} \boldsymbol{\mathcal{J}}_F \cdot \mathbf{n} dS = \int_{\mathcal{V}} \dot{W}_F d\mathcal{V}, \tag{7.3}$$

just like the flux term and the production term. In theory, a knowledge of the laws of state (Chap. 2), the complementary laws (Chap. 3: phenomenological relations, laws of chemical kinetics), and the balance equations (Chap. 4)

$$\begin{cases} \rho\, dv/dt = \boldsymbol{\nabla}\cdot\mathbf{v} \\ \rho\, dY_j/dt + \boldsymbol{\nabla}\cdot\boldsymbol{\mathcal{J}}_{Dj} = \dot{W}_j, \\ \rho\, d\mathbf{v}/dt + \boldsymbol{\nabla}\cdot\mathbf{P} = \sum_j \rho_j \mathbf{f}_j, \\ \rho\, de/dt + \boldsymbol{\nabla}\cdot\mathbf{q} = r - \mathbf{D} : \mathbf{P} + \sum_j \boldsymbol{\mathcal{J}}_{Dj}\cdot\mathbf{f}_j. \end{cases} \tag{7.4}$$

is sufficient to solve this problem when the boundary conditions and initial conditions are well specified. An examination of the entropy production rate highlights the dissipative terms, which are sources of irreversibility:

$$\begin{aligned} \dot{W}_S &= -\textstyle\sum_j (g_j/T)\dot{W}_j - (1/T)\mathbf{D} : (\mathbf{P} - p\mathbf{1}) + \mathbf{q}\cdot\boldsymbol{\nabla}(1/T) \\ &- \textstyle\sum_j \boldsymbol{\mathcal{J}}_{Dj}\cdot(\boldsymbol{\nabla}(g_j/T) - \mathbf{f_j}/T). \end{aligned} \tag{7.5}$$

Note in particular that the fluxes $\boldsymbol{\mathcal{J}}_{Dj}$ and $\mathbf{q}$ induce dissipation (mass diffusion and heat conduction), whereas the momentum flux $\mathbf{P}$ includes a nondissipative part $p\mathbf{1}$ and a dissipative part $\mathbf{P} - p\mathbf{1}$ (viscosity). The chemical kinetics only result in entropy production out of equilibrium.

We know that many couplings will take place in a flow: between parameters; between dissipative or nondissipative phenomena. These couplings can result from the balance equations, the state relations, and the phenomenological relations. They can also come from the boundary conditions of the problem. We now provide some elementary examples of coupled phenomena in fluids. More complex interactions are examined in subsequent sections and chapters.

7.1.2 Incompressible Nonviscous Fluid

An incompressible fluid conforms to

$$\boldsymbol{\nabla}\cdot\mathbf{v} = 0, \tag{7.6}$$

$$\rho\frac{d\mathbf{v}}{dt} + \boldsymbol{\nabla}\cdot\mathbf{P} = \rho\mathbf{f}. \tag{7.7}$$

If $\mathbf{f} = -\boldsymbol{\nabla}V$, where V is a potential, and if $\boldsymbol{\nabla}\times\mathbf{v} = 0$ (irrotational flow) initially, then $\mathbf{v} = \boldsymbol{\nabla}\phi$ at any time, where ϕ is the velocity potential. The velocity vector can therefore be obtained from the potential $\phi(\mathbf{x}, t)$, which conforms to

$$\Delta\phi = 0. \tag{7.8}$$

The velocity field is thus independent of the other fields. The pressure can be deduced from the modulus v of the velocity vector using the Bernoulli equation

$$\frac{\partial\phi}{\partial t} + \frac{v^2}{2} + \frac{p}{\rho} + V = C(t), \tag{7.9}$$

which results from the momentum equation in this irrotational case. Only boundary conditions relating to p can make the field depend on the velocity of the pressure field and thus induce pressure–velocity coupling.

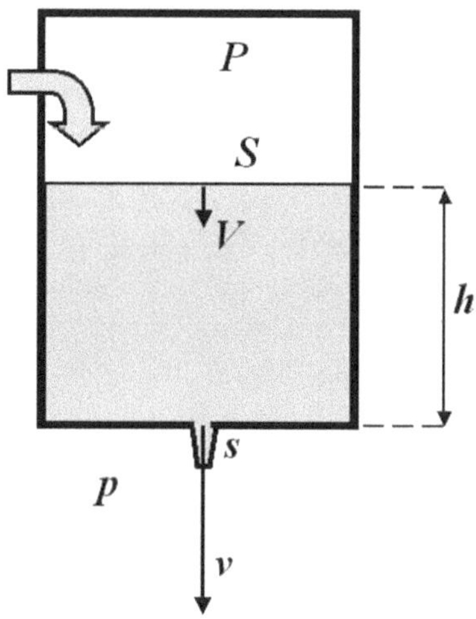

Fig. 7.1. Vessel with a hole in the bottom

We will highlight such an interaction using an example. Let us consider a cylindrical vessel of cross-section S that has a hole in the bottom of cross-section s. A liquid is fed into the vessel and steadily runs out through the hole such that the level of liquid in the vessel is constant (see Fig. 7.1).

When $s \ll S$, the velocities are written

$$V \cong \sqrt{2(s^2/S^2)(gh + (P - p)/\rho)},\ SV = sv,$$

where P is the pressure at the level of the free surface and p is the external pressure. Speed and pressure thus interact if $p \neq P$. At the limit, $V = 0$ if $P = p - \rho gh$.

7.1.3 Incompressible Viscous Fluid

In the case of an incompressible viscous fluid (Eq. 7.6 remains valid), the momentum equation indicates that there is a direct interaction between the inertial terms and the viscous terms,

$$\rho \frac{d\mathbf{v}}{dt} + \boldsymbol{\nabla}\mathbf{p} = \mu \Delta \mathbf{v}, \tag{7.10}$$

in the absence of external forces. The relative importance of the inertial compared to the viscous terms is characterized by the Reynolds number:

$$Re_L = \rho V L / \mu, \tag{7.11}$$

where L is a characteristic length. Moving a plate horizontally at a certain speed induces motion in the fluid (which is initially at rest), whereas this motion is nonexistent in a perfect fluid. Real fluids—unlike perfect fluids—are viscous, and this viscosity transmits the motion of the plate to the fluid (Fig. 7.2).

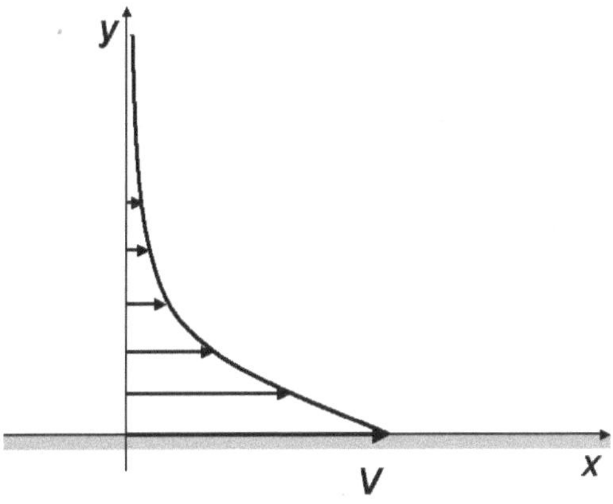

Fig. 7.2. Velocity profile in an unsteady boundary layer

Let us now consider 2D flow over a flat plate. We have

$$\begin{cases} \partial u/\partial x + \partial v/\partial y = 0, \\ \rho(\partial u/\partial t + u\partial u/\partial x + v\partial u/\partial y) = \mu(\partial^2 u/\partial x^2 + \partial^2 u/\partial y^2), \\ \rho(\partial v/\partial t + u\partial v/\partial x + v\partial v/\partial y) = \mu(\partial^2 v/\partial x^2 + \partial^2 v/\partial y^2), \end{cases} \tag{7.12}$$

with the boundary conditions

$$t \leq 0: \; u = v = 0; \; t \geq 0, \, y = 0: \; u = V, \, v = 0. \tag{7.13}$$

The solutions for $v = 0$ can be written as

$$u(y,t) = V(1 - \operatorname{erf}(y/\sqrt{4\nu t}))$$

where $\nu = \mu/\rho$. This solution is analyzed in more detail in Sect. 9.1.1.[2]

7.1.4 Reactive Incompressible Viscous Flow

Let us assume that Newton's viscosity law and Fick's diffusion law hold. If the temperature is constant, like D and μ, we have

$$\begin{cases} \boldsymbol{\nabla} \cdot \mathbf{v} = 0, \\ \rho\, d\mathbf{v}/dt + \boldsymbol{\nabla}\mathbf{p} = \mu \Delta \mathbf{v}, \\ \rho\, dY_j/dt = \rho D \Delta Y_j + \dot{W}_j. \end{cases} \tag{7.14}$$

In this case, the velocity field is independent of the chemistry, but it influences the chemistry via the convection term $\rho\mathbf{v} \cdot \boldsymbol{\nabla} Y_j$. In other cases, the reverse interaction also occurs. Recall that the relative effects of the diffusion and viscosity are characterized by the Schmidt number

$$Sc = \mu/\rho D. \tag{7.15}$$

7.2 Coupling Between Chemical Kinetics and Nondissipative Flow: Compressible Reactive Fluid

Let us suppose that the dissipative fluxes $\mathcal{J}_{Dj}$, $\mathbf{P} - p\mathbf{1}$ and $\mathbf{q}$ are negligible, like r and $\mathbf{f}_i$. The entropy production rate is then

$$\dot{W}_S = -\sum_j \frac{g_j}{T} \dot{W}_j. \tag{7.16}$$

A knowledge of the laws of chemical kinetics allows us, as described in Chap. 3, to avoid having to use linearized laws to express the chemical production rate. An assumption that chemical evolution occurs near equilibrium is thus not necessary, and the balance equations provide the means to solve problems related to flows that are not in chemical equilibrium.

---

[2]In particular, we see in this section that, in this case, the Reynolds number associated with the thickness of the resulting unsteady boundary layer is a function of time: $Re_\delta = V\delta/\nu$ with $\delta = O(\sqrt{\nu t})$.

Basic Equations for a Relaxing Flow

The balance equations become

$$\rho \frac{d\vartheta}{dt} = \nabla \cdot \mathbf{v}, \tag{7.17}$$

$$\rho \frac{dY_j}{dt} = \dot{W}_j, \tag{7.18}$$

$$\rho \frac{d\mathbf{v}}{dt} + \nabla p = 0, \tag{7.19}$$

$$\rho \frac{de}{dt} = -p\nabla \cdot \mathbf{v}. \tag{7.20}$$

The elimination of $\nabla \cdot \mathbf{v}$ between (7.17) and (7.20) gives us

$$\frac{de}{dt} + p\frac{d\vartheta}{dt} = 0, \tag{7.21}$$

or, according to the Gibbs relation,

$$\rho \frac{ds}{dt} + \sum_j g_j \frac{dY_j}{dt} = 0. \tag{7.22}$$

Consider a steady flow: we have $d/dt = \mathbf{v} \cdot \nabla$. The equation for the total energy becomes

$$\frac{d(h + v^2/2)}{dt} = 0 \tag{7.23}$$

or

$$h + v^2/2 = h_0. \tag{7.24}$$

If, moreover, the chemical process involves only one reversible reaction; for example

$$\mathrm{A_2 + M \rightleftharpoons A + A + M}, \tag{7.25}$$

where M is a neutral species or any of species A or A_2, then

$$\dot{W}_A = 2\mathcal{M}_A(\dot{\zeta}_D - \dot{\zeta}_R), \tag{7.26}$$

$$\dot{W}_{A_2} = -\mathcal{M}_{A_2}(\dot{\zeta}_D - \dot{\zeta}_R), \tag{7.27}$$

where

$$\mathcal{M}_{A_2} = 2\mathcal{M}_A,\ \dot{\zeta}_D = k_D(T)C_{A_2}C,\ \dot{\zeta}_R = k_R(T)(C_A)^2C,\ C = \sum_j C_j. \tag{7.28}$$

Let us set

$$dY_j = \nu_j \mathcal{M}_j d\xi. \tag{7.29}$$

The result is

$$\rho \frac{d\xi}{dt} = \dot{\zeta}_D - \dot{\zeta}_R. \tag{7.30}$$

The set of equations to be solved then becomes

$$\nabla \cdot (\rho \mathbf{v}) = 0, \tag{7.31}$$

$$\rho \frac{ds}{dt} - A \frac{d\xi}{dt} = 0, \tag{7.32}$$

$$\rho \frac{d\xi}{dt} = \dot{\zeta}_D - \dot{\zeta}_R, \tag{7.33}$$

$$h + v^2/2 = h_0, \tag{7.34}$$

$$A = -\sum_j \nu_j \mathcal{M}_j g_j. \tag{7.35}$$

We obviously need to take into account the equations of state

$$p = p(T,\, \rho,\, \xi),\ h = h(T,\, \rho,\, \xi),\ s = s(T,\, \rho,\, \xi). \tag{7.36}$$

The simplest case is that of flow in planar sections (quasi-one-dimensional flow) [31, 217]. It is assumed that the parameters are uniform in each cross-section of the stream tube studied. This leaves only one variable: the coordinate x along the average streamline. If $\Sigma(x)$ is the cross-sectional area, we obtain the following system:

$$\begin{cases} \rho v \Sigma(x) = \dot{m}, \\ \rho\, ds/dx - A\, d\xi/dx = 0, \\ \rho v\, d\xi/dx = \dot{\zeta}_D - \dot{\zeta}_R = \dot{\zeta}(T,\, \rho,\, \xi), \\ h + v^2/2 = h_0, \\ p = p(T,\, \rho,\, \xi), \\ h = h(T,\, \rho,\, \xi), \\ s = s(T,\, \rho,\, \xi). \end{cases} \tag{7.37}$$

This system can be solved numerically, and the solution highlights the coupling between the flow and the chemistry present.

Subsonic–Supersonic Flow in a Nozzle

In a rocket propulsion nozzle, the flow is generally initially subsonic and then supersonic (see Fig. 7.3).

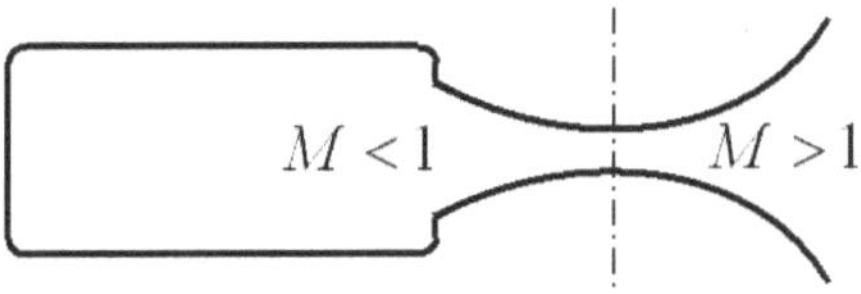

Fig. 7.3. A de Laval rocket nozzle

The temperature, pressure and concentration Y_A are decreasing functions of the abscissa x, whereas the concentration of A_2 increases because of the recombination process that occurs when the temperature decreases. The chemical production rates $\dot{\zeta}_D$ and $\dot{\zeta}_R$ take the forms shown in Fig. 7.4.

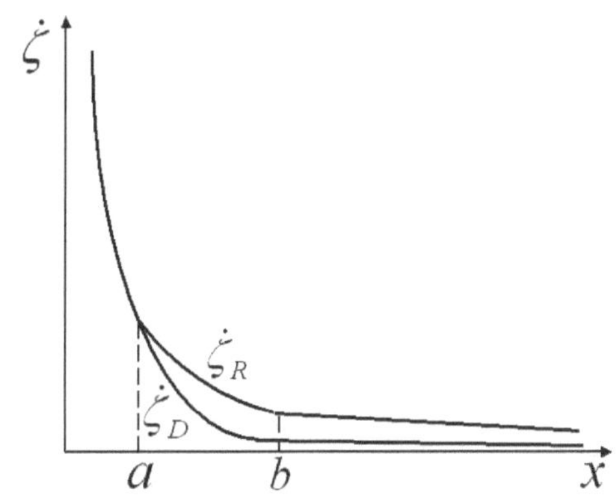

Fig. 7.4. Chemical production rates in a rocket nozzle

- For $x < a$, chemical equilibrium is obtained and $A \cong 0$. The flow is relatively slow, and at high temperature the gradients are weak: $ds/dx \cong 0$.
- For $a < x < b$, the evolution is clearly out of equilibrium, but the temperature is still high enough for the reactions to unfold, although they are slower than for $x < a$, $ds/dx > 0$.
- For $x > b$, the reaction rates are weak, but the tendency of the species A_2 to recombine remains stronger than its tendency to dissociate. We have $\dot{\zeta}_D/\dot{\zeta}_R \cong 0$ and $\rho v \, d\xi/dx \cong -\dot{\zeta}_R$: very weak. The flow is frozen and $ds/dx \cong 0$.

In all cases, the entropy production rate (Fig. 7.5) is equal to

$$\dot{W}_S = \dot{\zeta}_R A(e^{A/RT} - 1)/T. \tag{7.38}$$

(See Sect. 3.4.)

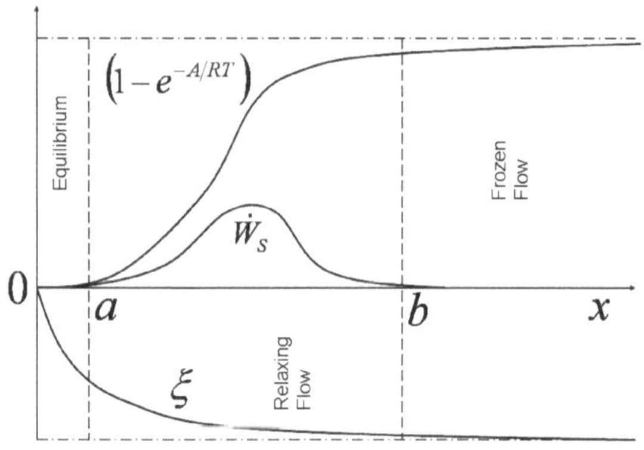

Fig. 7.5. Entropy production rate

In some cases, when the irreversible zone ab is confined, we can consider the evolution to take the form of a flow at chemical equilibrium and then a frozen flow. There is abrupt freezing at point a, where the rate $\dot{W}_S$ can be considered a Dirac peak: $\dot{W}_S = \dot{W}_{Sa}\delta(x-a)$. Chemical irreversibility (chemical relaxation) limits the recombination of the species A in the flow; since the reaction is exothermic, the energy released by recombination is thus smaller than at equilibrium. Therefore, the thrust of the rocket engine decreases due to relaxation. In order to evaluate the importance of the chemical irreversibility in advance, we make use of the first Damköhler parameter before we numerically resolve this system (see Sect. 3.1.1).

Near-Equilibrium Transonic Flow in a Nozzle

Singularities are often present in the throat zones of subsonic–supersonic nozzle flows.

In the case of a nonreactive flow, it is easy to see that the throat itself is a singular point, and that the condition that continuous flow must not present a velocity gradient of zero results from the fact that the flow velocity is equal to the characteristic speed there, $c = (\partial p/\partial \rho)_s^{1/2}$ (speed of sound, see Chaps. 2 and 10). System (7.37) then becomes (with $\vartheta = 1/\rho$)

$$v\Sigma = \dot{m}\vartheta, \; s = s_0, \; h + v^2/2 = h_0.$$

Assuming an ideal gas, we obtain

$$(\frac{\vartheta}{\vartheta^*})^2 = (\frac{\Sigma}{\Sigma^*})^2[\frac{\gamma+1}{\gamma-1} - \frac{2}{\gamma-1}(\frac{\vartheta}{\vartheta^*})^{1-\gamma}],$$

where the superscript (*) indicates critical conditions, and γ denotes the (presumably constant) isentropic coefficient.

The mass flow rate then reaches a critical value that cannot be exceeded in the steady regime, whatever the exit conditions (i.e., stagnation conditions).

We now use the subscript ($_0$) to denote the parameters of a basic uniform flow, which is assumed to be very similar to a real transonic flow. We divide the quantities relating to the real flow by the corresponding quantities relating to the uniform basic flow (ϑ_0, $v_0 = c_0$, s_0, Σ_0, $\dot{m}_0$) to obtain dimensionless quantities. If the notation used for the dimensionless parameters is not changed, the equations for the latter system remain the same. Using the small parameter ϵ, which characterizes the order of magnitude of the variation in nozzle cross-section in the vicinity of the throat, and n, a strictly positive number that needs to be determined, the expansions of the various parameters are then

$$\Sigma = 1 + \epsilon\Sigma_1, \; \dot{m} = 1 + \epsilon^n \dot{m}_1 + \epsilon^{2n}\dot{m}_2 + \ldots, \; s = 1 + \epsilon^n s_1 + \epsilon^{2n} s_2 + \ldots,$$

$$\vartheta = 1 + \epsilon^n \vartheta_1 + \epsilon^{2n}\vartheta_2 + \ldots, \; v = 1 + \epsilon^n v_1 + \epsilon^{2n} v_2 + \ldots$$

It can be shown that the only interesting choice is $n = 1/2$. By assuming a sonic basic flow ($v_0 = c_0^*$), we then obtain

$$\frac{\gamma+1}{2}v_1^2 - \Sigma_1 + \dot{m}_2 = 0, \; \vartheta_1 = v_1, \; s_1 = 0, \; \dot{m}_1 = 0. \tag{7.39}$$

The nozzle profile is given by $\Sigma_1 = \Sigma_1^* + ax^2$, meaning that $(\gamma+1)v_1^2/2 - ax^2 = \Sigma_1^* - \dot{m}_2$. In this case, the solutions form a family of hyperbolae. The critical mass flow rate is that which cancels out the right hand side; this corresponds to continuous subsonic–supersonic (or supersonic–subsonic) evolution. If the mass flow rate is lower than the critical mass flow rate, continuous evolution still occurs, but the regime (subsonic or supersonic) does not change (see Fig. 7.6).

In reactive flow, the singular point moves downstream from the throat. It is possible to study the throat zone analytically using the method of small singular disturbances if the cross-section of the nozzle does not vary too quickly and if the mixture is close to chemical equilibrium [218]. We will also assume one condition relating to the partial second derivatives; this condition results in the fact that the "frozen" speed of sound and the "equilibrium" speed of sound are very close to each other: $c_f^2/c_e^2 - 1 = e_{\vartheta\xi}^2/(e_{\vartheta\vartheta}e_{\xi\xi} - e_{\vartheta\xi}^2) \ll 1$ (see Sect. 2.5.3). Let us now indicate by the subscript ($_0$) the parameters of a basic uniform flow at chemical equilibrium that is very similar to a real transonic flow. We divide the quantities relating to the real flow by these quantities: ϑ_0, $v_0 = c_{f0}$, s_0, Σ_0, $\dot{m}_0$. We introduce the reference quantity

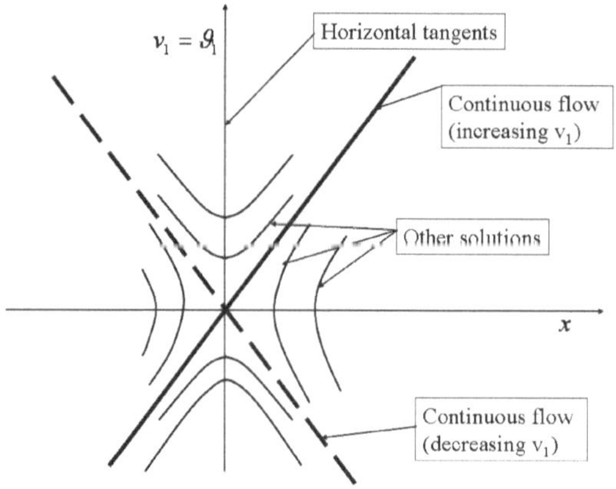

Fig. 7.6. Transonic behavior of a simple compressible fluid in a nozzle

$\xi_r = -(\partial\xi/\partial\vartheta)^0_{sA}$ as a progress variable. We also introduce the chemical time $\tau_c = (\tau_{s\vartheta})_0$ and the ratio $\theta = \tau_r/\tau_c$, where τ_r is a reference time, a reference length $l_r = v_0\tau_r$, and a chemical length $l_c = v_0\tau_c$. Without changing notation, the dimensionless system (7.37) becomes

$$v\Sigma = \dot{m}\vartheta, \;\; Tds/dx - Ad\xi/dx = 0, \;\; vd\xi/dx = \theta LA, \;\; h + v^2/2 = h_0.$$

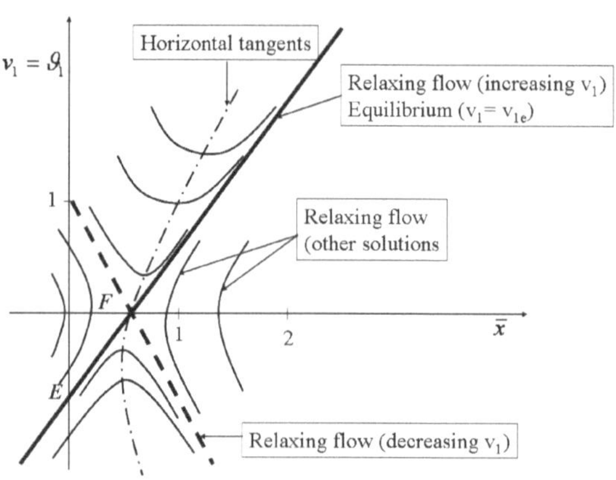

Fig. 7.7. Transonic behavior of a chemically reactive compressible flow in a nozzle: velocity

The phenomenological coefficient L results from the linearization of the chemical production rate in the vicinity of chemical equilibrium. The expansions of the various parameters considered are obtained as they were previously; for the additional parameters, $\xi = \xi_0 + \epsilon^n \xi_1 + \epsilon^{2n}\xi_2 + \ldots$, $A = \epsilon^n A_1 + \epsilon^{2n} A_2 + \ldots$.

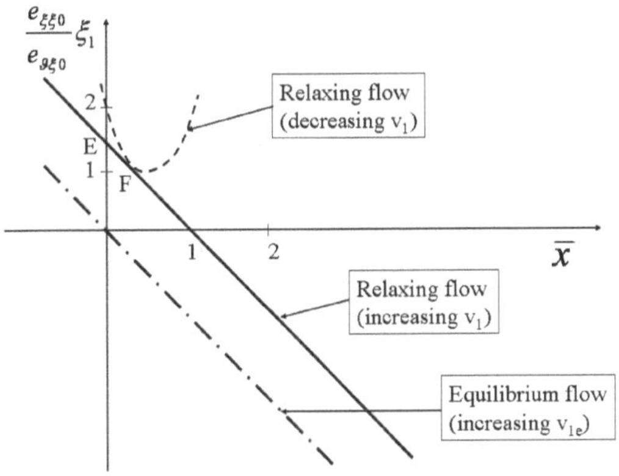

Fig. 7.8. Transonic behavior of a chemically reactive compressible flow in a nozzle: chemical progress variable

The other thermodynamic parameters yield expansions similar to those obtained previously. Then, knowing the fundamental energy law $e = e(s, \vartheta, \xi)$, we can deduce $A_1 = -e_{\xi s0}s_1 - e_{\xi\vartheta 0}\vartheta_1 - e_{\xi\xi 0}\xi_1$, $A_2 = -e_{\xi s0}s_2 - e_{\xi\vartheta 0}\vartheta_2 - e_{\xi\xi 0}\xi_2 + \frac{A_{ss0}}{2}s_1^2 + \frac{A_{\vartheta\vartheta 0}}{2}\vartheta_1^2 + \frac{A_{\xi\xi 0}}{2}\xi_1^2 + A_{s\vartheta 0}s_1\vartheta_1 + A_{s\xi 0}s_1\xi_1 + A_{\vartheta\xi}\vartheta_1\xi_1$.

We will characterize the ratio θ by a number δ such that $\theta = \tau_r/\tau_c = \epsilon^{\delta n}$, and we also introduce an arbitrary number α that makes it possible to change the x-coordinate scale at will. Thus, $\bar{x} = \epsilon^{\alpha n}x = \epsilon^{\alpha n}x^+/l_r = \epsilon^{(\delta-\alpha)n}x^+/l_c$, where x^+ indicates the natural abscissa. We will not write the system to be solved in the general case; we will restrict ourselves to presenting the results obtained. Applying the least degeneracy principle leads us to choose $n = 1/2$, $\delta = \alpha$, and so: $\bar{x} = x^+/l_c$. The system to be solved is then:

$$Z\xi_1 = \frac{\Gamma+1}{2}v_1^2 - \Sigma_1 + \dot{m}_2, \quad \frac{d\xi_1}{d\bar{x}} + \xi_1 + \vartheta_1 = 0, \quad \vartheta_1 = v_1, \quad s_1 = 0, \tag{7.40}$$

where $\Gamma = h_{\vartheta\vartheta 0}$, $c_{f0}^2/c_{e0}^2 = Z\epsilon^{1/2}$, $v_0 = c_{f0} = \vartheta_0 = 1$, and $\tau_c = (\tau_{s\vartheta})_0$.

The preceding system can be solved. By eliminating ξ_1, we obtain a nonlinear differential equation in v_1. The right hand side of this equation depends on the throat profile and mass flow rate data. Figures 7.7 and 7.8 [218] provide examples of solutions obtained with a parabolic cross-section and for fixed values of $\dot{m}_2$ and $\dot{m}_{2e}$: $\dot{m}_2 = \Sigma_1^* + 1.25$, $\dot{m}_{2e} = \Sigma_1^* + 0.25$.

7.3 Thermal Transfer and Mass Diffusion

Aside from couplings between transfer phenomena, inertial terms or external forces, intrinsic couplings can occur between transfer phenomena. For example, we have seen that there are intrinsic relations between generalized fluxes and generalized forces of the same tensorial order. This occurs in particular for species diffusion and thermal transfer [108]. We will initially study this phenomenon in a continuous medium.

Consider a continuous medium in which the only irreversible processes are thermal transfer and species diffusion. The entropy production rate is

$$\dot{W}_S = \mathbf{q} \cdot \boldsymbol{\nabla}(\frac{1}{T}) - \sum_j \boldsymbol{\mathcal{J}}_{Dj} \cdot \boldsymbol{\nabla}(\frac{g_j}{T}) \tag{7.41}$$

(the forces $\mathbf{f}_j$ are neglected here, which is a valid approach when only gravity has an effect, for example, because $\mathbf{f}_j = \mathbf{g}$, $\sum_j \boldsymbol{\mathcal{J}}_{Dj} \cdot \mathbf{f}_j = (\sum_j \boldsymbol{\mathcal{J}}_{Dj}) \cdot \mathbf{g} = 0$). We rearrange the right hand side of (7.41) to reveal the heating effects present in the term $\boldsymbol{\nabla}(g_j/T)$. We know that $g_j = \mu_j/\mathcal{M}_j$. For one mole of component j in the mixture, we have

$$d\mu_j = -\bar{S}_j dT + \bar{\mathcal{V}}_j dp + \sum_j (\partial \mu_j/\partial n_j)_{T,p} dn_j \tag{7.42}$$

or

$$d\mu_j = -\bar{S}_j dT + (d\mu_j)_T. \tag{7.43}$$

We then have

$$d(\frac{\mu_j}{T}) = \frac{1}{T}(d\mu_j)_T - \frac{\mu_j + T\bar{S}_j}{T^2} dT. \tag{7.44}$$

The partial molar quantities $\bar{S}_j$ and $\bar{H}_j$ conform to the relation (see Chap. 2)

$$\bar{H}_j = \mu_j + T\bar{S}_j. \tag{7.45}$$

It follows that

$$d(\frac{\mu_j}{T}) = \frac{1}{T}(d\mu_j)_T - \frac{\bar{H}_j}{T^2} dT \tag{7.46}$$

or, by introducing the partial enthalpy per unit mass $\bar{h}_j = \bar{H}_j/\mathcal{M}_j$,

$$d(\frac{g_j}{T}) = \frac{1}{T}(dg_j)_T - \frac{\bar{h}_j}{T^2} dT. \tag{7.47}$$

Therefore, the entropy production rate becomes

$$\dot{W}_S = \mathbf{q}' \cdot \boldsymbol{\nabla}(\frac{1}{T}) - \sum_j \boldsymbol{\mathcal{J}}_{Dj} \cdot (\boldsymbol{\nabla}(\frac{g_j}{T}))_T \tag{7.48}$$

where

$$\mathbf{q}' = \mathbf{q} - \sum_j \bar{h}_j \boldsymbol{\mathcal{J}}_{Dj}. \tag{7.49}$$

Let us now consider a binary mixture. We assume that the barycentric velocity is equal to zero and that the pressure is uniform. It follows that $(\boldsymbol{\nabla} g_j)_T = (\boldsymbol{\nabla} g_j)_{T,p}$ in this case. The Gibbs–Duhem equation leads to

$$Y_1(dg_1)_{T,p} + Y_2(dg_2)_{T,p} = 0. \tag{7.50}$$

Given that $Y_1 + Y_2 = 1$ and $\boldsymbol{\mathcal{J}}_{D1} + \boldsymbol{\mathcal{J}}_{D2} = 0$, we find that

$$\boldsymbol{\mathcal{J}}_{D1} \cdot (\boldsymbol{\nabla}(g_1/T))_T + \boldsymbol{\mathcal{J}}_{D2} \cdot (\boldsymbol{\nabla}(g_2/T))_T = (\boldsymbol{\nabla}(g_1)_{p,T}/T\,Y_2) \cdot \boldsymbol{\mathcal{J}}_{D1}. \tag{7.51}$$

There is only one independent concentration variable here, so

$$(dg_1)_{T,p} = g_{11}\, dY_1 \tag{7.52}$$

We then obtain

$$\dot{W}_S = -\mathbf{q}' \cdot \frac{\boldsymbol{\nabla}(T)}{T^2} - \boldsymbol{\mathcal{J}}_{D1} \cdot (\frac{g_{11}}{T\,Y_2} \boldsymbol{\nabla} Y_1). \tag{7.53}$$

Thus, the phenomenological relations are

$$\begin{cases} \boldsymbol{\mathcal{J}}_{D1} = -L_{11}(g_{11}/T\,Y_2)\nabla Y_1 - L_{12}\boldsymbol{\nabla} T/T^2 \\ \mathbf{q}' = -L_{21}(g_{11}/T\,Y_2)\nabla Y_1 - L_{22}\boldsymbol{\nabla} T/T^2, \end{cases} \tag{7.54}$$

and we have Onsager's symmetry relation $L_{21} = L_{12}$ and the conditions

$$L_{11} > 0,\; L_{22} > 0 \text{and} L_{11}L_{22} - L_{12}^2 > 0. \tag{7.55}$$

Let us now set

$$\begin{cases} L_{11} = \rho Y_2 T D/g_{11}, \\ L_{22} = \lambda T^2, \\ L_{21} = L_{12} = \rho Y_1 Y_2 T^2 D_T. \end{cases} \tag{7.56}$$

We are therefore highlighting the coefficients of thermal conductivity λ, species diffusion D, and thermal diffusion D_T. The expressions for $\mathbf{q}'$ and $\boldsymbol{\mathcal{J}}_{D1}$ reveal the coupling via the coefficient D_T:

$$\begin{cases} \boldsymbol{\mathcal{J}}_{D1} = -\rho D \nabla Y_1 - \rho Y_1 Y_2 D_T \boldsymbol{\nabla} T, \\ \mathbf{q}' = -\lambda \boldsymbol{\nabla}(T) - \rho Y_1 g_{11} T D_T \nabla Y_1. \end{cases} \tag{7.57}$$

A heat flux due to a concentration gradient, also know as the Dufour effect, occurs in the first equation. The thermal diffusion process occurs because a variation in temperature gives rise to a diffusive flux (the Soret effect). These coupling effects are not always negligible. These results are generalized to the case of N species with some restrictive assumptions in the Appendix (see Sect. A.6.3).

7.4 Shvab–Zel'dovich Approximation

The Shvab–Zel'dovich approximation leads to a simplified form of the balance equations for the flow [290]. It is used to solve many problems, some of which are treated in this work, such as the one-dimensional propagation of a planar premixed flame (Sect. 10.5), a boundary layer with chemical reactions above a planar plate (Emmons problem: Sect. 9.4.2), and the combustion of a droplet (Sect. 12.4.3). Within the framework of this approximation, the following assumptions are permitted:

- There is steady flow in a suitably chosen reference frame
- There is no thermal diffusion; $D_T = 0$
- External forces are negligible
- Viscosity is negligible
- To a first approximation, the static pressure is constant
- Fourier's law holds for thermal conduction
- Fick's law holds for diffusion, and there is a single diffusion coefficient for all species
- The Lewis number is close to unity
- There is only one chemical reaction
- A mixture of N perfect gases is present.

Mass conservation is then

$$\nabla \cdot (\rho \mathbf{v}) = 0. \tag{7.58}$$

The species chemical balance is given by

$$\nabla \cdot (\rho_j \mathbf{v}_j) = \dot{W}_j, \quad j = 1, \ \ldots, \ N. \tag{7.59}$$

Given that

$$\dot{W}_j = \nu_j \mathcal{M}_j \dot{\zeta}, \tag{7.60}$$

and by setting

$$\beta_j = Y_j / \nu_j \mathcal{M}_j, \tag{7.61}$$

we obtain

$$\nabla \cdot (\rho \mathbf{v} \beta_j - \rho D \nabla \beta_j) = \dot{\zeta}. \tag{7.62}$$

The energy balance is

$$\nabla \cdot (\rho \mathbf{v} e + \mathbf{q}) = -p \nabla \cdot \mathbf{v}. \tag{7.63}$$

Noting the assumptions listed above, we have

$$\mathbf{q} = -\lambda \nabla T + \sum_j \bar{h}_j \boldsymbol{\mathcal{J}}_{Dj},$$

and so

$$\nabla \cdot (\sum_j \rho_j \mathbf{v}_j \bar{h}_j - \lambda \nabla T) = 0. \tag{7.64}$$

We have (see Sect. 2.4.1)

$$\bar{h}_j = (q_f^0)_j + \int_{T^0}^{T} c_{p,\,j} dT, \tag{7.65}$$

so that

$$\sum_j \nabla \cdot (\rho_j \mathbf{v}_j)(q_f^0)_j + \nabla \cdot (\sum_j \rho_j \mathbf{v}_j \int_{T^0}^{T} c_{p,\,j} dT - \lambda \nabla T) = 0 \tag{7.66}$$

or

$$\begin{aligned}&(\textstyle\sum_j \nu_j \mathcal{M}_j (q_f^0)_j)\dot{\zeta} + \nabla \cdot (\rho \mathbf{v} \sum_j Y_j \int_{T^0}^{T} c_{p,\,j} dT \\ &-\rho D \textstyle\sum_j \nabla Y_j \int_{T^0}^{T} c_{p,\,j} dT - \lambda \nabla T) = 0.\end{aligned} \tag{7.67}$$

Upon noting that

$$\sum_j \nu_j \mathcal{M}_j (q_f^0)_j = -\Delta H, \tag{7.68}$$

the enthalpy of the reaction defined by (2.114), and setting

$$\beta_T = \sum_j Y_j \int_{T^0}^{T} c_{p,\,j} dT / \Delta H, \tag{7.69}$$

we obtain

$$\nabla \cdot (\rho \mathbf{v} \beta_T - \rho D \sum_j \nabla Y_j \int_{T^0}^{T} c_{p,\,j} dT / \Delta H - \lambda \nabla T / \Delta H) = \dot{\zeta}. \tag{7.70}$$

Differentiating β_T yields

$$\nabla T/\Delta H = (\nabla \beta_T - \sum_j \nabla Y_j \int_{T^0}^{T} c_{p,\,j} dT/\Delta H)/c_{p,\,f} \tag{7.71}$$

where

$$c_{p,\,f} = \sum_j Y_j c_{p,\,j}. \tag{7.72}$$

Finally,

$$\nabla \cdot (\rho \mathbf{v} \beta_T - \lambda \nabla \beta_T / c_{p,\,f} + \rho D(Le - 1) \sum_j \nabla Y_j \int_{T^0}^{T} c_{p,\,j} dT/\Delta H) = \dot{\zeta}. \tag{7.73}$$

The assumption

$$Le = \lambda/\rho D c_{p,\,f} \cong 1 \tag{7.74}$$

leads to the final equation

$$\nabla \cdot (\rho \mathbf{v} \beta_T - \lambda / c_{p,\,f} \nabla \beta_T) = \dot{\zeta}, \tag{7.75}$$

which is identical in form to the species balance equation (7.62) (note that we obtain the same result if $Le \neq 1$ but $c_{p,\,j}$ is independent of j). $N+1$ equations with the right hand side obtained in this way can therefore be replaced with N equations with zero right hand sides in $(\beta_i - \beta_j)$ or $(\beta_j - \beta_T)$. There is then only one equation with a right hand side of $\dot{\zeta}$, which is the most difficult to solve considering the nonlinear character of this function of β_j and β_T.

7.5 Phase Change of a Pure Constituent in a Gaseous Mixture

Let us now consider a surface of discontinuity for a phase change between a pure substance H and its vapor in a gaseous mixture. The general balance equations at an interface were established in Sect. 4.9.

The following assumptions are allowed:

- There is no thermal diffusion; $D_T = 0$
- Viscosity is negligible
- Fourier's law holds for thermal conduction
- Fick's law holds for diffusion, and there is a single diffusion coefficient for all species
- A mixture of perfect gases is present
- There is only one species (H) in the condensed phase

- The density of the gaseous mixture is smaller than that of the condensed phase
- There are no vapor recoil phenomena,[3] so relative velocity squared terms can be neglected
- There is no surface tension.

Let us define the relative velocity component normal to the interface as $u = (\mathbf{v} - \mathbf{W}) \cdot \mathbf{N}$. The balance equations at the discontinuity can then be written successively for the total mass, species, momentum and energy as

$$\begin{cases} [\rho u]_-^+ = 0, \text{ or } \rho^+ u^+ = \rho^- u^- = \dot{m}, \\ [\mathcal{J}_{Dj\perp} + \dot{m} Y_j]_-^+ = 0, \\ [p + \dot{m} u]_-^+ = 0, \\ \left[q_\perp + v_\perp p + \dot{m}(e + v^2/2)\right]_-^+ = 0. \end{cases} \tag{7.76}$$

The momentum equation can be written

$$[p]_-^+ = -\dot{m}[u]_-^+ = -[\rho u^2]_-^+ \cong 0.$$

The pressure does not vary across the interface: $p^+ = p^-$. On the other hand, the energy equation gives

$$[q_\perp + \dot{m}(h + v^2/2) + wp]_-^+ = 0,$$

and we have the identity $v^2/2 \equiv (w^2 + u^2)/2 + wu$, which is valid for $[\ ]_-^+$. Therefore, the energy equation can be rewritten as

$$[q_\perp + \dot{m}(h + u^2/2)]_-^+ = 0.$$

Ignoring terms in $u^2/2$, we obtain

$$[q_\perp + \dot{m} h]_-^+ = 0.$$

From (7.49), we can deduce that

$$q_\perp = -\lambda \frac{\partial T}{\partial N} + \sum_j \mathcal{J}_{Dj\perp} h_j.$$

[3]Vapor recoil results from the presence of velocity squared terms in the momentum equation. If such terms are non-negligible, a pressure jump occurs across the evaporation surface, even for a flat interface without mass, and the pressure is stronger on the side with the greater mass density (see [194, 94]). This phenomenon can occur when the pressure is much smaller than the saturating vapor pressure, leading to large evaporation rates. It can also occur near the saturation vapor pressure in slightly subcritical fluids.

Finally, the balance equations at the interface are

$$\begin{cases} \rho^+ u^+ = \rho^- u^- - \dot{m}, \\ \mathcal{J}^+_{Dj\perp} = \dot{m}(Y_j^- - Y_j^+), \\ p^+ = p^-, \\ -\lambda^+(\frac{\partial T}{\partial N})^+ + \lambda^-(\frac{\partial T}{\partial N})^- = \dot{m}(h_H^- - h_H^+) = -\dot{m}l, \end{cases} \tag{7.77}$$

where l is the latent heat of the phase change per unit mass of species H.

7.6 Thermal Osmosis: Minimum Entropy Production

Let us return to the example of Sect. 4.7 for the case of a mixture and a discrete system (see Fig. 7.9). The fluxes through the porous wall into subsystem 1 are the mass flux $\dot{m}_j$ of the species j, the internal energy flux q, and the entropy flux $\delta_e\dot{S} = (q - \sum_j g_j \dot{m}_j)/T$. By setting $\Delta\varphi = \varphi_2 - \varphi_1$, the entropy production rate at the level of the porous wall for an unspecified parameter φ becomes

$$\delta_i\dot{S} = q\,\Delta T/T^2 + \sum_j \dot{m}_j\,\Delta(g_j/T), \tag{7.78}$$

where we assumed that the variations Δ are relatively small.

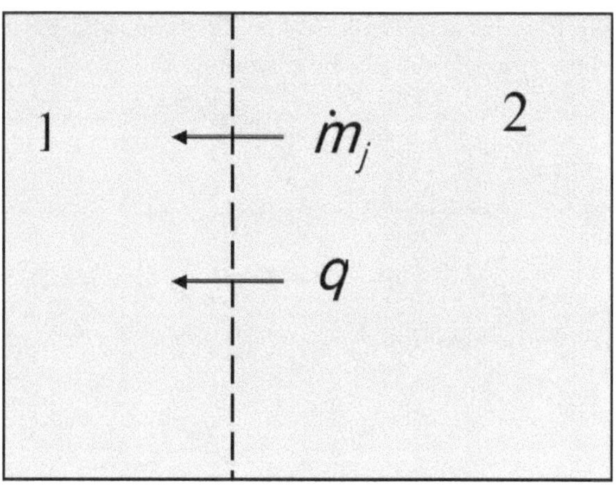

Fig. 7.9. Thermal osmosis

Just as we did previously (see Eq. 7.47), we can write

$$\Delta(\frac{g_j}{T}) = \Delta(\frac{g_j}{T})_T - \frac{\bar{h}_j}{T^2}\Delta T, \tag{7.79}$$

so that

$$\delta_i \dot{S} = q' \, \Delta T/T^2 + \sum_j \dot{m}_j (\Delta(g_j/T))_T \tag{7.80}$$

with

$$q' = q - \sum_j \bar{h}_j \dot{m}_j. \tag{7.81}$$

For a binary mixture, and assuming that a device that equalizes pressures p_1 and p_2 is present, we also obtain the following from (7.57):

$$\delta_i \dot{S} = q' \, \Delta T/T^2 + (g_{11}/TY_2) \, \dot{m}_1 \Delta Y_1. \tag{7.82}$$

By setting

$$q' = J_{th}, \; \Delta T/T^2 = F_{th}, \; \dot{m}_1 = J_m, \; (g_{11}/TY_2) \, \dot{m}_1 \Delta Y_1 = F_m, \tag{7.83}$$

we obtain the phenomenological relations

$$\begin{cases} J_m = L_{11} \, F_m + L_{12} \, F_{th}, \\ J_{th} = L_{21} \, F_m + L_{22} \, F_{th}, \end{cases} \tag{7.84}$$

where

$$\begin{cases} L_{12} = L_{21}, \\ L_{11} > 0, \; L_{22} > 0, \\ L_{11} \, L_{22} - L_{12}^2 > 0. \end{cases} \tag{7.85}$$

The entropy production rate then becomes

$$\delta_i \dot{S} = F_{th} \, J_{th} + F_m \, J_m = L_{11} F_m^2 + 2 L_{12} F_m F_{th} + L_{22} F_{th}^2. \tag{7.86}$$

The coupling coefficient L_{12} leads, for a given $F_{th} = \Delta T/T^2$, to a species flux even when $\Delta Y_1 = 0$. This is known as thermal osmosis.

The steady state (out of equilibrium, with a given F_{th}) corresponds to $\dot{m}_1 = J_m = 0$. We will now show that, if the coefficients L_{ij} are constant, this steady state corresponds to a minimum in the entropy production per unit time. If T_1 and T_2 are maintained constant, it is enough to consider two evolutions that are close to one another, with differences that are characterized by δJ_{th}, δJ_m, δF_{th}, δF_m. In this case, we obtain

$$\delta(\delta_i \dot{S}) = 2 L_{12} F_{th} \, \delta F_m + 2 L_{11} F_m \, \delta F_m. \tag{7.87}$$

Thus, $\delta(\delta_i \dot{S})/\delta F_m = 0$ in the steady state characterized by $J_m = 0$. We also find that $\delta^2(\delta_i \dot{S})/(\delta F_m)^2 = 2 L_{11} > 0$. It is thus a minimum (see Fig. 7.10).

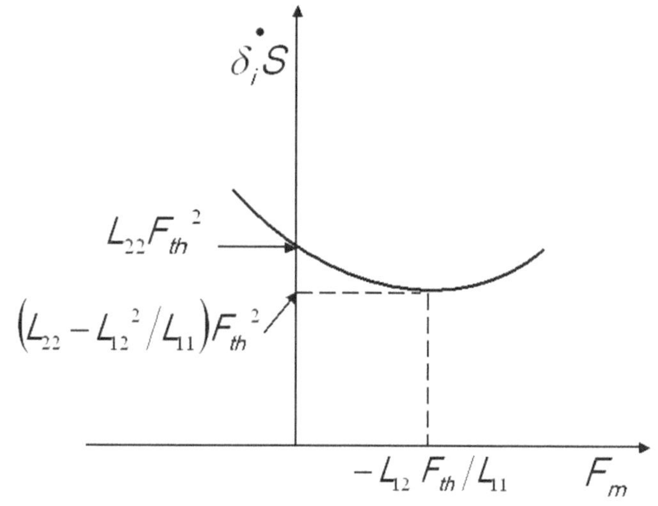

Fig. 7.10. Minimum entropy production

This is known as Prigogine's theorem [213], which is applicable to coupled systems with constant phenomenological coefficients.[4]

Note: Thermal osmosis leads to concentration changes due to temperature differences. It can therefore be used as separation process in chemical engineering. Another process is based on a difference in pressure across a porous wall rather than a difference in temperature. To illustrate this, assume that the pressure p is relatively high (a few atmospheres) on the left hand side of the wall, and so thermal equilibrium is ensured by Brownian motion (see Fig. 7.11).

For each gas species,

$$\frac{1}{2}M_1\bar{c}_1^2 = \frac{3}{2}kT,\ \frac{1}{2}M_2\bar{c}_2^2 = \frac{3}{2}kT, \tag{7.88}$$

so that

$$\bar{c}_1/\bar{c}_2 = \sqrt{M_2/M_1}. \tag{7.89}$$

There is a vacuum on the right hand side of the wall, so Brownian collisions do not dominate the molecular motions, which are characterized by a mean velocity $\bar{U}'$. The conservation of species fluxes yields

---

[4]In the presence of only one gas species, a jump in temperature across a wall produces a difference in pressure. This is another coupled effect, and is called thermomolecular pressure. It is similar to thermal osmosis because it depends on three independent phenomenological coefficients. Prigogine's theorem is also applicable here.

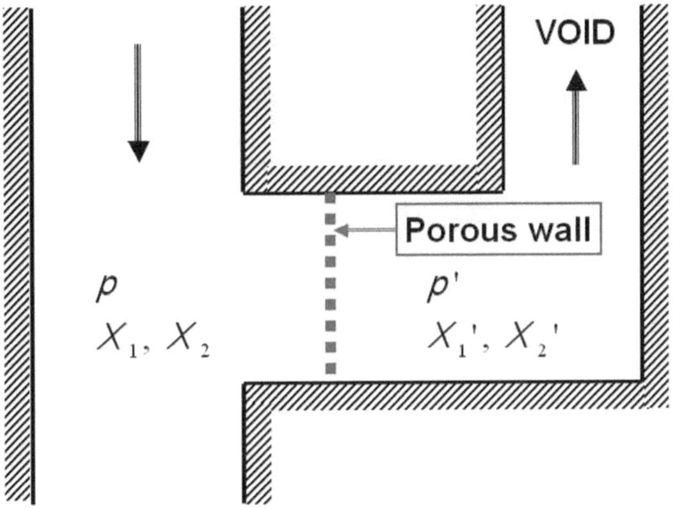

Fig. 7.11. Separation process in a rarefied medium

$$N_1\bar{c}_1 = N_1'\bar{U'},\ N_2\bar{c}_2 = N_2'\bar{U'}. \tag{7.90}$$

Eliminating $\bar{c}_1$, $\bar{c}_2$ and $\bar{U'}$ between these relations gives

$$N_1'/N_2' = \sqrt{M_2/M_1}\ N_1/N_2 \tag{7.91}$$

or, with the usual notation,

$$X_1'/X_2' = \sqrt{\mathcal{M}_1/\mathcal{M}_2}\ X_1/X_2. \tag{7.92}$$

Thus, when the two species cross the porous wall, the lighter species will be enriched compared to the other species (i.e., this phenomenon can be used to separate out species). Note that separation is not possible if the molar masses of thee species are equal.

7.7 Coupling Between Chemical Kinetics and Dissipative Flow: Laminar Flames

In this section we examine phenomena where the chemical kinetics is coupled with the flow, and dissipative processes occur (in contrast to the case treated in Sect. 7.2). Flames are particular examples of exothermal chemical reactions where the pressure is nearly uniform. The flame is localized in a relatively thin zone (the higher the activation energy of the reaction, the thinner the flame).

The case where the fuel and the oxidizer are initially premixed should be distinguished from the case where the combustible and the oxidizer are initially separated. The first case is termed premixed combustion, and the second case is known as a diffusion flame.

Several examples of these types of combustion are treated in this book. Premixed flames are examined in Chap. 10 (Sects. 10.4 and 10.5), and several diffusion flame examples are given in Chaps. 9 and 12 (see Sects. 9.4.2 and 12.4.3 for example). We can also get triple flames, which are a combination of diffusion flames and premixed flames [183].

Laminar *premixed flames* result from interactions between chemical processes and the phenomena of mass diffusion and thermal diffusion. For deflagration waves with high activation energies, we find in Sect. 10.5 that the burning velocity is an eigenvalue of the problem involving interactions between diffusion processes and chemical kinetics.

Diffusion flames develop at the contact surface between fuel and oxidizer. The burning velocity results from the diffusion of the active components at this contact surface. The contact surface between the two components is very important, and the larger surface, the more intense the combustion.

7.7.1 Example of a Laminar Premixed Flame: The Bunsen Thin Flame

The coupled problem of a normal planar premixed flame with a high activation energy is solved in Sect. 10.5, where the mass flow rate and the burning speed s_L in the laminar steady regime are expressed.

Let us now consider a thin flame over a plane 2D burner. We assume that the main part of the flame is a dihedron (or a cone in the case of a cylindrical Bunsen burner), the axis of which is the axis of the burner. The base of the flame is near the lip of the burner and the flame tip is smooth. Upstream, the density and velocity of the unburned fresh fluid are uniform (see Fig. 7.12). The change in velocity upon crossing from one side of the thin flame to the other obeys interfacial condition (4.97). Downstream, the heat released means that the burned gases are hotter $T_b > T_u$ and less dense $\rho_b < \rho_u$. The normal component v_{uN} for the unburned gases is equal to the burning speed s_L^0, which is deduced from the constant coefficient Λ, as defined by (10.166) and calculated by via (10.170). Thus, the normal velocity of the burned gases can be deduced using the gas density ρ_b; we have $v_{bN} = (\rho_u/\rho_b)v_{uN} = (\rho_u/\rho_b)s_L^0$.

For constant, uniform pressure, the momentum equation (4.99) leads to the conservation of the tangential velocity: $v_{bT} = v_{uT}$. Consequently, the flow velocity of the burnt gases is known, except at the flame tip, where the profile of the flame deviates from a planar (or conical) profile, becoming smooth. Along the axis of symmetry, the normal speed is v_u, which is higher than the previous burning speed $v_{uN} = s_L$. The increase in the normal velocity from s_L to v_u is due to the curvature of the front, which stretches the flame (as described in Chap. 11); note that linear relations such as (10.177) are not valid for high values of curvature (see [207, 206] for the solution to this complex problem).

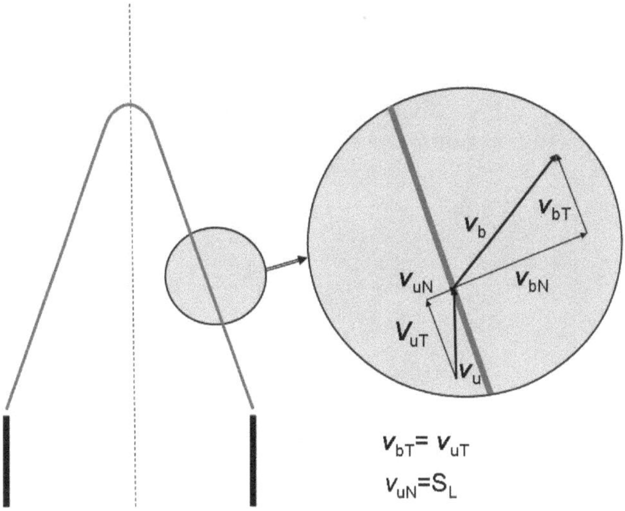

Fig. 7.12. Velocities for a steady oblique premixed flame

Figure 7.13 shows a real bunsen flame (seeded with oil particles that are vaporized at the diffusion zone boundary; the upper luminous surface corresponds to the reaction zone) sliced by a laser. The increased diffusion thickness is due to the low pressure (3×10^4 Pa).

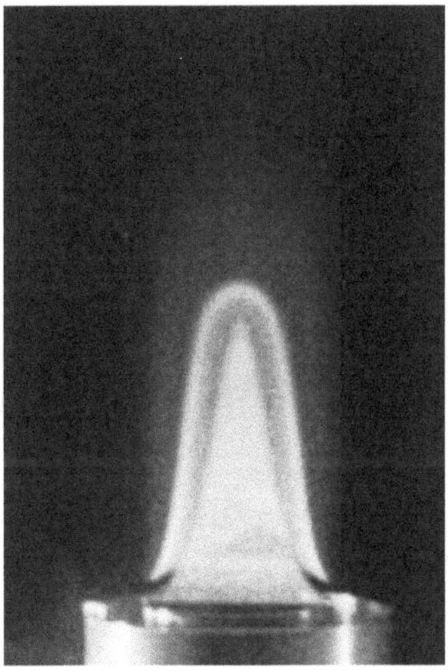

Fig. 7.13. A laser tomogram of a methane–air premixed flame at low pressure (photo from the Laboratoire d'Aérothermique du CNRS, private communication, 1985)

Note that the calculated downstream flow will depend on the physical hypotheses made, such as whether gravitational effects [142] (buoyancy forces) are present. Gravity results in deviations in the streamlines in the burned gases (Fig. 7.14), as well as flame flickering [77] due to Kelvin–Helmholtz instabilities produced by friction between the external air flow and the internal flow of the burned gases due to buoyancy forces. Gravity also acts on air–propane flame stability, as shown in [76].

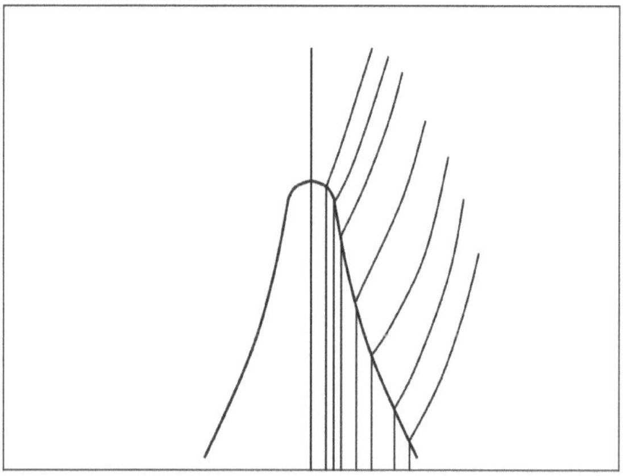

Fig. 7.14. Trajectories of particles through the inner cone of a flame

Indeed, premixed flames are very sensitive to the thermodynamic conditions present. In particular, temperature variations have a significant influence on the reactive zone because of their effects on the exponential factor of the reaction rate. Temperature and concentration fluctuations can induce instabilities, as shown in Sect. 8.2.1.

Other Examples of Laminar Stretched Premixed Flames

Examples of laminar stretched premixed flames are given in [206]. Figure 7.15 shows two of these flames: 1) a planar flame in front of a stagnation plane in the case of two steady flows directed towards each other (fresh gas directed towards combustion products); 2) an unsteady spherical flame (see Sect. 10.7.2 and [128] on cellular instabilities of spherical premixed flames).

The G-Equation

The G-equation introduced by Williams [291] is applicable to thin-flame structures that propagate with a well-defined burning velocity. Consider the

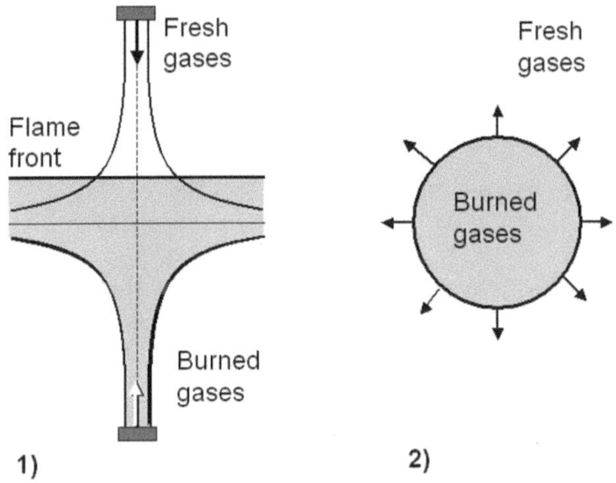

Fig. 7.15. Two examples of laminar stretched premixed flames: **1** a planar steady flame in front of a stagnation plane in the case where a fresh gas flow is directed towards an oncoming flow of combustion products); **2** an unsteady spherical flame

isoscalar surface[5] defined by an implicit equation analogous to (A.1) in the Appendix, with no chemical reactions,

$$\mathcal{F}(\mathbf{x},\, t) = G(\mathbf{x},\, t) - G_0 = 0.$$

The unburnt mixture corresponds to $G(\mathbf{x},\, t) < G_0$ and the burnt gas to $G(\mathbf{x},\, t) > G_0$. The normal to the flame front in the direction of the unburnt mixture is defined as (see Eq. A.2):

$$\mathbf{n} = -\boldsymbol{\nabla} G / \left| \boldsymbol{\nabla} G \right| . \tag{7.93}$$

If $\mathbf{v}_f$ is the flow velocity at the front and s_L is the burning velocity in the normal direction, the flame velocity $\mathbf{w}_f$ is then

$$\mathbf{w}_f = \mathbf{v}_f + s_L \mathbf{n}$$

[5]In combustion, the state of the mixture depends on the velocity vector $\mathbf{v}(\mathbf{x}, \mathrm{t})$ as well as the mass fractions Y_j of the species and temperature T, which are scalars. In a reactive mixture, these scalars obey balance equations that account for chemical production, and they are called reactive scalars. If there is no chemical reaction, or if we consider combinations of scalars that obey balance laws with no chemical production, such as $\beta_i - \beta_j$ or $\beta_T - \beta_j$ in Sect. 7.4, we term them nonreactive or passive scalars.

and so the quantity G obeys the following balance equation:

$$\partial G/\partial t + \mathbf{v}_f \cdot \nabla G = s_L \left|\nabla G\right|, \tag{7.94}$$

which is called the G-equation.

This equation plays an important role in premixed turbulent combustion, just as the "Z-equation" (presented in Sect. 7.7.2) does for nonpremixed turbulent combustion.

7.7.2 An Example of a Laminar Diffusion Flame

A simple example of a diffusion flame is given by the Burke–Shumann [37] problem. This models a diffusion flame that occurs between two coaxial flows with equal velocities, where one is a fuel flow and the other an oxidizer flow (see [142, 290]). The chemical reaction is of the form

$$\nu'_F \, \mathrm{F} + \nu'_O \, \mathrm{O} \rightarrow \nu''_P \, \mathrm{P}$$

The following assumptions are made:

- The Schvab–Zel'dovich approximation is valid (see Sect. 7.4).
- At the level of the outlets of the coaxial ducts, the velocities of the fuel and the oxidizer[6] (air for example) flows are equal (v_z) and uniform. We can change the proportion of fuel and air in the downstream flow by varying the radii of the tubes and the molar fuel ratio.
- The diffusion in the axial direction is negligible compared to that in the radial direction: $\partial^2 Y_j/\partial r^2 \ll \partial^2 Y_j/\partial z^2$.
- Mixing is caused by diffusion only, and the radial velocity component is equal to zero: $v_r = 0$.
- The reaction takes place at the surface of the flame, where the equivalence ratio is equal to unity.

The equations for this problem are deduced from (7.62), where the definition of β_j is given by (7.61). The reaction rate is eliminated by utilizing the variable $\beta = \beta_F - \beta_O$. In cylindrical coordinates, the equation to solve is

$$\partial\beta/\partial z - (D/v_z r)\, \partial(r\partial\beta/\partial r)/\partial r = 0, \tag{7.95}$$

with the boundary conditions

$$\begin{cases} z = 0, 0 \le r \le r_F : \; \beta = \beta_{F0} = -Y_F|_{z=0}/\nu'_F M_F, \\ z = 0, r_F \le r \le r_O : \; \beta = \beta_{O0} = -Y_O|_{z=0}/\nu'_O M_O, \\ r = 0, r_F, \; z > 0 : \; \partial\beta/\partial r = 0. \end{cases} \tag{7.96}$$

At this point it is convenient to define the dimensionless coordinates

[6] An oxidizer is a chemical needed by a fuel to burn.

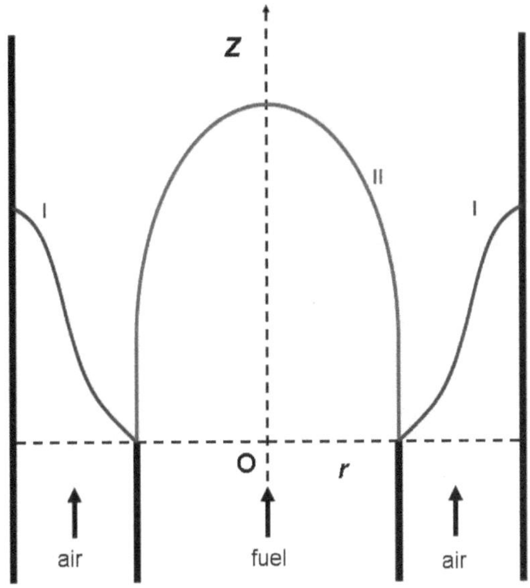

Fig. 7.16. Diffusion flame shapes for coaxial cylindrical tubes: *I*, underventilated; *II*, overventilated

$$\xi = r/r_O;\ \eta = Dz/v_r r_O^2$$

and the reduced parameters

$$C = r_F/r_O;\ \nu = \beta_{F0}/\beta_{O0};\ \gamma = -\beta/\beta_{F0}.$$

We then make use of the separation of variables method, and the ξ-dependent part of the solution involves the Bessel functions J_0 and J_1. We obtain (see [142] for detailed calculations)

$$\gamma = (1+\nu)C^2 - \nu + 2(1+\nu)C\sum_{n=1}^{\infty}\frac{1}{\phi_n}\frac{J_1(C\phi_n)}{[J_0(\phi_n)]^2}J_0(\phi_n\xi)e^{-\phi_n^2\eta}, \qquad (7.97)$$

where the ϕ_n are the successive roots of the equation $J_1(\phi) = 0$ (with the ordering convention $\phi_n > \phi_{n-1}$; $\phi_0 = 0$).

We assume that the reaction occurs at the flame surface (so we also assume an infinitely fast reaction leading to a thin flame). The flame profile $\eta = f(\xi)$ can be deduced from setting $\beta = 0$ and so $\gamma = 0$ in (7.97). The flame height is then obtained by setting $\xi = 0$. Neglecting all of the terms except $n = 1$, the following approximation is obtained for the dimensionless flame height of an underventilated flame:

$$\eta = (1/\phi_L^2) \ln 2(1+\nu)CJ_1(C\phi_1)/[\nu - (1+\nu)C^2]\phi_1 J_0(\phi_1). \tag{7.98}$$

The flame height and flame shapes obtained in this manner by Burke and Schumann (Fig. 7.16) are in good agreement with those obtained experimentally.

In contrast to the case for a thin premixed flame, the flame shape is found without using the energy equation. This equation would, however, be used if we wanted to find the temperature field.

The Z-Equation

Z is the mixture fraction variable. It is a combination of the mass fractions of the fuel and of the oxidizer, and its stoichiometric value Z_{st} corresponds to the flame position in the same manner as in the previous Burke–Shumann flame problem. For the single reaction

$$\nu'_F \,\mathrm{F} + \nu'_O \,\mathrm{O} \to \nu''_P \,\mathrm{P},$$

the variations in Y_F and Y_O due to chemistry are $dY_F/\nu'_F\mathcal{M}_F = dY_O/\nu'_O\mathcal{M}_O$. If there are two flows 1 and 2 with mass flow rates $\dot{m}_1$ and $\dot{m}_2$, where the first flow contains the species F (mass fraction $Y_{F,1}$) along with a diluent and combustion products but without species O, and the second flow contains the species O (mass fraction $Y_{O,2}$) along with a diluent and combustion products but without species F, we have

$$Z = \dot{m}_1/(\dot{m}_1 + \dot{m}_2).$$

Defining the stoichiometric ratio as

$$r_{st} = \nu'_O\mathcal{M}_O/\nu'_F\mathcal{M}_F, \tag{7.99}$$

we have

$$Z = \frac{r_{st}Y_F - Y_O + Y_{O,2}}{r_{st}Y_{F,1} + Y_{O,2}}. \tag{7.100}$$

At the flame level, $r_{st}Y_F - Y_O = 0$ and $Z = Z_{st} = (1 + r_{st}Y_{F,1}/Y_{O,2})^{-1}$. For one binary diffusion coefficient only, the mixture fraction is a nonreactive scalar that obeys the conservative balance equation

$$\rho\partial Z/\partial t + \rho\mathbf{v}\cdot\boldsymbol{\nabla} Z = \boldsymbol{\nabla}(\rho D\boldsymbol{\nabla} Z), \tag{7.101}$$

which is called the Z-equation. Just as the G-equation (7.94) does for premixed combustion, this equation plays an important role in nonpremixed turbulent combustion.

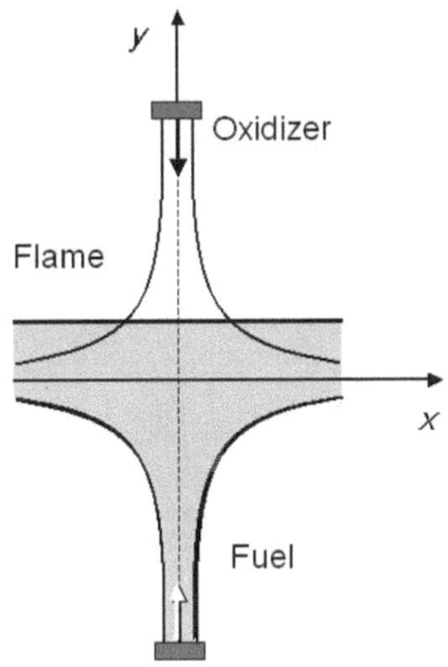

Fig. 7.17. An example of a laminar strained diffusion flame

An Example of a Laminar Strained Diffusion Flame: The Planar Counterflow Diffusion Flame

Figure 7.17 shows a counterflow diffusion flame for gaseous fuels. A constant diffusion coefficient is assumed. For dilute mixtures, such a flame is relatively simple to model because the density can be assumed to be constant and there is no effect of the flame on the velocity field of the strained flow considered [206]. If y is the coordinate along the flow axis (i.e., normal to the flame), the velocity components of the fluid flow are

$$u = ax, \; v = -ay$$

on the flame side, which we assume to be located in the $y > 0$ half plane a priori, as shown in Fig. 7.17. The Z-equation can be written (we also assume that Z do not depends on x and z)

$$-ay\partial Z/\partial y = D\partial^2 Z/\partial y^2. \tag{7.102}$$

Setting $\eta = y\sqrt{a/D}$, we obtain

$$\partial^2 Z/\partial \eta^2 + \eta \partial Z/\partial \eta = 0,$$

and the solution

$$Z = (1/2)(1 - \mathrm{erf}(\eta/\sqrt{2})). \tag{7.103}$$

The flame location is given by $Z_{st} = (1/2)(1 - \mathrm{erf}(\eta_f/\sqrt{2}))$. It is apparent that for $Z_{st} < 1/2$ the flame is located on the oxidizer side of the stagnation plane (the studied case), and conversely for $Z_{st} > 1/2$ the flame is located on the fuel side of the stagnation plane.

When density variations occur due to heat release, we also obtain a planar flame with a strained rate a that is imposed by the flow (but this is far more complex to demonstrate, so we will not do it here) and the following approximate value for the mixture fraction variable [202]:

$$Z = (1/2)(1 - \mathrm{erf}(\eta/\sqrt{2})) \tag{7.104}$$

where this time $\eta = \sqrt{a/D_\infty} \int_0^y (\rho/\rho_\infty) dy$, assuming that $\rho^2 D = \rho_\infty^2 D_\infty$.

7.8 Coupling Between Heat and Momentum Transfer in the Presence of Gravity

7.8.1 Historical Considerations

The Couette flow of a compressible flow and the structure of a shock wave are examples of situations where coupling occurs between heat and momentum transfer (see Chaps. 9 and 10) [99, 262]

For buoyancy-driven flows, the Boussinesq approximation is traditionally applied: the fluid is assumed to be incompressible, but dilation effects appear only on the right hand side of the momentum equation, through the gravity term, with temperature depending on density [27, 165].

Heat and momentum transfer coupling often causes instability. Two examples of such instabilities in liquids are presented in Sects. 7.8.2 and 7.9.2. The first involves gravity and the second the dependence of surface tension on temperature.

The first quantitative experiments on natural convection were performed by Henri Bénard around year 1900 [12]. Bénard studied the stability of a thin fluid layer that was open to the air and permitted a vertical temperature gradient. He accurately determined properties such as the spatial periodicity of the hexagonal convection cell pattern, how this pattern varied, and the profile of the interface. Later, in 1916, Lord Rayleigh [230] developed a complete linear stability analysis assuming stress-free conditions for the velocity and good heat-conducting plates. This mechanism was considered to explain Bénard's results until the role of the thermal Marangoni effect was pointed out, in particular by Pearson [197]. In this effect, a temperature fluctuation at the surface induces tangential stresses that can be amplified by hot fluid arriving from the interior [165].[7]

This theory is presented in Sect. 7.9.2.

[7] Later, Nield studied the influences of both gravity and the surface tension gradient on a fluid layer heated from below [186]. Scriven and Sternling [250] added the effect of free surface deformations to the Pearson theory.

Fig. 7.18. Henri Bénard, Professor in Paris (1874–1936), here when he left ENS and started his thesis on convection cells (E. Wesfreid, private communication, 2009; see also [287])

7.8.2 Rayleigh–Bénard Instability

Consider a fluid layer heated from below. This layer is confined between two horizontal planes at constant and uniform temperatures, where the temperature at the bottom is higher than the temperature at the top (see [43] for an extensive study of this problem). The bulk equations are those described in Chap. 4 with Boussinesq's approximation and with constant viscosity, specific heats, and thermal diffusivity. We have

$$\begin{cases} \boldsymbol{\nabla} \cdot \mathbf{v} = 0, \\ \rho_1(\partial \mathbf{v}/\partial t + \mathbf{v} \cdot \boldsymbol{\nabla} \otimes \mathbf{v}) + \boldsymbol{\nabla} p - \mu \Delta \mathbf{v} = \rho \mathbf{g}, \\ \partial T/\partial t + \mathbf{v} \cdot \nabla T - \kappa \Delta T = 0, \end{cases} \tag{7.105}$$

where $\mathbf{g} = -g\mathbf{k}$ and $\rho = \rho_1[1 + \alpha(T_1 - T)]$ (α is a constant). The constants T_1 and ρ_1 refer to the bottom ($z = 0$), and T_2 is used for the upper boundary.

The boundary conditions are as follows for a rigid upper wall:[8]

$$\begin{cases} z = 0,\ T = T_1,\ z = h,\ T = T_2,\ T_2 < T_1, \\ z = 0,\ z = h,\ u = v = w = 0. \end{cases} \tag{7.106}$$

The reference configuration is motionless and corresponds to a purely conductive situation. Denoting reference quantities with the subscript "$_0$", we easily obtain

[8] For a free horizontal upper boundary, the condition $u = v = 0$ at $z = h$ is valid assuming zero horizontal friction $(\mathbf{P} - p\mathbf{1}) \cdot \mathbf{k} = \mathbf{0}$.

$$\begin{cases} T_0(z) = T_1 - G\,z,\; G = (T_1 - T_2)/h, \\ p_0(z) = p_1 - \rho_1\, g(z + \alpha\, G\, z^2/2). \end{cases} \tag{7.107}$$

We want to study the fluid motion resulting from an experimentally observed instability. It is appropriate to study the linearized problem in this case, setting

$$T' = T - T_0(z),\; \rho' = \rho - \rho_0(z),\; p' = p - p_0(z),\; \mathbf{v}' = \mathbf{v}. \tag{7.108}$$

The small perturbation equations are (neglecting second-order perturbations):

$$\begin{cases} \partial u/\partial x + \partial v/\partial y + \partial w/\partial z = 0, \\ \rho_1(\partial u/\partial t + \partial p'\partial x) = \mu(\partial^2 u/\partial x^2 + \partial^2 u/\partial y^2 + \partial^2 u/\partial z^2), \\ \rho_1(\partial v/\partial t + \partial p'\partial y) = \mu(\partial^2 v/\partial x^2 + \partial^2 v/\partial y^2 + \partial^2 v/\partial z^2), \\ \rho_1(\partial w/\partial t + \partial p'\partial z) = \mu(\partial^2 w/\partial x^2 + \partial^2 w/\partial y^2 + \partial^2 w/\partial z^2) - \rho_1 g\alpha T', \\ \partial T'/\partial t = \kappa(\partial^2 T'/\partial x^2 + \partial^2 T'/\partial y^2 + \partial^2 T'/\partial z^2) + w\,G. \end{cases} \tag{7.109}$$

The boundary conditions are

$$z = 0,\, h :\; T' = 0,\; u = v = w = 0. \tag{7.110}$$

This linear problem is solved using the normal mode method,[9] where the perturbation f' of the f function is given in the form

$$f' = F(z,\, t)\exp(iK_x x + iK_y y + \lambda t)$$

and the solution that gives marginal stability (i.e., $\lambda = 0$) is controlled by the Rayleigh number, which was introduced in Chap. 5 as $Ra_{th} = Gr_{th}\, Pr$, with $Pr = \nu/\kappa$ and $Gr = \Delta\rho\, g\, L^3/\nu^2$. The present Rayleigh number can be written

$$Ra = \alpha G\, g\, h^4/\kappa\nu,$$

with $\nu = \mu/\rho_1$.

This leads to a marginal stability curve of Ra as a function of K, where K is the horizontal wavenumber modulus $K = \sqrt{K_x^2 + K_y^2}$. This curve (see Fig. 7.19), which characterizes the limits of the Rayleigh–Bénard instability, presents a minimum that defines the critical values Ra_C and K_C. For $Ra < Ra_C$, we have the purely conductive mode, and for $Ra \geq Ra_C$ we have the unstable regime. Their actual values are $Ra_C = 1707.76$, $h\, K_C = 3.12$ for our

[9]We will not give the complex, detailed calculations that can be found in several cited works here.

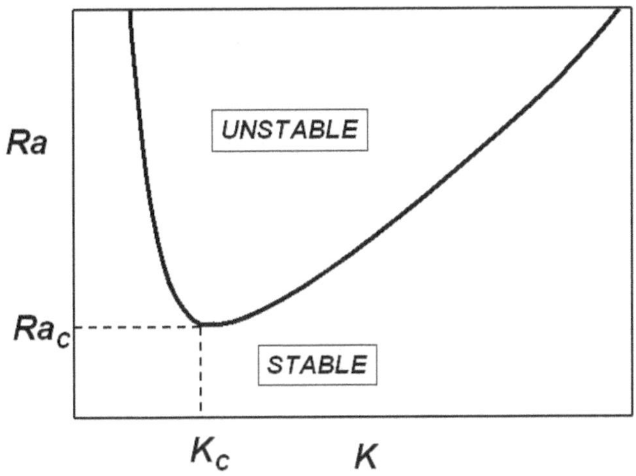

Fig. 7.19. Marginal stability curve for the Rayleigh–Bénard problem

case of no-slip boundary conditions and isothermal plates. Other cases[10] have been studied by several authors [43].

This example shows the complexity of heat and momentum transfer coupling under the action of gravity.

7.9 Surface Tension and Viscosity

Liquid/gas surface tension is a function of the temperature, which we will assume to be the same at equilibrium and during the motion of the fluid.

7.9.1 Marangoni Effect in a Highly Conducting Fluid Layer

We now consider a horizontal liquid layer of thickness h and length $l \gg h$, with a free upper limit.

A constant temperature gradient is imposed on the liquid/gas interface [159] (see Fig. 7.20), where $dT/dx = (T_2 - T_1)/l = G$. This is possible if the movements induced by the surface tension gradient do not affect the temperature field of the liquid. This is the case for small values of the Prandtl number (for liquid metals for example).

[10]For stress-free velocity upper boundary conditions and for no slip at the lower boundary conditions as well as for isothermal plates, we get $Ra_C = 1100.65$ and $h\,K_C = 2.68$.

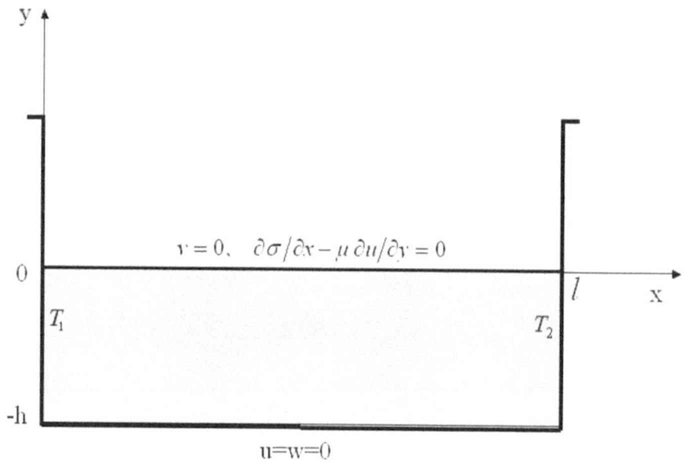

Fig. 7.20. Marangoni effect for a fixed temperature gradient

We propose to determine the velocity field and the pressure field by neglecting the vertical component of the velocity and by assuming steady motion. By assumption, we have

$$\sigma = \sigma(T_1) + \sigma_T G x, \ \sigma_T = d\sigma/dT < 0. \tag{7.111}$$

We set

$$-\sigma_T G = -\frac{T_2 - T_1}{l}\frac{d\sigma}{dT} = a. \tag{7.112}$$

a is negative if $T_2 < T_1$. The equations for the liquid give

$$\begin{cases} \partial u/\partial x = 0, \ \text{so } u = u(y), \ v = 0, \\ \partial p/\partial x = \mu \partial^2 u/\partial x^2, \\ \partial p/\partial y = 0. \end{cases} \tag{7.113}$$

We then find that $\partial p/\partial x$ is constant and that

$$\begin{cases} p = (dp/dx)x + p_0, \\ u = (dp/dx)y^2/2\mu + Ay + B. \end{cases} \tag{7.114}$$

At the bottom of the container $u = 0$, so

$$u = (dp/dx)(y^2 - h^2)/2\mu + A(y + h). \tag{7.115}$$

At the surface of the liquid, there is equality between the surface tension gradient and the viscous stress. Now, the balance of the forces must be evaluated for a surface element with negligible mass, which therefore does not have any inertia (see Fig. 7.21). In this case, the tensor for the viscous pressures is

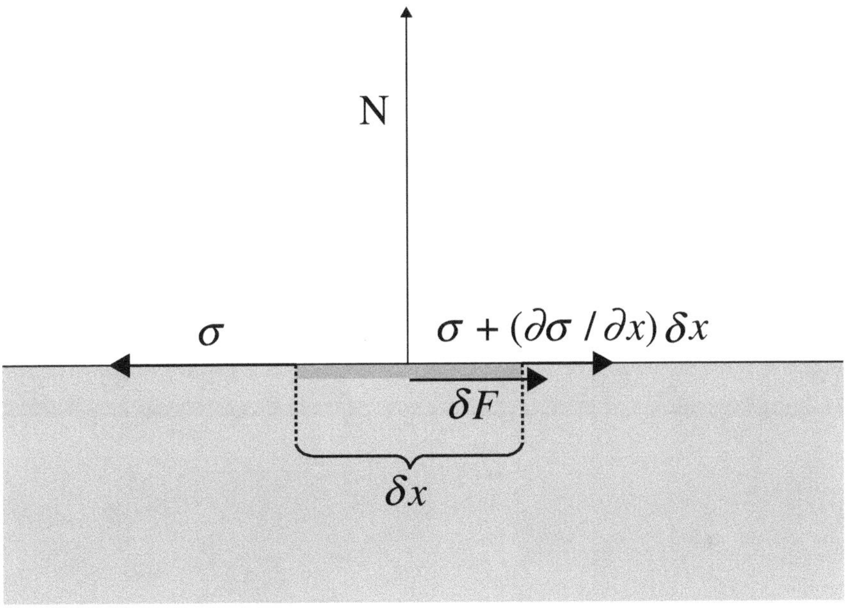

Fig. 7.21. Forces at the surface of the liquid

$$\mathbf{P} - p\mathbf{1} = \left\| \begin{matrix} 0 & -\mu du/dy \\ -\mu du/dy & 0 \end{matrix} \right\| . \tag{7.116}$$

The viscous force $\delta\mathbf{F}$ exerted on the element is thus

$$\delta\mathbf{F} = (\mathbf{P} - p\mathbf{1}) \cdot \mathbf{N}\delta x = \left\| \begin{matrix} -\mu du/dy \\ 0 \end{matrix} \right\| , \tag{7.117}$$

or, along x (i.e., along the surface),

$$\delta F = -\mu(du/dy)\, \delta x. \tag{7.118}$$

The force due to the surface stresses is $\sigma(x+\delta x) - \sigma(x)$; i.e., $(\partial\sigma/\partial x)\delta x$ if we neglect the second-order term. The sum of these forces is zero, so

$$\partial\sigma/\partial x - \mu du/dy = 0 \text{ for } y = 0. \tag{7.119}$$

However,

$$\partial\sigma/\partial x = \sigma_T G = -a. \tag{7.120}$$

We therefore obtain

$$du/dy = -a/\mu \text{ for } y = 0. \tag{7.121}$$

This condition makes it possible to deduce the integration constant

$$A = -a/\mu. \tag{7.122}$$

The flow velocity field is thus

$$u = (dp/dx)(y^2 - h^2)/2\mu - a(y + h)/\mu. \tag{7.123}$$

The pressure gradient is determined by writing a conservation condition for the total mass. It is assumed that, for any x-coordinate, the flow rate vanishes; i.e.,

$$\int_{-h}^{0} u\, dy = 0. \tag{7.124}$$

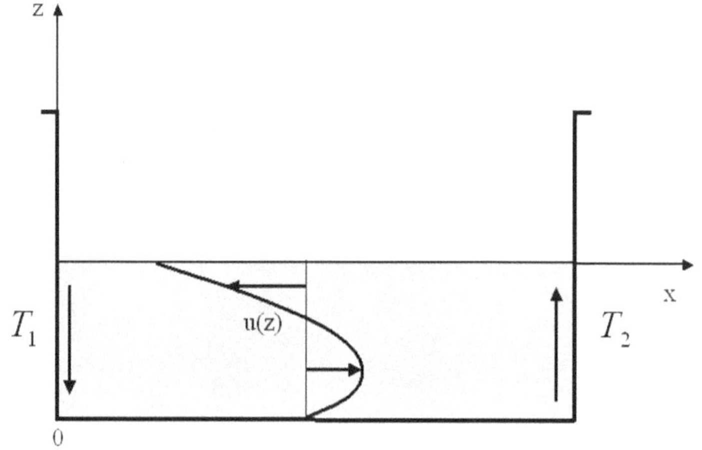

Fig. 7.22. Marangoni effect for a fixed temperature gradient (when $T_2 > T_1$; i.e., $\sigma_2 < \sigma_1$): velocity profile far from the extremities

This assumption leads to the results (see Fig. 7.22, from [159])

$$\begin{cases} dp/dx = -3a/2h, \\ u = -a(3y^2 + 4hy + h^2)/4\mu h = -a(3y + h)(y + h)/4\mu h. \end{cases} \tag{7.125}$$

This treatment of the motion of the surface layer and of the liquid is open to criticism: the motions of the liquid at the ends of the container have been ignored, and the assumptions lead to a thermodynamic pressure gradient that does not vanish at the surface, whereas in theory the pressure of the gas phase is uniform. These defects come from our simplifying assumptions. However, the main aspects of the phenomenon are well described.

7.9.2 Bénard–Marangoni Instability

The effect observed in Sect. 7.9.1 appears as soon as a longitudinal heat gradient is imposed, however small it is. This is no longer the case in the presence of a single transverse gradient when the liquid is not very conductive (the Prandtl number is then not small like it was in Sect. 7.9.1). When this transverse temperature gradient is lower than a threshold value, the regime is purely conductive. Above the critical gradient, an instability known as the Bénard–Marangoni instability occurs; fluctuations due to Brownian motion tend to develop, which give rise to convective swirls. Those can be coupled to the Rayleigh–Bénard convective structures associated with gravity and thermal dilation. In the absence of gravitational effects (a thin layer), with no curvature of the free surface [197], and assuming a two-dimensional configuration, the motion is controlled by the following equations (see Fig. 7.23):

$$\begin{cases} \partial u/\partial x + \partial v/\partial y = 0, \\ \rho \partial u/\partial t + \rho(u\partial u/\partial x + v\partial u/\partial y) + \partial p/\partial x = \mu(\partial^2 u/\partial x^2 + \partial^2 u/\partial y^2), \\ \rho \partial v/\partial t + \rho(u\partial v/\partial x + v\partial v/\partial y) + \partial p/\partial y = \mu(\partial^2 v/\partial x^2 + \partial^2 v/\partial y^2), \\ \partial T/\partial t + u\partial T/\partial x + v\partial T/\partial y = \kappa(\partial^2 T/\partial x^2 + \partial^2 T/\partial y^2), \end{cases} \tag{7.126}$$

with the boundary conditions

$$\begin{cases} y = -h \;\rightarrow\; u = v = 0,\; T = T_1, \\ y = 0 \;\rightarrow\; \sigma_T \partial T/\partial x - \mu \partial u/\partial y = 0,\; \partial T/\partial y + \Lambda(T - T_0) = 0, \end{cases} \tag{7.127}$$

if we assume a linear law with a coefficient Λ for the heat exchange between the free surface and the gas above. The purely conductive solution is

$$u = v = 0,\; T = T_1 - \frac{\Lambda(T_1 - T_0)}{1 + \Lambda h}(y + h) = T_1 + G(y + h). \tag{7.128}$$

We then set $T = T_1 + G(y+h) + T'$, $p = p_0 + p'$, $u = u'$, $v = v'$, with the disturbances considered to be small. Substitution into the preceding system leads

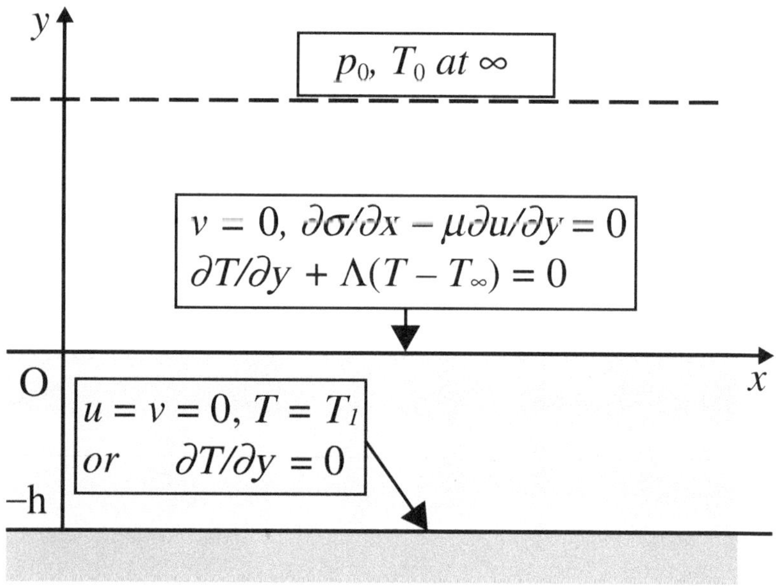

Fig. 7.23. Boundary conditions for the Marangoni instability problem

us to the desired system of equations for small disturbances. Perturbations p' and u' are then eliminated by successive differentiation, and we set

$$v' = V(y,\, t)\cos Kx,\ T' = \theta(y,\, t)\cos Kx, \tag{7.129}$$

where K is an unknown constant wavenumber. The following system is obtained:

$$\begin{cases} (\partial/\partial t)(\partial^2 V/\partial y^2 - K^2 V) - \nu(\partial^4 V/\partial y^4 - 2K^2\partial^2 V/\partial y^2 + K^4 V) = 0,\ \nu = \frac{\mu}{\rho}, \\ \partial\theta/\partial t - \kappa(\partial^2\theta/\partial y^2 - K^2\theta) + GV = 0, \\ y = -h,\ V = 0,\ \partial V/\partial y = 0,\ \theta = 0 \text{ or } \partial\theta/\partial y = 0, \\ y = 0,\ V = 0,\ \mu\,\partial^2 V/\partial y^2 + \sigma_T K^2\theta = 0,\ \partial\theta/\partial y + \Lambda\theta = 0. \end{cases} \tag{7.130}$$

The steady solutions are of the form

$$\begin{cases} V = (Ay + B)\exp(Ky) + (Cy + D)\exp(-Ky), \\ (4\kappa K/G)\theta = [Ay^2 + (2B - A/K)y - E]\exp(Ky) \\ +[Cy^2 - (2D + C/K)y - F]\exp(-Ky). \end{cases} \tag{7.131}$$

In the conducting case, respecting the boundary conditions leads to

$$\frac{\sigma_T G h^2}{\kappa \mu} = \frac{8Kh(Kh\cosh Kh + \Lambda h \sinh Kh)(Kh - \sinh Kh \cosh Kh)}{(Kh)^3 \cosh Kh - \sinh^3 Kh}. \tag{7.132}$$

The left hand side of this equation is known as the Marangoni number:

$$Ma = \sigma_T G h^2 / \kappa \mu, \tag{7.133}$$

and the critical value Ma_c (i.e., the minima of Ma as a function of α, as shown in Fig. 7.24 for several values of the Nusselt number Nu) [197]. We will not give a more detailed analysis of this problem here; we will limit ourselves to the traditional techniques used for instabilities in fluids. Also note that (i) dimensional analysis can give the Marangoni number directly; (ii) in reality, this phenomenon is generally three-dimensional; and (iii) these couplings between the surface tension and the fluid flow also occur in the presence of concentration gradients, and a similar treatment is justified in such cases.

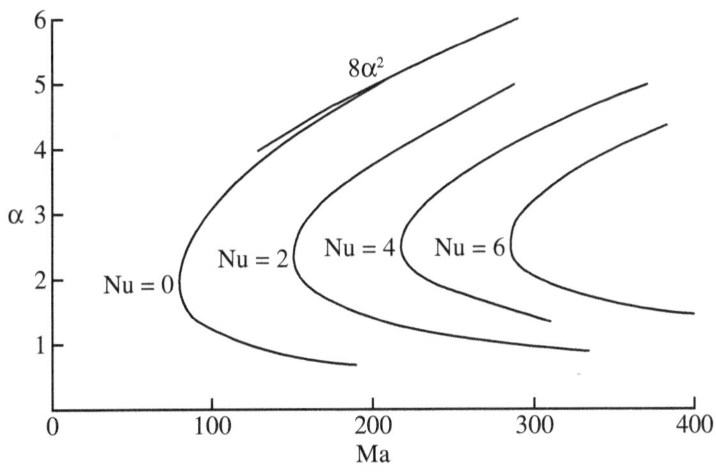

Fig. 7.24. Neutral stability curves for an isothermal bottom (reprinted, with permission, from [197]). Here, $\alpha = Kh$ and $Nu = \Lambda h$

Finally, it is possible to account for the surface deformation, which was previously assumed to be planar [94]. Scriven and Sternling [250] have studied the Marangoni instability in this more complex situation, taking into account surface deformations and interface viscosity coefficients κ_a and ϵ_a (see Eq. 11.41).

8

Turbulent Flow Concepts

This chapter deals with turbulent flow, as opposed to laminar (nonchaotic) flow. The reference is always a laminar flow, and this can become turbulent flow as a result of external destabilization. Turbulent flows are characterized by high complexity. Turbulent processes can be modeled using a variety of numerical simulations associated with computational fluid dynamics (CFD).

In Sect. 8.1, we present experimental evidence of turbulent flows occurring in tubes (an experiment performed by Reynolds in the nineteenth century) or close to solid walls. We show that friction coefficients are significantly modified by the onset and development of turbulence. Aerodynamicists were among the first to calculate friction coefficients in turbulent conditions, where they were used to investigate and design airplane wing profiles. Specifically, they were applied to assess changes in drag and lift, as well as the onset of aerodynamic stall (stall can be suppressed but cannot be avoided completely).

Section 8.2 deals with turbulence onset and decay. We investigate—analytically, if possible—examples of laminar flow destabilization, which can lead to random chaos and turbulence if the constraints applied to the system of interest are increased. We describe Rayleigh–Taylor, Kelvin–Helmholtz and shear flow instabilities, as well as the von Kármán vortex street.[1] Once the onset of unstable eddies occurs, the process leading to their possible decay by dissipation at small scales is initiated.

Classical turbulence theory (Reynolds-averaged Navier–Stokes or "RANS" theory) is discussed in Sect. 8.3, mainly for single-component flows. We introduce the mean values of all quantities (dilatable and compressible fluids require the use of Favre averaging), and turbulent transfer coefficients (by analogy with molecular transfer coefficients). Statistical modeling (using the well-known k–ϵ method for example) and spectral analysis, which have long been the main tools for analyzing (often incompressible and nonreactive) turbulent flows, are then presented.

[1]Other types of instability will also be addressed in subsequent chapters.

R. Prud'homme, *Flows of Reactive Fluids*,
DOI 10.1007/978-0-8176-4659-2_8, © Springer Science+Business Media, LLC 2010

A new approach to investigating compressibility and chemical reactivity is presented in Sect. 8.4 in relation to RANS simulations. Studies of combustion reveal the existence of nonlinear chemical production terms. Regime diagrams are presented for premixed and nonpremixed flames. Some turbulent combustion models are introduced in varying levels of detail.

The difficulties encountered when using statistical methods and the continual growth of computer power have resulted in a temptation to treat turbulent flows through direct numerical simulation (DNS). Indeed, for most phenomena, turbulent flows are governed by the same equations as laminar flows, but they are characterized by fluctuations in velocities and state variables in space and time.[2]

As a result, the costs of DNS calculations and the times required for them quickly become prohibitive. An attempt was made to solve this problem by combining direct numerical simulation (for large eddies) and statistical modeling (for smaller vortices). This combination of approaches is called the large eddy simulation (LES) method, and it is described in Sect. 8.5. Although the LES method was initially developed for incompressible flows, it is now also being applied to combustion.

8.1 Experimental Evidence for Turbulence

One characteristic of turbulent flows is rapid and random fluctuations in the measured quantity f (see Fig. 8.12) as a function of time, whereas the same quantity gives a more regular, quieter signal in laminar flows. There is known to be a threshold above which certain flows present significant and disordered fluctuations over time, and we now present several cases of flows that are laminar in an upstream zone but become turbulent downstream.

We pay attention to the fact that this change in flow regime has an important effect on the coefficients that characterize the forces acting on the surfaces of a body. We start this section by discussing Reynolds' experiment regarding the turbulent transition in a duct, and then consider viscous flow over a flat plate. Finally, we investigate the effect of micromixtures on chemical processes.

---

[2] "The Navier–Stokes equations, combined with a proper set of initial and boundary conditions, are deterministic in that they are believed to possess a unique solution. However, at high Re, the solution is strongly sensitive throughout most of the flow field to the initial and boundary conditions, and in reality the conditions cannot be specified accurately enough to obtain the deterministic solution, either theoretically or experimentally. It then becomes reasonable to introduce statistical methods and to seek only probabilistic aspects of the solutions, that is, to treat turbulence as a random process." [290].

Fig. 8.1. Professor Osborne Reynolds (1842–1912). Reprinted, with permission, from [121]

Reynolds' Experiment

This experiment, performed by Reynolds in 1883 [236], relates to *water run-off in a tube with a circular cross-section* [249].

The turbulence is visualized using a dye. As long as the Reynolds number $Re = \rho VD/\mu$ is lower than 2000, the flow is laminar, and we have already seen (see Sect. 5.1.4) that the head-loss coefficient is then

$$\Lambda = \frac{D}{\rho V^2}\frac{\Delta p}{\Delta \rho} = \frac{64}{Re}. \tag{8.1}$$

After a transition zone, increasing the Reynolds number leads to an other relation for Λ, even in the absence of surface roughness (see Fig. 8.2, based on data from Schlichting [249]):

$$\Lambda \cong 0.3164 Re^{-1/4}. \tag{8.2}$$

For turbulent flow, the velocities of interest are averaged values of the measured instantaneous velocities (see Sect. 8.3).

A more satisfying formula is provided by Schlichting [249] (curve 3 of Fig. 8.2):

$$1/\sqrt{\Lambda} = 2\log[(VD/\nu)\sqrt{\Lambda}] - 0.8. \tag{8.3}$$

The experiment performed by Reynolds was the first to clearly show the occurence of two flow regimes (laminar and turbulent flow). It showed that there was a correlation between the random velocity fluctuations in the flow and the macroscopic effect on the head-loss coefficient as a function of the Reynolds number.

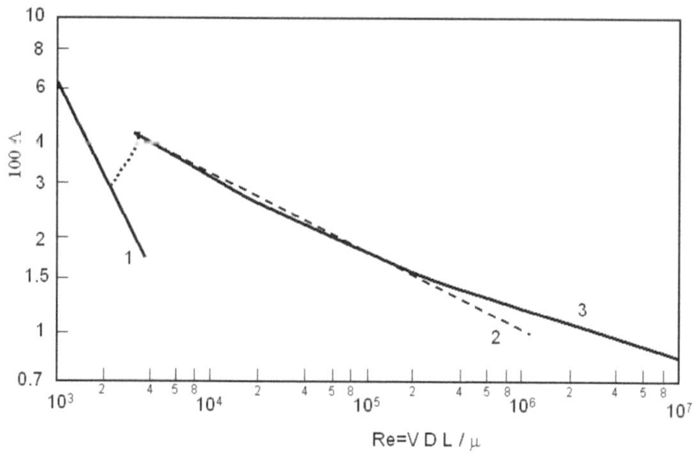

Fig. 8.2. Laminar–turbulent transition in a tube. *Curve 1*, laminar experimental results and formula (8.1). *Curve 2*, turbulent experimental results and empirical formula (8.2). *Curve 3*, experimental results and relation (8.3)

Viscous Flow Over a Smooth, Flat Plate

In the case of a *planar plate and a parallel flow* of mean velocity $\mathbf{V}$, a boundary layer develops (see Sects. 9.3 and 9.6), starting from the leading edge of the plate. Beyond this layer above the plate (such as far upstream of the leading edge), the flow remains inviscid [34, 62]. We define the coefficient of local friction on the wall via

$$C_f(x) = 2\,\tau(x)/\rho\,V^2, \tag{8.4}$$

where $\tau(x)$ is the local friction stress, and the coefficient of the average friction between $x = 0$ (the leading edge) and $x = L$ (downstream of the leading edge) is

$$C_{fm} = \frac{1}{L}\int_0^L C_f(x)dx.$$

Between the abscises $x = 0$ and x_T, the boundary layer is laminar, and we find that $C_{fm} = 1.328(Re)^{-1/2}$, where $Re = \rho V L/\mu$. Beyond x_T, the boundary layer becomes turbulent (see Sect. 9.6 for the theory behind this). Actually, there is no precise abscise x_T, but rather a laminar–turbulent transition zone.[3]

[3]The order of magnitude of x_T is given by $Re = \rho V x_T/\mu \approx 2 \times 10^5$.

In the case of a smooth wall and for a uniform pressure, and for the onset of turbulence at abscise x_T, the following empirical law can be used:

$$C_{fm} = 0.074(Re)^{-0.2} - (x_T/L)[0.074(Re_T)^{-0.2} - 1.328(Re_T)^{-0.5}], \quad (8.5)$$

where $Re_T = \rho V x_T/\mu$ (if the boundary layer was turbulent downstream of the leading edge, we would have $C_{fm} = 0.074(Re)^{-0.2}$). The results differ because of the width of the transition zone and the uncertainty in the value of x_T. In addition, for high Reynolds numbers, the formula with $(Re)^{-0.2}$ does not remain valid; instead, we get the Schultz–Grunow law that depends on $(\ln Re)^{2.584}$ for the coefficient of local friction, so the most suitable formula is that of Prandtl–Schlichting, $C_{fm} = 0.455(\ln Re)^{-2.58} - 2A(Re)^{-1}$, which is valid up to $Re \cong 10^9$. The coefficient A depends on the Reynolds number of transition Re_T as shown in Table 8.1 [91].

Re_T	1×10^5	3×10^5	5×10^5	1×10^6	3×10^6
A	150	525	850	1650	4350

Table 8.1. Dependence of the coefficient A on Re_T

Effect of the Rugosity of the Flat Plate

For a *rough plate* where the roughness is homogeneous and has a thickness of e, the relative roughness ϵ_δ will be $\epsilon_\delta = e/\delta$, where δ is the thickness of the boundary layer. According to Brun et al. [34], for a relative roughness of less than 1/3, the wall is aerodynamically smooth; it can be considered rough if $\epsilon_\delta < 6$; and it is semi-rough for intermediate values. Since the boundary layer thickness δ increases with the distance x from the leading edge, it then follows that the plate will be rough at the beginning but will then be able to appear smooth. In the rough zone, we can use Droblenkov's law, $C_f = 0.0139(\epsilon)^{1/7}$, $\epsilon = e/L$.

The boundary layer problem is studied in detail in Chap. 9. Other developments in relation to this problem are given in [249] (p. 529).

Effect of Micromixing on Chemical Reactions

Turbulence affects chemical reactions. Turbulence at small scales is responsible for what is known as micromixing in chemical engineering. The following example shows the importance of micromixing in chemical reactions. Consider the precipitation of barium sulfate by H^+ ions from a stable complex in a basic medium. There are two concurrent reactions:

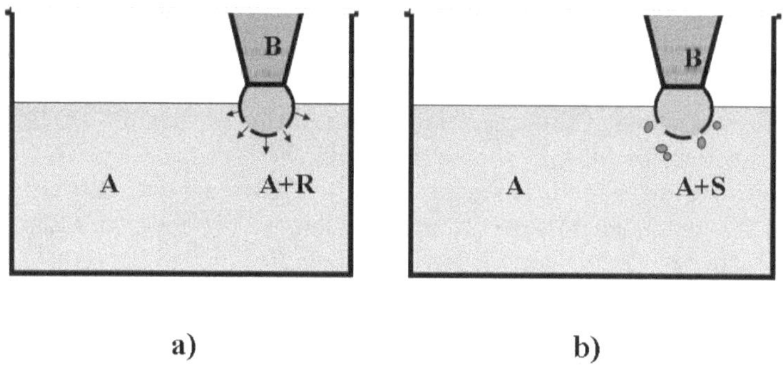

Fig. 8.3. Effect of micromixing: **a** Efficient micromixing at a molecular scale at the point when a drop of B is injected into liquid A. Species B is always in contact with A, but not with R. The precipitate S cannot form. **b** Imperfect micromixing. B is surrounded by R, R reacts with B to form the precipitate S

$$\begin{cases} \mathrm{A} + \mathrm{B} \longrightarrow \mathrm{R} \text{ (very fast)} \\ \mathrm{R} + \mathrm{B} \longrightarrow \mathrm{S} \downarrow \text{ (fast).} \end{cases} \tag{8.6}$$

Here, species A and B are liquids, as is the intermediate product R, while species S is barium sulfate.

A drop of B is injected into liquid A. Figure 8.3 indicates the two extreme situations that can occur depending on the quality of the turbulent stirring of liquid A [283].

The quantity of S produced makes it possible to measure the segregation index for the mixing. This index increases as the micromixing becomes more imperfect. Turbulence at large scales does not act directly on the reaction, but does influence the residence time of the species in an open reactor (see Chap. 6), leading to a residence time distribution $E(t_e)$.

8.2 Turbulence Onset and Damping

8.2.1 Turbulence Onset Mechanisms: Laminar Flow Instabilities

Many situations can lead to laminar flow instabilities. It is known that Rayleigh–Bénard rolls start to fluctuate at high Rayleigh numbers [43], and that this gradually leads to a chaotic regime upon increasing the Rayleigh number further [82, 110]. This instability was presented in Sect. 7.8, and another type of instability that occurs in a fluid that was initially at rest was introduced in Sect. 7.9.2.[4] Random fluctuations also occur at the bifurcation point between two stable modes. There are many sources of such instabilities. In the following sections we consider, for simple fluid flows, the Rayleigh–Taylor and Kelvin–Helmholtz instabilities, the Strouhal instability (which gives rise to von Kármán vortex streets and to the onset of turbulence at high Reynolds numbers), shear flow instability, and finally wrinkled flame instability (to illustrate the reactive case).

Rayleigh–Taylor and Kelvin–Helmholtz Instabilities

Two kinds of instability are presented here as particular cases of the more general problem of two superposed fluids with a planar surface of separation. The first, called the Rayleigh–Taylor instability, derives from the specific character of the equilibrium of the planar interface between two heavy fluids with differing densities in the presence of gravity. The second type of instability, called the Kelvin–Helmholtz instability, arises when two superposed fluids flow over one another with a relative horizontal velocity [43]. Surface tension has a stabilizing effect by reducing the domain of unstable wavelengths. For inviscid fluids of density ρ_1 and velocity U_1 for the lower fluid and ρ_2, U_2 for the upper fluid, and for constant surface tension σ, the amplification factor ω_i is obtained by analyzing the disturbance as a set of normal modes [94, 145]:

$$\omega_i^2 = \frac{\sigma}{\rho_1 + \rho_2} k^3 + \frac{\rho_1 \rho_2}{(\rho_1 + \rho_2)^2} (U_1 - U_2)^2 k^2 + g \frac{\rho_2 - \rho_1}{\rho_1 + \rho_2} k, \tag{8.7}$$

where k is the wavenumber and g is the acceleration due to gravity.[5]

---

[4] Note that we presented a linear analysis for these instabilities. This cannot be done for stronger instabilities, which require (generally numerical) nonlinear system resolution.

[5] The previous result was obtained by assuming inviscid fluids and semi-infinite layers. The velocity potential ϕ obeys the Laplace equation. The potential perturbation ϕ' can be decomposed into normal modes. The integration constants are determined using the interfacial conditions. The equation for the interface is obtained by assuming small perturbations near the planar reference surface, characterized by a function η. The boundary conditions at the interface (the Laplace law for the pressure jump due to surface tension and the impermeability conditions) are linearized, and the equations of motion for the fluids can be expressed via the Bernoulli laws. We obtain an homogeneous system of perturbation equations involving time and space derivatives of ϕ' and the interfacial function η. Finally, after eliminating the functions, we get the dispersion equation

$$(\omega - k\frac{\rho_1 U_1 + \rho_2 U_2}{\rho_1 + \rho_2})^2 = \frac{\sigma}{\rho_1 + \rho_2} k^3 - \frac{\rho_1 \rho_2}{(\rho_1 + \rho_2)^2} (U_1 - U_2)^2 k^2 - g \frac{\rho_2 - \rho_1}{\rho_1 + \rho_2} k.$$

Rayleigh–Taylor Instability

A Rayleigh–Taylor instability can occur for $\sigma = 0$, $U_1 = U_2$ and $\rho_2 > \rho_1$. The planar surface is then unstable at any wavenumber (i.e., at any propagation velocity).

In contrast, if $\sigma \neq 0$, $U_1 = U_2$ and $\rho_2 > \rho_1$, stability analysis gives a cut-off value k_c, where

$$k_c^2 = g(\rho_2 - \rho_1)/\sigma,$$

and stability is ensured for small wavelengths ($k > k_c$). The maximum ω_i appears for a particular value k_m of the wavenumber (see Fig. 8.4) given by [94].

$$k_m^2 = g(\rho_2 - \rho_1)/3\sigma.$$

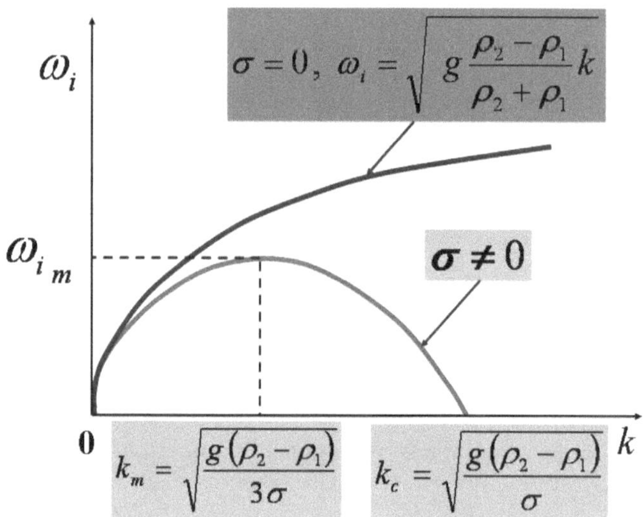

Fig. 8.4. Growth curves for the Rayleigh–Taylor instability in inviscid fluids when $U_2 = U_1$

Kelvin–Helmholtz Instability

A Kelvin–Helmholtz instability can occur for $g = 0$ and $\sigma = 0$. Instability can occur for any value of k. If $\sigma \neq 0$ and $g = 0$, a cut-off value (k_c) appears, and

---

For real values of the wavenumber k, we set $\omega = \omega_r + \omega_i$. The real value $\omega_r = k(\rho_1 U_1 + \rho_2 U_2)/(\rho_1 + \rho_2)$, and the imaginary part is given by (8.7).

we get the maximum perturbation growth rate for a particular wavenumber k_m (see Fig. 8.5) [94]. We have

$$k_c = \frac{\rho_1 \rho_2}{(\rho_1 + \rho_2)^2} \frac{(U_1 - U_2)^2}{\sigma}; \; k_m = \frac{2}{3} \frac{\rho_1 \rho_2}{(\rho_1 + \rho_2)^2} \frac{(U_1 - U_2)^2}{\sigma}.$$

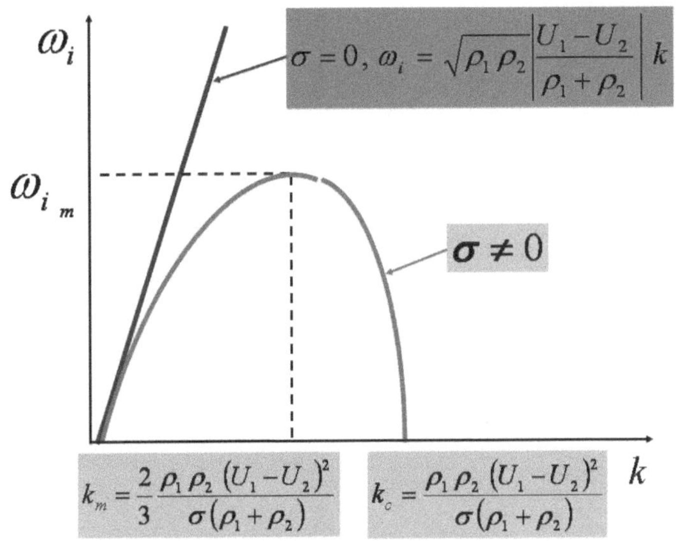

Fig. 8.5. Growth curves for a simple Kelvin–Helmholtz instability when $g = 0$

The Kelvin–Helmholtz instability gives rise to "rolled up" structures such as those shown in Fig. 8.6a.

It is possible to determine a convective speed U_c for the large structures observed. The streamlines of the flow observed in a reference frame moving at the speed U_c are represented in Fig. 8.6b, and exhibit a "neck" structure [195] similar to those seen for the planar counterflow diffusion flame of Fig. 7.17 and the planar stretched premixed flame of Fig. 7.15. This justifies the local use of these Lagrangian structures when studying and modeling turbulent flames.

To determine the speed U_c of a continuous flow of a compressible fluid, we note that the dynamic pressures p_0 of the two flows are equal at the stagnation point; i.e., the ratios $p_0/p = [(\gamma - 1)M^2/2 + 1]^{\gamma/(\gamma-1)}$, where p is the static pressure, γ is the isentropic coefficient and M is the Mach number, are the same. If the Mach numbers are not too high, and assuming isentropy, we obtain

$$\rho_1 (U_1 - U_c)^2 = \rho_2 (U_c - U_2)^2$$

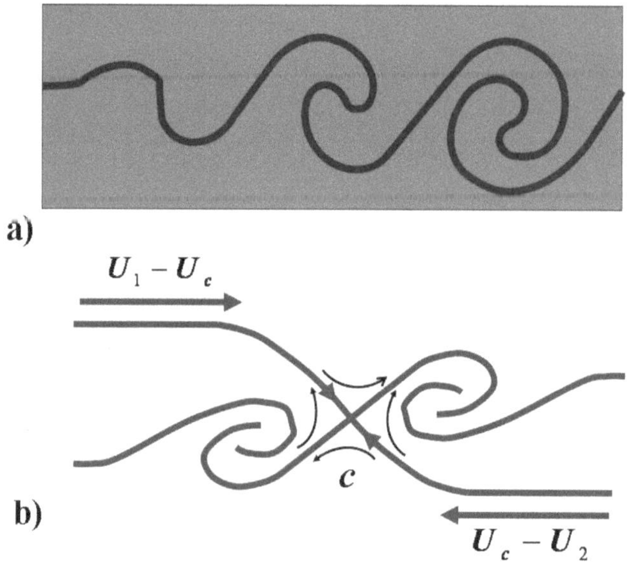

Fig. 8.6. **a** Example of the Kelvin–Helmholtz instability phenomenon that causes "rolling up" at the interface. **b** Convective frame and streamlines in this reference frame

if U_1 is the highest velocity, and we deduce

$$U_c = U_1 \frac{1 - U_2/U_1\sqrt{\rho_2/\rho_1}}{1 + \sqrt{\rho_2/\rho_1}}. \tag{8.8}$$

This formula is no longer valid if the flow does not stay isentropic (shock waves attached to the structures can appear at high velocities). For $M_c > 0.6$, the experimentally observed mixing layer becomes turbulent and exhibits 3D structures.

Kármán Vortex Streets

A Kármán vortex street is a repeating pattern of swirling vortices caused by the unsteady separation of flow over bluff bodies. They are named after the engineer and fluid dynamicist Theodore von Kármán.

If we consider a long circular cylinder,[6] the frequency of vortex shedding is given, for the range $250 < Re < 2 \times 10^5$, by the empirical formula

[6]This case illustrates the Strouhal instability and the particular type of wake known as the Karman vortex street. This is a succession of eddies that are created close to the cylinder and break away alternately from both sides of the cylinder. Vortices are emitted regularly and rotate in opposite senses.

$$fd/V = 0.198(1 - 19.7/Re),$$

where f is the vortex-shedding frequency, d is the cylinder diameter, V is the flow speed, and $Re = Vd/\nu$ is the Reynolds number ($\nu = \mu/\rho$ is the kinematic viscosity). The dimensionless parameter fd/V is known as the Strouhal number.

Up to the critical Reynolds number (about 40), the oscillations are so strong that one of the two vortices breaks away from the cylinder. This is the Strouhal instability. The second vortex is shed while the first is reshaped. The vortices appear and are shed alternately at a constant frequency (see Fig. 8.7 and [281]). At a Reynolds number of about 200, the structure of the Karman vortex street becomes three-dimensional. When the Reynolds number moves above 400, turbulence develops and the Karman vortex street disappears.

Fig. 8.7. Vortex street behind a circular cylinder at $Re = 138$. Hydraulic analogy visualization in the ONERA hydrodynamic tunnel. (ONERA, private communication, 2009)

The Shear Flow Instability

Let us now consider cases where the viscous terms are in competition with the inertial terms, and where the reference laminar flow is bidimensional with parallel stream lines.

The Orr–Sommerfeld Equation

We suppose that the laminar flow is characterized by the velocity profile (with no disturbances) $u = U(y)$, the velocity component v is zero, and the flow is 2D planar. In this case, the pressure P is a linear function of the abscises x, and that it obeys

$$\frac{dP}{dx} = \nu \frac{d^2U}{dy^2}. \tag{8.9}$$

The disturbed flow will be

$$\begin{cases} u = U(y) + u', \\ v = v', \\ p = P(x) + p'. \end{cases} \tag{8.10}$$

The fluid is assumed to be incompressible, so the mass conservation equation leads to

$$u' = \partial\psi/\partial y, \; v' = -\partial\psi/\partial x. \tag{8.11}$$

Let us study disturbances of the type

$$\psi = \phi(y)e^{ik(x-ct)}. \tag{8.12}$$

The momentum equation becomes

$$\begin{cases} \partial^2\psi/\partial t\partial y + U(y)\,\partial^2\psi/\partial x\partial y - (\partial\psi/\partial x)\,dU/dy + (1/\rho)\,\partial p'/\partial x \\ = \nu\,\partial(\Delta\psi)/\partial y, \\ -\partial^2\psi/\partial t\partial x - U(y)\,\partial^2\psi/\partial x^2 + (1/\rho)\,\partial p'/\partial y \\ = -\nu\,\partial(\Delta\psi)/\partial x. \end{cases} \tag{8.13}$$

After eliminating p', we have

$$\frac{\partial(\Delta\psi)}{\partial t} + U(y)\frac{\partial(\Delta\psi)}{\partial x} - \frac{\partial\psi}{\partial x}\frac{d^2U}{dy^2} = \nu\Delta(\Delta\psi). \tag{8.14}$$

As

$$\Delta\psi = (\frac{d^2\phi}{dy^2} - k^2\phi)e^{ik(x-ct)}, \tag{8.15}$$

we obtain

$$\phi^{(4)} - 2k^2\phi'' + k^4\phi = \frac{ik}{\nu}[(U-c)(\phi'' - k^2\phi) - U''\phi]. \tag{8.16}$$

This is the Orr–Sommerfeld equation (see, for example, Landau and Lifschitz [147]).

The Nonviscous Shear Flow Case

Studying the same problem for a perfect fluid ($\nu = 0$) leads to Rayleigh's equation:

$$(U - c)(\phi'' - k^2\phi) - U''\phi = 0. \tag{8.17}$$

We can show that one requirement for the development of instability in a perfect fluid is that the velocity profile $U(y)$ presents a point of inflection.

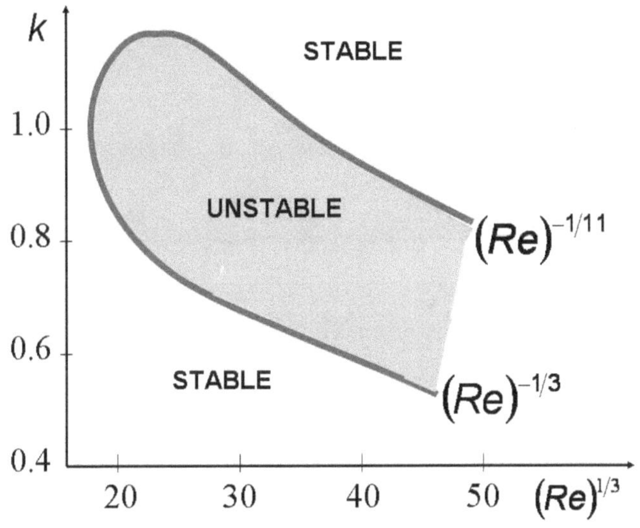

Fig. 8.8. Stability diagram for the Orr–Sommerfeld equation

Demonstration for a particular case: In the case of a flow between two parallel flat surfaces, we have

$$\begin{cases} k \in \Re,\ c = c_r + ic_i,\ \phi = \phi_r + i\phi_i, \\ \psi = \phi\, e^{kc_i t}\, e^{ik(x - c_r t)}. \end{cases} \tag{8.18}$$

Instability occurs if $c_i > 0$. We define the real and imaginary parts of the quantity $(U - c)^{-1}$:

$$(U - c)^{-1} = \gamma_r + i\gamma_i,\ \gamma_r = \frac{U - c_r}{(U - c_r)^2 + c_i^2},\ \gamma_i = \frac{c_i}{(U - c_r)^2 + c_i^2}, \tag{8.19}$$

$$\phi'' - k^2\phi - U''\phi(\gamma_r + i\gamma_i) = 0. \tag{8.20}$$

Separating the real and imaginary parts, we have:

$$\begin{cases} \phi_r'' - k^2\phi_r - U''(\phi_r\gamma_r - \phi_i\gamma_i) = 0, \\ \phi_i'' - k^2\phi_i - U''(\phi_r\gamma_i + \phi_i\gamma_r) = 0. \end{cases} \tag{8.21}$$

By combining these equations, we obtain $\phi_i\phi_r'' - \phi_r\phi_i'' + U''(\phi_i^2 + \phi_r^2)\gamma_i = 0$, or

$$\frac{d}{dy}(\phi_i\phi_r' - \phi_r\phi_i') + U''(\phi_i^2 + \phi_r^2)\gamma_i = 0$$

Integration leads to

$$[\phi_i\phi_r' - \phi_r\phi_i']_{-h}^{+h} + \int_{-h}^{+h} U''(\phi_i^2 + \phi_r^2).\gamma_i \, dy = 0 \tag{8.22}$$

As $v = -\partial\psi/\partial x = -ik\phi(y)\, e^{kc_it}\, e^{ik(x-c_rt)}$, and with $\phi(h) = \phi(-h) = 0$, the jump $[\phi_i\phi_r' - \phi_r\phi_i']_{-h}^{+h}$ is zero. Also,

$$\int_{-h}^{+h} U''(\phi_i^2 + \phi_r^2)\gamma_i \, dy = 0. \tag{8.23}$$

If $c_i > 0$, so $\gamma_i > 0$, this integral is only canceled out if the sign of U'' changes between $y = -h$ and $y = +h$. Therefore, the velocity field $U(y)$ must present an inflection point.

The General Case for Viscous Shear Flow

The problem becomes trickier to tackle when $\nu \neq 0$ (the Orr–Sommerfeld equation). For a flow between two parallel planes, we can proceed analytically or numerically for each value of ν and k. We obtain the condition $c_i = F(k, \nu)$, meaning that there are no trivial solutions that obey $u(\pm h) = v(\pm h) = 0$.

The sign of c_i is thus that of $\Im(F(k, v))$. We find that there is always stability for $Re = 2\bar{U}h/\nu < 5800$, where $\bar{U}$ is the mean velocity of the flow (Fig. 8.8). Instabilities are possible for certain wavenumbers with large values of Re. Lastly, stability is obtained for infinite Re (the two branches meet), which proves the limits of this linearized theory.

Instability of a Premixed Laminar Flame

Premixed flame instability is easily obtained by varying the equivalence ratio of the flame of Bunsen burner. Beyond a critical value of the chemical composition of the mixture, the cone is instantly transformed into a polyhedron. The spontaneous appearance of cells is only observed for mixtures that are rich in heavy hydrocarbons (propane) or that have only a small fraction of light combustibles (hydrogen); i.e., when the species that limits the reaction is the lightest one. Indeed, the flame can exhibit several types of instability (see for instance [198]).

We present two types of instability in a laminar premixed flame below: hydrodynamic instability and thermodiffusive instability.

Hydrodynamic Instability of a Planar Laminar Premixed Flame

The hydrodynamic instability of a planar laminar premixed flame is known as the Darrieus–Landau instability [22, 147]. Here, we investigate the stability of a planar flame front with respect to infinitesimal disturbances. The flame of interest is reduced to a discontinuity, as described in Sect. 10.4, and it is a deflagration wave.

We assume the plane of discontinuity to be the yz-plane, with unperturbed gas velocities in the positive x-direction: v_1 for the unburned gas and v_2 for the burned gases. We have $\rho_1 v_1 = \rho_2 v_2$ and $\rho_1 > \rho_2$ (due to the heat release), as well as $p_1 = p_2$ (squared velocity terms are negligible).

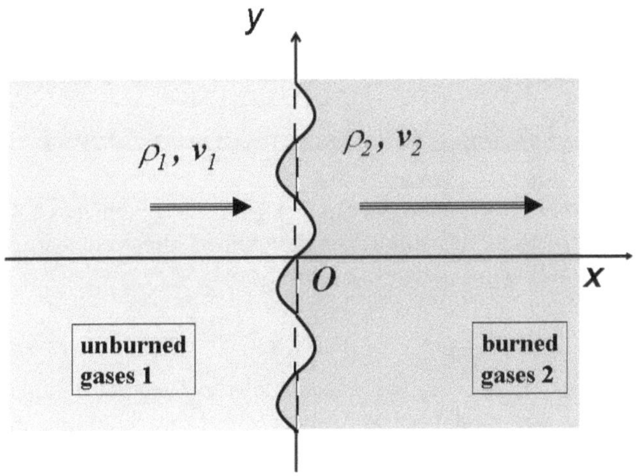

Fig. 8.9. A planar flame front with infinitesimal disturbances

The perturbation equations on each side (1 and 2) of the discontinuity are those of incompressible nonviscous fluids:

$$\nabla \mathbf{v}' = 0, \;\; \partial \mathbf{v}'/\partial t + (\mathbf{v} \otimes \nabla) \cdot \mathbf{v}' + \nabla p'/\rho = 0. \tag{8.24}$$

The perturbed flame surface is defined by $x = \eta(y, t)$ (2D case, Fig. 8.9). On the surface of discontinuity (i.e., for x =0), the following conditions must be satisfied:

- The pressure continuity equation, $p_1' = p_2'$
- The condition that the velocity component tangential to the surface is continuous, $v_{1y}' + v_1 \partial\eta/\partial y = v_{2y}' + v_2 \partial\eta/\partial y$

- The condition that *the gas velocity normal to the surface of discontinuity is unchanged,* $v'_{1x} - \partial\eta/\partial t = v'_{2x} - \partial\eta/\partial t$.

In the region $x < 0$ (the unburned gas, region 1), the solution to (8.24) is

$$\begin{cases} v'_{1x} = A\exp(iky + kx - i\omega t),\ v'_{1y} = iA\exp(iky + kx - i\omega t), \\ p'_1 = A\rho_1(i\omega/k - v_1)\exp(iky + kx - i\omega t). \end{cases} \tag{8.25}$$

In the region $x > O$ (the combustion products, region 2), we can write [147]

$$\begin{cases} v'_{2x} = B\exp(iky - kx - i\omega t) + C\exp(iky + i\omega x/v_2 - i\omega t), \\ v'_{2y} = -iB\exp(iky - kx - i\omega t) - (\omega/kv_2)C\exp(iky + i\omega x/v_2 - i\omega t), \\ p'_2 = -B\rho_2(i\omega/k + v_2)\exp(iky - kx - i\omega t). \end{cases} \tag{8.26}$$

Setting

$$\eta(y, t) = D\exp(iky - i\omega t),\ \omega = \omega_r + i\omega_i, \tag{8.27}$$

an unstable solution will give $\omega_i > 0$.

We obtain four homogeneous equations for the coefficients A, B, C, D. A simple calculation gives the condition that allows these equations to be compatible; i.e., the following dispersion equation:

$$\omega_i^2(v_1 + v_2) + 2\omega_i k v_1 v_2 + k^2 v_1 v_2 (v_1 - v_2). \tag{8.28}$$

Here $v_1 < v_2$ because $\rho_1 > \rho_2$ (on account of the considerable heating during combustion) and $\rho_1 v_1 = \rho_2 v_2$; however, the roots are real and have opposite signs, and the original motion is unstable.

Thermodiffusive Instability of a Premixed Laminar Flame

Let us now consider a reactive mixture of species A and B diluted in a neutral gas, where the composition is far from being stoichiometric, and the lightest species A is also the rarest [52]. In this case, the course of the reaction only affects species A appreciably. The coefficient of diffusion is therefore large (see Chap. 3) because of the low molar mass, and so the Lewis number $Le_A = \lambda/\rho c_p D_A$ is small compared to the Lewis number of (the heavier) species B. The thermal diffusivity $\kappa = \lambda/\rho c_p$ is independent of the proportions of A and B because of the strong dilution of the mixture.

The combustion velocity of a planar premixed laminar flame is well defined (Sect. 10.5). Locally, a stable Bunsen flame is comparable with such a planar flame. Let us suppose that a local disturbance induces a deformation that results in hollows and swells (Fig. 8.10). The transverse diffusion of species A tends to decrease the concentration C_A at a point M on the reaction surface, which is located in a hollow. In contrast, heat from the burned gases diffuses towards the interior. However, in this case, where Le_A is small, the effect of the diffusion of species A dominates, and the reaction rate decreases even though the temperature rises in M, leading to a drop in the speed of the combustion wave; i.e., the celerity of the flame relative to the fresh gases. The zone considered will become increasingly hollow. The opposite occurs at point N along the cut $y'y$. There is flame instability.

If the Lewis number was greater than 1, the effects of thermal conduction would be stronger than those of molecular diffusion, resulting in stability.

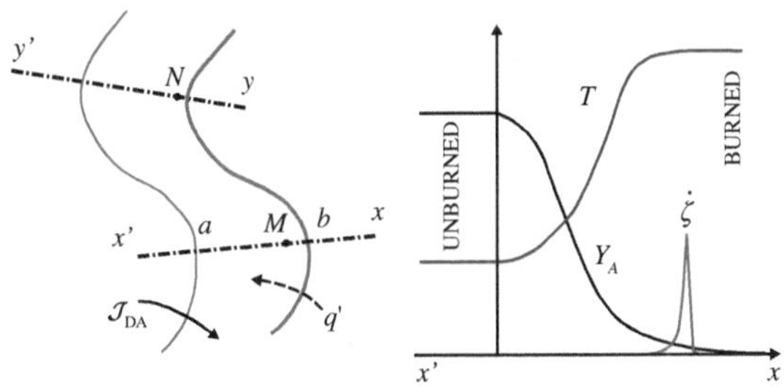

Fig. 8.10. Thermodiffusive instability of a laminar flame

These instability effects give rise to thermodiffusive autoturbulence through the amplification of defects. If the flow upstream of the fresh gases is already turbulent, the two phenomena will interact.

Motion studies of turbulent wrinkled flames were a focus of research thirty years ago [288]. Clavin and Joulin [53] showed that, within the framework of certain assumptions, just one scalar that measures the stretching of the flame front controls the form and dynamics of the flame locally (Sect. 11.4.2). This scalar consists of two terms that represent the contribution of the flame geometry (the curvature of the front, advancing with a prescribed normal velocity) and that of the inhomogeneity of the flow, as characterized by the tensor of the strain rates. Law, Jomaas and Bechtold [128, 152] recently studied the appearence of cellular instabilities in expanding spherical flames.

8.2.2 Turbulence Decay Mechanisms: Vortex Dissipation

In Sect. 8.2.1, we presented some instability mechanisms that can give rise to turbulent structures. In the following sections, we will see that the turbulent structures evolve by exchanging energy. These exchanges can be more or less inertial in nature, and dissipation can occur.

The small structures are generally damped by molecular diffusion. For pure fluids, we can compare them to small eddies where the rotation is slowed by the viscosity. Species diffusion also intervenes in reactive flows, just as it does in chemical reactions.

Viscous Damping of a Vortex

To illustrate the viscous dissipation of turbulent small structures, we now present the classical study of the viscous dissipation of a vortex. The laminar character assumed for the flow is justified by the fact that at scales smaller than the Kolmogorov scale introduced in Sect. 8.3.4, the rotation is damped by molecular viscosity in the same manner as in laminar flow.[7] Consider the flow of pure, incompressible, viscous fluid. The momentum balance equation is

$$\frac{d\mathbf{v}}{dt} + \frac{1}{\rho}\boldsymbol{\nabla} p = \nu \Delta \mathbf{v} + \mathbf{f}, \tag{8.29}$$

which can also be written

$$\frac{\partial \mathbf{v}}{\partial t} + \boldsymbol{\nabla}(\frac{v^2}{2}) + (\boldsymbol{\nabla} \times \mathbf{v}) \times \mathbf{v} + \frac{1}{\rho}\boldsymbol{\nabla} p = \nu \Delta \mathbf{v} + \mathbf{f}. \tag{8.30}$$

Applying the operator $\boldsymbol{\nabla}\times$ to each side of this equation gives

$$\partial \boldsymbol{\omega}/\partial t + \boldsymbol{\nabla} \times (\boldsymbol{\omega} \times \mathbf{v}) = \nu \Delta \boldsymbol{\omega}, \tag{8.31}$$

where $\omega = \frac{1}{2}\boldsymbol{\nabla}\times\mathbf{v}$, and where we have assumed that $\mathbf{f}$ derives from a potential.
If the fluid is nonviscous, (8.31) reduces to

$$\partial \boldsymbol{\omega}/\partial t + \boldsymbol{\nabla} \times (\boldsymbol{\omega} \times \mathbf{v}) = \mathbf{0}. \tag{8.32}$$

In 2D planar flow, where $\boldsymbol{\omega}$ is parallel to the Oz axis ($\boldsymbol{\omega} = \omega \mathbf{e}_z$), we have

$$\frac{\partial \omega}{\partial t} = \frac{\nu}{r}\frac{\partial}{\partial r}(r\frac{\partial \omega}{\partial r}), \tag{8.33}$$

and, for an inviscid flow,

$$\partial \omega/\partial t = 0. \tag{8.34}$$

In this latter case, we know the solution, which is generally presented in the complex plane:

$$v_r = 0,\ v_\theta = \Gamma/2\pi r. \tag{8.35}$$

This corresponds, for $v \neq 0$, to irrotational flow ($\boldsymbol{\omega} = \mathbf{0}$). This refers to an irrotational (or free) vortex that has a singularity at its center O. The stream lines are concentric circles. This solution is also a particular solution of (8.31) for the viscous fluid.

---

[7]Note that the concept of a turbulent eddy is definitely not reducible to that of free vortex. Turbulent eddies are more or less coherent and interacting structures that are frequently used in turbulence theory in relation to spectral analysis, as we will see later.

Let us now search for a more general solution [96] to (8.31) where the stream lines are still concentric circles and which is identical to the free vortex at time $t = 0$. For a self-similar solution, we can choose the variable η such that

$$\eta = r^2/4\nu t, \ \omega = f(\eta), \tag{8.36}$$

using a similar approach to that employed in Chap. 5 for the problem of thermal transfer (Sect. 5.3). The equation

$$d\omega/d\eta = (A/\eta)e^{-\eta} \tag{8.37}$$

is obtained, where A is an integration constant.

The function

$$\omega(r,t) = (A/t)e^{-\eta}$$

obeys (8.37), and the corresponding velocity v_θ is

$$v_\theta = (4\nu A/r)(1 - e^{-r^2/4\nu t}). \tag{8.38}$$

As the time t tends to zero, we must obtain an irrotational vortex. The integration constant A is thus equal to $A = \Gamma/8\pi\nu$. The (presumably single) solution to this problem is thus[8]

$$v_\theta = (\Gamma/2\pi r)(1 - e^{-r^2/4\nu t}). \tag{8.39}$$

The coherent structure to choose for modeling depends on the type of turbulent flow being studied (i.e., mixing layer or some other type), and is not always easy to identify [126].

Growth of a Flame in a Vortex Field

To illustrate the coupling between eddies and chemical reactions and species diffusion through the use of examples, we will now consider a vortex damped by viscosity at the same time that it is introduced at the interface between a fuel and an oxidizer, where a diffusion flame is burning.

For the planar injection of parallel jets, one of fuel F and the other of oxidizer O, the shear flow gives rise to a swirling alley (Kelvin–Helmholtz instabilities, Sect. 8.2.1). By following the motion of the center of the vortex (Fig. 8.6), we note that a rolled-up layer gradually develops, as a double spiral

---

[8] In turbulence, vortices are generally three-dimensional, and so a more appropriate model to explain the evolution is Burgers' vortex [36]. Moreover, we use Burgers' vortex sheets to model turbulent dissipation. Burgers' vortex is defined by the equations: $v_r = -\sigma r$, $v_\theta = (\Gamma/2\pi r)(1 - e^{-\sigma r^2/2\nu})$, $v_z = 2\sigma z$, where σ denotes the strain and Γ represents the circulation of the vortex. The quantity $\delta = \sqrt{\nu/\sigma}$ characterizes the core size of the vortex.

that contains alternate layers of the fluids F and O. These layers become increasingly thin, which contains burnt gases produced by the diffusion flame (the chemical reaction is assumed to occur infinitely quickly). A core of burnt gases is then formed, and the diameter of this core increases over time.

This process, which occurs at the same time as and interacts with viscous damping, was studied by Marble and Broadwell [168], Marble [166], and by Karagozian and Marble [132].

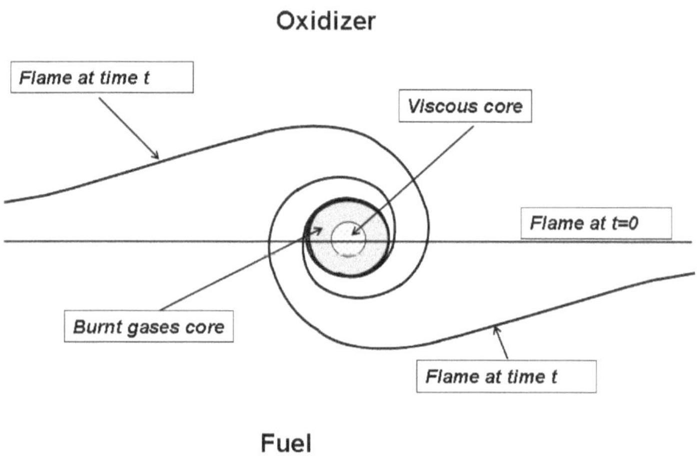

Fig. 8.11. Viscous and burnt gas cores for a diffusion flame that is initially horizontal but gradually rolls up into a vortex for $Re\sqrt{Sc} > 50$

Modeling the initial irrotational vortex with (8.35), we find that the viscous core radius is derived from (8.32):

$$r^* \approx \sqrt{\nu t},$$

and for $1 < Re = \Gamma/2\pi\nu < 10^3$ the radius of the diffusion core (which is filled with burned gases in the case of a fast reaction with an initially horizontal flame) r_b can increase more rapidly than the viscous core. For example, if $Re\sqrt{Sc} > 50$, we have the asymptotic law

$$r_b^* \approx \sqrt{\Gamma^{2/3}\, D^{1/3}\, t},$$

where Γ is the circulation of the vortex for large radii, and reactant consumption is enhanced by a factor proportional to $\Gamma^{2/3}\, D^{1/3}$ [168] (see Fig. 8.11).

Premixed flames that roll up into vortex structures have also been studied [149], and the dynamics of flame/vortex interactions were studied by Renard et al. [234]. The liquid case has also been examined [167].

8.3 Classical Turbulence Theory

8.3.1 Turbulent Transfer Coefficients and Chemical Kinetics (Simplified Statistical Theory, Incompressible Case)

Averaged Quantities

Let us now consider the fluctuating parameter $f(\mathbf{x}, t)$ (which could be pressure, temperature, velocity, concentration, ...) of a turbulent flow [117], as shown in Fig. 8.12. We can define the time average as

$$\bar{f}(\mathbf{x}, t) = \frac{1}{T} \int_{t-T/2}^{t+T/2} f(\mathbf{x}, \tau)\, d\tau, \tag{8.40}$$

where T is the sampling time, which is chosen to be large enough for f to be independent of it. The parameter f then comprises an average value of $\bar{f}$ and a random part f'.

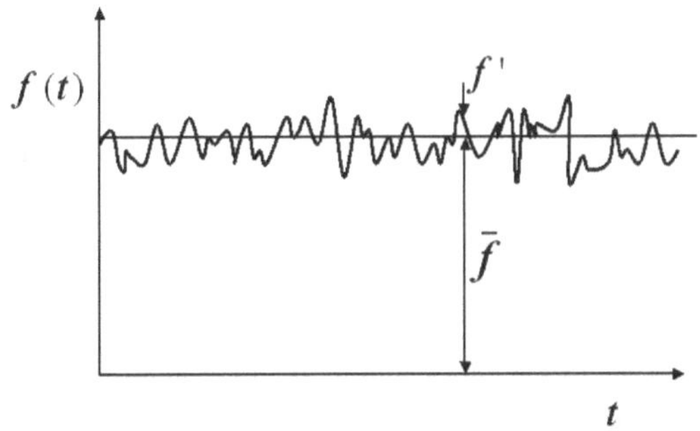

Fig. 8.12. Instantaneous turbulent quantity f versus time

We now suppose that the following Reynolds axioms [10] hold, where k =const.

$$\begin{cases} f = \bar{f} + f',\ \bar{f'} = 0, \\ \overline{kf} = k\bar{f}, \\ \overline{fg} = \bar{f}\bar{g} + \overline{f'g'}, \\ \overline{\partial f/\partial s} = \partial \bar{f}/\partial s,\ \overline{\int f\, ds} = \int \bar{f}\, ds. \end{cases} \tag{8.41}$$

For two points M_1 and M_2 separated by $\xi = |\boldsymbol{\xi}|$, where the velocity is $u_1(\mathbf{x})$ and $u_2(\mathbf{x} + \boldsymbol{\xi})$ at time t, respectively, we can define a correlation coefficient

$$R(\xi) = \overline{u_1' u_2'} / \sqrt{\overline{u_1'^2}}\sqrt{\overline{u_2'^2}} \tag{8.42}$$

and a correlation length L that gives the average eddy size:[9]

$$L = \int_0^\infty R(\xi)\, d\xi. \tag{8.43}$$

If $R(\xi)$ is independent of the direction of M_1M_2 and the position $\mathbf{x}$, the turbulence is known to be *homogeneous and isotropic* (see Sect. 8.3.2).

Coefficients of Turbulent Exchange

Turbulence results in new *flux terms* that must be added to fluxes of the various quantities that have an effect following transfer processes between molecules. For the incompressible case, the balance equation

$$\partial(\rho f)/\partial t + \nabla \cdot (\rho f \mathbf{v} + \boldsymbol{\mathcal{J}}_F) = \dot{W}_F \tag{8.44}$$

becomes, by taking the average,

$$\rho \partial \bar{f}/\partial t + \rho \nabla \cdot (\bar{f}\bar{\mathbf{v}}) + \nabla \cdot \overline{\boldsymbol{\mathcal{J}}_F} + \rho \nabla \cdot \overline{f'\mathbf{v}'} = \overline{\dot{W}_F}. \tag{8.45}$$

Thus, the momentum equation yields the Reynolds tensor

$$\overline{\rho \mathbf{v}' \otimes \mathbf{v}'}, \tag{8.46}$$

which is added to the viscous pressure tensor (the opposite of the viscous strain tensor) to give the sum of the molecular and turbulent viscosities

$$-2\mu \mathbf{D} + \rho \overline{\mathbf{v}' \otimes \mathbf{v}'}. \tag{8.47}$$

We must now express the additional terms that appear in the averaged balance equations. This is the closure problem. This initially focuses on the Reynolds tensor, and more generally the tensors $\rho \overline{f'\mathbf{v}'}$, which, as we will see below, can be interpreted as turbulent fluxes. They are then the production terms $\overline{\dot{W}_F}$. For chemical production, these latter terms are not equal to those computed using the averages of the quantities on which they depend.

[9] We assume that the basic turbulent structures are eddies.

A simplifying assumption (the Boussinesq hypothesis) is to assume that the turbulent fluxes are proportional to the corresponding generalized forces (first gradient closure). We then have

$$\rho\,\overline{u'v'} = -K_u \partial u/\partial y, \tag{8.48}$$

where K_u is a coefficient of turbulent exchange. In contrast to molecular transfer coefficients, coefficients of turbulent exchange depend not only on the local state of the fluid but also the flow characteristics. We can only rarely assume them to be constant. For a turbulent Poiseuille flow, we can assume for example that K_u is related to y. This leads to semi-empirical coefficients that generally have very restricted ranges of validity, but which have the advantages of ensuring the closure of the system of equations and providing qualitative information. For the quantity F, we can introduce coefficient K_F such that

$$\rho\,\overline{f'v'} = -K_F\,\partial \bar{f}/\partial y. \tag{8.49}$$

We therefore define the three *coefficients of turbulent exchange* μ_t, $\lambda_t/\overline{c_p}$ and $\bar{\rho}D_t$, where[10]

$$\begin{cases} \overline{(\rho\mathbf{v})'\otimes\mathbf{v}'} = -\mu_t\boldsymbol{\nabla}\otimes\bar{\mathbf{v}}, \\ \overline{(\rho\mathbf{v})'h'} = -\lambda_t/\bar{c}_p\boldsymbol{\nabla}\bar{h}, \\ \overline{(\rho\mathbf{v})'Y_j'} = -\rho\,D_t\boldsymbol{\nabla}\bar{Y}_j. \end{cases} \tag{8.50}$$

We can then introduce turbulent Schmidt and Prandtl numbers:

$$Sc_t = \mu_t/\rho\,D_t,\ Pr_t = \mu_t\bar{c}_p/\lambda_t. \tag{8.51}$$

By analogy with the kinetic theory of gases, we can now define the "mean length of the mixture" (similar to the mean free path). As an example, consider the transfer of a quantity F in direction y. We assume that the transfer is performed by random jets of length l'_F in the Oy direction, and that these jets have an average value $\bar{f}(y - 1'_F)$. After this free path, there is a sudden change, with the average value becoming $\bar{f}(y)$. Therefore, at coordinate y, the fluctuations f' will be (Fig. 8.13)

$$\begin{cases} f' = \bar{f}(y - l'_F) - \bar{f}(y), \\ f' \cong -l'_F\,\partial f/\partial y. \end{cases} \tag{8.52}$$

By accounting for all possible random jets and taking the average we get

$$\overline{(\rho\mathbf{v})'f'} = -\overline{(\rho\mathbf{v})'l'_F}\,\partial f/\partial y. \tag{8.53}$$

[10]These relations, in which counter-gradient diffusion occurs (see Sect. 8.4.4), are also valid for compressible fluids provided that the Favre averages of Sect. 8.3.2 are used.

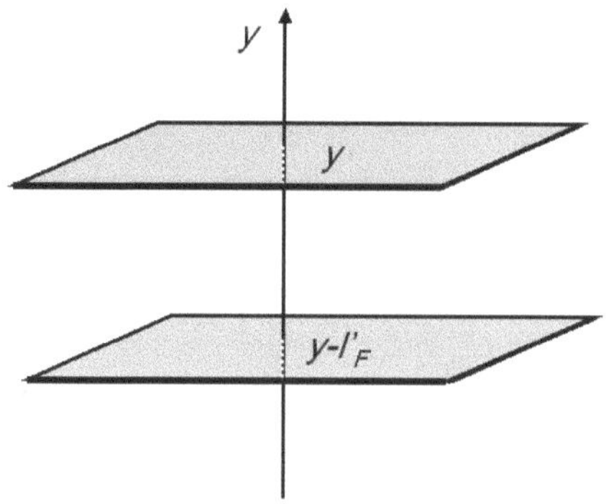

Fig. 8.13. The turbulent fluctuations

Therefore,

$$K_F = \overline{(\rho \mathbf{v})' l'_F}. \tag{8.54}$$

Let us suppose that ρ is constant and that the fluctuations u' are of same order of magnitude as the fluctuations v'. By setting $f = u$, we get

$$\begin{cases} u' \cong -l'_u \, \partial \bar{u}/\partial y \cong v', \\ K_u = \mu_t = \overline{(\rho \mathbf{v})' l'_u} = \rho \, \overline{l'^2_u} \, \partial \bar{u}/\partial y, \end{cases} \tag{8.55}$$

or $K_u/\rho = L_P^2 \, \partial \bar{u}/\partial y$ if we set $L_P^2 = \rho \, \overline{l'^2_u}$. L_P is the Prandtl mixing length. It can be interpreted as the most effective eddy size for mixing.

Turbulence and Chemical Kinetics

Now let us now study the *turbulent terms resulting from the chemical kinetics.* We consider the chemical reaction A + B $\rightarrow$ C. The production rates of the species are

$$\dot{W}_j = \nu_j \mathcal{M}_j \, k \rho^2 \, Y_A Y_B / \mathcal{M}_A \mathcal{M}_B = \pm k_j Y_A Y_B, \tag{8.56}$$

where the sign is (+) for j = C and (−) for j = A or B, and the specific reaction rate k is given by (3.52). The corresponding averaged formula is

$$\begin{aligned} \overline{\dot{W}_A} = & -\bar{k}_1 \, (\bar{Y}_A \bar{Y}_B + \overline{Y'_A Y'_B}) - \bar{Y}_A \, \overline{k'_1 Y'_B} \\ & -\bar{Y}_B \, \overline{k'_1 Y'_A} - \overline{k'_1 Y'_A Y'_B}. \end{aligned} \tag{8.57}$$

We note that, even when k_1' is negligible,[11] the average production rate is the sum of two terms and can therefore be very different from the production rate calculated from the average concentrations:

$$\overline{\dot{W}_A} = -\bar{k}_1\,(\bar{Y}_A\bar{Y}_B + \overline{Y_A'Y_B'}). \tag{8.58}$$

By introducing a characteristic diffusion time τ_D and a chemical time τ_c, it can be shown that
a) If $\tau_D \ll \tau_c$ (i.e., if the reactions are slow or the system is perfectly stirred), the turbulent chemical term is negligible compared to the turbulent diffusion term. Therefore $\overline{Y_A'Y_B'} \cong 0$ and

$$\overline{\dot{W}_A} = -\bar{k}_1\,\bar{Y}_A\bar{Y}_B. \tag{8.59}$$

b) If $\tau_c \ll \tau_D$, the chemical reaction is much faster than the diffusion phenomenon. The species are slightly stirred and in the extreme case they do not meet. Any particular point in space sees the species A or B alternately, but never both at the same time. This means that the production rate vanishes, which leads to

$$\bar{Y}_A\bar{Y}_B = -\overline{Y_A'Y_B'}. \tag{8.60}$$

c) Between these extreme cases, the correlation modulus $|\overline{Y_A'Y_B'}|$ takes values ranging between 0 and 1.

Note that this very simplified presentation of turbulent chemically reacting mixtures focuses on mixing processes and do not takes the heat release effects that generally occur during combustion into account. This approach is justified for dilute gaseous mixtures. To study turbulent combustion, we will use the Favre average defined below. The averaged balance equation (8.45) that was established for the incompressible case is no longer valid in this case, and will be replaced by an other, as shown in Sect. 8.4.1.

8.3.2 Some Definitions Relating to Turbulence

Statistical Processing

The statistical processing of turbulence is generally more sophisticated than described in Sect. 8.3.1 [2]. The statistical average of a random function $f(\mathbf{x}, t)$ is obtained starting from N events $f_j(\mathbf{x}, t)$ so that

$$\langle f\rangle(\mathbf{x}, t) = lim_{N\to\infty}(\frac{1}{N}\sum_{i=1}^{N} f_i(\mathbf{x}, t)). \tag{8.61}$$

[11] This is possible if the temperature fluctuations are negligible, which is obviously not the case in combustion, where temperature changes due to heat release are generally significant. If k_1' is neglected, we are especially interested in the mixing effects.

More generally, we introduce a probability density $P(f|\mathbf{x},\, t)$, and the average becomes

$$\langle f\rangle(\mathbf{x},\, t) = \int_{-\infty}^{+\infty} f(\mathbf{x},\, t)P(f|\mathbf{x},\, t)\, df. \qquad (8.62)$$

This is the *ensemble average*. For pairs of random variables, we also use the density $P(f|\mathbf{x},\, t|\mathbf{x}',\, t')$, which gives access to the correlations. When the probability at a point $P(f|\mathbf{x},\, t)$ does not depend explicitly on time, the turbulence is *steady*. When it does not depend on spatial position, it is *homogeneous*. There is *isotropy* when there is invariance upon translation or rotation, and when there is symmetry about any arbitrary plane. In experiments, we have only one configuration along with variable times or positions. We must therefore recourse to space-time averages rather than ensemble averages. If the weight function is denoted by $\phi(\mathbf{x},\, t)$, we get

$$\langle f\rangle(\mathbf{x},\, t) = \int_{\mathbf{x},\, t} f(\boldsymbol{\xi},\, \tau)\, \phi(\mathbf{x}-\boldsymbol{\xi},\, t-\tau)\, d\boldsymbol{\xi}\, d\tau, \qquad (8.63)$$

with

$$\int_{\mathbf{x},\, t} \phi(\boldsymbol{\xi},\, \tau)\, d\boldsymbol{\xi}\, d\tau = 1. \qquad (8.64)$$

The ensemble averages and the space-time averages are equivalent provided that the assumption of *ergodicity* holds. According to the principle of ergodicity, an infinitely repeated experiment with a single drawing is equivalent—statistically speaking—to a single experiment with an infinite number of drawings.

We also introduce space-time averages at two points for couples between random variables.

Ergodicity was implicitly assumed in Sect. 8.3.1, where time averages were used. In this case, the average value of an unspecified function φ was $\langle\varphi\rangle = \overline{\varphi}$.

Favre Averages

For flows with large changes in density (e.g., in flows of compressible fluids or for combustion), we generally use density-weighted Favre averages [83], [84]. The Favre average $\tilde{f}$ of the quantity f is defined based on the classical average of ρf:

$$\bar{\rho}\tilde{f} = \overline{\rho f}. \qquad (8.65)$$

Splitting $f(\mathbf{x},t)$ into $\tilde{f}(\mathbf{x},t)$ and f'', we have $\widetilde{f''} = 0$ and $\overline{\rho f''} = 0$.

The use of the Favre average makes it simpler to formulate some products that depend on density. The classical average of the triple product $\rho u f$ then yields four terms:

$$\overline{\rho u f} = \bar{\rho}\,\bar{u}\,\bar{f} + \bar{\rho}\,\overline{u'f'} + \overline{\rho u'}\,\bar{f} + \overline{\rho f'}\,\bar{u} + \overline{\rho' u' f'}, \tag{8.66}$$

and the Favre average of the same product gives

$$\overline{\rho u f} = \bar{\rho}\,\tilde{u}\,\tilde{f} + \bar{\rho}\,\widetilde{u'' f''}, \tag{8.67}$$

which has the same formal structure as the conventional average of uf for constant density flows:

$$\overline{uf} = \bar{u}\bar{f} + \overline{u'f'}. \tag{8.68}$$

Difficulties that arise with Favre averaging in viscous and diffusive transport terms are relatively unimportant, since these terms are usually neglected for high Reynolds number turbulence [202].

The Taylor Hypothesis

The assumption called the Taylor hypothesis is valid only for weak rates of turbulence: $\overline{\nu_i'\nu_j'}^{1/2} \ll (\bar{\nu}_i\bar{\nu}_j)^{1/2}$. It is the assumption that, in the homogeneous case, the turbulent field is frozen and the fluctuations are transported by the motion with a mean velocity $\overline{\mathbf{v}}$ that is presumed to be locally constant. This assumes a unique advection velocity, which is not possible in the case of compressible flows (for example in acoustics), or for shear flows where $\bar{u}$ is scale dependent. We therefore obtain a relation between the space and time dependencies. Based on this assumption, it is therefore sufficient to experimentally measure the temporal fluctuations using a fixed probe. The spatial fluctuations can then be deduced using

$$f(t) = f(x/\bar{u}) \Longleftrightarrow \frac{d}{dt} \cong \bar{u}\frac{\partial}{\partial x}. \tag{8.69}$$

Scales of Turbulence

The *integral scale* ℓ characterizes large turbulent structures. We will see in Sect. 8.3.4 that the turbulent kinetic energy is distributed according to wavenumber. In the homogeneous case, the spectral distribution obtained by Fourier transformation is $E(k)$ per unit mass and wavenumber. It can be shown that the integral scale is equal to

$$\ell = \frac{\pi}{2}\frac{E(0)}{\overline{u'^2}}, \tag{8.70}$$

where $E(k)$ is the one-dimensional spectral distribution.

The small structure scale can be defined as the size at which the structures are strongly affected by the viscosity (dissipative area). We thus obtain the *Kolmogorov microscale* ℓ_K described in Sect. 8.3.4.

The *Taylor microscale* λ is intermediate between the integral scale and the Kolmogorov microscale. This makes it possible to connect the variances of the velocity fluctuation gradients to the variances of the velocity fluctuations, and we can define the Taylor microscale via

$$\lambda = (\overline{u'^2}/\overline{(\partial u'/\partial x)^2})^{1/2}. \tag{8.71}$$

Spectral analysis shows that it can be deduced from the energy spectrum $E(k)$:

$$\lambda^2 \int_0^\infty k^2 E(k)\, dk = \int_0^\infty E(k)\, dk. \tag{8.72}$$

8.3.3 k–ϵ Modeling (Closing the Transfer Terms)

We now present a turbulence model for the Navier–Stokes equations, which is first applied to the case of constant density flow, and then to the case of variable density. This model is frequently used in engineering design.

k–ϵ Modeling of Incompressible Fluids

Here, we will only study the case of an nonreactive incompressible fluid with a constant viscosity coefficient. Recall that [2]

$$\begin{cases} \nabla \cdot \overline{\mathbf{v}} = 0, \\ \partial \overline{\mathbf{v}}/\partial t + \nabla \cdot (\overline{\mathbf{v}} \otimes \overline{\mathbf{v}}) + \nabla \cdot \overline{\mathbf{v}' \cdot \mathbf{v}'} + \nabla p/\rho_0 - \nabla \cdot (2\nu\, \overline{\mathbf{D}}) = 0, \end{cases} \tag{8.73}$$

where $\nu = \mu/\rho$, and where the definition of $\mathbf{D}$ is given by (3.32). System closure is obtained when we know the coefficient of turbulent transfer μ_t and the average kinetic energy of turbulence. We can write (more precisely than in the preceding section)

$$\overline{\mathbf{v}' \cdot \mathbf{v}'} - (1/3)\overline{v'^2}\mathbf{1} = -2\nu_t \bar{\mathbf{D}} \tag{8.74}$$

in the case of isotropic turbulence, where the tensor on the left hand side is the tensor for turbulent stresses, which has zero trace. The coefficient $\nu_t = \mu_t/\bar{\rho}$ can be evaluated starting from the Prandtl mixing length L_P. However, a more sophisticated theoretical approach connects it to the average kinetic energy of turbulence per unit mass

$$\bar{k} = \overline{v'^2}/2 \tag{8.75}$$

and to the average rate of turbulence dissipation

$$\bar{\epsilon} = \overline{\nu(\nabla \otimes \mathbf{v}' + \widetilde{\nabla \otimes \mathbf{v}'}) : \nabla \otimes \mathbf{v}'}. \tag{8.76}$$

We then have

$$\nu_t = \overline{v'l'_u} \cong a_v L\sqrt{\bar{k}} \tag{8.77}$$

where a_v is a constant coefficient and L is a mixing length that can be different from L_P. It can be shown that, for high Reynolds numbers, the rate of dissipation is connected to L and $\bar{k}$ by the relation

$$\bar{\epsilon} = c_1 \bar{k}^{3/2}/L. \tag{8.78}$$

It follows that

$$\nu_t = c_\mu \bar{k}^2/\bar{\epsilon}, \tag{8.79}$$

where $c_v = a_v c_1$ is constant. By utilizing the local balance equations for mass and momentum, multiplying appropriately by v'_j and its derivatives, and taking the average, we obtain two balance equations in $\bar{k}$ and $\bar{\epsilon}$ that reveal a series of second- or third-order correlation terms and their derivatives. System closure is obtained by neglecting some of these terms and by providing approximations for the others. Neglecting molecular viscosity, the result (which will not be proven here) is as follows:

$$\begin{cases} \partial\bar{k}/\partial t + \bar{\mathbf{v}}\cdot\nabla\bar{k} = \nabla\cdot(\nu_t\nabla\bar{k}/\sigma_k) + \bar{P}(k) - \bar{\epsilon}, \\ \partial\bar{\epsilon}/\partial t + \bar{\mathbf{v}}\cdot\nabla\bar{\epsilon} = \nabla\cdot(\nu_t\nabla\bar{\epsilon}/\sigma_\epsilon) + c_{\epsilon 1}(\bar{\epsilon}/k)\bar{P}(k) - c_{\epsilon 2}\bar{\epsilon}^2/\rho_0\bar{k}. \end{cases} \tag{8.80}$$

On the right hand sides of these equations, the first terms are flux terms, σ_k is the Prandtl number for turbulent energy, while σ_ϵ is the Prandtl number for turbulent dissipation. The other terms are production and destruction terms. We have

$$\bar{P}(k) = -\nu_t\,\overline{\mathbf{v}'\otimes\mathbf{v}'}\mathbf{:}\nabla\otimes\bar{\mathbf{v}}, \tag{8.81}$$

where ν_t obeys (8.79).

The constants have been evaluated in many experiments (in the atmosphere and in oceans), and good choices appear to be[12]

$$c_\mu = 0.09,\ \sigma_k = 1,\ \sigma_\epsilon = 1.3,\ c_{\epsilon 1} = 1.44,\ c_{\epsilon 2} = 1.92 \tag{8.82}$$

The equations obtained do not then contain the correlations but simply the gradients of the average quantities. They are coupled with the average balance equations, and the whole system can be solved by numerical methods, taking into account suitable initial conditions and boundary conditions.

---

[12]These values are associated with a particular model of isotropic turbulence. Other models give different coefficients. For instance, Turpin [279] uses $c_v = 0.09$, $\sigma_k = 1$, $\sigma_\epsilon = 1.7$, $c_{\epsilon 1} = 1.5$, $c_{\epsilon 2} = 1.82$ in a study of an axisymmetric Masri flame.

k–ϵ Modeling of Variable Density Fluids

Neglecting the buoyancy terms and the bulk viscosity, the averaged Navier Stokes equations can be written as

$$\begin{cases} \partial\bar{\rho}/\partial t + \nabla\cdot(\bar{\rho}\tilde{v}) = 0, \\ \partial(\bar{\rho}\tilde{v})/\partial t + \nabla\cdot(\bar{\rho}\tilde{\mathbf{v}}\otimes\tilde{\mathbf{v}}) + \nabla\cdot(\bar{\rho}\widetilde{\mathbf{v}''\cdot\mathbf{v}''}) + \nabla\bar{p} - \nabla\cdot(2\bar{\rho}\nu\,\tilde{\mathbf{D}}) = 0. \end{cases} \tag{8.83}$$

This time, we have

$$\bar{\rho}\widetilde{\mathbf{v}''\otimes\mathbf{v}''} = -\bar{\rho}\nu_t(2\tilde{\mathbf{D}} - (2/3)\nabla\cdot\tilde{\mathbf{v}}) + (2/3)\bar{\rho}\tilde{k}\mathbf{1} \tag{8.84}$$

and

$$\tilde{k} = \widetilde{\mathbf{v}'\cdot\mathbf{v}'}/2,\ \nu_t = c_\mu\tilde{k}^2/\epsilon,\ c_\mu = 0.09. \tag{8.85}$$

The k–ϵ system of equations is then

$$\begin{cases} \bar{\rho}\partial\tilde{k}/\partial t + \bar{\rho}\tilde{\mathbf{v}}\cdot\nabla\tilde{k} = \nabla\cdot(\bar{\rho}\nu_t\nabla\tilde{k}/\sigma_k) + \tilde{P}(k) - \bar{\rho}\tilde{\epsilon}, \\ \bar{\rho}\partial\tilde{\epsilon}/\partial t + \bar{\rho}\tilde{\mathbf{v}}\cdot\nabla\tilde{\epsilon} = \nabla\cdot(\bar{\rho}\nu_t\nabla\tilde{\epsilon}/\sigma_\epsilon) + c_{\epsilon1}(\bar{\rho}\tilde{\epsilon}/\tilde{k})\tilde{P}(k) - c_{\epsilon2}\bar{\rho}\tilde{\epsilon}^2/\tilde{k}, \end{cases} \tag{8.86}$$

with the same coefficient values as in (8.82), and

$$\tilde{P}(k) = -\bar{\rho}\,\nu_t\,\widetilde{\mathbf{v}'\otimes\mathbf{v}'}{:}\nabla\otimes\tilde{\mathbf{v}}. \tag{8.87}$$

Comments Regarding k–ϵ Modeling

k–ϵ models do have their weaknesses. For example, they give a trivial solution in the homogeneous (without an average velocity) and isotropic case.

One of their most important limitations is connected to the viscosity-related decrease in the kinetic energy of homogeneous and isotropic turbulence.[13] To treat this case, we must use spectral dynamics (this topic is addressed to some degree in Sect. 8.5.4), which takes into account the interactions between energy modes. Solution methods such as the k–ϵ method are based on a "one-point" formalism and so do not contain as much spectral information as those based on a two-point formalism [2, 158].

However, k–ϵ models are frequently utilized, including for reactive flows, where we must also determine the average chemical production terms $\overline{\dot{W}_j}$. They give good results for inhomogeneous flows, such as some flows of engineering interest, boundary layers, jets, wakes, and mixing layers.

---

[13] It should be noted that the k–ϵ model is not compatible with the energy decay laws of the spectral theory (see Sect. 8.3.4, [2]). In this case, the k–ϵ model reduces to $\partial k/\partial t = -\bar{\epsilon}$; $\partial\bar{\epsilon}/\partial t = -c_{\epsilon 2}\bar{\epsilon}^2 k$, which leads to $\partial^2 k/\partial t^2 = c_{\epsilon 2}/h(\partial k/\partial t)^2$ or $k = (At+B)^{-1/(c_{\epsilon 2}-1)}$. In the k–ϵ model, the energy decreases as $t^{-\alpha_e}$, where the superscript α_e does not depend on the shape of the initial spectrum, in contrast to what is actually observed.

8.3.4 Spectral Analysis and Kolmogorov's Theory

Applying the Fourier transformation to any function $f(\mathbf{x})$:

$$f(\mathbf{k}) = \frac{1}{8\pi^3} \int_{\Re^3} f(\mathbf{x}) e^{-i\mathbf{k}\cdot\mathbf{x}}\, d\mathbf{x} \tag{8.88}$$

makes it possible to study the energy spectrum for turbulent structures. The inverse of the wavenumber k defines the turbulence scale; i.e., the eddy size.

We will only outline the spectral theory here. Note, however, that energy spectra are easily obtained experimentally these days. We will assume here that the turbulence is homogeneous and isotropic (in particular, $\langle \mathbf{v} \rangle = 0$) and that the fluid is incompressible and viscous. Moreover, the evolution is unconstrained, and the effects of the walls are neglected. Finally, we assume that the initial conditions are random.

It can then be shown that the spectrum of turbulent energy $E(k)$ includes two distinct zones separated by a transition zone (Fig. 8.14):

- A zone with low values of k; i.e., a *zone of large eddies* (size k_e^{-1}) where energy is steadily fed in (by instabilities, interaction with the average motion, external forces, etc.) at a rate $\dot{e}_1$ and concentrated
- A *zone of small structures* of size $1/k_d$ (k_d is the Kolmogorov wavenumber) where energy is dissipated by the viscosity at a rate of $\dot{e}_2$
- A transition or *inertial zone* between the two other zones where the advection terms in $\partial(u_i u_j)/\partial x_j$ ensure the transfer of energy between the different scales in a conservative way.

If the transfer zone has a non-negligible extent, then $k_d \gg k_e$, which implies that there is a condition on the Reynolds number. We define a Reynolds number for turbulence starting from the eddy size k^{-1} and a speed based on the energy $E(k)$ (which is the energy of the velocity fluctuations per unit mass and unit wavenumber). The quantity $(k\, E(k))^{1/2}$ is thus the mean velocity of structures of size k^{-1}. Its associated Reynolds number is

$$Re = (1/\nu k)(k\, E(k))^{1/2}. \tag{8.89}$$

For $k = k_d$, the dissipative phenomena are around the same order of magnitude as the inertial phenomena, so that $(Re)_d \approx 1$.

$$(Re)_d = (1/\nu k_d)(k_d\, E(k_d))^{1/2}. \tag{8.90}$$

For large structures, we obtain $(Re)_e = (1/\nu k_e)(k_e\, E(k_e))^{1/2}$ with $E(k_e) > E(k_d)$ and $k_e \ll k_d$. It follows that

$$(Re)_e \gg (Re)_d \approx 1. \tag{8.91}$$

This inequality summarizes the condition required for the existence of a transfer zone where the motions are independent of large energy scales, sources of turbulence, and the viscosity.

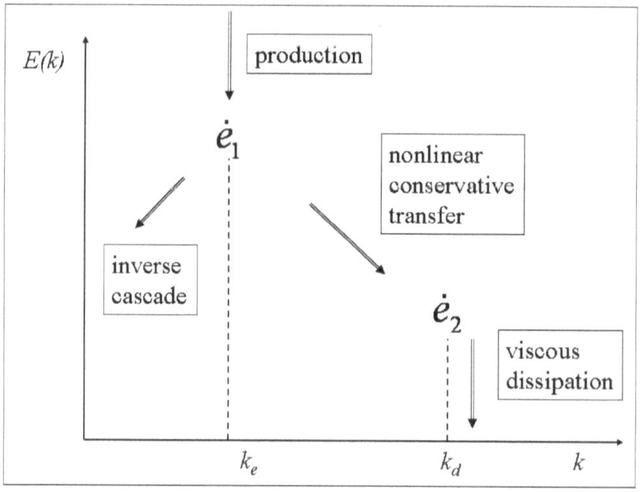

Fig. 8.14. Characteristic zones of the spectrum of turbulent energy

Kolmogorov's theory postulates: 1) the existence of small scales, and 2) the existence of an inertial zone (the "eddy cascade hypothesis").[14]

This makes it possible to determine the $E(k)$ spectrum there using dimensional analysis. We assume a quasi-steady regime; the rates $\dot{e}_1$ and $\dot{e}_2$ then have a common value ϵ that is the energy produced per unit mass per unit time (dimensions: L^2T^{-3}). We can relate ϵ directly to the turnover velocity and the length scale ℓ of integral-scale eddies:

$$\epsilon \approx v'^3/\ell. \tag{8.92}$$

In the same manner, the turnover velocity v_n of any inertial-range eddy is related to its size ℓ_n by

$$\epsilon = {v_n}^3/\ell_n \tag{8.93}$$

where the rate ϵ is the same as before. We can take k, $\bar{\epsilon}$, ν, $E(k)$ as characteristic parameters of the problem. It is useful to replace ν with a combination of $\bar{\epsilon}$ and ν that has the dimensions of length:

$$\ell_k = (\nu^3/\bar{\epsilon})^{1/4}. \tag{8.94}$$

[14]It is possible to define, as done by Peters [202], a discrete sequence of eddies within the inertial subrange by $\ell_n = \ell/2^n \geq \ell_K$.

This is known as the Kolmogorov length.[15]

We then choose k and $\bar{\epsilon}$ as basic quantities. The following Π ratios are then obtained from Table 8.2:

	k	$\bar{\epsilon}$	$\ell_{\mathbf{K}}$	**E(k)**
L	-1	2	1	3
M	0	0	0	0
T	0	-3	0	-2

Table 8.2. Dimensions of the quantities of interest

$$\begin{cases} \Pi_l = k\ell_k, \\ \Pi_E = k^{5/3}\,\bar{\epsilon}^{-2/3}\,E(k). \end{cases} \tag{8.95}$$

The Vashi–Buckingham theorem gives us

$$\Pi_E = \Psi(\Pi_l) \tag{8.96}$$

or

$$E(k) = \bar{\epsilon}^{\,2/3}\,k^{-5/3}\,\Psi(k\,\ell_K). \tag{8.97}$$

In the inertial zone, viscosity does not intervene, and we have $\Psi(k\,\ell_K) = c_K$, the Kolmogorov constant, and

$$E(k) = c_K\,\bar{\epsilon}^{2/3}\,k^{-5/3}. \tag{8.98}$$

This law has been confirmed experimentally for the nonreactive media.[16] The constant c_K is on the order of 1.5. Figure 8.15 relates to the one-dimensional spectrum, but a similar law applies in the three-dimensional case, although with the constant $(48/55)\,c_K$.

The inertial zone is limited by the wavenumber k_d, which, since $Re_d \approx 1$, is equal to

$$k_d \approx E(k_d)/\rho\nu^2. \tag{8.99}$$

Therefore, in the inertial zone, for $k = k_d$, we have

---

[15]The Kolmogorov eddy size and turnover velocity obey the relation $\epsilon = {v_K}^3/\ell_K$, similar to (8.93) and (8.92).

[16]In the presence of chemical reactions, and in particular combustion, the $k^{-5/3}$ law does not always hold. This is undoubtedly due to the fact that the turbulence is not homogeneous and isotropic in this case. In addition, other parameters come into play in addition to those accounted for in dimensional analysis, in particular terms related to chemical production, diffusion and thermal transfer, as well as geometric parameters. Heat release leads to a drop in the Reynolds number and then flow relaminarization [235].

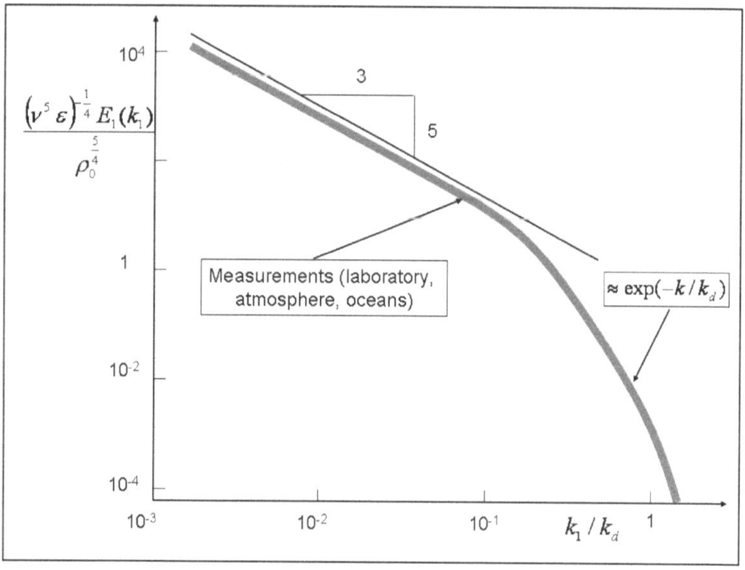

Fig. 8.15. One-dimensional energy spectrum showing the universality of the Kolmogorov law

$$k_d \approx \bar{\epsilon}^{\,2/3} k_d^{-5/3} / \nu^2 \tag{8.100}$$

or

$$k_d^4 \approx \bar{\epsilon}\,\nu^3 = \ell_K^{-4}, \; k\ell_K = k/k_d. \tag{8.101}$$

It follows from this and (8.97) that in the dissipative zone

$$E(k) = \bar{\epsilon}^{2/3}\, k^{-5/3}\, \Psi(k/k_d). \tag{8.102}$$

In addition, the Reynolds number for large eddies is

$$\begin{cases} (Re)_e = (1/\nu k_e)\,(k_e\, E(k_e))^{1/2}, \\ (Re)_e \approx (k_e^{-4/3}/\nu)\, \bar{\epsilon}^{\,1/3} = (k_d/k_e)^{4/3} \gg 1. \end{cases} \tag{8.103}$$

8.4 Turbulent Combustion

We can distinguish between several types of combustion that are laminar or turbulent a priori [8, 206].

A distinction was made between premixed flames (where the fuel and the oxidizer are initially premixed) and diffusion flames (where the fuel and oxidizer are initially separated) in Sect. 7.7. Several sections in this book refer to these types of flames.

Another distinction is based on the rapidity of the chemical kinetics of the reactions. This leads to two cases: infinitely fast chemistry and finite-rate chemistry.

A lot of models have been developed to study turbulent combustion. A classification of turbulent combustion models in terms of chemistry and mixing is provided by Peters [202]:

1. For infinitely fast chemistry and premixed combustion, we have the Bray–Moss–Libby model and the "coherent flame model;" for infinitely fast chemistry and nonpremixed combustion we have "conserved scalar equilibrium model"
2. For finite rate chemistry, we have the "pdf transport equation model" and the "linear eddy model" in the premixed and the nonpremixed cases, as well as the "flamelet model based on the G-equation" for premixed combustion and the "flamelet model based on the mixture fraction" and "conditional moment closure" for nonpremixed combustion.

Turbulence generally increases the contact surface between the fuel and the oxidizer and that between the unburned and burned gases through stretching and crumpling. The combustion of a homogeneous mixture can also begin in the bulk when the temperature and pressure are sufficient to cause self-ignition and the explosion of the mixture (as in diesel engines). Turbulence supports premixed combustion by increasing the flame surface (but excessive stretching can also result in extinction) and by supporting diffusive phenomena (turbulent transfer is often more effective than molecular transfer).

We will not go into all of these models in detail, but we will delve into some of them. For more information on them, refer to the books on combustion (or, more precisely, turbulent combustion) cited at the end of Chap. 8.

8.4.1 Averaged Balance Equation for Turbulent Combustion

In turbulent combustion, the equations relating to the velocity can be separated from those relating to the scalar quantities: the mass fractions of the species and temperature. An important assumption is that the cascade hypothesis of Sect. 8.3.4 remains valid. Before we study the scaling of and present regime diagrams for turbulent premixed combustion and nonpremixed combustion, and investigate some combustion models, we will first establish the balance equations for reactive scalars. This method will lead us to consider the Favre scalar dissipation rate and closure problems.

Let us again consider the balance equation (8.44) of a property F (assumed to be a scalar: mass fraction Y_j or temperature). Suppose that the flux $\mathcal{J}_F$ corresponds to only one diffusion coefficient D. The balance equation becomes

$$\rho\,\partial f/\partial t + \rho\mathbf{v}\cdot\boldsymbol{\nabla} f = \boldsymbol{\nabla}\cdot(\rho D\boldsymbol{\nabla} f) + \dot{W}_F. \tag{8.104}$$

This time, we consider a variable density ρ and make use of Favre averaging: $f = \tilde{f} + f''$. Setting $\dot{W}_F = \rho S_F$, we obtain

$$\bar{\rho}\,\partial\tilde{f}/\partial t + \bar{\rho}\tilde{\mathbf{v}}\cdot\boldsymbol{\nabla}\tilde{f} = \boldsymbol{\nabla}\cdot(\overline{\rho D\boldsymbol{\nabla} f}) - \boldsymbol{\nabla}\cdot(\bar{\rho}\widetilde{\mathbf{v}''f''}) + \bar{\rho}\tilde{S}_F. \tag{8.105}$$

At high Reynolds numbers, the molecular diffusion term $\boldsymbol{\nabla}\cdot(\overline{\rho D\boldsymbol{\nabla} f})$ is negligible.

The source term is proportional to $S_T = B(T_b - T)\exp(-E_a/RT)$ and can be written

$$S_T(T) = S_T(\tilde{T})(1 - T''/(T_b - T))\exp(E_a)T''/R\tilde{T}^2. \tag{8.106}$$

This causes enhanced fluctuations of the chemical source term around its mean value (see Sects. 10.5 and 11.4.2.)

For nonreacting scalars, it is possible to write $\widetilde{\mathbf{v}''f''} = -D_t\boldsymbol{\nabla}\tilde{f}$ [202]. Let us now write the balance equation for $\widetilde{f''^2}$,

$$\bar{\rho}\,\partial\widetilde{f''^2}/\partial t + \bar{\rho}\tilde{\mathbf{v}}\cdot\boldsymbol{\nabla}\widetilde{f''^2} = -\boldsymbol{\nabla}\cdot(\bar{\rho}\widetilde{\mathbf{v}''f''^2}) - 2\bar{\rho}\widetilde{\mathbf{v}''f''}\cdot\boldsymbol{\nabla}\tilde{f} - \bar{\rho}\widetilde{\chi_F} + 2\bar{\rho}\widetilde{f''S''_F}, \tag{8.107}$$

along with the Favre scalar dissipation rate associated with the scalar F,

$$\widetilde{\chi_F} = 2D(\widetilde{\boldsymbol{\nabla} f''})^2. \tag{8.108}$$

We define an integral scalar timescale $\tau_F = \widetilde{f''^2}/\chi_F$. In the nonreactive case, τ_F is proportional to the time $\tau = \tilde{k}/\tilde{\epsilon}$: $\tau = c_\chi\tau_F$, with $c_\chi \approx 2$, so $\chi_F = c_\chi(\tilde{\epsilon}/\tilde{k})\widetilde{f''^2}$. As production = dissipation (quasi-steady regime of the inertial cascade), we have

$$-2\bar{\rho}\widetilde{\mathbf{v}''f''}\cdot\boldsymbol{\nabla}\tilde{f} = \bar{\rho}\widetilde{\chi_F}$$

and

$$-2\widetilde{\mathbf{v}''f''}\cdot\boldsymbol{\nabla}\tilde{f} = c_\chi(\tilde{\epsilon}/\tilde{k})\widetilde{f''^2}.$$

Multiplication of both sides of this equation by $D_t \propto \tilde{k}^2/\tilde{\epsilon}$ leads to

$$-D_t\widetilde{\mathbf{v}''f''}\cdot\boldsymbol{\nabla}\tilde{f} \approx c_\chi\tilde{k}\widetilde{f''^2}.$$

Assuming isotropy, and thus a proportionality between these two quantities, we deduce the validity of the gradient transport assumption:

$$-\widetilde{\mathbf{v}''f''} \approx c_\chi^{-1}D_t\boldsymbol{\nabla}\tilde{f}. \tag{8.109}$$

This property is also valid for linearly reacting scalars (see [202]).

8.4.2 Turbulent Regimes for Premixed Combustion

The classification shown in Fig. 8.16, provided by Peters [202], relates to premixtures. This classification supposes that the turbulence presents the inertial range described in Sect. 8.3.4, and that the flames are thin; i.e., they have deflagration wave structure (see Sect. 10.5 and Sect. 11.4.2 for stretched flames), and that the reaction has a high activation energy.

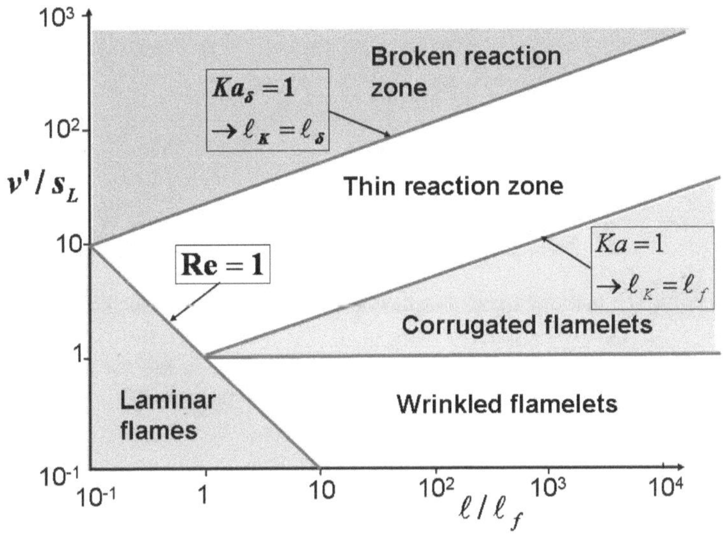

Fig. 8.16. Peters' regime diagram for premixed turbulent combustion. (Redrawn after [202]; reprinted with the permission of Cambridge University Press)

In Fig. 8.16, the lines of separation between the domains are defined based on characteristic quantities and dimensionless numbers. Some of these quantities are related to classical turbulence, such as:

- ℓ: the integral scale defined previously in Sect. 8.3.2
- $v' = \sqrt{u'^2}$: the intensity of the turbulence
- ℓ_K: the Kolmogorov length scale defined by (8.101).

Other quantities are related to the combustion:

- $\ell_f = D/s_L$: the flame diffusion thickness; for scaling purposes, it is assumed that the diffusion coefficient is the same for all species and that the Schmidt number $S_c = \nu/D = 1$
- $\ell_\delta = \delta\ell_f$: the flame reaction thickness; for a chemical reaction with a high activation energy, we have $\delta \ll 1$, as in Sect. 11.4.2 (see also Sect. 10.5)

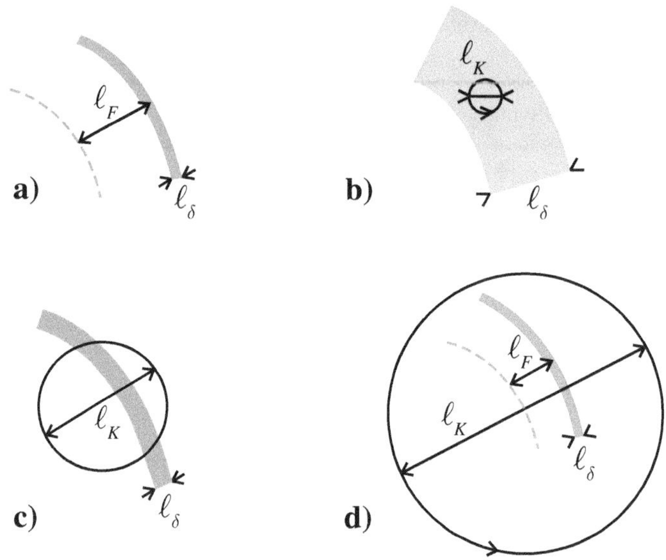

Fig. 8.17. Premixed turbulent combustion regimes. **a** Typical premixed flamelet structure: $\ell_\delta \ll \ell_f$. **b** Broken reaction zone: Kolmogorov eddies can break up the reacting zone, $\ell_K \ll \ell_\delta$. **c** Thin reaction zone: reactive layers are included in the Kolmogorov eddy, $\ell_\delta \ll \ell_K$. **d** Corrugated and wrinkled flamelets: flamelet inside a Kolmogorov eddy, $\ell_f \ll \ell_K$

- s_L: the normal combustion velocity, which is also deduced from the theory of laminar deflagration waves (see Sect. 10.5).

The Gibson scale $\ell_G = s_L^3/\epsilon$ is a quantity that couples turbulence and combustion.

The dimensionless numbers are:

- $Re = v'\ell/s_L\ell_f$: the turbulent Reynolds number, related to the flame thickness
- $Da = s_L\ell/v'\ell_f$: the turbulent Damköhler number
- $Ka = \ell_f^2/\ell_K^2$: the first Karlovitz number; we have $Re = Da^2Ka^2$
- $Ka_\delta = {\ell_\delta}^2/\ell_K^2 = \delta^2 Ka$: the second Karlovitz number.

We must now describe the different regimes of Fig. 8.16. Just before we do, it is important to explain the *flamelet concept*, as discovered by Williams [289], which is currently used in all types of turbulent combustion (we should also mention Peters, Kuznetsov and Bray for their extensions of this concept). Flamelets are thin reactive–diffusive layers embedded within an otherwise non-reacting turbulent flow field. Poinsot and Veynante [206] give a less restrictive definition of the flamelet regime that corresponds to a continuous flame front without quenching,[17] and where no reference is made to thin flames.

The laminar domain is limited to the region where $Re \ll 1$. The other regions of the diagram can be explained by comparing the previously defined scales, especially the Kolmogorov length scale ℓ_K and the characteristic thicknesses of the premixed flamelet, as can be seen in Fig. 8.17. In the cases (b) and (c), the reaction zone can be considered a well-stirred reactor (see Fig. 6.5 in Sect. 6.2): the combustion starts when the upper branch is obtained under the effect of heating; extinction can occur if their is too little heating. In case (d), the zone of corrugated flamelets is distinguished from the zone of wrinkled flamelets using the Gibson scale, where the flame speed s_L^0 is equal to the turbulent velocity v' (see [202]). For $v' < s_L$, the flamelet shape is relatively undisturbed: the speed of the turbulent motion is too low to wrinkle the flame front sufficiently to allow flame interactions. For $v' > s_L$, turbulent motions are able to wrinkle the flame front up to flame interactions, leading to the formation of pockets of fresh and burnt gases; this is the "thin flame regime with pockets" or the "corrugated flamelet regime."

Some deficiencies in the previous classification of premixed flame regimes were pointed out by Poinsot and Veynante, who used the results of direct numerical simulations (DNS) and experimental results to justify modifying combustion diagrams [206]. They provided several arguments for their refinements to Peters' regime diagram:

- The analysis assumes homogeneous and isotropic turbulence that is unaffected by the heat release defined by (2.116)
- Regime limits are based only on order-of-magnitude estimations, not on precise derivations
- The Kolmogorov scale ℓ_K is too small or has insufficient velocity to affect the flame front
- Kolmogorov vortices are effective at inducing strain, but they have short lifetimes due to viscous dissipation, and are probably unable to effectively quench the flame
- Scales smaller than ℓ_f induce local curvature and then thermodiffusive effects that may counteract the influence of strain
- The interaction between a given vortex and a flame front is essentially unsteady.

The modified combustion diagrams take these considerations, as well as the heat loss due to radiation, large flame stretching due to the action of pairs of vortices, and so on, into account. They are not given here, but are shown in [206].

[17] Flame quenching occurs when a flame front is submitted to external perturbations such as heat losses or aerodynamic stretching that is strong enough to decrease the reaction rate to a negligible value, and in some cases to completely suppress the combustion process.

An Example of a Wrinkled Flame

As an example of a wrinkled flame, we can cite the case of a premixed Bunsen flame under the influence of gravity. The flame is obtained using a burner that provides relatively low rates of turbulence. In certain situations a rippled flame is observed. The turbulence of the upstream flow (fresh gas) is modified when the flame crosses it. If we measure the fluctuation velocity at various points using a laser anemometer, we can deduce the turbulent energy spectrum $E(k)$ by treating the signal appropriately. The integral scale corresponds to small values of the wavenumber k, and can be deduced directly from the energy spectrum at each point.

The following result shows the complexity of the phenomenon: we observe a strong increase in the integral scale ℓ_z (Sect. 8.3.2), where z is the vertical distance from the burner exit, at the level where the flame crosses. This phenomenon originates with the fluctuations in the position of the unstable flame front (Sects. 10.4 and 10.5). Thus, the passage of the deflagration wave in the zone of measurement is observed. The observed effect is the intermittency of the flame, not the turbulence of the flow (which is incompressible upsteam and downstream of the flame). Downstream, the integral scale returns to its initial value, and then grows under the effect of turbulence damping at small scales, accelerated by the increase in viscosity ν, which is related to the rise in temperature.

8.4.3 Turbulent Regimes for Nonpremixed Combustion

The scaling is more complex in the case of nonpremixed combustion than it was for premixed combustion. Indeed, there is no simply defined flame thickness as described in Sect. 8.4.2, and turbulent scales are evaluated partially within the space of the mixture fraction Z introduced in Sect. 7.7.2. We have:

- A flame thickness $\ell_D = \sqrt{D_{st}/a}$, defined as the ratio of the diffusion coefficient taken at the location of the flame when $Z = Z_{st}$ and the strain rate a.
- $\Delta Z_f = |\nabla Z|_{st}\, \ell_D$, the corresponding diffusion thickness in the mixture fraction space. This diffusion thickness can be rewritten as a function of the strain rate a and the instantaneous local value of the scalar dissipation rate of (8.108) for a stoichiometric mixture, $\chi_{st} = 2D_{st}\, |\nabla Z|_{st}^2$. Thus, we have $\Delta Z_f = \sqrt{\chi_{st}/2a}$.
- A reaction thickness $\Delta Z_R = \epsilon \Delta Z_f$ that is equal to the oxidation layer thickness, which is proportional to $\chi_{st}^{1/4}$ for a four-step methane–air diffusion flame [202]. Finally, we obtain $\Delta Z_R = \epsilon_q(\chi_{st}/\chi_q)^{1/4}\Delta Z_f$, where ϵ_q and χ_q are values corresponding to the extinction limit.
- For turbulent diffusion flames, ΔZ_f and ΔZ_R must be compared to the mixture fraction fluctuation intensity $Z' = (\widetilde{Z''^2})^{1/2}$, and more precisely to $Z'_{st} = (\widetilde{Z''^2_{st}})^{1/2}$ at the flame location where $\tilde{Z}(\mathbf{x}, t) = Z_{st}$.

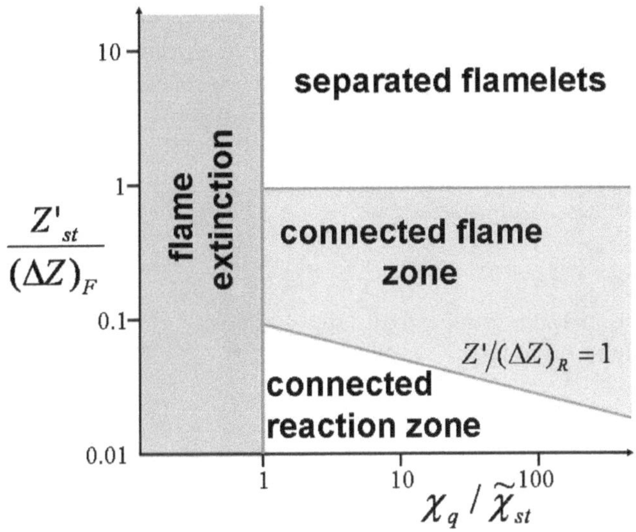

Fig. 8.18. Simplified Peters' regime diagram for nonpremixed turbulent combustion. (Redrawn after [202]; reprinted with the permission of Cambridge University Press)

The coordinates chosen for the Peters diagram (Fig. 8.18) [202] are then χ_q/χ_{st} and $Z'_{st}/\Delta Z_f$. We observe four regions in this diagram:

- A flame extinction zone for $\chi_q/\chi_{st} < 1$.
- A zone of separated flamelets for $Z'_{st}/\Delta Z_f > 1$.
- Two connected combustion zones separated by a straight line of slope $(-1/4)$ in logarithmic coordinates, corresponding to $Z'_{st} = \Delta Z_f$. For $Z'_{st} > \Delta Z_f$, the fluctuations of the mixture fraction are sufficient to separate the diffusion zones surrounding the reaction zone; for $Z'_{st} < \Delta Z_f$, the fluctuations of the mixture fraction are small and there is intense mixing.

8.4.4 Combustion Models

Authors such as Peters [202] and Poinsot and Veynante [206] have presented the main turbulent combustion models (TCMs). For infinitely fast chemistry, the Damköhler parameter (defined based on a characteristic turbulence time τ_t and the chemical time τ_{chim}) $Da = \tau_t/\tau_{chim} \gg 1$, and combustion is controlled solely by the turbulent stirring. However, as soon as finite-rate chemistry is introduced, the composition of the mixture is no longer solely dependent on the flow variables; it depends on the Damköhler number too. Some of these models are presented hereafter.

The Bray–Moss–Libby Model

The Bray–Moss–Libby (BML) model applies to a premixed flame with a very high reaction activation energy. Thus, the flame thickness is infinitely small, and the turbulence and combustion are separated. It combines a statistical approach using the probability density function with physical analysis of the flamelet regime.

Let us consider the progress variable c, which can be

$$c = (T - T_u)/(T_b - T_u) \text{ or } c = Y_P/Y_{P,b}. \tag{8.110}$$

We assume that the normalized probability density function of c is defined by two Dirac delta functions

$$\begin{cases} P(c; \mathbf{x}, t) = \alpha(\mathbf{x}, t)\delta(c) + \beta(\mathbf{x}, t)\delta(1 - c), \\ \int_0^1 P(c; \mathbf{x}, t)dc = 1. \end{cases} \tag{8.111}$$

Thus, we have

$$\alpha(\mathbf{x}, t) + \beta(\mathbf{x}, t) = 1 \tag{8.112}$$

As the pressure is constant and uniform to a first approximation, and the mean molar mass is the same for fresh and burnt gases, we can write

$$\rho/\rho_u = T_u/T = (1 - \gamma)/[1 - \gamma(1 - c)]; \ \gamma = 1 - \rho_b/\rho_u. \tag{8.113}$$

The following results can be deduced for the average values of ρ and c:

$$\begin{cases} \bar{\rho}/\rho_u = \int_0^1 \int_0^1 (\rho/\rho_u)P(c; \mathbf{x}, t)dc = 1 - \beta(\mathbf{x}, t)\gamma, \\ \tilde{c} = \overline{\rho c}/\bar{\rho} = (\rho_u/\bar{\rho}) \int_0^1 \rho c P(c; \mathbf{x}, t)dc = \beta(1 - \gamma)/(1 - \beta(\mathbf{x}, t)\gamma), \\ \bar{c} = \int_0^1 cP(c; \mathbf{x}, t)dc = \tilde{c}/[1 - \gamma)(1 - \tilde{c})], \\ \bar{\rho}/\rho_u = (1 - \gamma)/[1 - \gamma(1 - \tilde{c})]. \end{cases} \tag{8.114}$$

Let us now consider a velocity component u in addition to c. We obtain

$$\begin{cases} P(u, c; \mathbf{x}, t) = \alpha(\mathbf{x}, t)\delta(c)P(u_u; \mathbf{x}, t) + \beta(\mathbf{x}, t)\delta(1 - c)P(u_b; \mathbf{x}, t), \\ \int_0^1 P(u, c; \mathbf{x}, t)dudc = 1, \\ \tilde{u}(\mathbf{x}, t) = (1 - \tilde{c})\bar{u}_u(\mathbf{x}, t) + \tilde{c}\bar{u}_b(\mathbf{x}, t). \end{cases} \tag{8.115}$$

For the Favre correlation $\widetilde{u''c''} = \overline{\rho(u - \tilde{u})(c - \tilde{c})}/\bar{\rho}$, we get

$$\widetilde{u''c''} = \tilde{c}(1 - \tilde{c})(u_b - u_u). \tag{8.116}$$

This relation can be compared with (8.50) obtained by applying classical gradient theory to the Favre correlation (Sect. 8.3.2) and to (8.109), which can be written

$$\widetilde{u''c''} = -D_t \partial c/\partial x. \tag{8.117}$$

Note that, for a steady planar turbulent flame, there is a contradiction between (8.116), which is experimentally verified, and (8.117), where the gradient $\partial c/\partial x$ takes the place of the jump $(1-\tilde{c})$. Indeed, as $u_b > u_u$, then (8.116) indicates that $\widetilde{u''c''} > 0$, and this leads to a negative concentration gradient in (8.117). This result is called *countergradient diffusion*. Actually, it consists of two combined phenomena:

- Turbulent mixing
- Gas expansion

The countergradient diffusion is due to the expansion of gas at the flame front (see [282] for a more refined analysis of this problem). We must now calculate the turbulent source term arising from the chemical reaction, which appears in the averaged balance equation of c:

$$\bar{\rho}(\partial \tilde{c}/\partial t + \tilde{\mathbf{v}} \cdot \nabla \tilde{c}) + \nabla \cdot (\bar{\rho}\widetilde{\mathbf{v}''c''}) = \tilde{\dot{W}}_c. \tag{8.118}$$

The scalar dissipation rate $\tilde{\dot{W}}_c$ is determined using the flame crossing frequency concept or via flame surface density models [202, 206].

Averaged G-Equation

We introduced the G-equation for laminar premixed flames in Sect. 7.7.1. This equation is convenient when studying the corrugated flamelet regime for example [202].[18] Let us now define the probability density function $P(G; \mathbf{x}, t)$ and define the Favre average of G and its variance:

$$\begin{cases} \bar{\rho}\tilde{G} = \int_{-\infty}^{+\infty} \rho G P(G; \mathbf{x}, t) dG, \\ \bar{\rho}\widetilde{G''^2} = \int_{-\infty}^{+\infty} \rho G''^2 P(G; \mathbf{x}, t) dG. \end{cases} \tag{8.119}$$

Peters gives the balance equations for the mean value of G and for the variance. After analyzing the different terms of these equations and making approximations in order to establish the required relation closure, he obtains the following system:

$$\begin{cases} \bar{\rho}(\partial \tilde{G}/\partial t + \tilde{\mathbf{v}} \cdot \nabla \tilde{G}) = (\bar{\rho} s_T^0)|\nabla \tilde{G})| - \bar{\rho} D_t \tilde{\kappa} |\nabla \tilde{G})|, \\ \bar{\rho}(\partial \widetilde{G''^2}/\partial t + \tilde{\mathbf{v}} \cdot \nabla \widetilde{G''^2}) = \nabla_{//} \cdot (\bar{\rho} D_t \nabla_{//} \widetilde{G''^2}) + 2\bar{\rho} D_t (\nabla \tilde{G})^2, \\ -c_s \bar{\rho}(\tilde{\epsilon}/\tilde{k})\widetilde{G''^2}, \end{cases} \tag{8.120}$$

[18]This method is called the "level set approach" by Peters.

where s_T^0 is the turbulent burning velocity, $\tilde{\kappa}$ the mean curvature of the front, D_t is the turbulent diffusivity, and c_s is a modeling constant (Peters suggests that this should be assigned a value of 2.0). The turbulent flame speed is not a well-defined quantity,[19] and it depends on the turbulent combustion regime in particular. An interesting review of the theoretical [20] and empirical results and formulae associated with this topic is given in [202].

Eddy Break-Up Models

In an eddy break-up (EBU) model, chemistry does not play any explicit role; turbulent motions control the reaction rate. The reaction zone is considered a collection of fresh and burnt gaseous pockets transported by turbulent eddies. This model was originated by Spalding [265]. The k–ϵ model is used to describe the turbulence. The mean rate of consumption of fuel F in the premixed turbulent flame is expressed as

$$\widetilde{\dot{W}_F} = -C_{EBU}\bar{\rho}\,\frac{\epsilon}{k}\sqrt{\widetilde{Y_F''^2}}, \tag{8.121}$$

where C_{EBU} is a constant that is adjusted according to the considered mixture.

For diffusion flames, we model the term $\sqrt{(\widetilde{Y_F^2})''}$ with

$$\sqrt{\widetilde{Y_F''^2}} = \min(\widetilde{Y_F}, \widetilde{Y_O}/r_{st}), \tag{8.122}$$

where r_{st} is the stoichiometric mass ratio, defined by $r_{st} = \nu_O \mathcal{M}_O / \nu_F \mathcal{M}_F$.

For premixed flames, we can use any of the reduced variables defined in (8.110), such as $c = (T - T_u)/(T_b - T_u)$, in place of Y_F. The corresponding averaged production rate is then $\widetilde{\dot{W}_c} = -C_{EBU}\bar{\rho}\,(\epsilon/k)\sqrt{(\widetilde{c''^2})}$. Estimating the fluctuation $(\widetilde{c''^2})$ for an infinitely thin flame leads to $\rho\,(\widetilde{c''^2}) = \overline{\rho(c-\tilde{c})^2} = \rho\,[(\widetilde{c^2}) - \tilde{c}^2] = \rho\,\tilde{c}(1-\tilde{c})$ because temperature can only take either of the two values $c = c_u = 0$ and $c = c_b = 1$, and the mean reaction rate is then [206]

$$\widetilde{\dot{W}_c} = -C_{EBU}\bar{\rho}\,\frac{\epsilon}{k}\,\tilde{c}(1-\tilde{c}). \tag{8.123}$$

---

[19] Having noted that this formulation is not particularly well suited to closing Favre-averaged transport equations, certain authors consider it to be more appropriate for large eddy simulations [206].

[20] Damköhler was the first to present theoretical expressions for the turbulent burning velocity [61]. Starting from the equivalence between the mass flux of the laminar flow through the instantaneous flame surface A_T and the mass flux of the turbulent flow through the cross-sectional area A, he derived the following relation between the laminar and turbulent combustion velocities for a steady premixed flame in a duct: $\dot{m} = \rho_u s_L A_T = \bar{\rho_u} s_T A$.

This model, which has been compared to the Arrhenius model, is sometimes combined with it in industrial codes. Some adjustments of the constant C_{EBU} have been suggested in order to incorporate certain features [243].

Passive Scalar Models

When combining the balance equations to eliminate the chemical production term (see Sect. 7.4), we introduce a parameter $\phi = r_{st}Y_F - Y_O$ (called the Shvab–Zel'dovich variable), or the corresponding reduced variable[21]

$$Z = \frac{\phi - \phi_{min}}{\phi_{max} - \phi_{min}}. \tag{8.124}$$

We get back to concentrations by using the concept of the probability density function (we will discuss this function further shortly):

$$\widetilde{Y_F} = \int_0^1 Y_F(\phi)P(\phi)d\phi. \tag{8.125}$$

Models derived from this include the CRAMER model [75] and an extension of it, the CLE model [229].

Coherent Flame Model

The coherent flame model utilizes the concept of flame stretch (Sect. 11.2.1) as applied to the flame area density [162]. Marble and Broadwell originated this model, which has been adapted to diffusion flames [40]. It is called the flame surface density model in [206] (p. 224). Let us write the production term of (8.118), valid for a thin premixed flame, as [202]

$$\tilde{\dot{W}}_c = \rho_u \, s_L^0 \, I_0 \, \Sigma, \tag{8.126}$$

where ρ_u is the fresh gas density, s_L^0 is the laminar burning velocity, Σ is the flame surface density (the flame surface per unit volume), and I_0 is a stretch factor. I_0 can be deduced from the formula [206]

$$I_0 = \frac{1}{s_L^0} \int_0^\infty s_c(\kappa)\, p(\kappa)\, d\kappa.$$

The stretch factor I_0 is on the order of 1, s_c is the consumption speed, κ is the stretch rate, and the probability $p(\kappa)$ is assumed (in most practical implementations) to be a Dirac function, $p(\kappa) = \delta(\kappa - \bar{\kappa})$.

[21]The reason for considering Z rather than ϕ is that Z always lies between 0 and 1. In the case of a ramjet configuration with an air inlet and a (nonpremixed) fuel inlet, ϕ_{min} and ϕ_{max} are the values of ϕ in the air inlet and fuel inlet, respectively. Thus, $\phi_{min} = -Y_{O,\,air\,inlet} = -0.233$, $\phi_{max} = r_{st}Y_{F,\,fuel\,inlet}$ [279].

A convenient formulation (that gives satisfying results in comparison to the results from DNS) for the balance surface density equation is [277]

$$\frac{\partial \Sigma}{\partial t} + \boldsymbol{\nabla} \cdot (\tilde{\mathbf{v}} \Sigma) = \boldsymbol{\nabla} \cdot (D_t \boldsymbol{\nabla} \Sigma) + C_1 \frac{\epsilon}{k} \Sigma - C_2 s_L^0 \frac{\Sigma^2}{1 - \bar{c}}. \tag{8.127}$$

Arrhenius Model

The simplest way to take the chemical production terms into account is to write them as we did for a perfectly stirred reactor (see Sect. 6.2), where we considered average values for the concentrations and the temperature, as in (8.59). We then have

$$\widetilde{\dot{c}} = -B\,\tilde{\rho}\,(1 - \tilde{c})\exp(-\frac{T_a}{T_u + (T_b - T_u)\tilde{c}}). \tag{8.128}$$

This model is only relevant for low turbulent Damköhler numbers; i.e., if $\tau_t / \tau_{ch} \ll 1$.

Production Terms and the pdf

The chemical production terms are nonlinear. We cannot treat them in a similar manner to turbulent transfer coefficients, and they cannot be interpreted simply via Prandtl-type theories. The probability density function (pdf) is defined for a phase space that includes the state parameters of the flow T, v, Y_j as well as the time and space coordinates [2]. If $\boldsymbol{\zeta}$ is a vector that has these parameters as its components, we can define the pdf as

$$P(\boldsymbol{\zeta}, \mathbf{x}, t) \tag{8.129}$$

in a similar way as was done in Sect. 4.10.2.

The average quantities become

$$\bar{\varphi}(\mathbf{x}, t) = \int_{\mathcal{V}} \varphi\, P\, d\boldsymbol{\zeta} \tag{8.130}$$

where we know that

$$\int_{\mathcal{V}} P\, d\boldsymbol{\zeta} = 1. \tag{8.131}$$

Knowing the pdf, we can deduce (for example) the average species production rate:

$$\left\langle \dot{W} \right\rangle = \int_{\mathcal{V}} \dot{W}(\boldsymbol{\zeta}) P(\boldsymbol{\zeta}, \mathbf{x}, t)\, d\boldsymbol{\zeta}. \tag{8.132}$$

ζ will have the concentrations and the temperature as components, but for turbulent diffusion fluxes it will be necessary to use a pdf that depends on the concentrations and velocity components.[22]

The pdf obeys a balance equation in the phase space that is similar to population balance (see Sect. 4.10.2).

To obtain the closure conditions, which are needed to solve the equations of turbulent flow, we must know the pdf balance equation or provide a suitable form for the pdf in advance (this is known as the presumed pdf method). Among the various pdf balance equations we could choose, we will select the simplest, where the phase space includes only one concentration coordinate and does not depend on $\mathbf{x}$:

$$\begin{aligned}&\tfrac{\partial P(C,t)}{\partial t} + \alpha[P(C,t) - P_0(C,t)] + \tfrac{\partial}{\partial C}[\dot{W}(C)\,P(C,t)] = \\ &2\beta[\textstyle\int P(C',t)\,P(C'',t)\,\delta(\tfrac{C'+C''}{2} - C)\,dC'\,dC'' - P(C,t)].\end{aligned} \tag{8.133}$$

The first term on the left hand side is the nonsteady variation of P; the second corresponds to the convection; α is the inverse of the average residence time; and P_0 is the pdf at the reactor inlet. The third term corresponds to the chemical production; it is a flux term in the phase space. Finally, the right hand side accounts for the interactions between fluid particles by assuming that two particles of concentration C' and C'' meet to form a third particle of concentration $(C' + C'')/2$ with a coalescence–redispersion frequency of β (presumed constant). This equation, where we assume that $\rho =$const., does not take inhomogeneities into account.

Other models are more sophisticated. Some of them utilize the "two-point pdf." In these cases, we need to find good closure conditions for the pdf balance and then solve the equation.

The pdf can be measured. In combustion involving a mixture of burned gas and fresh gas, the pdf of temperature will generally be bimodal, but it can evolve in form during the reaction.

Example: Complete Solution of a Turbulent Reactive Flow Problem

Such a problem can only be solved numerically with the aid of approximations. For example, Peters [201] treated the case of a premixed flame by assuming a single reaction

$$\nu'_A\,\mathrm{A} + \nu'_B\,\mathrm{B} \longrightarrow \mathrm{C} \tag{8.134}$$

---

[22] In theory, the pdf is used to calculate the chemical production terms in particular, whereas turbulent flows require other closing methods (k–ϵ methods for example). However, we can also determine the fluxes using this method so long as we know the corresponding pdf.

with a high activation energy. If T_u is the temperature of the fresh gases and T_b is that of the burned gases under conditions of complete adiabatic combustion, we set

$$\epsilon = RT_b^2/E_a(T_b - T_u) \ll 1 \tag{8.135}$$

where E_a is the activation energy for the reaction. The reduced temperature

$$c = (T - T_u)/(T_b - T_u) \tag{8.136}$$

is written in the form

$$c = 1 - \epsilon y, \tag{8.137}$$

and all of the terms that depend on T are expanded asymptotically as a function of the small parameter ϵ. As the flow is compressible, Favre averages are used. Therefore, for the c variable defined by (8.110), we have

$$\overline{c} = \overline{\rho}\,\tilde{c}, \tag{8.138}$$

where $\overline{c}$ indicates the classical average and $\tilde{c}$ the Favre average, weighted by the mass. Thus

$$\begin{cases} c = \tilde{c} + c'', & \overline{\rho\, c''} = 0 \\ \rho = \bar{\rho} + \rho', & \bar{\rho'} = 0. \end{cases} \tag{8.139}$$

The energy equation for $Le = 1$ (see Chap. 6) becomes

$$\rho\frac{\partial c}{\partial t} + \rho\mathbf{v}\cdot\boldsymbol{\nabla} c - \boldsymbol{\nabla}\cdot(\rho D\boldsymbol{\nabla} c) = \frac{\Delta H}{C_p(T_b - T_u)}\dot{\zeta} \tag{8.140}$$

with

$$\dot{\zeta} = k(T)\,Y_A^{\nu'_A}\,Y_B^{\nu'_B} = B\,Y_A^{\nu'_A}\,Y_B^{\nu'_B}\,\exp -E_a/RT. \tag{8.141}$$

We can show that for $Le = 1$, Y_A and Y_B are linear functions of c. We set $S = \nu'_A\mathcal{M}_A\,\dot{\zeta}/\rho$, $J_{1\alpha} = \widetilde{v''_\alpha c''}$, $J_{2\alpha} = \widetilde{v''_\alpha (c'')^2}$, and define A as the rate of fuel consumption, yielding

$$\begin{aligned} &\bar{\rho}\tilde{v}_\alpha\partial\tilde{c}/\partial x_\alpha = \partial/\partial x_\alpha(-\bar{\rho}J_{1\alpha}) + \bar{\rho}[-\Delta H/\nu'_A\mathcal{M}_A C_p(T_b - T_u)]\tilde{S}, \\ &\bar{\rho}\tilde{v}_\alpha\partial\widetilde{(c")^2}/\partial x_\alpha = \partial/\partial x_\alpha(-\bar{\rho}J_{2\alpha}) + 2\bar{\rho}J_{1\alpha}\partial\tilde{c}/\partial x_\alpha, \\ &-2\bar{\rho}\tilde{\epsilon_c} + 2\bar{\rho}[-\Delta H/\nu'_A\mathcal{M}_A C_p(T_b - T_u)]\widetilde{c''S''}. \end{aligned} \tag{8.142}$$

The flux terms are modeled as follows:

$$\begin{cases} J_{1\alpha} = -D_t\partial\tilde{c}/\partial x_\alpha, \\ J_{2\alpha} = -D_t\partial\widetilde{(c'')^2}/\partial x_\alpha, \end{cases} \tag{8.143}$$

and the dissipation is given by

$$\tilde{\epsilon_c} = \frac{c_1}{2} \frac{\rho \widetilde{(v_{t,f})^2}}{\overline{\rho}^2 D_t}, \tag{8.144}$$

where $\widetilde{(v_{t,f})^2}$ is the square of the turbulent flame speed and c_1 is assumed to be constant. The production terms $\tilde{S}$ and $\widetilde{c''S''}$ are calculated using the following probability density $P(x)$, which takes the form of a β function:

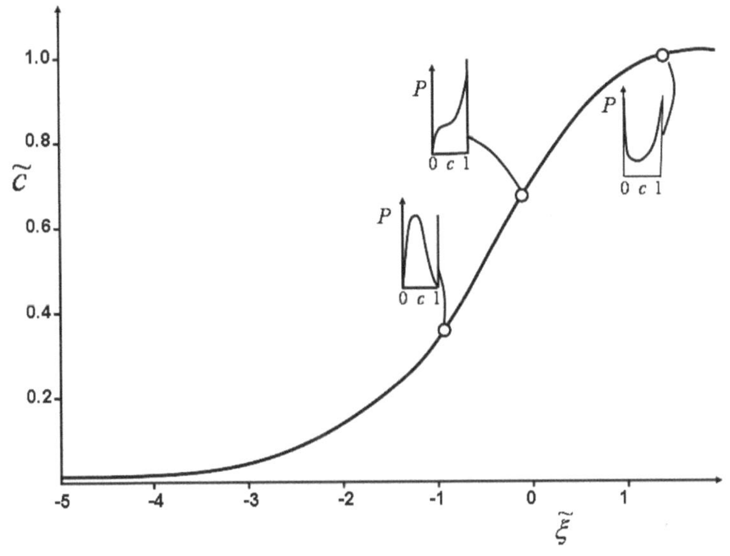

Fig. 8.19. Example of turbulent reactive flow (after Peters; reprinted with permission from [201])

$$P(x) = x^{\alpha-1}(1-x)^{\beta-1}\frac{\Gamma(\gamma)}{\Gamma(\alpha)\Gamma(\beta)}. \tag{8.145}$$

This pdf has the advantage of evolving in both α and β, and it is compatible with experimental observations. In addition, α and β are related to the average $\bar{x}$ and the variance $\overline{x'^2}$:

$$\begin{cases} \alpha = \gamma\bar{x}, \ \beta = (1-\bar{x})\gamma, \\ \gamma = \alpha + \beta = \bar{x}(1-\bar{x})/\overline{x'^2} - 1, \\ \alpha, \ \beta, \ \gamma > 0. \end{cases} \tag{8.146}$$

We then have

$$\begin{cases} \tilde{S} = \int_0^1 S(c)\tilde{P}(c)\,dc, \\ \widetilde{c''S''} = \int_0^1 cS(c)\tilde{P}(c)\,dc - \tilde{c}\tilde{S}, \end{cases} \tag{8.147}$$

with

$$\tilde{P} = \frac{\Gamma(\gamma)}{\Gamma(\alpha)\Gamma(\beta)}(1-\hat{\alpha}y)(\epsilon y)^{\beta-1}, \; \hat{\alpha} = (\alpha-1)\epsilon. \tag{8.148}$$

These terms are calculated via asymptotic expansions in ϵ.

Numerical calculations make it possible to solve the equations associated with this problem with respect to the boundary conditions. Figure 8.19 presents the solution in $c(\xi)$, where ξ is a dimensionless space variable that is normal to the flame (which is assumed to be planar). This figure shows how the pdf evolves up to the flame crossing.

Flamelet Models

As previously noted (see Sect. 8.4.2), flamelets are thin reactive–diffusive layers that are embedded within an otherwise nonreacting turbulent flow field [202]. This concept focuses on the location of the flame surface: the flame front for premixed flames and the stoichiometric surface for nonpremixed flames. Scalar quantities are considered in both cases.

The balance equations for these quantities are the G-equation (7.94) for premixed flames and the Z-equation (7.101) for nonpremixed flames. These equations were presented in Sects. 7.7.1 and 7.7.2.

We now present a model equation for the premixed flame surface area ratio. The flame surface area ratio $\bar{\sigma}$ is the mean gradient defined by the relation

$$(\bar{\rho} s_T^0)|\nabla \tilde{G}| = (\rho s_L^0)\bar{\sigma}, \tag{8.149}$$

where s_T^0 is the turbulent burning velocity described at the end of the section devoted to the averaged G-equation. It can be proven that the surface area ratio is proportional to the wrinkled flame front defined by the area $G = G_0$ after filtering (see [202]). As derived by Rutland et al. [240], assuming constant values for the density ρ, laminar combustion speed s_L^0 and diffusion coefficient D, we have

$$\partial\sigma/\partial t + \mathbf{v}\cdot\nabla\sigma = -\mathbf{n}\cdot\nabla\otimes\mathbf{v}\cdot\mathbf{n} + s_L^0(\kappa + \nabla^2 G) + D\,\mathbf{n}\cdot\nabla(\kappa\sigma) \tag{8.150}$$

where $\mathbf{n}$ is defined by (7.93). The first term on the right hand side accounts for flow field strain, which represents flame surface area production. The next term is a kinematic restoration term. The last term is a scalar dissipation term (see the variance equation for G (8.120)). At large Reynolds numbers, and for large values of v'/s_L, it is possible to write $\bar{\sigma} = |\nabla\tilde{G}| + \bar{\sigma}_t$. The equation obtained by Peters is written

$$\begin{aligned}\bar{\rho}\partial\bar{\sigma}_t/\partial t + \bar{\rho}\tilde{\mathbf{v}}\cdot\nabla\bar{\sigma}_t &= \nabla_{//}\cdot(\bar{\rho}D_t\nabla_{//}\sigma_t) - c_0\bar{\rho}\widetilde{\mathbf{v}''\otimes\mathbf{v}''}{:}\nabla\otimes\tilde{\mathbf{v}}\bar{\sigma}_t/\tilde{k} \\ &+ c_1\bar{\rho}D_t(\nabla\tilde{G})^2\sigma_t/(\widetilde{G''^2}) - c_2\bar{\rho}s_L^0\overline{\sigma_t}^2/((\widetilde{G''^2}))^{1/2} - c_3\bar{\rho}D\overline{\sigma_t}^3/(\widetilde{G''^2}).\end{aligned} \tag{8.151}$$

The constants c_1, c_2, c_3 were determined by Wenzel and Peters: $c_1 = 4.63$, $c_2 = 1.01$, $c_3 = c_1$.

The relation between $\overline{\sigma}$ and Σ has been discussed by Candel and Poinsot [39].

8.5 Concepts of Large Eddy Simulation

The direct numerical simulation (DNS) method, which involves numerically solving the balance equations for the flows, is most exact. It is particularly appropriate for the flows with low Reynolds numbers. However, it becomes prohibitive for large Reynolds numbers because current levels of computing power are not sufficient (even in the case of parallel computing).

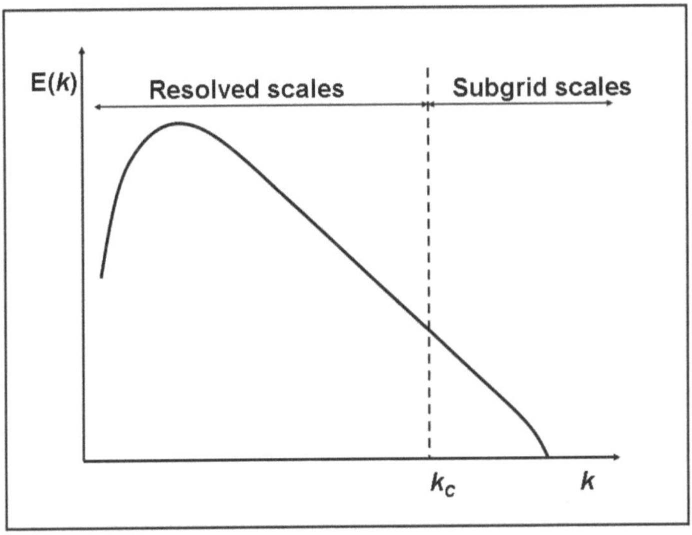

Fig. 8.20. Cut-off wavenumber

The inaccuracies of averaged methods and the inability to use DNS for flows with large Reynolds numbers has resulted in the use of a mixed method, large eddy simulation (LES), since the 1960s [261].

The aim of large eddy simulation is to explicitly compute the largest turbulent structures of the flow field (typically structures that are larger than the computational mesh size), while the effects of the smallest ones are modeled instead.

We will not give a complete account of this method—which is used in many modeling studies—here, since it is beyond the scope of this book (for a discussion of this method in relation to incompressible fluids, refer to [241]; for the combustion field, refer to Poinsot and Veynante [206]). We will only provide some guidelines below.

8.5.1 Filtering

In LES, variables are filtered in spectral space or in physical space. In spectral space, we consider a cut-off wavenumber k_c in the turbulent kinetic energy spectrum $E(k)$ (Fig. 8.20).

The aim is to treat, as much as possible, small scales of turbulence (*or unresolved scales, or subgrid scales, or those modeled in LES*, as defined by $k > k_c$) in a statistical way. We then establish equations that are valid for large scales (*or resolved scales, or those computed in LES*) and solve these equations directly.

We must take into account the interactions between modes at various scales and try to re-introduce the concept of turbulent viscosity. The first task is to establish a system of equations that are to be solved for the various scales and are based on the classical balance laws.

We call $G = G(\mathbf{x}, t)$ a filter in physical space. When applied to an unspecified quantity $f(\mathbf{x}, t)$, this filter provides the resolved part $\bar{f}(\mathbf{x}, t)$:[23]

$$\bar{f}(\mathbf{x}, t) = \int_{\Re^3} f(\mathbf{x}, t)\, G(\mathbf{x} - \boldsymbol{\xi}, t - t')dt'\, d^3\boldsymbol{\xi}, \tag{8.152}$$

or, symbolically, $\bar{f} = Gf$. Conversely, the unresolved part will be $f' = (1 - G)f$.
Transposed into Fourier space, these definitions give

$$\bar{f} = G(k)\bar{f}, f' = [1 - G(k)]\, f$$

where $G(k) = G(\mathbf{k}, \omega)$ and $f = f(\mathbf{k}, \omega)$ this time.

Let us now present three common filters used in LES (the shapes of these filters in physical space and in spectral space are shown in Fig. 8.21):

- In spectral space, the **sharp cut-off filter** has $G(k) = 1, |\mathbf{k}| \leq k_c$, $G(k) = 0, |\mathbf{k}| > k_c$, with $k_c = \pi/\Delta$ (Δ is the filter size; the filter keeps length scales larger than 2Δ). In physical space, the corresponding filter is $G(x) = \sin(k_c x)/k_c x$.
- A **box (or top-hat) filter** has $G(x) = 1/\Delta$ for $|x| \leq \Delta/2$ and 0 otherwise. It becomes $\hat{G}(k) = \sin(k\Delta/2)/k\Delta/2$ in spectral space. In the present case, this filter corresponds to averaging over a one-dimensional box of size Δ.

[23] In the absence of ambiguity, we will retain the same notation for used filtered quantities defined in (8.152) and for classically averaged quantities (see Sect. 8.3.1). This will also be true of the Favre-filtered quantities defined by (8.153).

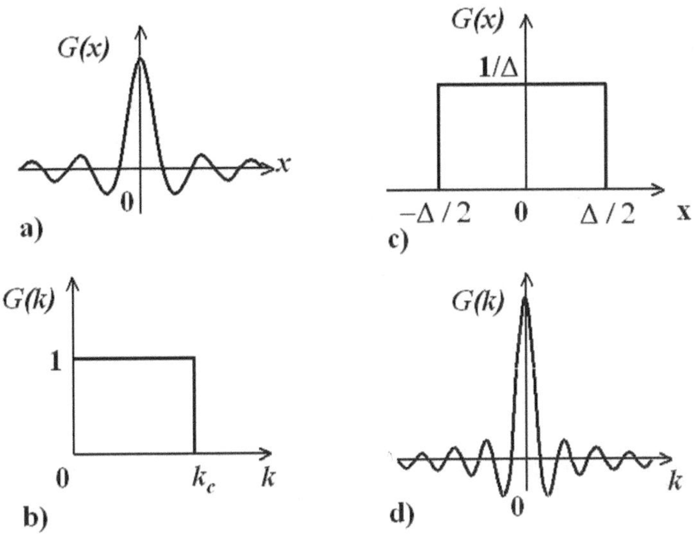

Fig. 8.21. Filter examples (after [241]): **a** cut-off filter in physical space; **b** cut-off filter in spectral space; **c** box filter in physical space; **d** box filter in spectral space

- A **Gaussian filter** $G(x) = (\gamma/\pi\Delta^2)^{1/2}\exp(\gamma x^2/\Delta^2)$ in physical space, where γ is a constant, gives another Gaussian filter $G(k) = \exp(-\Delta^2)k^2/4\gamma$ in spectral space.

All of these filters are normalized: $\int_{-\infty}^{+\infty} G(x)\,dx = 1$ in the 1-D case. Favre averaging can also be defined for turbulent flows with variable density via

$$\bar{\rho}\,\tilde{f}(\mathbf{x},t) = \int_{\Re^3} (\rho f)\,(\mathbf{x},t)\,G(\mathbf{x}-\boldsymbol{\xi},t-t')dt'\,d^3\boldsymbol{\xi}. \tag{8.153}$$

We now quote the properties, as stated by Ghosal and Moin [103], which indicate that a filter is acceptable in the homogeneous one-dimensional case: (1) it is symmetrical; (2) the constants are conserved; (3) it decreases rapidly; (4) it exhibits quasi-locality in physical space.

Note also that [206]:

- In contrast to RANS averaging, the filtered value of a LES perturbation is not zero: $\bar{f}' \neq 0$. Thus, filtered and double-filtered values are generally not equal: $\bar{\bar{f}} \neq \bar{f}$. Also, $f'' = f - \tilde{f}$, $\tilde{f''} \neq 0$, and $\tilde{\tilde{f}} \neq \tilde{f}$.
- To derive balance equations for the filtered quantities it is necessary to exchange filter and derivative operators.

8.5.2 Filtered Balance Equations

Filtered NS Equations for a Nonreactive Incompressible Fluid

After filtering, the Navier–Stokes equations for a nonreactive incompressible fluid with constant viscosity become

$$\begin{cases} \partial \bar{u}_i/\partial t + \partial \overline{u_i u_j}/\partial x_j = -\partial \bar{p}/\partial x_i + \nu(\partial x_j)(\partial \bar{u}_i/\partial x_j + \partial \bar{u}_j/\partial x_i), \\ \partial \bar{u}_i \partial x_i = 0. \end{cases} \tag{8.154}$$

Unlike in (8.45), $\bar{f}$ are averages of filtered quantities this time.

Filtered Balance Equations for Combustion

The equations of (8.154) are not valid for combustion. We must consider compressible flows and use Favre averaging, as well as introduce concentration variables, chemical production rates, and the heat released. The resulting equations are formally similar to (8.83) for mass and momentum and to (8.105) for scalar quantities. We will write them (as Poinsot and Veynante [206] do) in the following form:

$$\begin{cases} \partial \bar{\rho}/\partial t + \boldsymbol{\nabla} \cdot (\bar{\rho}\tilde{v}) = 0, \\ \partial(\bar{\rho}\tilde{v})/\partial t + \boldsymbol{\nabla} \cdot (\bar{\rho}\tilde{\mathbf{v}} \otimes \tilde{\mathbf{v}}) + \boldsymbol{\nabla}\bar{p} = -\boldsymbol{\nabla} \cdot [\bar{\boldsymbol{\Pi}} + \bar{\rho}(\widetilde{\mathbf{v} \otimes \mathbf{v}} - \tilde{\mathbf{v}} \otimes \tilde{\mathbf{v}})], \\ \bar{\rho}\partial \tilde{Y}_j/\partial t + \bar{\rho}\tilde{\mathbf{v}} \cdot \boldsymbol{\nabla}\tilde{Y}_j = \boldsymbol{\nabla} \cdot (\overline{\rho\, D_j \boldsymbol{\nabla} Y_j}) \\ -\boldsymbol{\nabla} \cdot [\bar{\rho}(\widetilde{\mathbf{v}Y_j} - \tilde{\mathbf{v}}\tilde{Y}_j)] + \bar{\dot{W}}_j, \; j = 1, \dots N, \\ \bar{\rho}\partial \tilde{h}_s/\partial t + \bar{\rho}\tilde{\mathbf{v}} \cdot \boldsymbol{\nabla}\tilde{h}_s = \partial \bar{p}/\partial t + \overline{\mathbf{v} \cdot \boldsymbol{\nabla} p} + \boldsymbol{\nabla} \cdot (\overline{\lambda \boldsymbol{\nabla} T}) \\ -\boldsymbol{\nabla} \cdot (\bar{\rho}[\widetilde{\mathbf{v}h_s} - \tilde{\mathbf{v}}\tilde{h}_s] - \overline{\Pi{:}\nabla \otimes \mathbf{v}} - \boldsymbol{\nabla} \cdot \overline{(\rho \sum_{j=1}^{N} \mathbf{V}_j Y_j h_{s,j})} + \bar{\dot{W}}_T, \end{cases} \tag{8.155}$$

where $h_{s,j} = \int_{T^0}^{T} c_{p,\,j} dT$ and $h_s = h - \sum_{j=1}^{N} Y_j (q_f^0)_j$. The quantity h_s is the sensible enthalpy corresponding to the quantity β_T of Sect. 7.4 if we assume only one chemical reaction: $h_s = \sum_j Y_j \int_{T^0}^{T} c_{p,\,j} dT = \Delta H \beta_T$. We will assume the approximation $\overline{\mathbf{v} \cdot \boldsymbol{\nabla} p} \cong \tilde{\mathbf{v}} \cdot \boldsymbol{\nabla}\bar{p}$. Filtered molecular diffusion fluxes are neglected or modeled through a simple gradient assumption such as

$$\overline{\rho\, D_j \boldsymbol{\nabla} Y_j} = \bar{\rho}\, \bar{D}_j\, \boldsymbol{\nabla}\tilde{Y}_j, \quad \overline{\lambda \boldsymbol{\nabla} T} = \bar{\lambda}\, \boldsymbol{\nabla}\tilde{T}. \tag{8.156}$$

LES is utilized for the determination of various flows—compressible or incompressible, more or less confined and homogeneous [64, 73, 269]. Some combustion problems can then be solved, as done in France by the CERFACS team [238, 252]. Poinsot and Veynante [206] provide simple closures for chemical production rates that are assumed to apply to both premixed and diffusion flames.

8.5.3 Closure Relations for Filtered Balance Equations

We will now summarize the main approaches for modeling the unresolved transport terms (unresolved Reynolds stresses $\mathcal{T} = \widetilde{\mathbf{v} \otimes \mathbf{v}} - \tilde{\mathbf{v}} \otimes \tilde{\mathbf{v}}$, unresolved scalar fluxes $\widetilde{\mathbf{v}Y_j} - \tilde{\mathbf{v}}\tilde{Y}_j$ and $\widetilde{\mathbf{v}h_s} - \tilde{\mathbf{v}}\tilde{h}_s$), then the filtered laminar diffusion fluxes, and after that the filtered chemical reaction rates $\bar{\dot{W}}_j$. The subsequent models can only be validated by comparison with DNS analysis and experimental data (if possible).

Models for unresolved Reynolds stresses were first presented in [206], where constant density was assumed for the sake of simplification.

Models for Unresolved Reynolds Stresses

We will use Poinsot–Veynante symbols and tensorial notation (see Sect. A.1).

Smagorinsky's Model

Just as for classical averaging, the turbulent fluxes are assumed to be proportional to the turbulent forces. This leads to the introduction of a turbulent subgrid scale viscosity ν_t that is different from the turbulent viscosity defined by (8.79).

Writing

$$\mathcal{T} - \frac{1}{3}\,\mathrm{tr}(\mathcal{T})\,\mathbf{1} = -\nu_t\, \nabla \otimes \mathbf{v} = -2\nu_t \mathbf{S}, \tag{8.157}$$

where $\mathbf{S}$ is the symmetric part of the $\nabla \otimes \mathbf{v}$ tensor, we have

$$\nu_t = C_S^2\, \Delta^{4/3}\, \ell^{2/3}\, |\bar{S}|, \tag{8.158}$$

where ℓ is the integral scale of the turbulence, C_S is a constant, and $\bar{S} = (2\mathbf{S}{:}\mathbf{S})^{1/2}$ is the resolved shear stress. Equation 8.158 is simplified by assuming that the integral scale is on the order of the grid size; i.e., $\ell \approx \Delta$.

Scale Similarity Model

The scale similarity model is based upon a double-filtering approach and on the idea that unresolved stresses are mainly controlled by the largest unresolved structures (see Fig. 8.22a), which are similar to the smallest resolved structures. The subgrid tensor is then evaluated using

$$\mathcal{T} = \overline{\bar{\mathbf{v}} \otimes \bar{\mathbf{v}}} - \bar{\bar{\mathbf{v}}} \otimes \bar{\bar{\mathbf{v}}}. \tag{8.159}$$

Since this closure is insufficiently dissipative, it is generally combined with an eddy viscosity model such as Smagorinsky's model, producing a mixed model [158].

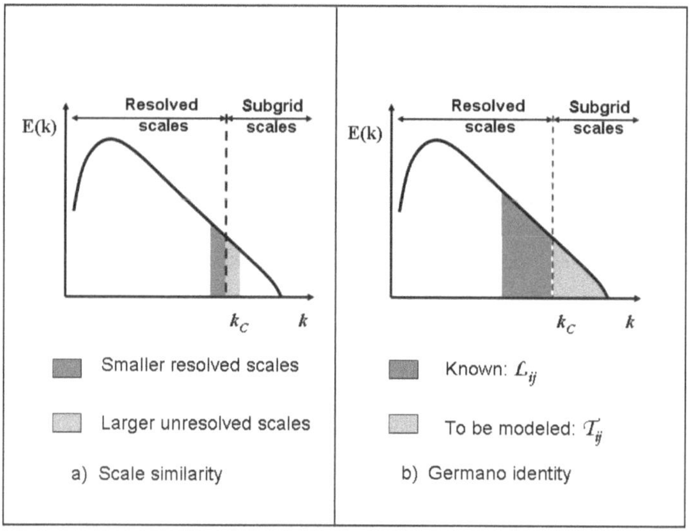

Fig. 8.22. Scale similarity and the Germano dynamic model

Germano's Dynamic Model

A test filter that is $\hat{\Delta}$ larger than the LES filter (size $\bar{\Delta}$) is now introduced [100] (see Fig. 8.22b)). For the unresolved momentum fluxes, we have

$$\boldsymbol{\mathcal{T}} = \overline{\mathbf{v} \otimes \mathbf{v}} - \bar{\mathbf{v}} \otimes \bar{\mathbf{v}}, \tag{8.160}$$

and for the unresolved fluxes at the test level we have

$$\widehat{\boldsymbol{\mathcal{T}}} = \widehat{\mathbf{v} \otimes \mathbf{v}} - \hat{\bar{\mathbf{v}}} \otimes \hat{\bar{\mathbf{v}}}. \tag{8.161}$$

We then introduce the double-filter "bar hat:"

$$\mathbf{T} = \overline{\widehat{\mathbf{v} \otimes \mathbf{v}}} - \hat{\bar{\mathbf{v}}} \otimes \hat{\bar{\mathbf{v}}}. \tag{8.162}$$

Now we consider the following resolved turbulent stress corresponding to the test filter applied to the field $\bar{\mathbf{v}}$:

$$\widehat{\boldsymbol{\mathcal{L}}} = \widehat{\bar{\mathbf{v}} \otimes \bar{\mathbf{v}}} - \hat{\bar{\mathbf{v}}} \otimes \hat{\bar{\mathbf{v}}}. \tag{8.163}$$

Adding (8.163) and (8.161) yields, via (8.162),

$$\widehat{\boldsymbol{\mathcal{L}}} = \mathbf{T} - \widehat{\boldsymbol{\mathcal{T}}}, \tag{8.164}$$

which is called Germano's identity. The Reynolds stress tensors $\boldsymbol{\mathcal{T}}$ and $\mathbf{T}$ can be estimated from the Smagorinsky model (Eqs. 8.157 and 8.158 with $\ell = \Delta$):

$$\begin{cases} \mathcal{T} - \frac{1}{3}\,\mathrm{tr}(\mathcal{T})\,\mathbf{1} = -2C\bar{\Delta}^2|\bar{\mathbf{S}}|\bar{\mathbf{S}} = -2C\boldsymbol{\alpha}, \\ \mathbf{T} - \frac{1}{3}\,\mathrm{tr}(\mathbf{T})\,\mathbf{1} = -2C\bar{\Delta}^2|\widehat{\bar{\mathbf{S}}}|\widehat{\bar{\mathbf{S}}} = -2C\boldsymbol{\beta}, \end{cases} \tag{8.165}$$

where C is a parameter that needs to be determined. The Germano identity (8.164) is then rewritten as

$$\mathcal{L} - \frac{1}{3}\,\mathrm{tr}(\mathcal{L})\,\mathbf{1} = 2C(\widehat{\boldsymbol{\alpha}} - \boldsymbol{\beta}). \tag{8.166}$$

To determine $C = C(\mathbf{x}, t)$, we also have—in the incompressible case—the two mass conservation equations $\mathrm{tr}(\bar{\mathbf{S}}) = \mathrm{tr}(\widehat{\bar{\mathbf{S}}}) = 0$. Determining C computationally leads to instability, so a least squares approach is used to minimize the error.

This model has also been extended to compressible turbulence [177]. It is a very efficient model for a large number of applications.

Structure Function Models

These models are based on a theoretical analysis of turbulence in spectral space and on the concept of subgrid-scale viscosity. The subgrid-scale dynamic viscosity is

$$\nu_t(\mathbf{x}, \Delta) = 0.105\, c_K^{-3/2}\, \Delta\, \sqrt{F_2(\mathbf{x}, \Delta)}, \tag{8.167}$$

where $c_K = 1.4$ is the Kolmogorov constant (see Eq. 8.98) and F_2 is the structure function

$$F_2(\mathbf{x}, \Delta) = [\bar{\mathbf{v}}(\mathbf{x} + \mathbf{r}) - \bar{\mathbf{v}}(\mathbf{x})]^2, \text{ with } \Delta = |\mathbf{r}|.$$

A high-pass filter is used to eliminate the large eddies (which cause high dissipation) with the corresponding filtered structure function $\bar{F}_2$. The resulting subgrid viscosity is then

$$\nu_t(\mathbf{x}, \Delta) = 0.0014\, c_K^{-3/2}\, \Delta\, \sqrt{\bar{F}_2(\mathbf{x}, \Delta)}. \tag{8.168}$$

Unresolved Scalar Flux Models

The simplest model for LES unresolved species flux is a gradient relation, as in classical turbulence theory with (8.50) and (8.51)

$$\widetilde{\mathbf{v}Y_j} - \tilde{\mathbf{v}}\tilde{Y}_j = -\frac{\nu_t}{Sc_j}\,\boldsymbol{\nabla}\tilde{Y}_j \tag{8.169}$$

Filtered Reaction Rate Closure Models

We describe two simple models for the LES filtered reaction rate that are assumed to apply to all types of flames. More specific models that are required for premixed flames and for nonpremixed flames will not be presented here; they are described in [206].

Arrhenius Model for Filtered Quantities

This model corresponds to the Arrhenius model of Sect. 8.4.4, but filtered quantities are considered here. Perfect mixing is assumed at the subgrid-scale level and subgrid-scale fluctuations are neglected. It is assumed that $\tau_t \ll \tau_c$, where τ_t is the turbulent subgrid timescale and τ_c is the chemical timescale. We then have, for a second-order chemical reaction,

$$\bar{\dot{W}}_j = \nu_j' B\bar{\rho}^2 \tilde{Y}_F \tilde{Y}_O \tilde{T}^s \exp(T_a/\tilde{T}), \tag{8.170}$$

where B and s are the constant coefficients in Arrhenius law (3.52) and $T_a = E_a/R$ is the activation temperature of the reaction.

If the fuel and the oxidizer are not perfectly mixed at the subgrid level, we can write the following formula with a segregation factor $(\widetilde{Y_F Y_O} - \tilde{Y}_F \tilde{Y}_O)/\tilde{Y}_F \tilde{Y}_O$:

$$\bar{\dot{W}}_j = \nu_j' B\bar{\rho}^2 \tilde{Y}_F \tilde{Y}_O \tilde{T}^s \exp(T_a/\tilde{T}) \left[1 + \frac{\widetilde{Y_F Y_O} - \tilde{Y}_F \tilde{Y}_O}{\tilde{Y}_F \tilde{Y}_O}\right]. \tag{8.171}$$

Reaction Rate Modeling with Scale Similarity Assumptions

Scale similarity assumptions have been extended to reaction rate modeling [101].

8.5.4 Spectral Analysis and LES

In this section we briefly present an approach based on the concepts of spectral eddy viscosity and spectral eddy diffusivity. We will work in Fourier space and use the sharp cut-off filter defined in Sect. 8.5.1 [158, 176]. We will also only consider incompressible fluids.

Aspects of Spectral Dynamics

In the two-point formalism [2], we must account for the interactions between modes of energy using triads of wavenumber vectors $(\mathbf{k}, \mathbf{p}, \mathbf{q})$, such as $\mathbf{k} + \mathbf{p} + \mathbf{q} = \mathbf{0}$. We then write the evolution equation of the energy spectrum for isotropic turbulence using the Fourier transformation defined by (8.88):

$$(\frac{\partial}{\partial t} + 2\nu k^2) E(k) = T(k) \tag{8.172}$$

(see [158], p. 172), where $T(k)$ is the transfer term corresponding to triple-velocity correlations arising from nonlinear interactions of the NS equations, which are often divided into two parts ([158], p. 234):

$$T(k) = T_{NL}(k) + T_L(k), \tag{8.173}$$

where the indices NL and L refer to nonlocal (due to triad interactions) and local transfers, respectively.

In spectral space, the filtered NS equations lead to the energy transfer equation for resolved scales:

$$(\frac{\partial}{\partial t} + 2\nu k^2)\bar{E}_r(k,t) = \bar{T}(\mathbf{k}), \tag{8.174}$$

with the decomposition

$$\bar{T}(\mathbf{k}) = \bar{T}_r(\mathbf{k}) + \bar{T}_{sg}(\mathbf{k}).$$

The terms on the right hand side represent energy exchanges between the mode $\mathbf{k}$ and all of the other modes associated with the calculable terms, respectively, starting from the resolved modes and the subgrid terms.

$E(k)$ is the energy present in a sphere of radius k; $\bar{E}_r(k) = \hat{G}^2(k)E(k)$ is the kinetic energy of the resolved modes (this is different from the filtered portion of the kinetic energy [241]). The conservation of kinetic energy for inviscid fluids (in the case of zero viscosity) implies that

$$\int_{\Re^3} (\bar{T}_r(\mathbf{k}) + \bar{T}_{sg}(\mathbf{k}))d^3\mathbf{k} = 0.$$

Models of Effective Viscosity

There are many ways to model the transfer terms. As an example, we now present a method that utilizes an "effective viscosity" that depends on the wavenuumber and the cut-off wavenumber [241]. It involves modeling the direct energy cascade process. The spectral models utilize an effective viscosity. They are derived from the analysis of Kraichnan [139]. Let us now summarize one of the most recent [158, 176]. Let us assume fully developed, isotropic, homogeneous turbulence and focus initially on the subgrid term $\bar{T}_{sg}(\mathbf{k})$ by assuming that $k \ll k_c$ and $E(k_c) \gg E(k)$. We have

$$(\frac{\partial}{\partial t} + 2\nu k^2)\bar{E}(k,t) = \bar{T}_{<k_c}(\mathbf{k},t) + \bar{T}_{sg}(\mathbf{k},t) \tag{8.175}$$

where $\bar{T}_{sg}(\mathbf{k}) = -2\nu_t^{\infty} k^2 \bar{E}(k,t)$. The other term $\bar{T}_{<k_c}(\mathbf{k})$ corresponds to triad interactions involving small wavenumbers (i.e., large wavelengths/sizes). This term will be calculated directly by LES. Nonlocal transfers from large scales to subgrid scales are considered negligible due to the assumptions used.

For an inertial zone that exhibits a $k^{-5/3}$ law, we find that

$$\nu_t^{\infty} = 0.441 c_K^{-3/2}[E(k_c)/k_c]^{1/2}, \; c_K \approx 1.4, \tag{8.176}$$

where c_K is the Kolmogorov constant.

For a spectrum that exhibits a k^{-m} ($m \leq 3$) dependence,

$$\nu_t^\infty = \frac{1}{15a_1}\frac{5-m}{m+1}\sqrt{3-m}[E(k_c)/k_c]^{1/2}. \quad (8.177)$$

If a_1 is close to 1, we get the preceding result for $m = 5/3$.

For $m > 3$, we get $\nu_t^\infty \propto E(k_c)/k_c$.

If k is close to k_c, we define $\nu_t(k|k_c)$, so that

$$\bar{T}_{sg}(\mathbf{k}) = -2\nu_t(k|k_c)k^2\bar{E}(k,t), \quad (8.178)$$

where $\nu_t(k|k_c) = K(k/k_c)\nu_t^\infty,\ K \cong 1$, except for $k = k_c$.

The modeling above can be extended to the case of a passive scalar—the temperature here. For $k \ll k_c$ and $E(k_c) \ll E(k)$, the equations then become

$$\begin{cases} (\partial/\partial t + 2\nu k^2)\bar{E}_T(k,t) = \bar{T}^T_{<k_c}(\mathbf{k},t) + \bar{T}^T_{sg}(\mathbf{k},t), \\ \bar{T}_{sg}(\mathbf{k}) = -2\nu_t^\infty k^2\bar{E}(k,t), \\ m \le 3:\ \kappa_t^\infty = \nu_t^\infty/Pr^{(t)};\ \Pr^{(t)} = 0.18(5-m), \end{cases} \quad (8.179)$$

and for k close to k_c they are

$$\begin{cases} \bar{T}^T_{sg}(\mathbf{k}) = -2\kappa_t(k|k_c)k^2\bar{E}(k,t), \\ \kappa_t(k|k_c) = C(k/k_c)\kappa_t^\infty,\ C \cong 1 \text{ except for } k = k_c. \end{cases} \quad (8.180)$$

8.6 Conclusion

The discussions of turbulence provided in this chapter illustrate the complexity of the problem. Fluid turbulence is currently regarded as one of the most difficult problems to solve. "One does not know, for instance, how to predict the critical Reynolds number of transition to turbulence in a pipe, nor how to compute precisely, the drag of a car or an aircraft, even with today's largest computers" (Lesieur's foreword in Sagaut's book [241]).

We have only introduced some of the methods used in this field: spectral analysis; modeling in k–ϵ; pdf; LES. There are others. Current research often focuses on improving the existing models and comparing theoretical results to experimental measurements. Other research relates to the behavior of the large stable or unstable structures in the flow. Lastly, the interaction between dynamic and chemical effects has not been completely clarified yet. Nevertheless, much of our current knowledge in this field is put to practical use, and even if there is still much progress to be made, we are now able to accurately model many practical situations.

A comparison can be made between the computational approaches used for turbulent combustion. Poinsot and Veynante distinguish three levels: RANS, LES and DNS (see Sect. 4.4 in [206]; also note the section "Comparing Large Eddy Simulation and Experimental Data" on p. 175 of that work). For DNS, in order to calculate all of the scales of the Kolmogorov cascade, we must have $Re_t < N^{4/3}$, where Re_t is a turbulent Reynolds number defined using the integral scale of turbulence ℓ and the Kolmogorov scale ℓ_K such that $Re_t = (\ell_K/\ell)^{4/3}$. There is a less restrictive condition for LES: $Re_t < (q\,N)^{4/3}$ with $q = l_c/\ell_K > 1$, where ℓ_c is the cut-off length scale. Therefore, in terms of cost, LES is definitely more expensive than RANS, but it is also more precise. On the other hand, it is less expensive than DNS.

This chapter uses ideas and results from recent books, such as [158, 241, 192, 242] for general turbulence-related methods, and [22, 23, 142, 150, 202, 206, 290] for combustion.

9

Boundary Layers and Fluid Layers

The presence of bodies and walls in a fluid flow results in the transfer of mass, momentum and heat, which occurs in addition to convection. These processes sometimes develop next to the wall in a layer (called the "boundary layer") which has a thickness that is much smaller than the characteristic size of the obstacle. The boundary layer can be either laminar or turbulent. Moreover, each transfer process can have its own particular boundary layer thickness. Chemical reactions can take place in the bulk of the fluid phase (in a flame for example) or at the wall (in heterogeneous reactions).

In certain situations we can assume that the presence of diffusion and chemical reactions does not disturb the viscous flow, which can therefore be determined separately. In other situations, such as in the Emmons problem for example, strong coupling effects exist due to the sublimation of a wall.

In Sect. 9.1 we will look at unsteady laminar flows with balance equations that have exact solutions, including a laminar viscous boundary layer that develops on a flat plate moving through an incompressible fluid at rest, and diffusion boundary layers caused by heat conduction or mass diffusion in a semi-infinite space limited by a wall.

Section 9.2 is devoted to the steady flow of a viscous incompressible fluid between two rotating coaxial cylinders. In the laminar case there is a classical Couette flow velocity profile rather than a boundary layer. After we show the effects of wall blowing and aspiration on the Couette flow, we investigate how it is destabilized by increasing the relative rotational velocity of the cylinders. This leads to Taylor–Couette instability.

In subsequent sections, we consider flow situations where approximations are required. The classical problem of the steady incompressible laminar boundary layer above a flat plate is the subject of Sect. 9.3. Self-similar solutions are obtained for the basic approximated equations, and the effects of wall blowing and aspiration on the flow are also studied.

Section 9.4 deals with steady laminar boundary layers with chemical reactions above a flat plate. Assuming that the diffusion processes do not modify the velocity profile of the incompressible flow, the concentration profile is

R. Prud'homme, *Flows of Reactive Fluids*,
DOI 10.1007/978-0-8176-4659-2_9, © Springer Science+Business Media, LLC 2010

found to be dependent on the Schmidt number. Next we study the Emmons problem, which is a steady compressible boundary layer with a diffusion flame that is induced by the sublimation of a flat plate of a solid combustible in a flow of oxidizer. Here, we find the velocity, temperature and concentration profiles and the mass flow rate of sublimation with respect to the Shvab–Zel'dovich approximation. Microgravitational behavior and the role of soot in this context are discussed.

Laminar and turbulent boundary layers and the diffusion of species across the chemically reacting wall of a rotating disc are studied in Sect. 9.5.

Finally, Sect. 9.6 is devoted to the turbulent boundary layer over a flat plate. The transition from laminar to turbulent flow is discussed with the aid of multiple-scale dimensional analysis.

9.1 Unsteady Boundary Layers

9.1.1 Viscous Boundary Layer in the Laminar Flow of an Incompressible Fluid

The problem of a viscous boundary layer in the laminar flow of an incompressible fluid was discussed in Sect. 7.1.3 in order to illustrate coupled phenomena. It will now be treated in more detail.

The momentum equation (4.47) is written

$$\rho \frac{d\mathbf{v}}{dt} + \boldsymbol{\nabla} \cdot \mathbf{P} = \sum_j \rho_j \mathbf{f}_j. \tag{9.1}$$

In the case of an incompressible pure fluid with a constant viscosity coefficient and no gravitational effects, this equation becomes

$$\rho \frac{d\mathbf{v}}{dt} + \boldsymbol{\nabla} p = \mu \Delta \mathbf{v}. \tag{9.2}$$

Let us now study the case of a plane flow with rectilinear and parallel trajectories with respect to the direction Ox. The mass balance shows that, in this case, the only nonzero velocity component is a function of only y and t:

$$u = u(y,\, t), \;\; v = w = 0. \tag{9.3}$$

The momentum equation gives

$$\begin{cases} \partial u/\partial t + (1/\rho)\partial p/\partial x = \nu \partial^2 u/\partial y^2, \\ \partial p/\partial y = 0, \;\; \nu = \mu/\rho. \end{cases} \tag{9.4}$$

It follows that

$$\partial u/\partial t - \nu \partial^2 u/\partial y^2 = (1/\rho)\partial p/\partial x = C'(t) \tag{9.5}$$

where $C(t)$ is an arbitrary function of time and $C'(t) = dC/dt$. We obtain $p = -\rho(C'(t)x + D(t))$, where $D(t)$ is an arbitrary function of time resulting from the integration of the second equation of (9.5), and setting $U(y,\, t) = u(y,\, t) - C(t)$ yields

$$\partial U/\partial t - \nu\partial^2 U/\partial y^2 = 0. \tag{9.6}$$

We restrict the search to self-similar solutions and then write

$$\eta = y/\sqrt{\nu t},\ U = f(\eta). \tag{9.7}$$

The partial differential equation (9.6) thus becomes a differential equation in $f(\eta)$:

$$\eta f'(\eta) + 2f''(\eta) = 0. \tag{9.8}$$

The integration of this differential equation gives

$$f(\eta) = A\,\mathrm{erf}(\eta/2) + B, \tag{9.9}$$

where A and B are constants. The velocity field is thus

$$U(y,\, t) = A\,\mathrm{erf}(y/\sqrt{4\nu t}) + E(t),\ E(t) = B + C(t). \tag{9.10}$$

This class of solutions allows us to treat the case of a flow that is initially at rest in the half-plane $y > 0$ above an infinite plate that is located at $y = 0$ and moves along Ox at a constant velocity V.
The boundary conditions

$$u(\infty,\, t) = 0\ \forall t, \quad u(0,\, t) = V,\ t > 0 \tag{9.11}$$

lead immediately to the solution (see Fig. 9.1)

$$u(y,\, t) = V(1 - \mathrm{erf}(y/\sqrt{4\nu t})). \tag{9.12}$$

The velocity u only has a significant value in a layer of thickness δ, where

$$\delta = O(\sqrt{\nu t}). \tag{9.13}$$

More precisely, by fixing the layer boundary at $\eta = \eta_0 = O(1)$, we have

$$\delta = 2\eta_0\sqrt{\nu t}. \tag{9.14}$$

Note that [99]:

- The thickness of the boundary layer grows with time as $\sqrt{\nu t}$.
- As soon as $t > 0$, the disturbance due to the motion of the plate exists throughout the half-plane $y > 0$. The propagation is thus instantaneous. In reality, propagation does not occur instantaneously, but this model of incompressible fluid cannot take this into account. An analogous remark was made in Sect. 5.3 about thermal propagation.

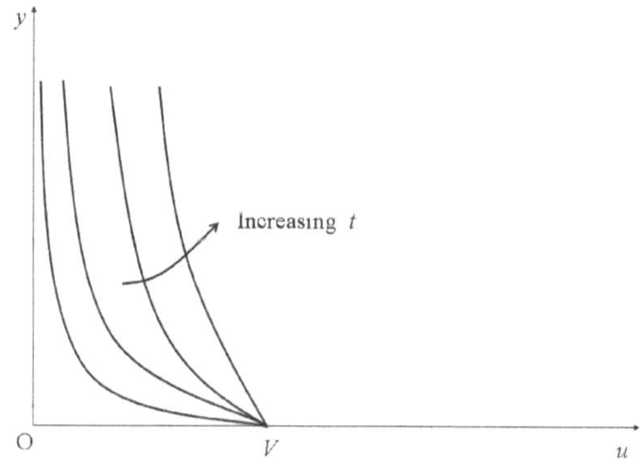

Fig. 9.1. Velocity of an unsteady boundary layer

- When ν tends to zero for given values of $y > 0$ and $t > 0$, the velocity u also tends to zero, but the fluid velocity remains nonzero and equal to V at $y = 0$. However, we find that $u = 0$ for $y > 0$, $\forall t$ in the case of a nonviscous fluid. This contradiction indicates singular behavior, which is characteristic of a boundary layer.

9.1.2 Diffusional Boundary Layers

The same phenomenon can be observed at rest when thermal conduction is present (see Chap. 5). In this case, we have a thermal boundary layer (Fig. 9.2) of thickness

$$\delta_T = 2\eta_0\sqrt{\kappa t} \tag{9.15}$$

with a thermal diffusivity of $\kappa = \lambda/\rho\, c_p$. In the same way, the diffusion of a substance A through a substance B according to Fick's law will provide a boundary layer thickness of

$$\delta_C = 2\eta_0\sqrt{Dt}. \tag{9.16}$$

Imagine a planar porous wall that is normal to $x'x$ at point O (Fig. 9.3). The part $x < 0$ contains liquid A at a concentration of $Y_A = 1$, which is kept constant over time since there is an adequate supply of A. A diffuses through the motionless part $x > 0$ in such a way that $Y_A(0, t) = 1$ and $Y_A(\infty, t) = 0$. The obtained solution is then, for $x > 0$,

$$Y_A = 1 - \mathrm{erf}(x/\sqrt{4Dt}). \tag{9.17}$$

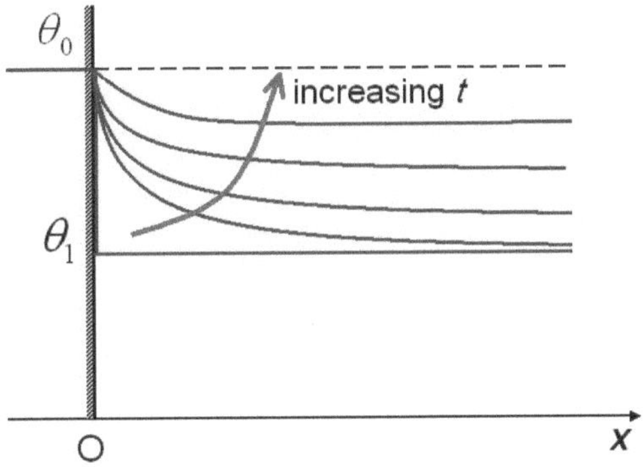

Fig. 9.2. Unsteady thermal boundary layer

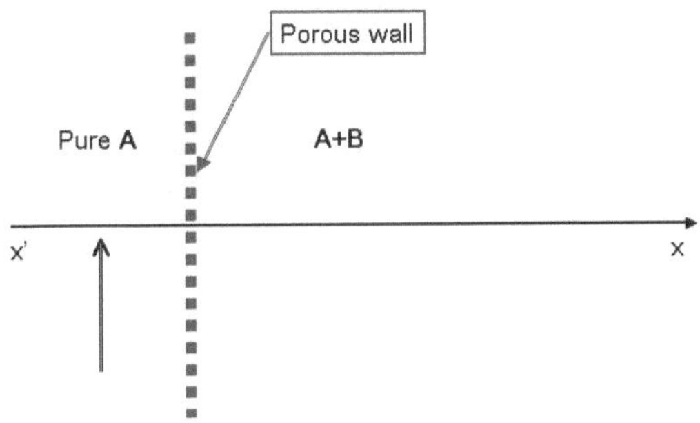

Fig. 9.3. Configuration of a boundary layer of concentration

9.2 Steady Flow of a Viscous Incompressible Fluid Between Two Coaxial Cylinders

9.2.1 Laminar Couette Flow

Two coaxial cylinders with radii of R_1 and R_2, respectively, where $R_2 > R_1$, are rotated with angular velocities of ω_1 and ω_2, respectively, where $\omega_2 \neq \omega_1$ [43].

Between these two infinitely long cylinders (the results remain valid at a sufficient distance from the extremities of cylinders that are sufficiently long compared to R_2) is an incompressible viscous fluid.

Assuming a zero axial velocity v_z, we look for solutions that give a zero radial velocity v_r. It is assumed that μ =const. The continuity equation shows that the tangential velocity v_θ is solely a function of the distance r to the common axis of the cylinders (see Fig. 9.4).

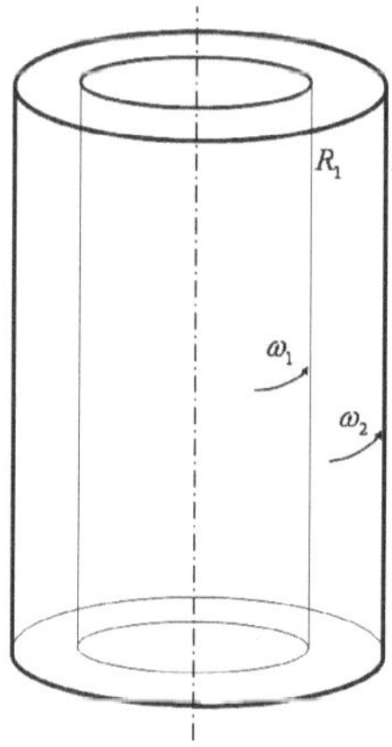

Fig. 9.4. Rotating coaxial cylinders

The momentum equation gives the relations

$$\begin{cases} -v_\theta^2/r^2 + (1/\rho)\,\partial p/\partial r = 0, \\ (1/\rho r)\,\partial p/\partial\theta = \nu[(1/r)\,(\partial/\partial r)(r\partial v_\theta/\partial r) - v_\theta/r^2], \\ \partial p/\partial z = 0, \end{cases} \tag{9.18}$$

with the boundary conditions

$$r = R_1 : \; v_\theta = \omega_1 R_1; \;\; r = R_2 : \; v_\theta = \omega_1 R_2. \tag{9.19}$$

We will also assume that $\partial p/\partial\theta = 0$. We now apply the identity

$$(1/r)\,(\partial/\partial r)(r\partial v_\theta/\partial r) - v_\theta/r^2 = (\partial/\partial r)[(1/r)\,\partial(rv_\theta)/\partial r]$$

to the right hand side of the second equation of (9.18) in order to deduce the solution, which takes the form $v_\theta = Ar + B/r$. The integration constants A and B are calculated from the boundary conditions. We finally obtain

$$v_\theta = \frac{\omega_2 R_2^2 - \omega_1 R_1^2}{R_2^2 - R_1^2} r + \frac{R_1^2 R_2^2(\omega_1 - \omega_2)}{R_2^2 - R_1^2}\frac{1}{r}. \tag{9.20}$$

The velocity profile obtained does not depend on the viscosity coefficient, which is logical since the equation involving v_θ is simply $\Delta v_\theta = 0$. There is an analogy here with steady thermal conduction, which gives $\Delta T = 0$ and does not depend on the thermal diffusivity κ.

Thus, there is no laminar viscous boundary layer that we can define directly starting from the velocity profile and μ. Calculating the shear stress gives

$$\tau_{r\theta} = 2\mu \frac{R_2^2(\omega_2 - \omega_1)}{R_2^2 - R_1^2} \frac{R_1^2}{r^2}. \tag{9.21}$$

Therefore, $\tau_{r\theta}$ decreases as $1/r^2$ from $r = R_1$ to $r = R_2$. The shear stress is the greatest at the wall of the inner cylinder. The rotation vector $\boldsymbol{\omega}$ is constant and equal to

$$\omega = \frac{\omega_2 R_2^2 - \omega_1 R_1^2}{R_2^2 - R_1^2}. \tag{9.22}$$

The flow is irrotational if

$$\omega_2 R_2^2 = \omega_1 R_1^2. \tag{9.23}$$

Such a system permits the viscosity coefficient μ to be measured at a given temperature. To do this, we only need to know the torque required to rotate one of the cylinders, and to apply the formula that yields $\tau_{r\theta}$. If C_1 is the torque exerted on the cylinder of radius R_1 and height h, we obtain (neglecting the edge effects, $R_1 < R_2 \ll h$)

$$\mu = C_1 \frac{R_2^2 - R_1^2}{4\pi h R_1^2 R_2^2 \omega_1}. \tag{9.24}$$

9.2.2 Blowing and Aspiration at the Walls

The previous problem also has an exact solution in the presence of uniform blowing or aspiration at the walls. We still assume $p = p(r)$, $v_\theta = v_\theta(r)$ and $v_z = 0$, but now $v_r = v_r(r)$. In this case, the continuity equation

$$\frac{\partial \rho}{\partial t} + \frac{1}{r}\frac{\partial \rho r v_r}{\partial r} + \frac{1}{r}\frac{\partial \rho v_\theta}{\partial \theta} + \frac{\partial \rho v_z}{\partial z} = 0 \tag{9.25}$$

gives

$$r v_r = Q, \tag{9.26}$$

where $2\pi\, h\, Q$ is the volume flow rate for a porous cylinder of height h. The momentum equations are

$$\begin{cases} dv_r/dt - v_\theta^2/r + (1/\rho)\, \partial p/\partial r = \nu\, [\Delta v_r - (2/r^2)\, \partial v_\theta/\partial \theta - v_r/r^2], \\ dv_\theta/dt + v_r v_\theta/r + (1/\rho r)\, \partial p/\partial \theta = \nu\, [\Delta v_\theta + (2/r^2)\, \partial v_r/\partial \theta - v_\theta/r^2]. \end{cases} \tag{9.27}$$

Converting the expressions into cylindrical coordinates:

$$df/dt = \partial f/\partial t + v_r\, \partial f/\partial r + (v_\theta/r)\, \partial f/\partial \theta + v_z\, \partial f/\partial z,$$

$$\Delta f = (1/r)\, \partial(r\, \partial f/\partial r)/\partial r + (1/r^2)\, \partial^2 f/\partial \theta^2 + \partial^2 f/\partial z^2,$$

and using our assumptions, we obtain

$$\begin{cases} v_r\, \partial v_r/\partial r - v_\theta^2/r + (1/\rho)\, \partial p/\partial r = \nu[(1/r)\, \partial(r\, \partial v_r/\partial r)/\partial r - v_r/r^2], \\ v_r\, \partial v_\theta/\partial r + v_r v_\theta/r = \nu[(1/r)\, \partial(r\, \partial v_\theta/\partial r)/\partial r - v_\theta/r^2]. \end{cases} \tag{9.28}$$

The first momentum equation gives $p(r)$ if v_r and v_θ are known. The second equation, which takes into account the conservation of the mass flow rate, gives us

$$\frac{Q}{r}\left(\frac{\partial v_\theta}{\partial r} + \frac{v_\theta}{r}\right) = \nu \frac{\partial}{\partial r}\left(\frac{\partial v_\theta}{\partial r} + \frac{v_\theta}{r}\right) \tag{9.29}$$

or

$$\frac{\partial}{\partial \ln r}\left(\frac{1}{r}\frac{\partial(r v_\theta)}{\partial r}\right) = \frac{Q}{\nu}\left(\frac{1}{r}\frac{\partial(r v_\theta)}{\partial r}\right). \tag{9.30}$$

The solution takes the form

$$\begin{cases} v_\theta = A r^{1+Q/\nu} + B/r, \; Q/\nu \neq -2 \\ v_\theta = A \ln r/r + B/r, \; Q/\nu = -2. \end{cases} \tag{9.31}$$

Boundary conditions specifying that the velocity v_θ is equal to $\omega_1 R_1$ on the inner cylinder and $\omega_2 R_2$ on the external cylinder provide the solution (setting $Q/\nu = \alpha$)

$$\begin{cases} v_\theta = [(R_2^2\omega_2 - R_1^2\omega_1)/(R_2^{\alpha+2} - R_1^{\alpha+2})]\, r^{\alpha+1} \\ +[R_1^2 R_2^2 (R_2^\alpha \omega_1 - R_1^\alpha \omega_2)/(R_2^{\alpha+2} - R_1^{\alpha+2})]\, r^{-1}, \; \alpha \neq -2 \\ v_\theta = [(R_2^2\omega_2 - R_1^2\omega_1)/(\ln R_2 - \ln R_1)]\, \ln r/r \\ +[(R_1^2\omega_1 \ln R_2 - R_2^2\omega_2 \ln R_1)/(\ln R_2 - \ln R_1)]\, r^{-1}, \; \alpha = -2. \end{cases} \tag{9.32}$$

Note that, in the presence of blowing ($\alpha > 0$) or aspiration ($\alpha < 0$), the solution depends on the viscosity through the coefficient

$$\alpha = Q/\nu = r v_r/\nu, \tag{9.33}$$

which is simply a Reynolds number associated with the radial motion.

The flow rotation vector, which is collinear with Oz, has a component along this axis of

$$\begin{cases} \omega = (1/2r)\partial(rv_\theta)/\partial r \\ = ((\alpha+2)/2)[(R_2^2\omega_2 - R_1^2\omega_1)/(R_2^{\alpha+2} - R_1^{\alpha+2})]\, r^\alpha, \; \alpha \neq -2 \\ \omega = [(R_2^2\omega_2 - R_1^2\omega_1)/2\ln(R_2/R_1)]\, r^{-2}, \; \alpha = -2. \end{cases} \tag{9.34}$$

Therefore, for any α, we have

$$\omega(r)/\omega(R_1) = (r/R_1)^\alpha. \tag{9.35}$$

Figure 9.5 shows the evolution of $\omega(r)/\omega(R_1)$ for various values of α.

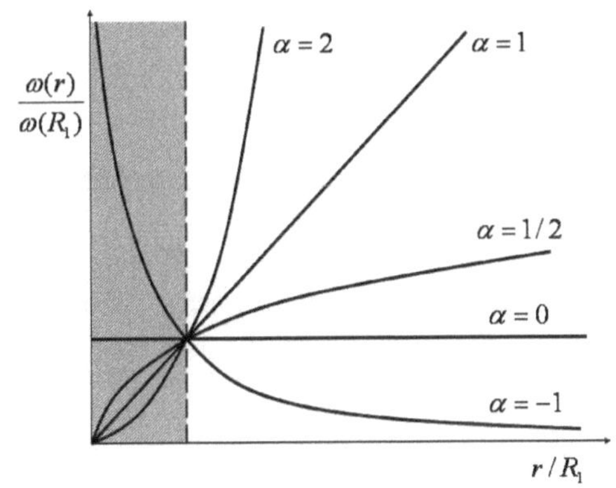

Fig. 9.5. Swirl as a function of the distance from the axis

For $\alpha > 0$, the flow becomes increasingly rotational when r increases from R_1 to R_2. This is the case for blowing. The opposite occurs for $\alpha < 0$. The aspiration tends to limit the rotational zone to the vicinity of the inner cylinder.

A boundary layer appears when $|\alpha|$ is large; i.e., when the Reynolds number is large. In the case of aspiration, for example, we can define a boundary layer thickness δ_ϵ in the vicinity of the inner cylinder by writing

$$r/R_1 = 1 + \delta_\epsilon/R_1, \; \delta_\epsilon/R_1 \ll 1, \tag{9.36}$$

and, for a given value of $\epsilon \ll 1$, using one of the v_θ expressions given in (9.32):

$$|v_\theta/R_1\omega_1| \leq \epsilon. \tag{9.37}$$

The uniform injection of a species A through the wall of the external cylinder, which leads to the chemical reaction

$$A \xrightarrow{k} B,$$

and the aspiration of the mixture by the inner cylinder are described by the equation

$$\frac{\partial}{\partial r}\left(r\frac{\partial Y_A}{\partial r}\right) - \frac{Q}{D}\frac{\partial Y_A}{\partial r} - \frac{k}{D} r\, Y_A = 0. \tag{9.38}$$

The concentration profile does not therefore depend on the rotational motion of the cylinders, only on the injection flow rate Q. The result would be the same with motionless cylinders.

9.2.3 Taylor–Couette Instability

In the absence of aspiration and blowing, the solution described in Sect. 9.2.1 gives

$$v_r = v_z = 0, \; v_\theta = Ar + B/r. \tag{9.39}$$

This is the only possible solution that gives $\mathbf{v} = \mathbf{v}(r)$. It is stable as long as $|B|/\nu$ is small.

The Taylor–Couette instability occurs for a sufficiently large value of $|B|/\nu$ and gives rise to a new solution involving parallel rolls (see Fig. 9.6a).

In turn, these Taylor vortices, which correspond to the solution $\mathbf{v} = \mathbf{v}(r, \theta, z)$ or $\mathbf{v} = \mathbf{v}(r, z)$, become unstable when $|B|/\nu$ is further increased by increasing the relative speed of rotation of the cylinders. We then obtain undulating vortices corresponding to a time-periodic mode (see Fig. 9.6b) [129].[1]

These instabilities are rather complex to study. We can consider steady solutions with axial symmetry for the linearized equations in u, v, w, where these quantities are defined like Couette flow velocity disturbances:

$$p = p(r) + p(r, z), \; v_r = u(r, z), \; v_\theta = v_\theta(r) + v(r, z), \; v_z = w(r, z). \tag{9.40}$$

After linearization, the equations of the steady regime become

$$\begin{cases} (1/r)\, \partial(ru)/\partial r + \partial w/\partial z = 0, \\ (1/\rho)\, \partial p/\partial r = \nu\, \mathcal{L}(u) + 2(A + B/r^2)v, \\ 2Au = \nu\, \mathcal{L}(v), \\ (1/\rho)\, \partial p/\partial z = \nu\left[(1/r)\, \partial(r\, \partial w/\partial r)/\partial r + \partial^2 w/\partial z^2\right], \end{cases} \tag{9.41}$$

with

[1]Spiral vortex flow and traveling waves can also be observed. The type of flow observed depends on the initial disturbance [104]. Moreover, turbulence appears for high values of the relative angular velocity of the cylinders.

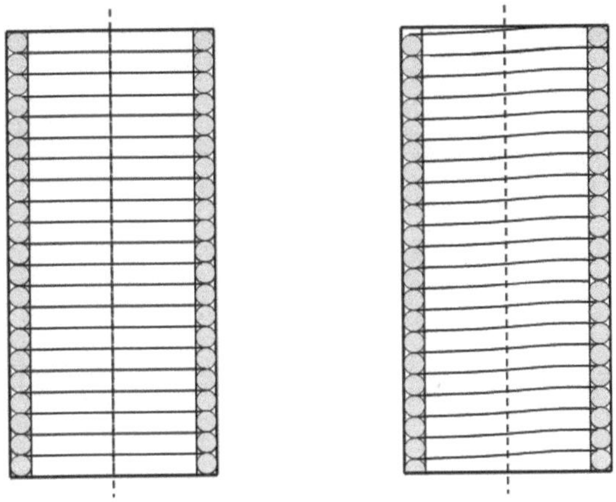

Fig. 9.6. Taylor–Couette instability

$$\mathcal{L}(f) = \frac{\partial}{\partial r}(\frac{1}{r}\frac{\partial(rf)}{\partial r}) + \frac{\partial^2 f}{\partial z^2}). \tag{9.42}$$

By eliminating w and p, we get

$$2(A + B/r^2)\partial^2 v/\partial z^2 + \nu\,\mathcal{L}^2(u) = 0,\ 2Au = \nu\,\mathcal{L}(v) \tag{9.43}$$

with the boundary conditions

$$r = R_1,\ R_2 :\ u = \partial u/\partial r = v = 0. \tag{9.44}$$

We then seek solutions of the type

$$u(r,\, z) = \hat{u}(r)\cos(kz),\ v(r,\, z) = \hat{v}(r)\cos(kz). \tag{9.45}$$

The preceding system becomes

$$\nu\, L^2(\hat{u}) = 2(A + B/r^2)k^2\hat{v},\ \nu\, L(\hat{v}) = 2A\hat{u} \tag{9.46}$$

with

$$L(\hat{f}) = \frac{d}{dr}(\frac{1}{r}\frac{d(r\hat{f})}{dr}) - k^2 \tag{9.47}$$

and

$$r = R_1,\ R_2 :\ \hat{u} = d\hat{u}/dr = \hat{v} = 0. \tag{9.48}$$

According to Chandrasekhar [43], when the radii R_1 and R_2 are similar we can set

$$d = R_2 - R_1,\ \mu = \omega_2/\omega_1,\ \zeta = (r - R_1)/d,\ k = a/d,\ \alpha = \mu - 1,\ D = d/d\zeta \tag{9.49}$$

and

$$T = -4A\omega_1 d^4/\nu^2,\ \hat{u} = 2d^2\omega_1 a^2 \hat{U}/\nu. \tag{9.50}$$

We then obtain

$$\begin{cases} (D^2 - a^2)^2\,\hat{U} = (1 + \alpha\zeta)\hat{v}, \\ (D^2 - a^2)\,\hat{v} = -Ta^2\,\hat{U}, \\ \hat{U} = D\hat{U} = \hat{v} = 0 \text{ for } \zeta = 0,\ 1. \end{cases} \tag{9.51}$$

The solution to this system can be found by expressing $\hat{v}$ in the form

$$\hat{v} = \sum_{m=1}^{\infty} C_m \sin \pi m\zeta. \tag{9.52}$$

After a rather complex calculation, and with the aid of an approximation, we find that the first Taylor–Couette rolls appear as soon as T exceeds the cutoff value

$$T_c = 3430/(1 + \mu) \tag{9.53}$$

obtained for $a = 3.12$.

The main reason that this problem is difficult to solve is that the first equation of system 9.51 includes a $1 + \alpha\zeta$ term. A similar solution is found by replacing this term with $(\mu + 1)/2$ (we set $r = (R_1 + R_2)/2$).

The differential system is then linear with constant coefficients. There are also known solutions to this problem for very different radii R_1 and R_2.

As mentioned previously, other instabilities such as undulating vortices can appear above the critical value.[2] These do not have axial symmetry, and in this case the velocity field, like the pressure field, depends on r, θ, z, t. Figure 9.7 (after [129]) gives the limits of stability deduced from the linearized theory (and corresponding to the experimental results) in the plane $(\omega_1 R_1^2/\nu,\ \omega_2 R_2^2/\nu)$.

We will limit our study here to flows between two coaxial cylinders in rotation. Various authors have also treated the case of an axial flow between such cylinders. In the presence of a radial flow, we can also envisage the existence

[2]According to Squire's theorem [62], 2D instabilities always occur before 3-D instabilities (this has also been established for a liquid film streaming across a plane plate under the action of gravity). This is the famous cascade of instabilities that leads to turbulence.

of instabilities that give rise to vortices. Unlike in Poiseuille flow (studied in Sect. 9.2.2), in this case the differentiated rotational motion influences the chemical processes beyond the stability threshold.

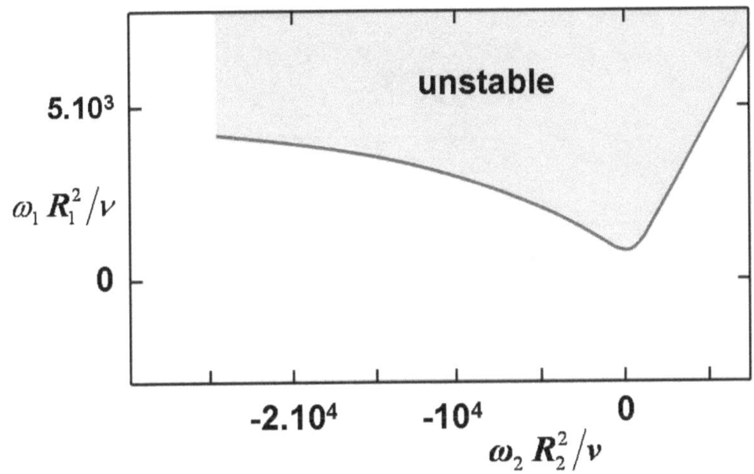

Fig. 9.7. Results for $R_2/R_1 = 1.135$

In the following sections, we will shift our focus away from the onset of instabilities and the laminar–turbulent threshold. We will calculate a fully laminar or a fully turbulent boundary layer. Approximations will be required to determine the motion close to walls with or without chemical reactions. Also note that certain results in relation to turbulent flows in the vicinity of a plane plate have already been given in Chap. 7.

9.3 Steady Incompressible Laminar Boundary Layer Above a Flat Plate

9.3.1 Basic Equations of the Boundary Layer

The equations of flow are as follows [10, 34, 249]:

$$\nabla \cdot \mathbf{v} = 0, \quad \rho \mathbf{v} \cdot \nabla \otimes \mathbf{v} + \nabla p = \mu \, \Delta \mathbf{v}. \tag{9.54}$$

Let us consider a 2D flow with a velocity U_∞ at infinity upstream, which evolves above a semi-infinite plate of equation $y = 0$ for $x > 0$ (Fig. 9.8).

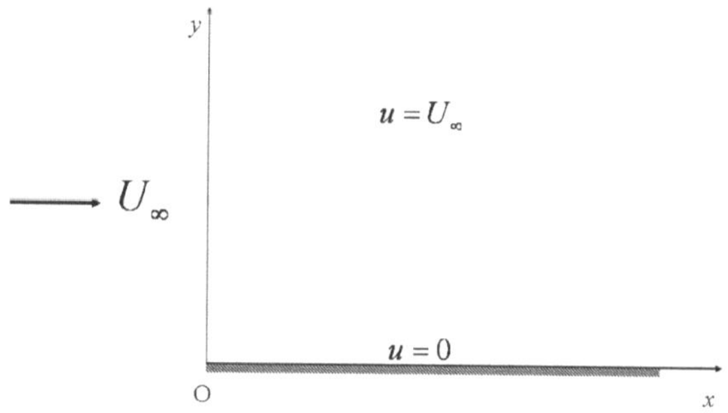

Fig. 9.8. Configuration of laminar boundary layer above a flat plate

We assume the presence of a boundary layer of thickness $\delta(x)$ (this is an experimental fact), and study what occurs inside this boundary layer; in other words, for $y = O(\delta(x))$. Outside of the boundary layer, $y/\delta(x) \to \infty$, $u = U_\infty$, $w = 0$. This problem has no solutions with streamlines parallel to Ox. Let us suppose that we have

$$v/u \approx \delta/x \ll 1 \tag{9.55}$$

inside the boundary layer (we will see the consequences of this assumption for the Reynolds number later), and that the first and second derivatives of the quantities have the same order of magnitude as the ratios of these quantities. Thus,

$$\partial^2 u/\partial x^2 \approx u/x^2,\ \partial p/\partial z \approx p/\delta,\ \dots \tag{9.56}$$

These assumptions mean that the terms in $\partial^2/\partial x^2$ can be ignored in comparison with those in $\partial^2/\partial y^2$. The equations thus become

$$\begin{cases} \partial u/\partial x + \partial v/\partial y = 0, \\ u\partial u/\partial x + v\partial u/\partial y + (1/\rho)\partial p/\partial x = \nu\partial^2 u/\partial y^2, \\ u\partial v/\partial x + v\partial v/\partial y + (1/\rho)\partial p/\partial y = \nu\partial^2 v/\partial y^2. \end{cases} \tag{9.57}$$

According to the last equation, the term $(1/\rho)\partial p/\partial y \approx (p/\rho)/\delta$ is of order $(u^2/x + \nu u/\delta^2)\delta/x$, so $(1/\rho)\partial p/\partial x \approx (u^2 + \nu u/\delta^2)(\delta/x)^2$ is much smaller than the other terms in the second equation of system (9.57).

These assumptions thus allow us to ignore the pressure gradient, which leads us to the simplified system

$$\begin{cases} \partial u/\partial x + \partial v/\partial y = 0, \\ u\partial u/\partial x + v\partial u/\partial y = \nu\partial^2 u/\partial y^2. \end{cases} \tag{9.58}$$

A More Rigorous Approach

We can establish (9.58) more rigorously via the asymptotic expansion method. The first step is to introduce dimensionless coordinates by dividing x and y by a reference length L. We now take ϵ, the inverse of the Reynolds number Re_∞, to be a small parameter:

$$\epsilon = (Re_\infty)^{-1} = \nu/U_\infty L \ll 1. \tag{9.59}$$

We then have

$$\begin{cases} u = U_\infty[\delta_1(\epsilon)u_1 + \delta_2(\epsilon)u_2 + \ldots] = O(U_\infty), \\ v = U_\infty \Delta(\epsilon)[\delta_1(\epsilon)v_1 + \delta_2(\epsilon)v_2 + \ldots] = O(U_\infty \Delta(\epsilon)), \\ p = p_\infty[\delta_1(\epsilon)p_1 + \delta_2(\epsilon)p_2 + \ldots], \\ Y = y/d(\epsilon), \end{cases} \tag{9.60}$$

where the quantities u_i, v_i and p_i are unknown functions of x and y, and the gage functions $\delta_i(\epsilon)$, $\Delta(\epsilon)$, $d(\epsilon)$ are unknown functions of ϵ. These gage functions δ_i also obey the condition

$$\lim_{\epsilon\to 0}[\delta_i(\epsilon)/\delta_{i-1}(\epsilon)] = 0. \tag{9.61}$$

The quantities u_i, v_i, p_i and their derivatives with respect to x and Y are assumed to be $O(1)$. For $Y \to \infty$ (or $y \gg \delta$) we have $u \to U_\infty$, so we will take $\delta_1(\epsilon) = 1$. The continuity equation becomes

$$\partial u_1/\partial x + O(\delta_2) + (\Delta(\epsilon)/d(\epsilon))(\partial v_1/\partial Y + O(\delta_2)) = 0. \tag{9.62}$$

We then obtain a first order nontrivial equation provided that

$$\Delta(\epsilon) = d(\epsilon). \tag{9.63}$$

It follows that

$$\begin{cases} u = U_\infty[u_1 + \delta_2(\epsilon)u_2 + \ldots], \\ v = U_\infty d(\epsilon)[v_1 + \delta_2(\epsilon)v_2 + \ldots], \\ p = p_\infty[p_1 + \delta_2(\epsilon)p_2 + \ldots], \\ Y = y/d(\epsilon). \end{cases} \tag{9.64}$$

The momentum equations can then be written

$$\begin{cases} u_1\partial u_1/\partial x + v_1\partial u_1/\partial Y + (p_\infty/\rho U_\infty^2)\partial p_1/\partial x + O(\delta_2) \\ = (\epsilon/d^2)(\partial^2 u_1/\partial Y^2 + O(\delta_2)), \\ u_1\partial v_1/\partial x + v_1\partial v_1/\partial Y + (1/d^2)[(p_\infty/\rho U_\infty^2)\partial p_1/\partial Y + O(\delta_2)] \\ = (\epsilon/d^2)(\partial^2 v_1/\partial Y^2 + O(\delta_2)). \end{cases} \tag{9.65}$$

The first-order equations depend on our choice of the function $d(\epsilon)$, which characterizes the scale of analysis in y.

For $d(\epsilon) = 1 :\ x = O(1),\ y = O(1)$, we obtain the nonviscous equations

$$\begin{cases} Y = y,\ \partial u_1/\partial x + \partial v_1/\partial Y = 0, \\ u_1\partial u_1/\partial x + v_1\partial u_1/\partial Y + (p_\infty/\rho U_\infty^2)\partial p_1/\partial x = 0, \\ u_1\partial v_1/\partial x + v_1\partial v_1/\partial Y + (p_\infty/\rho U_\infty^2)\partial p_1/\partial y = 0. \end{cases} \tag{9.66}$$

This system is valid outside the boundary layer when y and x are of the same order of magnitude. The solution is known to be:

$$\begin{cases} u_1 = 1,\ v_1 = 0,\ p_1 = 1, \text{ or} \\ u = U_\infty,\ v = 0,\ p = p_\infty. \end{cases} \tag{9.67}$$

For $d(\epsilon) = \epsilon^{1/2} :\ x = O(1),\ y = O(\epsilon^{1/2})$, we have

$$\begin{cases} Y = y/\sqrt{\epsilon},\ \partial u_1/\partial x + \partial v_1/\partial Y = 0, \\ u_1\partial u_1/\partial x + v_1\partial u_1/\partial Y + (p_\infty/\rho U_\infty^2)\partial p_1/\partial x = \partial^2 u_1/\partial Y^2, \\ u_1\partial v_1/\partial x + v_1\partial v_1/\partial Y + \epsilon^{-1}(p_\infty/\rho U_\infty^2)\partial p_1/\partial y = \partial^2 v_1/\partial Y^2. \end{cases} \tag{9.68}$$

The last equation shows us that

$$\partial p_1/\partial y = 0,\ p_1 = p_1(x). \tag{9.69}$$

However, for $y = O(1)$ (i.e., $Y \to \infty$), we know that $p = p_\infty$, so $p_1(x)$=const.= 1. The system reduces to

$$\begin{cases} \partial u_1/\partial x + \partial v_1/\partial Y = 0, \\ u_1\partial u_1/\partial x + v_1\partial u_1/\partial Y = \partial^2 u_1/\partial Y^2. \end{cases} \tag{9.70}$$

These equations are those of the boundary layer obtained previously (system 9.58), and the thickness of this boundary layer is

$$\delta = Ld(\epsilon) = L\epsilon^{1/2} = (\nu L/U_\infty)^{1/2}. \tag{9.71}$$

As x is assumed to be first order, the corresponding dimensional length x is of order L. It follows that the boundary layer thickness actually varies as

$$\delta = (\nu x/U_\infty)^{1/2}. \tag{9.72}$$

The assumption $\delta/x \ll 1$ thus results in

$$(\nu/U_\infty x)^{1/2} = (Re_x)^{-1/2} \ll 1 \text{ or } Re_x \gg 1, \tag{9.73}$$

and is only valid for rather large Reynolds numbers.

9.3.2 Self-Similar Solutions

We have

$$\begin{cases} \partial u/\partial x + \partial v/\partial y = 0, \\ u\partial u/\partial x + v\partial u/\partial y = \nu\partial^2 u/\partial y^2. \end{cases} \tag{9.74}$$

The self-similar solutions (see Chap. 5) will be such that

$$u^2/x \approx \nu u/y^2. \tag{9.75}$$

This suggests that the quantity

$$\eta = y(U_\infty/\nu x)^{1/2} \tag{9.76}$$

should be introduced. We find that the steady boundary layer thickness is

$$\delta = (\nu x/U_\infty)^{1/2}. \tag{9.77}$$

Let us seek a solution of the type

$$u = u(\eta). \tag{9.78}$$

The continuity equation enables us to introduce the stream function ψ such that

$$u = \partial\psi/\partial y, \ v = -\partial\psi/\partial x, \tag{9.79}$$

so that $\partial\psi/\partial y = u(\eta)$. Taking x and η as new variables, we obtain $\psi = \psi(x, \eta)$ and

$$u(\eta) = \frac{\partial\psi}{\partial y} = \frac{\partial\psi}{\partial\eta}\frac{\partial\eta}{\partial y} + \frac{\partial\psi}{\partial x}\frac{\partial x}{\partial y} = (U_\infty/\nu x)^{1/2}\frac{\partial\psi}{\partial\eta}, \tag{9.80}$$

or, by integration,

$$\psi(x, \eta) = (\nu x/U_\infty)^{1/2}\int u(\eta)\, d\eta + C(x). \tag{9.81}$$

Let us set $u(\eta) = U_\infty f'(\eta)$, where $f'(\eta)$ is the derivative of an unknown function $f(\eta)$ that can be shifted by any constant value. We then find that

$$\psi(x, \eta) = (\nu x U_\infty)^{1/2} f(\eta) + C(x). \tag{9.82}$$

Here, the velocity component v becomes

$$v = (\nu U_\infty/x)^{1/2}(\eta f'(\eta) - f(\eta))/2 - C'(x). \tag{9.83}$$

From the wall condition

$$f'(0) = 0, \; v = 0, \tag{9.84}$$

we deduce that

$$v = 1/2\,(\nu U_\infty/x)^{1/2}(f(0) + \eta f'(\eta) - f(\eta)) \tag{9.85}$$

and that

$$C'(x) = 1/2\,(\nu U_\infty/x)^{1/2} f(0). \tag{9.86}$$

The definition of $f(\eta)$ enables us to choose the constant $f(0)$ so as to cancel out $C(x)$. We can do this by taking $f(0) = 0$. Thus,

$$\begin{cases} \psi(x, \eta) = (\nu x U_\infty)^{1/2} f(\eta), \\ u = U_\infty f'(\eta), \\ v = 1/2\,(\nu U_\infty/x)^{1/2}(\eta f'(\eta) - f(\eta)). \end{cases} \tag{9.87}$$

Putting these expressions into the momentum equation, we obtain the classical equation of Blasius:[3]

$$2f''' + f\,f'' = 0 \tag{9.89}$$

with

$$f(0) = f'(0) = 0, \; f'(\infty) = 1. \tag{9.90}$$

Figure 9.9 gives the solutions obtained for the components of the velocity vector [34].

---

[3] $\eta = y(U_\infty/2\nu x)^{1/2}$ is usually chosen instead, so that Blasius' equation becomes

$$f''' + f\,f'' = 0 \tag{9.88}$$

with $f(0) = f'(0) = 0$, $f''(\infty) = 1$. We also have $\psi(x, \eta) = (2\nu x U_\infty)^{1/2} f(\eta)$, $u = U_\infty f'(\eta)$, $v = (\nu U_\infty/2x)^{1/2}(\eta f'(\eta) - f(\eta))$.

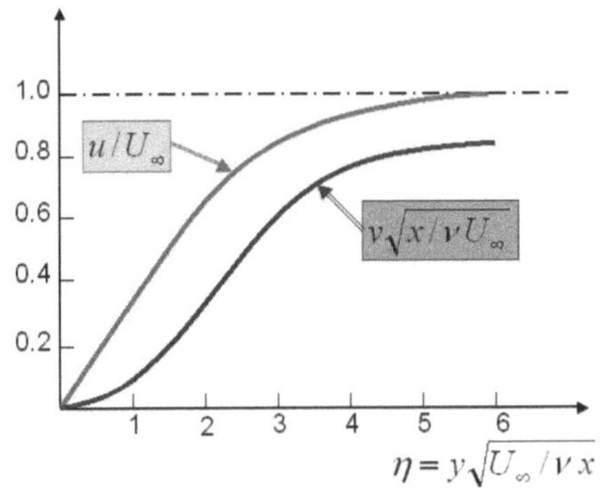

Fig. 9.9. Reduced velocity components as functions of the similarity variable

9.3.3 Blowing and Aspiration at the Wall

The preceding solutions of the Blasius equation also make it possible to describe the situation where there is blowing or aspiration of fluid through a wall. However, there are particular cases in this situation. Choosing $f(0) \neq 0$ results in a nonzero velocity component in the direction normal to the wall [249].

Moreover, the other conditions $f'(0) = 0$ (zero tangential velocity) and $f'(\infty) = 1$ remain valid. The mass flow rate for blowing at the wall will be

$$\dot{m} = (\rho v)_W = -(\rho \mu U_\infty / 2x)^{1/2} f(0). \tag{9.91}$$

Blowing occurs for $f(0) < 0$ and aspiration in the opposite case. However, this result is limited to cases where the injected or aspirated flow varies as $x^{-1/2}$, which limits its applicability. In particular, the assumption of similarity cannot be applied during uniform injection. To study the general case, we can take [161, 266] $\xi = \nu x$ and have η as a variable upon which the velocity component u depends, so that

$$u(\xi, \eta) = U_\infty f'(\xi, \eta) = U_\infty \partial f / \partial \eta \tag{9.92}$$

if we retain the same notation for successive derivatives of f with respect to η, which become partial derivatives here.

The Blasius equation (9.88) is then replaced with

$$f''' + f f'' = 2\xi (f' \partial f' / \partial \xi - f'' \partial f / \partial \xi), \tag{9.93}$$

with the boundary conditions

$$\begin{cases} f'(\xi, 0) = 0, \; f'(\xi, \infty) = 1, \\ (f + 2\xi \partial f / \partial \xi)_{\eta = 0} = -\dot{m} (2\xi / U_\infty)^{1/2} / \mu. \end{cases} \tag{9.94}$$

Setting

$$\chi = (2\xi)^{1/2}, \ G = \partial f/\partial \chi, \tag{9.95}$$

we obtain, by neglecting the terms in $\partial G/\partial \chi$,

$$\begin{cases} f''' + f\,f'' = \chi(f'G' - f''G), \\ G''' + fG'' + f''G = 2(f'G' - f''G), \end{cases} \tag{9.96}$$

with

$$\begin{cases} f(\chi, 0) = -\chi/2, \ f'(\chi, 0) = 0, \ f'(\chi, \infty) = 1, \\ G(\chi, 0) = -1/2, \ G'(\chi, 0) = 0, \ G'(\chi, \infty) = 0. \end{cases} \tag{9.97}$$

This system is solved numerically. This is the two-equation model.

If a new differentiation is carried out, we can introduce a new function H, and an additional differential equation is added as well as the corresponding boundary conditions for H. In this case, we ignore the $\chi\partial^2(f'G' - f''G)/\partial\xi^2$ term and we must then integrate a system of three ordinary differential equations. Thus, we have the three-equation model. Calculations show that the results become increasingly precise if we increase the number of equations for the model, ignoring terms that are increasingly small. Note that the results are already excellent with the two- or three-equation models (Fig. 9.10).

Also note that, in the case of aspiration, there is *an exact solution for the boundary layer equations if the aspiration is uniform* [249]. Consider the uniform aspiration speed $v_W < 0$. Now suppose that u and v are independent of the x-coordinate. We obtain, starting from the continuity equation

$$\partial u/\partial x + \partial v/\partial y = 0, \tag{9.98}$$

the result

$$v = v_W = \text{const.} \tag{9.99}$$

The momentum equation

$$u\partial u/\partial x + v\partial u/\partial y = \nu\partial^2 u/\partial y^2 \tag{9.100}$$

becomes

$$d^2u/dy^2 - (v_W/\nu)du/dy = 0, \tag{9.101}$$

which only permits a valid solution only if $v_W < 0$. We obtain

$$u = U_\infty[1 - \exp(v_W y/\nu)]. \tag{9.102}$$

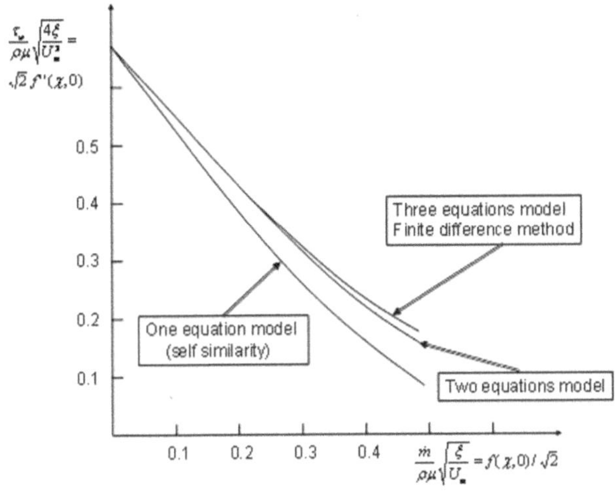

Fig. 9.10. Evolution of $f''(\chi, 0)$ according to $f(\chi, 0)$ in the case of uniform blowing at the wall, or tension from friction at the wall according to the unit mass flow rate of injection

The boundary layer thickness is then constant and equal to

$$\delta = \nu / v_W . \tag{9.103}$$

The solution, which is valid for an infinite planar plate, is only asymptotically suitable in the case of a semi-infinite plate ($x > 0$) or in the case of a plate of finite length.

9.4 Steady Laminar Boundary Layers with Chemical Reactions Above a Flat Plate

9.4.1 Boundary Layers with Diffusion

Let us assume that the diffusion does not modify the velocity profile of the flow [159]. If it has a Blasius profile, we have

$$\begin{cases} u = U_\infty f'(\eta), \ \eta = y(U_\infty / 2\nu x)^{1/2}, \\ v = (\nu U_\infty / 2x)^{1/2}(\eta f'(\eta) - f(\eta)), \\ f''' + f\, f'' = 0, \ f'(0) = f(0) = 0, \ f'(\infty) = 1. \end{cases} \tag{9.104}$$

The following four cases can be distinguished depending on the Schmidt number $Sc = \nu / D$:

If $Sc \gg 1$

Suppose that the diffusion initially takes place very close to the wall and that the diffusional boundary layer thickness is $\delta_C \ll \delta$. In this zone, we have

$$u \cong U_\infty \eta f''(0). \tag{9.105}$$

The species balance equation is

$$u\partial Y/\partial x + v\partial Y/\partial y - D\partial^2 Y/\partial y^2 = 0 \tag{9.106}$$

in the case of Fick's law with only one coefficient of diffusion and if we can ignore $\partial^2 Y/\partial x^2 = O(1/x^2)$ in comparison with $\partial^2 Y/\partial y^2 = O(1/\delta_C^2)$.

The first and the last terms are of the same order of magnitude, so that

$$u/x \cong D/\delta_C^2 \tag{9.107}$$

or

$$U_\infty f''(0)(\delta_C/x)(U_\infty/2\nu x)^{1/2} \approx D/\delta_C^2. \tag{9.108}$$

Since we have

$$\delta \approx (2\nu x/U_\infty)^{1/2}, \tag{9.109}$$

we deduce that

$$\delta_C/\delta \approx (D/2\nu f''(0))^{1/3} = O(Sc^{-1/3}), \tag{9.110}$$

where the Schmidt number $Sc = \mu/\rho D$. It is clear that this case corresponds to $Sc \gg 1$.

Let us now assume that a very fast chemical reaction A $\rightarrow$ B, corresponding to wall catalysis, is occurring. For $y = 0\,(\eta = 0)$, we have

$$Y_A = 0, \;\; Y_B = 1, \tag{9.111}$$

and for $y = \infty\,(\eta = \infty)$ we have

$$Y_A = 1, \;\; Y_B = 0. \tag{9.112}$$

In the vicinity of the wall, $y = O(\delta_C)$ and

$$\begin{cases} u = U_\infty \eta f''(0), \\ v = (1/2)(\nu U_\infty/2x)^{1/2} f''(0)\,\eta^2. \end{cases} \tag{9.113}$$

Let us write $Y_B = Y(\eta)$. We obtain

$$d^2Y/d\eta^2 + (\nu f''(0)\eta^2/2D)dY/d\eta = 0. \tag{9.114}$$

The solution to this equation is

$$Y = 1 - I(\eta)/I(\infty), \tag{9.115}$$

with

$$I(\eta) = \int_0^\eta e^{-f''(0)Sc\xi^3/6} d\xi. \tag{9.116}$$

Of course, we find that the order of magnitude of the thickness of the boundary layer of concentrations $\delta_C/\delta = O(Sc^{-1/3})$.

If $Sc \ll 1$

Let us now suppose that the Schmidt number is small. The diffusion occurs in a layer of thickness $\delta_C \gg \delta$, and the concentrations hardly vary in the viscous boundary layer, so that we have $u = U_\infty$ almost everywhere. The diffusion equation becomes

$$U_\infty \partial Y/\partial x - D\partial^2 Y/\partial y^2 = 0, \tag{9.117}$$

and the thickness δ_C is thus

$$U_\infty/x \approx D/\delta_C^2, \tag{9.118}$$

which gives

$$\delta_C/\delta \approx (D/2\nu)^{1/2} = O(Sc^{-1/2}), \; Sc \ll 1 \tag{9.119}$$

For the abovementioned reaction, we obtain the solution

$$Y = 1 - \mathrm{erf}(\sqrt{Sc/2}\,\eta). \tag{9.120}$$

If $Sc = 1$

If the Schmidt number is equal to one, we have

$$\begin{cases} u\partial u/\partial x + v\partial u/\partial y - \nu\partial^2 u/\partial y^2 = 0, \\ u\partial Y/\partial x + v\partial Y/\partial y - \nu\partial^2 Y/\partial y^2 = 0, \end{cases} \tag{9.121}$$

since $D = \nu$. The viscous and diffusion thicknesses of the boundary layer are the same, and in the particular case of a fast reaction at the wall, we obtain the self-similar solution

$$u/U_\infty = 1 - Y = f'(\eta), \tag{9.122}$$

where $f(\eta)$ is the solution of the Blasius equation with $f'(0) = f(0) = 0, \; f'(\infty) = 1$.

If $Sc = O(1)$

Finally, if Sc is of the same order of magnitude 1 but is not exactly equal to 1, the solution is not easily found (except in the case of a fast reaction). It is then necessary to solve the diffusion equation knowing the velocity profile.

Note that we have only considered here, as an example, a situation where diffusion controls the chemical process. In reality, there are very often interactions between the diffusion, the chemical kinetics and the thermal transfer. The following example illustrates such interactions.

9.4.2 The Emmons Problem

An ablative plane plate is sublimated with a latent heat ΔL in a combustive gas flow [80, 290]. The combustible plate has a density ρ_s and releases the gas H through sublimation (see Fig. 9.11). The chemical exothermic reaction $\mathrm{H} + \mathrm{O} \to \mathrm{P}$, in the region of contact with the oxidizer O. In this case, the standard heat of reaction is (see Eq. 2.114 of Sect. 2.4.1)

$$\Delta H = -\sum_j \nu_j \mathcal{M}_j (q_f^0)_j. \tag{9.123}$$

The phenomenon includes thermal conduction as well as the diffusion of the species involved and viscosity. Gravitational effects are neglected (Fig. 9.12). It is also assumed that the pressure gradient is negligible and that the Schvab–Zel'dovich approximation can be applied (see Sect. 7.4). The flow is two-dimensional planar and steady. It is easily shown that the terms in $\partial^2/\partial x^2$ are negligible compared to the terms in $\partial^2/\partial y^2$ because of the thinness of the layer: $\delta(x) \ll x$ for sufficiently large x (i.e., located sufficiently far from the leading edge, but with the flow remaining laminar).

Using the notation of Sect. 7.4, the equations for compressible and reactive fluid flow are as follows:

$$\begin{cases} \partial(\rho u)/\partial x + \partial(\rho v)/\partial y = 0, \\ \rho u \partial u/\partial x + \rho v \partial u/\partial y - (\partial/\partial y)(\mu \partial u/\partial y) = 0, \\ \rho u \partial \beta_j/\partial x + \rho v \partial \beta_j/\partial y - (\partial/\partial y)(\rho D \partial \beta_j/\partial y) = \dot{\zeta}, \\ \rho u \partial \beta_T/\partial x + \rho v \partial \beta_T/\partial y - (\partial/\partial y)((\lambda/c_{p,\,f})\partial \beta_T/\partial y) = \dot{\zeta}. \end{cases} \tag{9.124}$$

The conditions at the wall are obtained by applying the discontinuity balance equation (Sect. 4.9)

$$[\boldsymbol{\mathcal{J}}_F + \rho f(\mathbf{v} - \mathbf{W})]_-^+ \cdot \mathbf{N} = 0 \tag{9.125}$$

(with side $(-)$ corresponding to the solid wall and side $(+)$ to the fluid) to mass, momentum, chemical species (there is no surface reaction), and total energy. We get (v_a: ablation velocity, τ_{xy}: unknown friction stress, s: solid phase

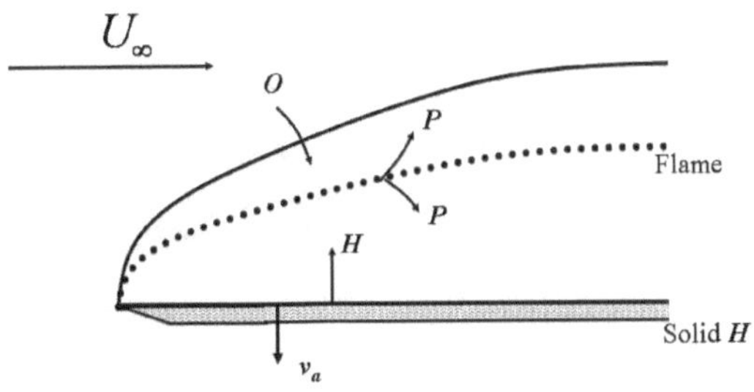

Fig. 9.11. The Emmons problem: the thin flame case

assumed to be homogeneous in both density, $\rho_{s\,W} = \rho_s$, and in concentration, $\beta_{j\,s\,W} = \beta_{j\,s}$):

$$\begin{cases} (\rho v)_W = \rho_s v_a, \\ \tau_{xy} = (\mu \partial u/\partial y)_W, \\ -(\rho D \partial \beta_j/\partial y)_W + (\rho v)_W\, \beta_{j\,W} = \rho_s v_a\, \beta_{j\,s}, \\ -((\lambda/c_{p,\,f})\partial \beta_T/\partial y)_W + (\rho v)_W\, \beta_{T\,W} = -\rho_s v_a\, l/\Delta H. \end{cases} \tag{9.126}$$

The gaseous medium is compressible. This difficulty is easily overcome by performing the Howarth variable change:

$$\xi = \int_0^x \rho\,\mu\,dx, \tag{9.127}$$

which gives $d\xi = \rho\,\mu\,dx$ if we assume that the quantity $\rho\,\mu$ depends only on the x-coordinate[4]

$$\zeta = \int_0^y \rho\,dy \text{ or } d\zeta = [\int_0^y (\partial\rho/\partial x)\,dy]\,dx + \rho\,dy. \tag{9.128}$$

Finally, we set

[4]This assumption (that $\rho\,\mu$ does not vary across the boundary layer) is often reasonable for gases; if changes in the average molecular weight are negligible, then—because the pressure is constant—the ideal gas law implies that $\rho \propto 1/T$, in which case the constant nature of $\rho\,\mu$ leads to $\mu \propto T$: a dependence that is close to the predictions of kinetic theory [290].

$$g = (1/\rho\mu)[\rho v + u \int_0^y (\partial\rho/\partial x]\, dy). \tag{9.129}$$

The resulting equations are then

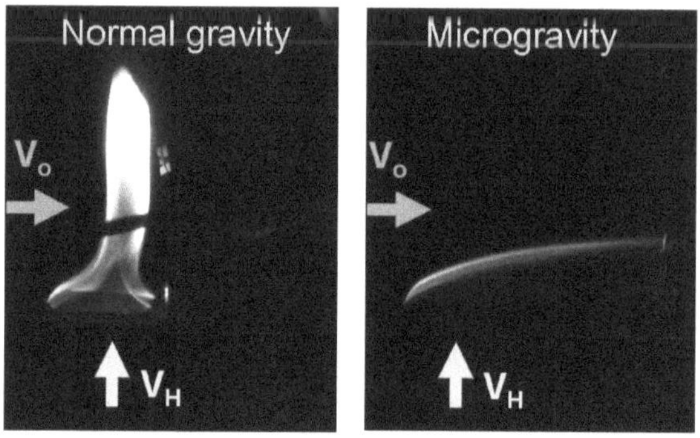

Fig. 9.12. Experimental observations of a diffusion flame under normal gravity and under microgravity (G. Legros, private communication, 2009)

$$\begin{cases} \partial u/\partial\xi + \partial g/\partial\zeta = 0,\ d\psi = u\, d\zeta - g\, d\xi, \\ u\partial u/\partial\xi + g\partial u/\partial\zeta - \partial^2 u/\partial\zeta^2 = 0, \\ u\partial\beta_j/\partial\xi + g\partial\beta_j/\partial\zeta - (\partial/\partial\zeta)(Sc^{-1}\partial\beta_j/\partial\zeta) = \dot{\zeta}/\rho^2\mu, \\ u\partial\beta_T/\partial\xi + g\partial\beta_T/\partial\zeta - (\partial/\partial\zeta)(Pr^{-1}\partial\beta_T/\partial\zeta) = \dot{\zeta}/\rho^2\mu, \end{cases} \tag{9.130}$$

with the following system applicable at the wall:

$$\begin{cases} (\rho\mu g)_W = \rho_s v_a \\ \tau_{xy} = \rho\mu(\partial u/\partial\zeta)_W \\ -Sc^{-1}(\partial\beta_j/\partial\zeta)_W + g\,(\beta_j - \beta_{js}) = 0 \\ -Pr^{-1}(\partial\beta_T/\partial\zeta)_W + g\, l/\Delta H = 0 \end{cases} \tag{9.131}$$

(see Sect. 4.9 for the general equations for discontinuities, and Sect. 7.5 to determine the precisions of the first and the last two equations). For $\xi = \infty$, we have

$$u = U_\infty,\ \beta_H = 0,\ \beta_O = \beta_{O\infty},\ \beta_T = \beta_{T\infty}. \tag{9.132}$$

The system of two equations in u and g is identical to that of the incompressible boundary layer, and permits self-similar solutions. The thickness of this boundary layer in the direction ζ is

$$\delta(\xi) = (\xi/U_\infty)^{1/2}, \tag{9.133}$$

and we set

$$\eta = \zeta(U_\infty/2\xi)^{1/2}. \tag{9.134}$$

The velocity components can be written

$$u = U_\infty f'(\eta),\ g = (U_\infty/2\xi)^{1/2}(\eta f'(\eta) - f(\eta)), \tag{9.135}$$

and the stream function becomes

$$\psi = \sqrt{2\xi U_\infty}\, f(\eta). \tag{9.136}$$

The conditions on the wall give

$$\begin{cases} -\rho\mu\sqrt{U_\infty/2\xi}\, f(0) = \rho_s\, v_a, \\ \tau_{xy} = \rho_W\, \mu_W \sqrt{U_\infty/2\xi}\, f''(0). \end{cases} \tag{9.137}$$

Note that the speed of ablation depends on $\xi^{-1/2}$. The value of $f(0)$ will be determined by the other conditions relating to β_j and β_T. Moreover, we have

$$f'(\infty) = 1,\ f'(0) = 0. \tag{9.138}$$

When the Schmidt and Prandtl numbers are both equal to one, the quantity $(\beta_j - \beta_T)$ obeys the same equation as u. The only solution to the problem is the linear relation of Crocco:

$$\beta_T - \beta_O = \frac{\beta_{T\infty} - \beta_{TW} - \beta_{O\infty}}{U_\infty} u + \beta_{TW}. \tag{9.139}$$

At the wall, we have

$$(\partial\beta_T/\partial\zeta)_W = g(0)\, l/\Delta H = -\sqrt{U_\infty/2\xi}\, f(0)\, l/\Delta H, \tag{9.140}$$

so

$$(\partial\beta_T/\partial\eta)_W = -f(0)\, l/\Delta H. \tag{9.141}$$

Crocco's relation gives

$$(\partial\beta_T/\partial\eta)_W = (\beta_{T\infty} - \beta_{TW} - \beta_{O\infty})\, f''(0), \tag{9.142}$$

because $\beta_{OW} = 0$ if it is assumed that all of the oxidizer is consumed inside the flame.

Let us now define the Spalding parameter:

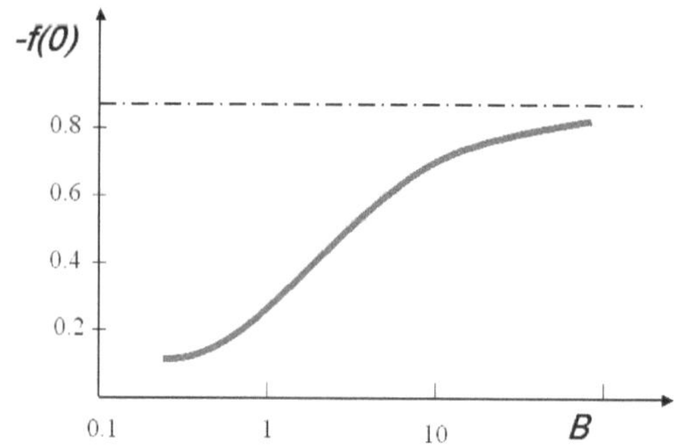

Fig. 9.13. The quantity $-f(0)$ as a function of B

$$B = (\beta_{T\infty} - \beta_{TW} - \beta_{O\infty})\, \Delta H/l. \tag{9.143}$$

The three boundary conditions

$$f'(\infty) = 1,\ f'(0) = 0,\ f(0) = -B\, f''(0) \tag{9.144}$$

enable us to solve the Blasius equation $f''' + f\, f'' = 0$.

The value of B (Fig. 9.13) thus provides the speed at which the fuel regresses:

$$v_a = -\sqrt{U_\infty/2\xi}\, f(0)\, (\rho\mu)_W/\rho_H, \tag{9.145}$$

where $\rho_H = \rho_s$ is the density of solid fuel, the friction stress on the plate

$$\tau_{xy} = U_\infty \sqrt{U_\infty/2\xi}\, f''(0)\, (\rho\mu)_W, \tag{9.146}$$

and the temperature gradient to the wall

$$(d\beta_T/d\eta)_W = -f(0)\, l/\Delta H. \tag{9.147}$$

These results are independent of the chemical kinetics and the position of the flame.

If the flame is thin (Fig. 9.11), the chemical production rate $\dot{\zeta}$ vanishes everywhere except on the locus defined by the value η_f (which provides a parabola $\zeta^2 = \eta_f^2\, 2\xi/U_\infty$), where the mixture is stoichiometric; i.e., $(Y_O/Y_H)_f = r_{st}$.[5] We then have

[5]The stoichiometric ratio r_{st} is defined by (7.99), where the combustible is denoted F.

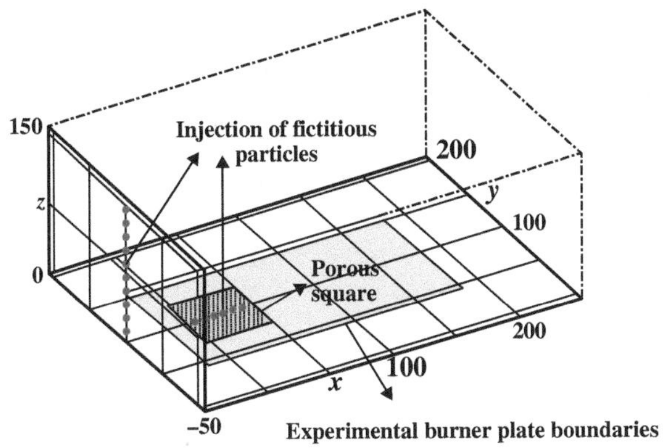

Fig. 9.14. Schematic view of the numerical domain (G. Legros, private communication, 2009; see also [157])

$$\begin{cases} Y_O = Y_{O\infty}(f'(\eta_f) - f'(\eta))/(f'(\eta_f) - 1), \; \eta > \eta_f, \\ Y_H = f(0)(f'(\eta_f) - f'(\eta))/(f(0)f'(\eta_f) - f''(0)), \; \eta < \eta_f. \end{cases} \tag{9.148}$$

The value of η_F is then determined by the equation $Y_{O\infty}/r_{st} = f(0)[f'(\eta_f) - 1]/[f(0)f'(\eta_f) - f''(0)]$, and the concentration and temperature profiles can be deduced. The coefficient of local friction is

$$C_F = \tau_{xy}/(\rho U_\infty^2/2) = (2\mu_\infty/\xi U_\infty)^{1/2} f''(0) = (2/B)\rho_s v_a/\rho_\infty U_\infty. \tag{9.149}$$

To be applicable, this theory requires the use of the Shvab–Zel'dovich assumption. However, two factors act very strongly on the flames:

- Gravity (Fig. 9.12), which makes hot gases rise
- The presence of soot, which induces radiative losses.

Heat is also lost at the wall. Torero et al. [275], who studied the flames associated with fires in space vehicles, proposed a modified mass transfer number in the absence of gravity, and this has been corroborated by many experiments performed under microgravity. This number is given by

$$B_T = \frac{(1-\chi)\Delta H_c Y_{O_{2\infty}} + c_{p\infty}(T_f - T_\infty)}{\Delta H_p + Q}, \tag{9.150}$$

where χ is the radiative loss to the surroundings, related to the radiative flow from the flame to the surface; ΔH_c is the enthalpy of combustion per unit mass of consumed oxygen; $Y_{O_{2\infty}}$ is the mass fraction of oxygen under ambient conditions; $c_{p\infty}$ is the heat capacity at constant pressure per unit mass under ambient conditions; T_f is the temperature of the flame; T_∞ is the temperature of the ambient gas; ΔH_p is the enthalpy of pyrolysis per unit mass of pyrolyzed fuel; and Q is the total surface enthalpic loss per unit mass of pyrolyzed fuel. This loss includes heat conduction across the ablative plate, re-emission from the ablative surface (which can be estimated from the temperature of pyrolysis), and radiative flux feedback from the flame.

In theory, we can evaluate χ and Q by simply juxtaposing the analytical solution of Emmons with the experimental profile.

Diffusion flame modeling is the subject of many theoretical, numerical (Fig. 9.14), and experimental studies [86]. One of the most difficult problems to model is the interaction of soot with the whole flow [285]. One mechanism of soot formation was provided by Bockhorn in 1994 [19]. It involves the following consecutive steps: a molecular stage containing fuel, oxidizer and reaction products (characteristic entity size: 0.5 nm); a particulate stage that can be subdivided into particle inception, surface growth, and coagulation (characteristic entity size: 50 nm).

To simplify the experimental configuration, we often utilize gaseous fuel injection through a porous square [156], and the two-dimensional assumption can be evaluated experimentally in order to assess its validity [155]. More recent studies of soots (numerical and experimental) were published by Legros et al. [157]; the authors integrated new and existing numerical modeling and experimental observations to provide a consistent explanation for observed flame lengths and soot volume fractions for laminar diffusion flames.

9.5 The Rotating Disc

This problem is of interest in the field of electrolysis, where a rotating electrode is commonly used [13, 159].

9.5.1 Viscous Boundary Layer in Laminar Flow

Let us study the steady velocity profile in a liquid above a disc rotating around the Oz axis. By assuming axial symmetry, the components u, v, w of the velocity vector in cylindrical coordinates are solutions of

$$\begin{cases} \partial(ru)/\partial r + \partial(rw)/\partial z = 0, \\ u\,\partial u/\partial r + w\,\partial u/\partial z - v^2/r = -(1/\rho)\,\partial p/\partial r + \nu\,(\partial^2 u/\partial r^2 \\ +\partial(u/r)/\partial r + \partial^2 u/\partial z^2), \\ u\,\partial v/\partial r + w\,\partial v/\partial z + uv/r = \nu\,(\partial^2 v/\partial r^2 + \partial(v/r)/\partial r + \partial^2 v/\partial z^2), \\ u\,\partial w/\partial r + w\,\partial w/\partial z = -(1/\rho)\,\partial p/\partial z + \nu\,(\partial^2 w/\partial r^2 \\ +(1/r)\,\partial w/\partial r + \partial^2 w/\partial z^2), \end{cases} \tag{9.151}$$

and the boundary conditions

$$\begin{cases} u = w = 0, \; v = r\Omega \text{ for } z = 0 \\ u = v = 0 \text{ for } z = \infty. \end{cases} \tag{9.152}$$

We will assume an infinite disc, but the results will also be valid for a disc with a finite diameter because we can generally ignore edge effects in most of the flow.

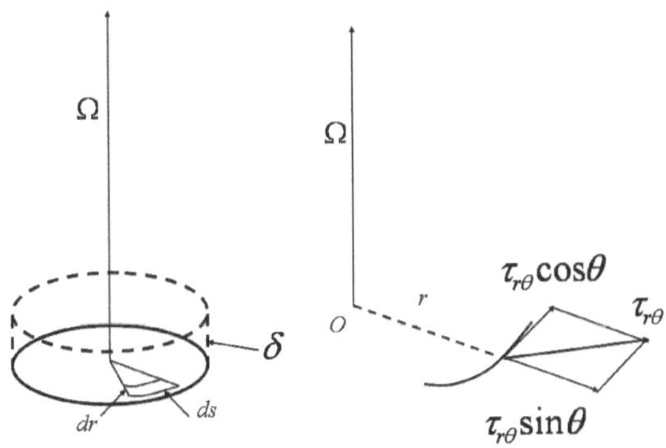

Fig. 9.15. The coordinate system

The boundary layer thickness δ is evaluated as done previously. Locally, the centrifugal force is $\rho r \Omega^2 \delta \, dr \, ds$. The frictional force has components of (see Fig. 9.15 for definitions)

$$\begin{cases} \tau_{r\theta} \sin\theta \, dr \, ds = \rho r \, \Omega^2 \delta \, dr \, ds, \\ \tau_{r\theta} \cos\theta \, dr \, ds = \mu r \, \Omega / \delta \, dr \, ds. \end{cases} \tag{9.153}$$

Therefore,

$$\tau_{r\theta} = \rho r \, \Omega^2 \delta / \sin\theta \cong \mu r \, \Omega / \delta \cos\theta, \; \delta^2 \cong \nu \tan\theta / \Omega. \tag{9.154}$$

If we assume that the sliding direction close to the wall does not depend on r, we obtain

$$\delta \cong \sqrt{\nu / \Omega}. \tag{9.155}$$

Therefore, we will now introduce the variable $\zeta = z/\delta = \sqrt{\Omega/\nu}\, z$.

We also introduce the von Kármán functions [133] to describe the velocity and pressure fields:

$$\begin{cases} u(r,\,z) = r\Omega F(\zeta),\ v(r,\,z) = r\Omega G(\zeta), \\ w(z) = (\nu\Omega)^{1/2} H(\zeta),\ p(z) = -\rho\nu\Omega P(\zeta). \end{cases} \tag{9.156}$$

We can deduce the following set of equations and boundary conditions:

$$\begin{cases} H' + 2F = 0 \\ F'' - HF' - F^2 + G^2 = 0, \\ G'' - HG' - 2FG = 0, \\ H'' - HH' + P' = 0, \\ F(0) = H(0) = F(\infty) = G(\infty) = 0,\ G(0) = 1. \end{cases} \tag{9.157}$$

We can also deduce

$$P(\zeta) = P_0 + H^2/2 - H' \tag{9.158}$$

and, after eliminating F between the remaining equations,

$$\begin{cases} G'' - HG' + H'G = 0, \\ H''' - HH'' + H'^2/2 - 2G^2 = 0, \\ H(0) = H'(0) = H'(\infty) = G(\infty) = 0,\ G(0) = 1. \end{cases} \tag{9.159}$$

Let us set $H(\infty) = -c$ with $c > 0$ and introduce Cochran's variable change [13] $\lambda = e^{-c\zeta}$. We will also set $G(\zeta) = c^2\, g(\lambda),\ H(\zeta) = -c + c\, h(\lambda)$. This gives

$$\begin{cases} \lambda g'' + hg' - h'g = 0, \\ \lambda h''' + 2\lambda^2 h'' + \lambda^2 hh'' + \lambda hh' - \lambda^2 h'^2/2 + 2g^2 = 0, \\ c^2 g(1) = 1,\ g(0) = h'(1) = 0,\ h(1) = 1. \end{cases} \tag{9.160}$$

The following expansions in powers of λ:

$$g(\lambda) = \sum_{i=1}^{n} a_i\, \lambda^i,\ h(\lambda) = \sum_{j=1}^{n} b_j\, \lambda^j, \tag{9.161}$$

as introduced by Benton, have the advantage of converging more quickly than traditional expansions in ζ. Recurrence relations are found between the a_i and the b_j.

$$\begin{cases} a_2 = 0, \qquad\qquad\qquad b_2 = -(b_1^2 + 4a_1^2)/8, \\ a_3 = (a_1 b_2 - a_2 b_1)/6,\ b_3 = -(3b_1 b_2 + 4a_1 a_2)/18, \\ a_4 = (a_1 b_3 - a_3 b_1)/6,\ b_4 = -(7b_1 b_3 + 4a_1 a_3 + 2b_2^2 + 4a_2^2)/48. \end{cases} \tag{9.162}$$

It is then necessary to determine a_1 and b_1 in order to satisfy the boundary conditions $h(1) = 1$, $h'(1) = 0$. The results obtained are $a_1 = 1.53678$ and $b_1 = 2.36449$, for which $c = 0.88447$.

We can therefore determine the functions F, G, H and $P - P_0$ and the derivatives F' and G' for all of the values of ζ with a high degree of accuracy by taking $n = 25$ (see Table 9.1 [13] and Fig. 9.16 [159]).

ζ	F	G	H	F′	G′	P − P₀
0.0	0.0000	1.0000	0.0000	0.5102	-0.6159	0.0000
0.5	0.1536	0.7076	-0.0919	0.1467	-0.5321	0.3115
1.0	0.1802	0.4766	-0.2655	-0.0157	-0.3911	0.3955
1.5	0.1559	0.3132	-0.4357	-0.0693	-0.2677	0.4066
2.0	0.1189	0.2033	-0.5732	-0.0739	-0.1771	0.4020
2.5	0.0848	0.1313	-0.6745	-0.0612	-0.1153	0.3970
3.0	0.0581	0.0845	-0.7452	-0.0455	-0.0745	0.3939
3.5	0.0389	0.0544	-0.7932	-0.0319	-0.0480	0.3924
4.0	0.0257	0.0349	-0.8251	-0.0216	-0.0309	0.3917
∞	0.0000	0.0000	-0.8845	0.0000	0.0000	0.3911

Table 9.1. Functions of the rotating disc

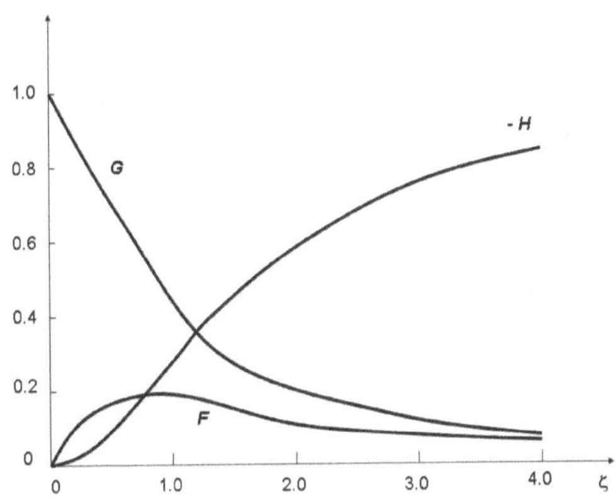

Fig. 9.16. F, G, H as functions of ζ

The results are displayed in Fig. 9.17. Traces can be obtained on the disc with an acid solution [159].

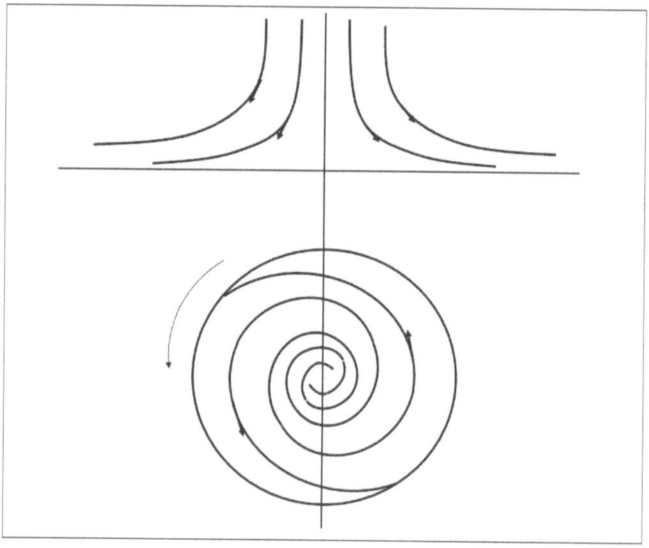

Fig. 9.17. Streamlines in planes $(r,\, z)$ and $(r,\, \theta)$

9.5.2 Diffusion in the Vicinity of a Rotating Disc

Let us now assume that a heterogeneous reaction occurs on the surface of the disc. This time, the liquid is an aqueous solution of concentration C (recall that $C_j = \rho Y_j/\mathcal{M}_j$). Since chemical reactions do not occur in the liquid, the chemical species balance equation reduces to

$$u\frac{\partial C}{\partial r} + \frac{v}{r}\frac{\partial C}{\partial \theta} + w\frac{\partial C}{\partial z} = D(\frac{\partial^2 C}{\partial r^2} + \frac{1}{r}\frac{\partial C}{\partial r} + \frac{1}{r^2}\frac{\partial^2 C}{\partial \theta^2} + \frac{\partial^2 C}{\partial z^2}) \qquad (9.163)$$

We will assume cylindrical symmetry. If C_∞ is the concentration at $z = \infty$ and C_0 is its value at $z = 0$ (presumed constant and determined by the surface reaction), let us assume that $\partial C/\partial r$ is negligible, and we will set

$$X = (C_\infty - C)/(C_\infty - C_0). \qquad (9.164)$$

We then obtain

$$w(z)dX/dz = Dd^2X/dz^2,\ X(\infty) = 0,\ X(0) = 1. \qquad (9.165)$$

The solution is (see Fig. 9.18)

$$X(z) = 1 - \frac{I(z)}{I(\infty)} \qquad (9.166)$$

where

$$I(z) = \int_0^z e^{\frac{1}{D}\int_0^\tau w(\eta)d\eta} d\tau.$$

Let us study the *case where the Schmidt number Sc is very large*:

$$Sc = \mu/\rho D = \nu/D \gg 1. \tag{9.167}$$

Diffusion only takes place on the inside of a thin layer of thickness δ_C in the vicinity of the disc:

$$\delta_C \ll \delta. \tag{9.168}$$

The integrals are calculated by noting that

$$\begin{cases} -\text{ for } z < \delta, \; H^{(1)}(\zeta) \cong -0.510\,\zeta^2 + \zeta^3/3 - 0.616\,\zeta^4/4 + \ldots \\ -\text{ for } z > \delta, \; H^{(2)}(\zeta) \cong -0.88. \end{cases} \tag{9.169}$$

Then

$$I(\infty) = \int_0^\delta e^{\frac{1}{D}\int_0^\tau w^{(1)}(\eta)d\eta} d\tau + \int_\delta^\infty e^{\frac{1}{D}\int_0^\tau w^{(2)}(\eta)d\eta} d\tau. \tag{9.170}$$

The last integral is zero to a first approximation. The first, calculated starting from series (1) of (9.169) and valid for $z < \delta$, is practically equal to

$$\int_0^\infty e^{\frac{1}{D}\int_0^\tau w^{(1)}(\eta)d\eta} d\tau \cong 1.61 D^{1/3}\nu^{1/6}\Omega^{-1/2}. \tag{9.171}$$

This $I(\infty)$ integral has the dimensions of length and is used as the definition of the concentration boundary layer thickness; i.e.,

$$\delta_C = I(\infty). \tag{9.172}$$

We can calculate $I(z)$ for $z < \delta$ using the same process ($I_{z>\delta} = I(\infty)$):

$$I(z) \cong 0.6 D^{1/3}\nu^{1/6}\Omega^{-1/2} \int_0^{(0.51\Omega^{3/2}/3D\nu^{1/2})z^3} e^{-\eta}\eta^{-2/3}\, d\eta. \tag{9.173}$$

By more precisely defining the viscous thickness of boundary layer δ as the value of z for which $G(\zeta) = 0.05$, we find that $\delta = 3.6\sqrt{\nu/\Omega}$ and

$$\delta_C/\delta = 0.45 Sc^{-1/3}. \tag{9.174}$$

The flux $\mathcal{J}_D = -(\mathcal{J}_{Dj})_z/\mathcal{M}_j$ of the solute towards the disc is, at $z = 0$,

$$\mathcal{J}_D = D(\frac{\partial C}{\partial z})_0 = -D(C_\infty - C_0)(\frac{\partial X}{\partial z})_0 \cong D(C_\infty - C_0)/\delta_C. \tag{9.175}$$

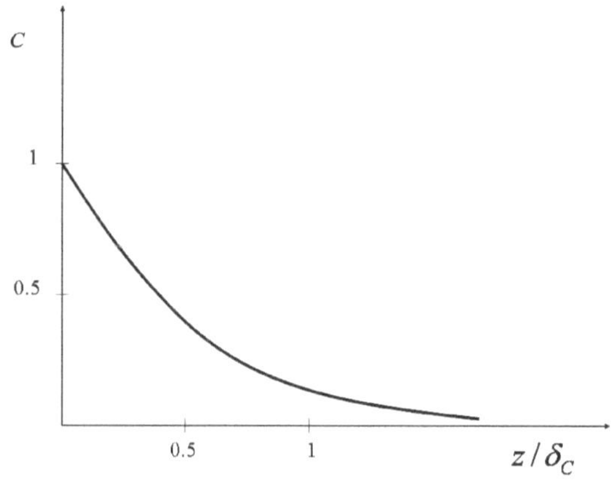

Fig. 9.18. Concentration profile

If the solute is consumed at the surface by a very fast reaction, we obtain

$$\mathcal{J}_{D\ \mathrm{lim}} = DC_\infty/\delta_C. \qquad (9.176)$$

If the solute is rejected at the surface (in a very fast reaction), we have another limiting flux,

$$\mathcal{J}_{D\ \mathrm{lim}} = -DC_0/\delta_C. \qquad (9.177)$$

For a disc of radius R, the total flux is

$$J = \pi R^2 D(C_\infty - C_0)/\delta_C, \qquad (9.178)$$

and the Nusselt number for diffusion Nu_C (or the Sherwood number Sh) is

$$Nu_C = JR/D(C_\infty - C_0) = 0.62\, Sc^{1/3} Re^{1/2}, \;\; Re \cong \Omega R^2/\nu. \qquad (9.179)$$

The convection coefficient is

$$\alpha_c = \rho J/(C_\infty - C_0) = -\pi R^2 \mathcal{J}_{Dj}/(Y_{j\infty} - Y_{j0}). \qquad (9.180)$$

Increasing the angular velocity Ω of the disc while the other quantities remain unchanged causes δ and δ_C to decrease and the mass fluxes to increase.

9.5.3 A Rotating Disc in Turbulent Flow

The tangential friction stress in laminar flow is

$$\tau_{z\theta} = \mu(\partial v/\partial z)_0 = r\rho\nu^{1/2}\Omega^{3/2}G'(0). \qquad (9.181)$$

For a disc of radius R, the torque is

$$M = \int_0^R r\tau_{z\theta} 2\pi r \, dr \cong -0.3\pi\rho\nu^{1/2}\Omega^{3/2}R^4 \tag{9.182}$$

and the corresponding friction coefficient is

$$C_M = -M/(\rho\Omega^2 R^5/2) \cong 1.885 Re^{-1/2}. \tag{9.183}$$

This result corresponds rather well to reality for Reynolds numbers of less than 3×10^5. Beyond this approximate value, the mode becomes turbulent.

Based on experimental results, von Kármán [133] derived the following formulae for the turbulent velocity profiles:

$$\begin{cases} u = \alpha r\Omega (z/\delta)^{1/7}(1 - z/\delta), \\ v = r\Omega(1 - (z/\delta)^{1/7}), \end{cases} \tag{9.184}$$

where α is a constant coefficient. The obtained boundary layer thickness is

$$\delta = 0.525 r(\nu/\Omega R^2)^{1/5}. \tag{9.185}$$

The turbulent boundary layer thickness thus depends on the distance r to the disc axis. The corresponding torque coefficient (Fig. 9.19) is given by the von Kármán formula [133, 249]

$$C_M = 0.146 Re^{-0.2}. \tag{9.186}$$

Landau and Levich [146] proposed an ideal model with several layers. The friction stress can be written as

$$\bar{\tau}_{z\theta} = (\mu + K_u)\partial v/\partial z. \tag{9.187}$$

We then have:

- $K_u \gg \mu$ and $K_u \approx \rho z$ "far" from the disc
- $K_u \approx \rho z^m$ with $m = 4$ closer to the disc
- A laminar mode very close to the disc (laminar underlayer): $K_u \ll \mu$.

9.6 Turbulent Boundary Layer and Dimensional Analysis

9.6.1 Turbulent Boundary Layer on a Flat Plate

We will now assume a smooth planar plate and determine the velocity profile in the vicinity of the wall using dimensional analysis for a simplified case. Let us suppose that the mean velocity of the flow is parallel to the plate in turbulent flow. We will also assume that the pressure is uniform, as in the laminar case [2]. Lastly, let us suppose that turbulence is homogeneous in the

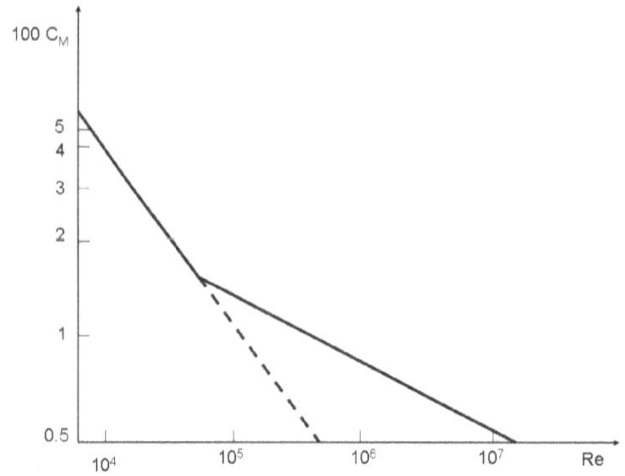

Fig. 9.19. Torque coefficient (turning moment) on a rotating disk in the laminar and turbulent regimes

directions parallel to the plate (i.e. that the correlations do not depend on x and z but on the normal distance y).

The continuity equation gives, with $\bar{v} = \bar{w} = 0$,

$$\partial \bar{u}/\partial x = 0. \tag{9.188}$$

Since the average flow is two-dimensional planar, we have

$$\bar{u} = \bar{u}(y), \tag{9.189}$$

so that $d\bar{u}/dt$ is equal to zero.

We then obtain the following for the momentum (ρ is assumed to be constant):

$$\begin{cases} \partial \overline{u'v'}/\partial y = \nu\, \partial^2 \bar{u}/\partial y^2, \\ \partial \overline{v'^2}/\partial y = \partial \overline{v'w'}/\partial y = 0. \end{cases} \tag{9.190}$$

On the wall, the velocity vector vanishes, so $\bar{\mathbf{v}}$ and $\mathbf{v}'$ vanish like $\bar{u}$, $\overline{u'v'}$, $\overline{v'^2}$ and $\overline{v'w'}$. Consequently, at any point

$$\overline{v'^2} = \overline{v'w'} = 0. \tag{9.191}$$

With these assumptions, the first momentum equation gives

$$\nu\, \partial u/\partial y - \overline{u'v'} = \text{const.} \tag{9.192}$$

This constant has the dimensions of the square of a velocity that we will call the speed of friction u^*, so that

$$\nu\, \partial u/\partial y - \overline{u'v'} = u^{*2} = \nu(\partial u/\partial y)_{y=0}. \tag{9.193}$$

We will also define the length of friction based on u^* and the viscosity coefficient ν:

$$y^* = \nu/u^* \tag{9.194}$$

The solution to the problem depends on y, y^* (or ν), u^* and $\bar{u}$. There are two Π ratios, $\bar{u}/u^*$ and y/y^*. According to the Vashi–Buckingham theorem (Sect. 5.1.2), we have

$$\bar{u}/u^* = \psi(y/y^*). \tag{9.195}$$

Very close to the wall, molecular flux dominates compared to turbulent fluxes, so that

$$\rho\nu\partial\bar{u}/\partial y \gg \rho\overline{u'v'}. \tag{9.196}$$

We therefore obtain

$$\nu\, \partial\bar{u}/\partial y = u^{*2}, \tag{9.197}$$

for which the solution is

$$\bar{u} = u^*\, y/\nu \Longrightarrow \bar{u}/u^* = y/y^*. \tag{9.198}$$

Thus, in this simplified case, the boundary layer includes a part with laminar character of thickness δ_v, which is on the order of magnitude of y^*. We can set

$$\delta_v = \alpha_v\, y^*, \tag{9.199}$$

where α_v is determined experimentally. In this zone,

$$\psi = y/y^*. \tag{9.200}$$

Farther from the wall, turbulent fluxes become more significant than molecular fluxes, so that

$$\rho\,\overline{u'v'} \gg \rho\,\nu\, \partial\bar{u}/\partial y \tag{9.201}$$

We then obtain

$$-\overline{u'v'} = u^{*2}. \tag{9.202}$$

In this zone, the velocity variations do not now depend on the molecular viscosity (i.e., on ν or y^*). The law $\bar{u}/u^* = \psi(y/y^*)$ must be such that $d\bar{u}/dy$ does not depend on y^*.

We have

$$\frac{\partial \bar{u}}{\partial y} = \frac{u^*}{y^*}\psi'(\frac{y}{y^*}). \tag{9.203}$$

The only solution is $\psi' = 1/(ky/y^*)$, where k is a constant. Thus, $d\bar{u}/dy = u^*/ky$. The solution to this equation is

$$\bar{u}/u^* = (1/k)\ln(y/y^*) + \text{const.} \tag{9.204}$$

This is the logarithmic sublayer obtained for $y \gg \delta_l > \delta_v$, with $\delta_l = \alpha_l y^*$. After obtaining α_l experimentally, we can use it to deduce the value of the constant as a function of $\bar{u}(\delta_l)$. Setting $\alpha_0 = \alpha_l e^{-k\bar{u}(\delta_l)/u^*}$, we then obtain

$$\bar{u}/u^* = \psi(y/y^*) = (1/k)\ln(y/\alpha_0 y^*). \tag{9.205}$$

This is therefore a model with two sublayers, a viscous sublayer for $0 < y < \delta_v$ and a logarithmic sublayer for $\delta_l < y$.

The coefficient k can be interpreted using the Prandtl turbulence model. We have

$$\rho\,\overline{u'v'} = -K_u\,\partial\bar{u}/\partial y \tag{9.206}$$

with the turbulent viscosity coefficient

$$K_u = \mu_t = \rho\,L_P^2\,\partial\bar{u}/\partial y. \tag{9.207}$$

We deduce

$$\overline{u'v'} = -L_P^2\,(\partial\bar{u}/\partial y)^2. \tag{9.208}$$

However, in the logarithmic sublayer,

$$\overline{u'v'} = -u^{*2},\ \partial\bar{u}/\partial y = u^*/ky. \tag{9.209}$$

It obviously follows that the Prandtl mixing length obeys the following expression close to the wall:

$$L_P = ky. \tag{9.210}$$

9.6.2 Method of Multiple Scales and Dimensional Analysis

The following method [239] makes it possible to find relations between dimensionless numbers that are valid for laminar flow and for turbulent flow. It applies to flows that can exhibit diffusion or thermal conduction. We will apply it to fluid layers.

We generally define the heat exchange coefficient α in the vicinity of a wall. Temperature changes are limited to a relatively thin layer through which the heat flow exchanged between the wall (of temperature T_W) and the undisturbed flow (of temperature T_∞) is given by the expression

$$q = \alpha\,(T_W - T_\infty). \tag{9.211}$$

At any point of the layer, we have

$$q_y = -\lambda\,\partial T/\partial y. \tag{9.212}$$

The Nusselt number is defined as

$$Nu = \alpha L/\lambda, \tag{9.213}$$

where L is a reference length.

In the presence of diffusion, the mass exchange between the wall (due to a chemical reaction, to dissolution...) and the flow will be

$$\mathcal{J}_D = \beta\rho\,(Y_W - Y_\infty), \tag{9.214}$$

where Y_W is the concentration at the wall and Y_∞ is the concentration in the main flow.

The diffusion flux of species j is

$$\mathcal{J}_{Dj} = -\rho D\,\partial Y_j/\partial y, \tag{9.215}$$

with a single diffusion coefficient D. We define the Sherwood number as

$$Sh = \beta L/D. \tag{9.216}$$

In both cases, L is a characteristic dimension of the system.

Relations between Nu, Re, Pr and between Sh, Re and Sc are found experimentally of the form

$$\begin{cases} Nu = A\,Re^m Pr^n, \\ Sh = A'\,Re^{m'} Pr^{n'}, \end{cases} \tag{9.217}$$

where A, A', m, m', n and n' are constants. Here, the method of multiple scales makes it possible, via modified dimensional analysis, to determine the values of m, m', n and n' in different zones corresponding to laminar and turbulent modes. We will also see that this method can be extended to other problems.

In classical dimensional analysis, the three universal basic quantities are length, mass and time.[6] In multiple-scale dimensional analysis, we will assume that there are two basic length quantities: L characterizes the macroscopic scale and l relates to molecular processes. In the same way, we have a macro-scale time T and a micro-scale time t.

In exchange via diffusion (the same reasoning is valid for thermal transfer) close to a wall, the parameters are: a characteristic length d (tube diameter for example), the flow velocity U_∞, ν, D and β. Table 9.2 shows the table of

[6] As explained in the footnote in Sect. 5.1.1, we do not consider temperature a primary quantity from a dimensional point of view.

dimensions for these parameters. β depends on four quantities but has the global dimension of a velocity:

$$\beta = L^u\, l^{1-u} / T^v\, t^{1-v}. \tag{9.218}$$

	d	U_∞	ν	D	β
L	1	1	0	0	u
l	0	0	2	2	$1-u$
T	0	-1	0	0	$-v$
t	0	0	-1	-1	$v-1$

Table 9.2. Dimensions of a diffusive boundary layer

By taking d, U_∞ and ν as the basic quantities (there are three basic independent quantities here), we can build the following ratios Π:

$$\begin{cases} \Pi_D = D/\nu, \\ \Pi_\beta = \beta / d^{u-v} U_\infty^v \nu^{1-v}, \end{cases} \tag{9.219}$$

with the relation

$$v = (1+u)/2 \Longrightarrow u = 2v - 1, \tag{9.220}$$

due to the implicit dimensionless character of Π_β.

The theorem of Vashi–Buckingham gives us

$$\Pi_\beta = \psi(\Pi_D) \tag{9.221}$$

or

$$\beta = d^{u-v}\, U_\infty^v\, \nu^{1-v}\, \psi(D/\nu). \tag{9.222}$$

Let us assume a power law for ψ: $\psi = A(D/\nu)^w$. Then, taking into account the relation between u and v, we have

$$\beta\, d/D = A\,(U_\infty d/\nu)^v\, (\nu/D)^{1-w} \tag{9.223}$$

or

$$Sh = A\, Re^v\, Sc^{1-w}. \tag{9.224}$$

The macro scales participate to the greatest extent in the *turbulent mode.* In extreme cases, for high Reynolds numbers, the micro scales do not intervene at all, and this fixes the exponents in the expression of β: $u = v = 1$.

The micro scales show the greatest participation in the *laminar mode* (although the macro scales still intervene), which can result in $v = 0$, $u = -1$ but also $u = 0$, $v = 1/2$.

Reality is not so definite, and in fact some zones of the plane (u, v) are valid in laminar flow and turbulent flow, as shown in Fig. 9.20.

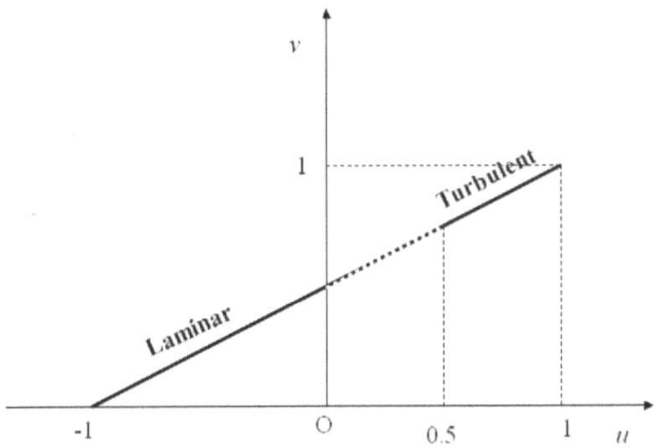

Fig. 9.20. Laminar and turbulent modes in the plane (u, v)

The exponent w is independent of the nature of the flow. We therefore find that $u < 0$ and $v < 0.5$ in laminar mode, and $0.5 < u < 1$ and $0.75 < v < 1$ in turbulent mode.

These results, and in particular the jump in the exponent v at the threshold of the transition from the laminar to the turbulent mode, have been confirmed experimentally.

The exponent w is the same in laminar flow and turbulent flow. It has been found that $(1 - w)$ is equal to 1/3 in the vicinity of a solid wall and to 1/2 at the fluid/fluid interface.

9.6.3 Turbulent Diffusion and First-Order Chemical Reaction in the Vicinity of a Wall

Let us suppose that the process of diffusion takes place very close to the wall, so the thickness of the boundary layer of diffusion is low compared to that of the viscous boundary layer. In this case, we can say that

$$\partial \bar{u}/\partial y \cong (\partial \bar{u}/\partial y)_0. \tag{9.225}$$

The velocity fluctuation u' is such that

$$u' = -l'_u\, \partial \bar{u}/\partial y \tag{9.226}$$

where l'_u is the characteristic turbulent length associated with u. We assume that this length (as well as its quadratic average L_W) is proportional to the distance y to the wall (see Sect. 8.1). Thus the fluctuation u' is proportional to y.

The continuity equation then tells us that v' is proportional to y^2. By definition, the turbulent diffusion coefficient $K_D = \rho\, D_t$ is equal (with ρ=const.) to

$$K_D = \rho\, \overline{v' l'_D}. \tag{9.227}$$

Let us also assume that l'_D is proportional to y. It follows that $K_D/\rho = D_t$ is proportional to y^3; that is to say

$$D_t = \gamma y^3. \tag{9.228}$$

Now we consider the concrete case of a wall that is slowly dissolving in a turbulent liquid flow in a tube in the presence of a first-order chemical reaction A $\longrightarrow$ B.

Since the phenomenon is occurring in the immediate vicinity of the wall, we can neglect any effect of curvature, so that the mass balance of species A is

$$\frac{d}{dy}[(D + D_t)\frac{dY}{dy}] = kY, \tag{9.229}$$

where k is the specific rate of the reaction. The boundary conditions are

$$\begin{cases} Y = Y_S \text{ in } y = 0, \\ Y = Y_\infty \text{ in } y \to \infty. \end{cases} \tag{9.230}$$

Evaluating the terms of the preceding equation gives

$$\frac{d}{dy}(D\frac{dY}{dy}) \cong D\, Y_S/\delta^2, \tag{9.231}$$

where δ is the thickness of the corresponding boundary layer,

$$\begin{cases} (d/dy)(D_t\, dY/dy) \cong \gamma\, \delta Y_S, \\ k\, Y \cong k\, Y_S. \end{cases} \tag{9.232}$$

The final linear relation will be of the type

$$AD/\delta^2 + B\gamma\delta = k. \tag{9.233}$$

We propose to determine A and B and then to deduce a valid relation for laminar and turbulent flows.

For $\gamma = 0$, the solution to the diffusion equation is

$$Y = Y_S \exp{(-\sqrt{k/D}\; y)}, \tag{9.234}$$

which gives the following relation for the thickness of the boundary layer δ:

$$\delta = \sqrt{D/k}. \tag{9.235}$$

We also have

$$AD/\delta^2 = k. \tag{9.236}$$

Thus $A = 1$. Let us call δ_1 the thickness of the boundary layer in the absence of a chemical reaction ($k = 0$). We then have

$$D/\delta_1^2 + B\,\gamma\,\delta_1 = 0, \tag{9.237}$$

which gives B as a function of δ_1, D and γ.

We finally obtain

$$(\delta_1/\delta)^2 - \delta/\delta_1 = k\,\delta_1^2/D. \tag{9.238}$$

This result, which is valid in laminar as well as turbulent flows, provides the ratio of boundary layer thicknesses with and without a chemical reaction. If we introduce the corresponding mass exchange coefficients β and β_1, we get

$$(\beta/\beta_1)^2 - \beta_1/\beta = kD/\beta_1^2. \tag{9.239}$$

This relation is in good agreement with the exact solution of the differential equation. Establishing linear algebraic relations between the orders of magnitude of the terms that appear in the differential equations (or in the partial differential equations) of a given problem frequently enables us to find very interesting experimentally accessible global relations. This method is a great aid in the study of turbulent flows.

10

Reactive and Nonreactive Waves

In this chapter, we will tackle various questions by referring to the thermodynamic discussions of Sect. 2.5.

We start in Sect. 10.1 with the notion of shock waves and their propagation. Nonreactive compressible media can exhibit macroscopic discontinuities of differing intensities. Thus, a supersonic fluid flow can give rise to shock waves, leading to a sharp increase in pressure.

Small disturbances near chemical equilibrium will be considered in Sect. 10.2, where we address the propagation of weak waves.

Then, in Sect. 10.3, situations associated with small stationary disturbances are examined, such as transonic flows near equilibrium.

Section 10.4 introduces both detonation and deflagration via the theory of Rankine–Hugoniot. Detonation and deflagration are combustion phenomena, which involve chemical reactions that are assumed to be exothermic. The distinction between the two types of wave is conveniently defined using the Clapeyron diagram. Detonation waves (which are a form of compression wave) can occur in a mixture of fuel and oxidizer.[1] Deflagrations are discontinuities that lead to weak decreases in pressure but to significant variations in concentration and temperature.

The structures of planar and curved deflagration waves will be studied successively in Sect. 10.5. The structure of a planar detonation wave will then be studied in Sect. 10.6.

We investigate various spherical wave scenarios in Sect. 10.7: the small movement case;[2] spherical deflagration; and finally blast waves that result from an explosion occurring at a particular point in space.

---

[1]Remember that when the fuel and the oxidizer are not premixed, there are significant concentration changes at the location of the flame, resulting in a so-called diffusion flame (see for example Sects. 7.7.2 and 9.4.2).

[2]Also consider the case of a shock wave crossing a mixture that is initially at chemical equilibrium, where the increase in internal energy does not derive from the heat released by the reaction (which is the case for detonation) but by the temperature jump of the shock wave. This causes the shock and the chemical reactions to

R. Prud'homme, *Flows of Reactive Fluids*,
DOI 10.1007/978-0-8176-4659-2_10, © Springer Science+Business Media, LLC 2010

10.1 Continuous and Discontinuous One-Dimensional Waves in a Barotropic Medium

10.1.1 Basic Equations of One-Dimensional Waves

Let us consider a gas moving in an adiabatic rectilinear tube, and assume nonstationary one-dimensional evolution. The variables are the abscissa x along the tube and the time t. Suppose that irreversible phenomena can only take place inside surfaces of discontinuity normal to the x-axis [42, 99].

In the continuous zone, which can vary in extent over time, the motion equations are the mass balance

$$\frac{\partial \rho}{\partial t} + u\frac{\partial \rho}{\partial x} + \rho\frac{\partial u}{\partial x} = 0, \tag{10.1}$$

the momentum balance

$$\frac{\partial u}{\partial t} + u\frac{\partial u}{\partial x} + \frac{1}{\rho}\frac{\partial p}{\partial x} = 0, \tag{10.2}$$

and the entropy balance (isentropic evolution)

$$\frac{\partial s}{\partial t} + u\frac{\partial s}{\partial x} = 0. \tag{10.3}$$

Equation 10.3 indicates that the entropy of each particle of fluid which has its motion followed is constant. Thus, if the entropy is uniform in the medium at a given initial time, it will remain uniform at later times. The state of the gas can be described by just one state variable: pressure or density or temperature or so on. This type of evolution is called barotropic evolution.

We define the characteristic speed c via the equality

$$c^2 = (\partial p/\partial \rho)_s. \tag{10.4}$$

It is known that $(\partial p/\partial \rho)_s$ is always positive (see Eq. 2.175).

Thus, for the flow studied here, we simply have

$$dp = c^2\, d\rho. \tag{10.5}$$

The mass balance becomes

$$\frac{1}{c^2}\frac{\partial p}{\partial t} + \frac{u}{c^2}\frac{\partial p}{\partial x} + \rho\frac{\partial u}{\partial x} = 0, \tag{10.6}$$

and the momentum balance can be rewritten as follows:

$$\frac{\partial u}{\partial t} + u\frac{\partial u}{\partial x} + \frac{c}{\rho c}\frac{\partial p}{\partial x} = 0. \tag{10.7}$$

---

interact. Such interaction issues also arise with detonation if it is generated by the passage of a shock wave.

The function P (see [99]) defined such that

$$dP = dp/\rho c \tag{10.8}$$

can be used to replace p, so that we have

$$\begin{cases} \partial P/\partial t + u\partial P/\partial x + c\partial u/\partial x = 0, \\ \partial u/\partial t + u\partial u/\partial x + c\partial P/\partial x = 0. \end{cases} \tag{10.9}$$

By combining these two equations, we obtain

$$\begin{cases} \partial(u+P)/\partial t + (u+c)\partial(u+P)/\partial x = 0, \\ \partial(u-P)/\partial t + (u-c)\partial(u-P)/\partial x = 0. \end{cases} \tag{10.10}$$

These equations mean simply that:

- The quantity $u+P$ is constant if we follow the motion velocity $u+c$ (i.e., $u+P=\alpha$) when $dx=(u+c)dt$ on the line (α)
- The quantity $u-P$ is constant if we follow the motion velocity $u-c$ (i.e., $u-P=\beta$) when $dx=(u-c)dt$ on the line (β).

The lines (α) and (β) are called characteristic lines.[3]

---

[3] *Note about the theory of characteristics:* The concept of a characteristic can be presented in a more mathematical and general manner than was done here. This is achieved by considering the problem of determining the partial derivatives in a situation governed by a known system of equations.

Let us consider as an illustration the case of the previous unsteady one-dimensional evolution. This is governed by the following system of partial differential equations with two variables t and x and two functions p and u:

$$\begin{cases} (1/c^2)\,\partial p/\partial t + (u/c^2)\,\partial p/\partial x + \rho\,\partial u/\partial x = 0, \\ \partial u/\partial t + u\,\partial u/\partial x + (1/\rho)\,\partial p/\partial x = 0. \end{cases} \tag{10.11}$$

Now consider the determination of the partial derivatives $\partial p/\partial t$, $\partial p/\partial x$, $\partial u/\partial t$ and $\partial u/\partial x$ knowing dt, dx, dp and du. We can add the following equations to the previous two:

$$\begin{cases} \partial p/\partial t\,dt + \partial p/\partial x\,dx = dp, \\ \partial u/\partial t\,dt + \partial u/\partial x\,dx = du. \end{cases} \tag{10.12}$$

The set of systems (10.11) and (10.12) forms a linear system of four equations with four unknowns. Solving this system (when possible) gives the four unknown partial derivatives by simple matrix inversion. The impossible cases are those that correspond to a vanishing determinent; i.e.,

$$(dx - u\,dt)^2 - c^2\,(dt)^2 = 0. \tag{10.13}$$

We then obtain the characteristic lines found previously.

At any point, the state and the motion of the fluid are perfectly defined by the thermodynamic state variable P and the material velocity u. The quantities $u+P$ and $u-P$ provide information about the fluid. Simultaneous knowledge of them at the same point of space completely defines the state of the fluid at this point at the time considered. The information $u+P$ is preserved along the characteristic (α) and is propagated at a speed of $u+c$. The information $u-P$ is preserved along (β) and is propagated at a speed of $u-c$. Based on these observations, we can interpret the characteristic speed c; in a medium initially at rest, a small disturbance will propagate at speeds of c and $-c$. The quantity c is thus the speed of sound in the medium considered.

10.1.2 Piston Moving in an Infinite Cylinder

Let us consider the classical problem of a piston moving in an infinite cylinder, starting from a position at rest (Fig. 10.1). We propose to initially describe the motion graphically in the reference plane (x,t) for the case of an ideal gas at rest. The piston motion is given by $X = X(\tau)$, with $d^2X/d\tau^2 \geq 0$.

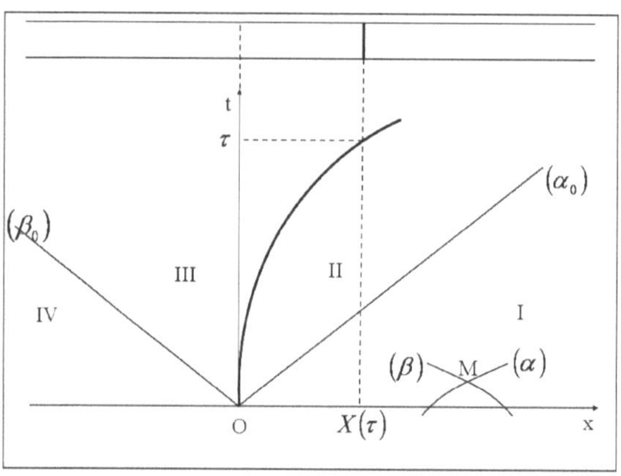

Fig. 10.1. Piston in an infinite cylinder

At times close to $t = 0$, two characteristic lines are arriving at a given point M. Provided that one is sufficiently far from the curve $x = X(t)$ characterizing the piston motion, these characteristic lines cross the axis Ox directly. We then have

$$\begin{cases} u + P = P_0 \quad (\alpha) \\ u - P = -P_0 \quad (\beta) \end{cases} \tag{10.14}$$

at M. We deduce that the state of the gas at M is the same as that of the fluid at rest $u = 0$, $P = P_0$, $c = c_0$. In this zone, the characteristic lines are straight lines with slopes of

$$\begin{cases} dt/dx = c_0^{-1} \ (\alpha), \\ dt/dx = -c_0^{-1} \ (\beta). \end{cases} \tag{10.15}$$

This region corresponding to fluid particles at rest is delimited on the right hand side by a characteristic (α_0), and on the left of the curve $x = X(t)$ by a characteristic (α_0), as indicated in Fig. 10.1, where we assume that $(dX/dt)^{-1} > c_0^{-1}$.

Apart from the zones of zero velocity (zones I and IV), it is easily shown that the characteristic lines (α) are straight lines in area II, and that it is the same for the characteristic lines (β) in area III. The diagram shown in Fig. 10.2 demonstrates a step that allows the conditions at an unspecified point N in area II to be determined.

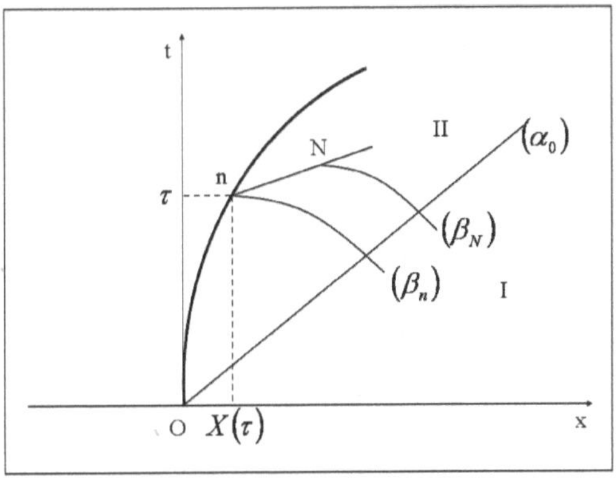

Fig. 10.2. The characteristic lines at a given point N in area II: $u + P = u_n + P_n$ (α), $u - P = -P_0$ (β_N), $u_n - P_n = -P_0$ (β_n), $u = dX/dt = \dot{X}$

It follows that

$$\begin{cases} u + P = 2\dot{X} + P_0 \ (\alpha) \\ u - P = -P_0 \ (\beta_N) \end{cases} \tag{10.16}$$

at point N, and that

$$\begin{cases} u = \dot{X}, \\ P = P_0 + \dot{X}. \end{cases} \tag{10.17}$$

At point N, the conditions u and P are entirely determined by P_0 and $\dot{X}$, which is the speed of the piston at point n (i.e., at a previous time τ). These conditions remain identical at the considered characteristic (α). It follows that c is also constant at (α) and so the slope $(u+c)^{-1}$ of this characteristic is constant. The characteristics (α) are thus straight lines in area II. This is also true of the characteristics (β) in area III.

In the case of an ideal gas, it is easy to show that

$$P = 2c/(\gamma - 1). \tag{10.18}$$

Therefore, at point N in area II, we have

$$\begin{cases} u = \dot{X}, \\ c = c_0 + \frac{\gamma-1}{2}\dot{X}. \end{cases} \tag{10.19}$$

Consequently, the slope of the characteristic (α) is defined by

$$\frac{dt}{dx} = (c_0 + \frac{\gamma+1}{2}\dot{X})^{-1}. \tag{10.20}$$

This slope is a decreasing function of time τ if $\dot{X}(\tau)$ is an increasing function of τ, which corresponds to the case shown in Fig. 10.1. In this case, the characteristics (α) of area II form a convergent bundle of straight lines. Conversely, the characteristics (β) of area III form a divergent bundle of straight lines. As soon as $\dot{X}$ becomes constant (uniform piston motion), the lines become parallel. There is a contradiction in zone II. The lines are cut at a finite distance. This means that there are several velocities at a point, which is not correct. This occurs in a zone of the plane (x, t) between (α_0) and the hull (e) of characteristics (α) in area II (Fig. 10.3).

For example, for $\dot{X} = \Gamma\tau$ (uniformly accelerated piston motion), the equation for hull (e) is

$$x = \frac{1}{2\gamma\Gamma}(\frac{\gamma+1}{2}\Gamma t - c_0)^2 + c_0 t. \tag{10.21}$$

This contradiction can only be overcome if a shock wave starts in front of the piston at time

$$t_c = 2c_0/(\gamma+1)\Gamma. \tag{10.22}$$

Before the onset of the shock, the system evolves as shown in Fig. 10.3.

For uniformly accelerated piston motion with zero initial acceleration, hull (e) exhibits the shape indicated in Fig. 10.4.

If we now compel the piston to have a speed V_P =const. from the initial moment onwards, the shock immediately appears because the zone of acceleration is concentrated at the origin.[4]

[4]The preceding theory is valid to the left and right of the piston, and on both sides of the shock. We still need to determine the motion of the shock wave. We

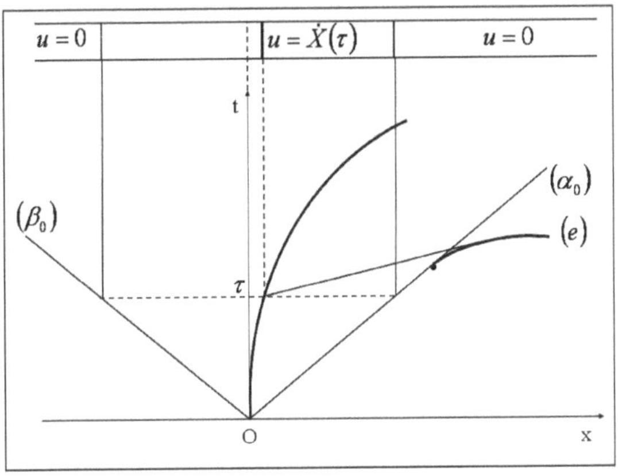

Fig. 10.3. Onset of a shock

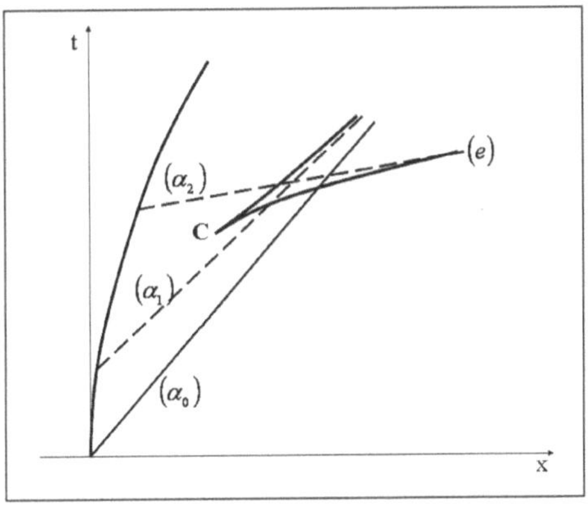

Fig. 10.4. Case of zero initial acceleration

10.1.3 Speed of a Normal Shock Wave Generated by a Piston

The balance laws through a surface of discontinuity were established in Sect. 4.9. We assume here that the fluid is nonviscous and barotropic[5] on both sides

---

will limit our study to the cases where the piston speed is constant, $\dot{X} = V_P > 0$. Experiments also show that the shock has a constant speed $w > V_P$.

[5]The state of a barotropic fluid depends on a single thermodynamic variable: the pressure. Fluids that can be considered incompressible and fluids undergoing isentropic evolution are then barotropic. In such cases we have s =const. on both sides of the shock wave.

of this surface and that the problem is one-dimensional. In the absence of the accumulation or production of wave energy and momentum, we have

$$\begin{cases} [\rho(u-w)]_-^+ = 0, \\ [p+\rho u(u-w)]_-^+ = 0, \\ \left[pu+\rho(e+u^2/2)(u-w)\right]_-^+ = 0, \\ \dot{W}_{aS} = [\rho s(u-w)]_-^+ \geq 0. \end{cases} \tag{10.23}$$

If indices 1 and 2 correspond to states immediately upstream and downstream of the shock, respectively, and $\dot{m}$ is the unit mass flow rate through the shock, we get

$$\begin{cases} \rho_1(u_1-w) = \rho_2(u_2-w) = \dot{m}, \\ p_1+\rho_1 u_1(u_1-w) = p_2+\rho_2 u_2(u_2-w), \\ p_1 u_1+\rho_1(e_1+u_1^2/2)(u_1-w) = p_2 u_2+\rho_2(e_2+u_2^2/2)(u_2-w). \end{cases} \tag{10.24}$$

By setting $v = u - w$, we can easily show that

$$\rho_1 v_1 = \rho_2 v_2 = \dot{m}, \tag{10.25}$$

$$p_1+\rho_1 v_1^2 = p_2+\rho_2 v_2^2, \tag{10.26}$$

$$p_1 v_1+\rho_1(e_1+v_1^2/2)v_1 = p_2 v_2+\rho_2(e_1+v_2^2/2)v_2, \tag{10.27}$$

which is the same as being placed in a reference frame associated with the wave at the time considered.

If we factorize the unit mass flow rate $\rho v = \dot{m}$, the energy equation is

$$h_1+v_1^2/2 = h_2+v_2^2/2. \tag{10.28}$$

In the following, we will assume that the fluid considered is an ideal gas. We then have

$$h = c_p T = c^2/(\gamma-1) \tag{10.29}$$

since

$$c^2 = \gamma r T = \gamma p/\rho, \tag{10.30}$$

so that the energy equation (10.27) is

$$\frac{c_1^2}{\gamma-1}+\frac{v_1^2}{2} = \frac{c_2^2}{\gamma-1}+\frac{v_2^2}{2} = \frac{\gamma+1}{2(\gamma-1)}c^{*2}, \tag{10.31}$$

where c^*, the critical normal relative velocity, is defined by the two left hand sides by making $c_1 = v_1 = c^*$ or $c_2 = v_2 = c^*$. This relation is *Hugoniot's equation*. The momentum equation (10.26) gives us

$$p_1/\rho_1 v_1 + v_1 = p_2/\rho_2 v_2 + v_2 \tag{10.32}$$

after dividing by $\dot{m} = \rho_1 v_1 = \rho_2 v_2$. This equation can also be written

$$c_1^2/\gamma v_1 + v_1 = c_2^2/\gamma v_2 + v_2. \tag{10.33}$$

Let us replace c_1 and c_2 by their values in terms of v_1, v_2 and c^* deduced from the double Hugoniot equation (10.31). After simplifying, we obtain

$$v_1 v_2 = c^{*2}. \tag{10.34}$$

This is the *Prandtl relation.*

These relations will enable us to calculate the speed w of a shock crossing a medium that is initially at rest. To do this, we assume that the velocities are uniform on both sides of shock. We will show then that the result obtained is coherent with the case of the uniformly moving piston. Here we have

$$\begin{cases} v_1 = -w \\ v_2 = u_2 - w. \end{cases} \tag{10.35}$$

From the relations of Hugoniot (10.31) and Prandtl (10.34), we can deduce that

$$\frac{c_1^2}{\gamma - 1} + \frac{v_1^2}{2} = \frac{\gamma + 1}{2(\gamma - 1)} v_1 v_2 \tag{10.36}$$

or, if we denote the upstream state by the index (0),

$$\frac{c_0^2}{\gamma - 1} + \frac{w^2}{2} = -\frac{\gamma + 1}{2(\gamma - 1)} w(u_2 - w). \tag{10.37}$$

This expression provides the sought-after relation between u_2 and w. It is also written

$$u_2 = \frac{2}{\gamma + 1} \frac{w^2 - c_0^2}{w}. \tag{10.38}$$

For a constant piston speed, the result can be represented as in Fig. 10.5.

Areas I and IV are at rest. In area II, velocity is constant and equal to the piston speed V_P. The speed w is determined by the previous equation by making $u_2 = V_P$:

$$V_P = \frac{2}{\gamma + 1} \frac{w^2 - c_0^2}{w}. \tag{10.39}$$

The characteristics (α) are straight lines, and the slopes of these lines can be determined knowing c_2. The formulae obtained for continuous isentropic flow cannot be used to find this slope because the flow is not isentropic during the crossing of a shock. The slope of (α) is of course given by

$$dt/dx = (c_2 + V_P)^{-1}. \tag{10.40}$$

The Hugoniot equation and the Prandtl relation provide the value of c_2:

$$c_2^2 = (w - V_P)(w + \frac{\gamma+1}{2} V_P). \tag{10.41}$$

In area III, we obtain an expansion centered at the origin.

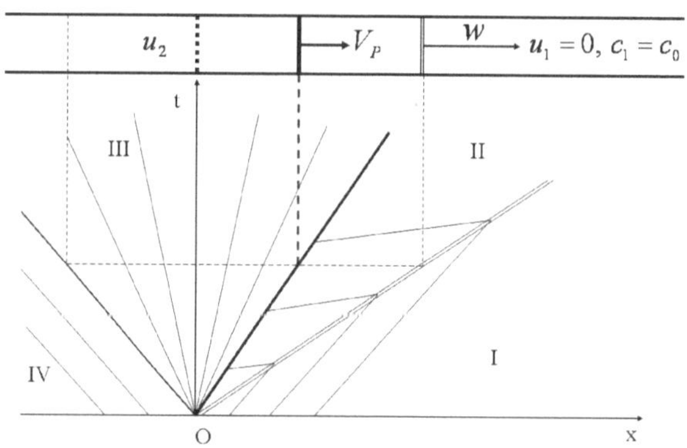

Fig. 10.5. Characteristics for a piston with constant velocity

When the speed of the piston is low compared to c_0, we obtain

$$\begin{cases} w = c_0 + \frac{\gamma+1}{4} V_P, \\ c_2 = c_0 + \frac{\gamma-1}{2} V_P, \\ dt/dx = \pi^{-1} = (c_0 + \frac{\gamma-1}{2} V_P)^{-1} \quad \text{(slope of the characteristics)}. \end{cases} \tag{10.42}$$

We can deduce from this that the speed of the weak shock wave is equal to half of the sum of the speed of sound c_0 and the inverse of the slope of the characteristics after the shock. In the coordinate system of (x, t), we can also write[6]

$$\frac{\pi^{-1} + c_0^{-1}}{2} \cong w^{-1}. \tag{10.43}$$

[6]If we ignore the first-order terms in V_P/c_0, we can see that the shock wave propagates at the speed of sound, so its intensity is very low and it is not then useful to regard it as a discontinuous phenomenon. In the following, we will study these weak waves, and consider the case of a reactive mixture that is at rest and chemical equilibrium in the reference state.

Thermoconvective Wave Near the Critical Point

There is another phenomenon that is relatively similar to a planar discontinuity wave. It occurs during the transport of heat in a pure fluid near the critical point. This phenomenon, called the "piston effect," which includes wave propagation at a certain scale, is presented in Sect. A.5.2.

10.2 Small Motions of a Fluid in Linearized Theory

10.2.1 Case of a Nonreactive Fluid

Let us establish the linearized equations of a moving nonreactive fluid (we will apply the same method to the near-equilibrium reactive mixture in Sect. 10.2.2 and study sound propagation in this case). We will set

$$\mathbf{v} = \epsilon \mathbf{v}_1, \; p = p_0 + \epsilon p_1, \; \rho = \rho_0 + \epsilon \rho_1, \; s = s_0 + \epsilon s_1, \tag{10.44}$$

where $\epsilon \ll 1$. Since the flow is isentropic (ideal fluid), we have

$$s_1 = 0, \tag{10.45}$$

so

$$p_1 = c_0^2 \rho_1, \tag{10.46}$$

where c_0 is the characteristic speed in the reference state indicated by the index (0). The linearized balance equations are as follows:

$$\begin{cases} \partial \rho_1/\partial t + \rho_0 \boldsymbol{\nabla} \cdot \mathbf{v}_1 = 0, \\ \rho_0 \partial \mathbf{v}_1/\partial t + \boldsymbol{\nabla} p_1 = 0. \end{cases} \tag{10.47}$$

We can set (if $\boldsymbol{\omega}_1 = \frac{1}{2} \boldsymbol{\nabla} \times \mathbf{v}_1 = 0$ at time $t = 0$)

$$\mathbf{v}_1 = \boldsymbol{\nabla} \varphi_1 \tag{10.48}$$

and eliminate p_1 and v_1 between these equations. We then obtain

$$\partial p_1/\partial t = -\rho_0 c_0^2 \Delta \varphi_1, \tag{10.49}$$

$$\rho_0 \boldsymbol{\nabla}(\partial \varphi_1/\partial t) + \boldsymbol{\nabla} p_1 = 0. \tag{10.50}$$

Equation 10.50 gives us

$$p_1 = -\rho_0 \partial \varphi_1/\partial t. \tag{10.51}$$

By replacing p_1 with its value in (10.51), we obtain

$$\partial^2 \varphi_1/\partial t - c_0^2 \Delta \varphi_1 = 0. \tag{10.52}$$

Solving this linear partial differential equation enables us to determine the evolution for the case of small motions.

We can take the general solution of this equation to be the sum of elementary solutions of the type

$$f = \hat{f} \exp i(\mathbf{K} \cdot \mathbf{x} - \omega t). \tag{10.53}$$

If we assume that ω is a real quantity, we find that $\mathbf{K}$, the wavenumber vector, is also a real quantity. We obtain

$$-\omega^2 \hat{\varphi}_1 + c_0^2 K^2 \hat{\varphi}_1 = 0; \tag{10.54}$$

that is to say,

$$K^2 = \omega^2 / c_0^2. \tag{10.55}$$

The length K of the real vector $\mathbf{K}$ is equal to ω/c_0. However, ω/K is by definition the propagation velocity of the wavy phenomenon. Checking it once more, the wavy phenomenon propagates at the characteristic speed c_0.

Let us now consider the particular case of one-dimensional motion where the velocity of the piston has a small amplitude:

$$dX(t)/dt = \epsilon dX_1(t)/dt = \epsilon V_{P1}. \tag{10.56}$$

We study the right part of the cylinder $x > X(t)$, presumed to be infinite. The equation to solve is

$$\partial^2 \varphi_1 / \partial t - c_0^2 \partial^2 \varphi_1 / \partial x^2 = 0. \tag{10.57}$$

Its general solution is

$$\varphi_1 = f(x - c_0 t) + g(x + c_0 t). \tag{10.58}$$

Setting

$$\xi = x - c_0 t, \ \eta = x + c_0 t, \tag{10.59}$$

we have

$$\begin{cases} \varphi_1 = f(\xi) + g(\eta), \\ u_1 = \partial \varphi_1 / \partial x = df/d\xi + dg/d\eta, \\ p_1 = -\rho_1 \partial \varphi_1 / \partial t = \rho_0 c_0 (df/d\xi - dg/d\eta). \end{cases} \tag{10.60}$$

We therefore note that

$$\begin{cases} u_1 + p_1/\rho_0 c_0 = \alpha = \text{const. for } \xi = \text{const.} = x - c_0 t, \\ u_1 - p_1/\rho_0 c_0 = \beta = \text{const. for } \eta = \text{const.} = x + c_0 t. \end{cases} \tag{10.61}$$

Our analysis of the motion then proceeds as in Sect. 10.1.

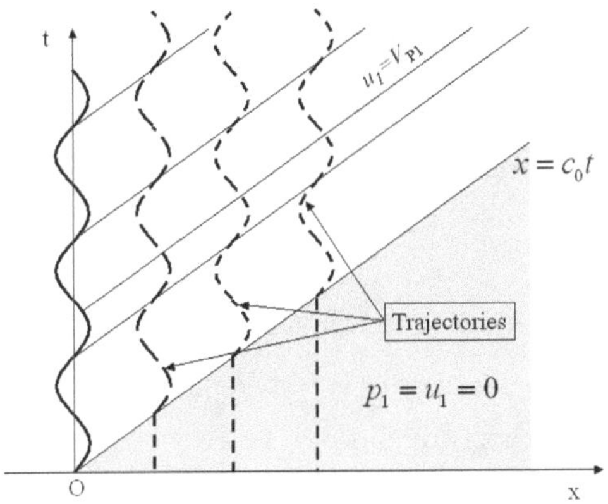

Fig. 10.6. Small periodic motion of a piston around $x = 0$

The characteristic lines (α) are now parallel straight lines of slope c_0^{-1} (see Fig. 10.6). When $x > c_0 t$, $t > 0$, we find that $\varphi_1 = 0$, $u_1 = p_1 = 0$, and for $c_0 t > x > X(t)$ we obtain

$$\epsilon\varphi_1 = -c_0 X(t - x/c_0), \ \epsilon u_1 = (dX/dt)(t - x/c_0) \tag{10.62}$$

and

$$p_1 = \rho_0 c_0 u_1. \tag{10.63}$$

10.2.2 Case of a Monoreactive Fluid

The chemical reaction considered here is assumed to be reversible and near to equilibrium [10, 78]. If ξ is its progress variable, then, in the absence of diffusion, we have

$$\rho\, d\xi/dt = L\, A, \tag{10.64}$$

where A is the chemical affinity and L is the phenomenological coefficient (see Sect. 3.4).

The other equations of this problem are

$$\begin{cases} d\rho/dt + \rho\boldsymbol{\nabla}\cdot\mathbf{v} = 0, \\ \rho\, d\mathbf{v}/dt + \boldsymbol{\nabla}p = 0, \\ \rho\, ds/dt = L\, A^2/T. \end{cases} \tag{10.65}$$

It is necessary to add to this system the equations of state for the mixture or its fundamental energy law (Chap. 2), for example in the form

$$e = e(s, \vartheta, \xi), \ \vartheta = 1/\rho. \tag{10.66}$$

The reference state is characterized by

$$\rho = \rho_0, \ p = p_0, \ \mathbf{v} = \mathbf{0}, \ s = s_0, \ A = 0, \ \xi = \xi_0, \tag{10.67}$$

and the slightly disturbed state by

$$\begin{cases} \rho = \rho_0 + \epsilon \rho_1, \ p = p_0 + \epsilon p_1, \ \mathbf{v} = \epsilon \mathbf{v}_1, \\ s = s_0 + \epsilon s_1, \ A = \epsilon A_1, \ \xi = \xi_0 + \epsilon \xi_1. \end{cases} \tag{10.68}$$

The linearized theory then gives us

$$\rho_0 \, \partial \xi_1 / \partial t = L_0 \, A_1, \tag{10.69}$$

$$\partial \rho_1 / \partial t + \rho_0 \boldsymbol{\nabla} \cdot \mathbf{v}_1 = 0, \tag{10.70}$$

$$\rho_0 \, \partial \mathbf{v}_1 / \partial t + \boldsymbol{\nabla} p_1 = 0, \tag{10.71}$$

$$s_1 = 0. \tag{10.72}$$

Before transforming these equations, it is useful to develop p_1 as a function of ρ_1 and ξ_1, and then as a function of ρ_1 and A_1. By choosing s, ρ and ξ as thermodynamic variables, we have

$$p_1 = c_{f0}^2 \, \rho_1 + a_0 \xi_1, \tag{10.73}$$

where $c_f = (\partial p / \partial \rho)_{s,\xi}^{1/2}$ is the local speed of sound of the medium, which has a fixed composition. The speed c_f is identical to the speed c of a non-reactive fluid. Moreover, as defined by (2.179), quantity $a = (\partial p / \partial \xi)_{s\rho}$ is a thermodynamic quantity.

If we now take s, ρ and A to be variables, we have

$$p_1 = c_{e0}^2 \, \rho_1 + b_0 \, A_1, \tag{10.74}$$

with $c_e = (\partial p / \partial \rho)_{sA}^{1/2}$, where c_e is the speed of sound in a mixture at chemical equilibrium (this is defined by $A = 0$), and where $b = (\partial p / \partial A)_{s\rho}$ is defined by (2.180).[7]

[7]It is known that c_f and c_e are positive real numbers. We do not know the signs of a and b a priori, but we do know that the ratio b/a is negative (see Eqs. 2.179 and 2.180). If we consider a chemical reaction occurring at constant volume, close to equilibrium we have the relations: $\rho_0 \, \partial \xi_1 / \partial t = L_0 \, A_1, \ p_1 = a_0 \, \xi_1, \ p_1 = b_0 \, A_1$, yielding $\partial \xi_1 / \partial t - (L_0 a_0 / \rho_0 b_0) \xi_1 = 0$. Thus, $\tau_{v0} = -\rho_0 b_0 / L_0 a_0 > 0$ represents the characteristic time of the reaction. In the same way, for a reaction occurring at constant pressure, we have $\tau_{p0} = -\rho_0 b_0 c_{f0}^2 / L_0 a_0 c_{e0}^2 > 0$.

We can deduce from (10.71) that

$$\mathbf{v}_1 = \nabla\varphi_1,\ p_1 = -\rho_0\partial\varphi_1/\partial t. \tag{10.75}$$

The continuity equation (10.70) gives us

$$(c_{f0})^{-2}(\partial p_1/\partial t - a_0\partial\xi_1/\partial t) + \rho_0\nabla\cdot\mathbf{v}_1 = 0, \tag{10.76}$$

or, since $\rho_0(\partial\xi_1/\partial t)$ is given by (10.69),

$$(c_{f0})^2\,\Delta\varphi_1 - \partial^2\varphi_1/\partial t^2 = L_0 a_0 A_1/\rho_0^2. \tag{10.77}$$

Let us differentiate the two sides of (10.77) with respect to t. We get

$$(\partial/\partial t)[(c_{f0})^2\,\Delta\varphi_1 - \partial^2\varphi_1/\partial t^2] = (L_0 a_0/\rho_0^2 b_0)(\partial/\partial t)(p_1 - c_{e0}^2\rho_1) \tag{10.78}$$

or

$$(\partial/\partial t)[(c_{f0})^2\,\Delta\varphi_1 - \partial^2\varphi_1/\partial t^2] - (L_0 a_0/\rho_0 b_0)[(c_{e0})^2\,\Delta\varphi_1 - \partial^2\varphi_1/\partial t^2] = 0. \tag{10.79}$$

Based on our previous discussion, this equation, which governs the motions of small disturbances, can be written in one of the two equivalent forms

$$\begin{cases} \tau_{v0}(\partial/\partial t)[\partial^2\varphi_1/\partial t^2 - (c_{f0})^2\,\Delta\varphi_1] + \partial^2\varphi_1/\partial t^2 - (c_{e0})^2\,\Delta\varphi_1 = 0, \\ \tau_{p0}(\partial/\partial t)[(c_{f0})^{-2}\,\partial^2\varphi_1/\partial t^2 - \Delta\varphi_1] + (c_{e0})^{-2}\,\partial^2\varphi_1/\partial t^2 - \Delta\varphi_1 = 0. \end{cases} \tag{10.80}$$

The partial differential equation in φ_1 is linear. We can seek elementary solutions of the type

$$f = \hat{f}\,\exp i(\mathbf{K}\cdot\mathbf{x} - \omega t), \tag{10.81}$$

representing free planar waves of frequency $\omega/2\pi$. We then obtain the following equation for $\mathbf{K}$ and ω:

$$i\omega\tau_{v0}(c_{f0}^2K^2 - \omega^2) - (c_{e0}^2K^2 - \omega^2) = 0, \tag{10.82}$$

or

$$K^2 = \frac{\omega^2(1 - i\omega\tau_{v0})}{c_{e0}^2 - i\omega\tau_{v0}c_{f0}^2}. \tag{10.83}$$

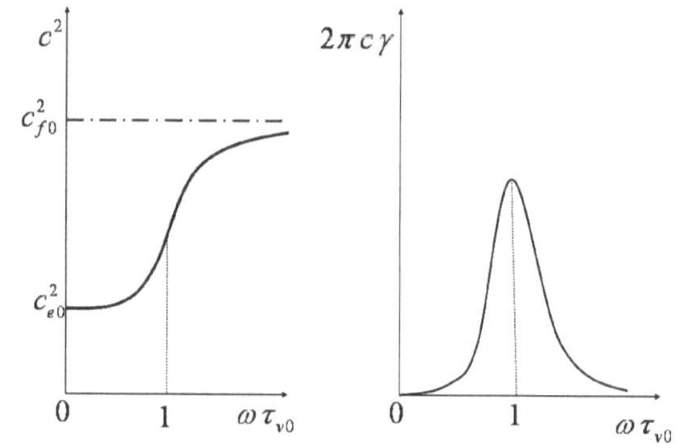

Fig. 10.7. Speed and attenuation of the wave

When there are two finite values of $\omega\tau_{v0}$, there is relaxation.[8] The exponent $i(\mathbf{K}\cdot\mathbf{x}-\omega t)$ is also written $i\omega(\mathbf{K}\cdot\mathbf{x}/\omega-t)$. If we decompose the wavenumber vector $\mathbf{K}$ into a real part $\mathbf{K}_r$ and an imaginary part $\mathbf{K}_i$, the exponent of (10.81) becomes

$$-\mathbf{K}_i\cdot\mathbf{x}+i\omega(\mathbf{K}_r\cdot\mathbf{x}/\omega-t). \tag{10.84}$$

Thus, K_i characterizes the space damping of the wave and K_r/ω is the inverse of its propagation velocity c. The damping of the wave and its speed will be functions of the pulsation ω. These functions are only easily determined if the characteristic speeds c_f and c_e are close to each other. According to (2.178), c_f is always greater than c_e. The condition is thus written

$$c^2_{f0}=c^2_{e0}(1+\epsilon'). \tag{10.85}$$

Let us write

$$K=K_r+iK_i,\ K_r/\omega=1/c(\omega),\ K_i/\omega=\gamma(\omega). \tag{10.86}$$

After linearizing in ϵ' and separating the real and imaginary parts, we obtain

[8]Two borderline cases are interesting. When $\omega\tau_{v0}$ is close to zero (very slow vibration or a very fast chemical reaction), evolution is slow compared to the chemical reaction, and chemical equilibrium can be attained at any time. We then find that $\omega^2/K^2\cong c^2_{e0}$, and the speed of propagation of the small motions is equal to c_e, which accurately reflects the speed of sound at chemical equilibrium. When $\omega\tau_{v0}$ is large (very fast vibration or a very slow reaction), the chemical reactions are frozen, and we find that $\omega^2/K^2\cong c^2_{f0}$, as in the nonreactive case.

$$\begin{cases} 1/c^2 - \gamma^2 = (c_{e0})^{-2}\left[1 - \epsilon'\omega^2\tau_{v0}^2/(1+\omega^2\tau_{v0}^2)\right], \\ 2\gamma/c = \epsilon'\,(c_{e0})^{-2}\,\omega\tau_{v0}/(1+\omega^2\tau_{v0}^2). \end{cases} \tag{10.87}$$

Solving this system by ignoring the ϵ'^2 terms gives us

$$\begin{cases} c^2(\omega) = (c_{e0}^2 + \omega^2\tau_{v0}^2 c_{f0}^2)/(1+\omega^2\tau_{v0}^2), \\ \gamma = \left[(c_{f0} - c_{e0})/c_{e0}^2\right]\left[\omega\tau_{v0}/(1+\omega^2\tau_{v0}^2)\right]. \end{cases} \tag{10.88}$$

The attenuation of the wave per unit wavelength is

$$2\pi c\gamma \cong 2\pi\frac{c_{f0} - c_{e0}}{c_{e0}}\frac{\omega\tau_{v0}}{1+\omega^2\tau_{v0}^2}. \tag{10.89}$$

The expression of $c(\omega)$ is compatible with the limits found for small $\omega\tau_{v0}$ and infinite $\omega\tau_{v0}$. The curves in Fig. 10.7 summarize these results [140].

In conclusion, the small motions of a reactive fluid are propagated at a speed that depends on the frequency of the (single-frequency) plane wave considered. They attenuate in space, with the attenuation coefficient per unit wavelength being a function of the frequency according to the laws indicated above.

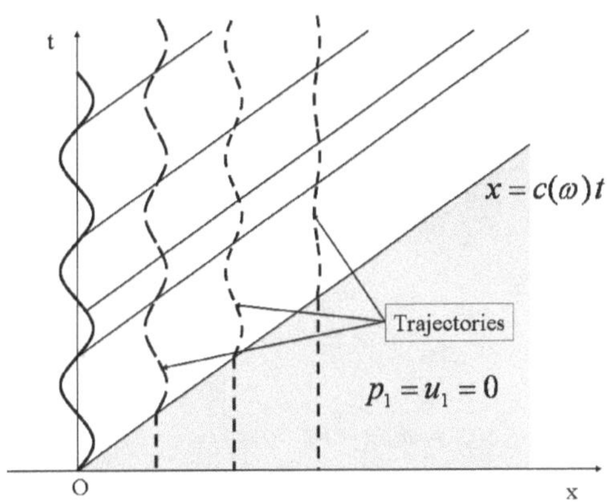

Fig. 10.8. Small perturbations in a mixture

Let us apply this result to a piston performing small motions around the position $x = 0$ (Fig. 10.8) with a frequency of $\omega/2\pi$. We obtain the same diagram as for a nonreactive fluid, but there is attenuation of the wave along the characteristics of slope $dt/dx = 1/c(\omega)$. The amplitude of the wave varies as $e^{-K_i x}$ and, if $(c_{f0}^2/c_{e0}^2) \ll 1$, K_i is equal to

$$K_i = \frac{(c_{f0} - c_{e0})\omega^2\tau_{v0}}{c_{e0}^2(1+\omega^2\tau_{v0}^2)}. \tag{10.90}$$

For a given frequency, the attenuation depends on τ_{v0}. This is maximum at $\tau_{v0} = 1/\omega$ for a given abscissa. It is proportional to x, the distance to the origin.[9]

This theory has been extended to the multireactive case. Interested readers can refer to [179] and [216] for more on this subject.

10.3 The Case of Small Stationary Disturbances

We will limit our study here to continuous flows. Some problems can be analyzed by the linearized theory of small disturbances, as described in Sect. 10.2. On the other hand, the study of transonic flows utilizes the method of small singular disturbances. In this section we discuss stationary two-dimensional planar flows where the reference flow proceeds along the axis Ox with velocity $\mathbf{v}_0 = U_0\mathbf{i}$.

10.3.1 Linearized Theory

The basic equations are similar to those of Sect. 10.2, with the difference that the temporal partial derivative $\partial/\partial t$ terms are replaced by terms in $\mathbf{v}\cdot\boldsymbol{\nabla} = U_0\,\partial/\partial x$. The order of magnitude of the small disturbance is now that of the velocity component normal to the direction of the uniform reference flow velocity $\mathbf{v}_0 = U_0\mathbf{i}$. Let us examine the consequences of this change for a fluid without a chemical reaction, and then for a fluid mixture with a single reaction.

Nonreactive Fluid

Let us write

$$\mathbf{v} = U_0\,\mathbf{i} + \epsilon\mathbf{v}_1,\ p = p_0 + \epsilon p_1,\ \rho = \rho_0 + \epsilon\rho_1,\ s = s_0 + \epsilon s_1.$$

The isentropy of the flow gives $s_1 = 0$ and $p_1 = c_0^2\,\rho_1$. The linearized equations become

$$\begin{cases} U_0\,\partial\rho_1/\partial x + \rho_0\boldsymbol{\nabla}\cdot\mathbf{v}_1 = 0, \\ \rho_0\mathbf{v}_0\cdot\boldsymbol{\nabla}\otimes\mathbf{v}_1 + \boldsymbol{\nabla}p_1 = 0, \end{cases} \tag{10.91}$$

and we can set $\mathbf{v}_1 = \boldsymbol{\nabla}\varphi_1$. Eliminating the quantities ρ_1, p_1, $\mathbf{v}_1$ leads to the result

[9] It should be noted that $K_i \to 0$ for $\omega \to 0$ (equilibrium), and that $K_i \to \epsilon'/2c_{e0}\tau_{v0}$ when $\omega \to \infty$ (freezing). The low-frequency waves are less attenuated and slower (speed c_{e0}), whereas the high-frequency waves undergo the most attenuation at fixed τ_{v0}. In the presence of small motions of the piston at a single frequency, there is a set of these waves, and the amplitude of each wave depends on its frequency.

$$U_0^2\,\partial^2\varphi_1/\partial x^2 - c_0^2\,\Delta\varphi_1 = 0, \tag{10.92}$$

which replaces (10.52) and can be written

$$M_0^2\,\partial^2\varphi_1/\partial x^2 - \Delta\varphi_1 = 0 \tag{10.93}$$

where M_0 is the Mach number of the unperturbed flow. For a two-dimensional planar flow, the system of characteristic equations for the components u_1 and v_1 of the velocity vector are

$$\begin{cases} (M_0^2-1)u_{1,x} - v_{1,y} = 0, \\ v_{1,x} - u_{1,y} = 0, \\ u_{1,x}dx + u_{1,y}dy = du_1, \\ v_{1,x}dx + v_{1,y}dy = dv_1. \end{cases} \tag{10.94}$$

Canceling the determinant gives the characteristic directions, which obey the equation

$$(M_0^2-1)(dx)^2 - (dy)^2 = 0. \tag{10.95}$$

These characteristic lines of the hyperbolic potential equation (10.93) are thus real, but only for supersonic flows. In particular, this theory makes it possible to study flows in the vicinity of thin aircraft wing profiles.

Fluid With One Chemical Reaction

Here, the reference flow is at chemical equilibrium, and we set

$$\mathbf{v} = U_0\mathbf{i} + \epsilon\mathbf{v}_1,\ p = p_0 + \epsilon p_1,\ \rho = \rho_0 + \epsilon\rho_1,$$
$$s = s_0 + \epsilon s_1,\ A = \epsilon A_1,\ \xi = \xi_0 + \epsilon\xi_1.$$

The linearized equations become

$$\begin{cases} U_0\,\partial\rho_1/\partial x + \rho_0\boldsymbol{\nabla}\cdot\mathbf{v}_1 = 0, \\ \rho_0\mathbf{U}_0\cdot\boldsymbol{\nabla}\otimes\mathbf{v}_1 + \boldsymbol{\nabla}p_1 = 0, \\ \rho_0\mathbf{U}_0\,\partial\xi_1/\partial x = L_0\,A_1, \\ s_1 = 0, \end{cases} \tag{10.96}$$

where L_0 is the unperturbed value of the phenomenological coefficient L in (10.64).

Using the same arguments as in Sect. 10.2, we derive the equation

$$\tau_{p0}U_0\,\frac{\partial}{\partial x}(M_{f0}^2\,\frac{\partial^2\varphi_1}{\partial x^2} - \Delta\varphi_1) + M_{e0}^2\,\frac{\partial^2\varphi_1}{\partial x^2} - \Delta\varphi_1 = 0. \tag{10.97}$$

Here, $M_{f0} = U_0/c_{f0}$ and $M_{e0} = U_0/c_{f0}$ are the "frozen" and "equilibrium" Mach numbers.

10.3.2 Small Singular Disturbances and Transonic Flow

Connecting the subsonic and supersonic zones of a flow with or without a chemical reaction is not possible directly in linearized theory. Indeed, the equations obtained are different for $M < 1$ (elliptic equation for the potential perturbation φ_1) and $M > 1$ (hyperbolic equation for the potential perturbation φ_1). To achieve this connection, for example in the vicinity of the throat of a de Laval nozzle with adapted flow, we turn to the method used for small singular disturbances of the "asymptotic expansion" type. This technique was discussed in Chap. 7 for a quasi-one-dimensional flow (Sect. 7.2); we apply it here to two-dimensional planar flow.

The order of magnitude of the small parameter is always that of the velocity component v, with the velocity component u being close to U_0. We set

$$u = U_0 + \epsilon^n u_1 + \epsilon^{2n} u_2 + \ldots, \quad v = \epsilon(v_1 + \epsilon^n v_2 + \ldots)$$

and consider different scales for x and y. If l_r is the mechanical reference length of the flow and x^+ and y^+ are dimensional coordinates, we can set

$$\bar{x} = \epsilon^{\alpha n} x^+/l_r, \; \bar{y} = \epsilon^{(\alpha-1)n} y^+/l_r. \tag{10.98}$$

Nonreactive Fluid

The basic equations are those of an isentropic steady flow of a compressible inviscid fluid:

$$\begin{cases} d\rho/dt + \rho \nabla \cdot \mathbf{v} = 0, \\ d(v^2/2)/dt + (1/\rho)\, dp/dt = 0, \\ ds/dt = 0 \;\Rightarrow s = s_0 \;\Rightarrow dp = c^2 d\rho. \end{cases} \tag{10.99}$$

By eliminating ρ and p between these equations, we obtain

$$\frac{d(\mathbf{v}^2/2)}{dt} - c^2 \nabla \cdot \mathbf{v} = 0, \tag{10.100}$$

and, in the case of a two-dimensional flow,

$$(u^2 - c^2)\frac{\partial u}{\partial x} + (v^2 - c^2)\frac{\partial v}{\partial y} + uv\,(\frac{\partial v}{\partial x} + \frac{\partial u}{\partial y}) = 0. \tag{10.101}$$

We choose reference quantities such that $l_r = v_r t_r$, $h_r = T_r s_r = p_r \vartheta_r = v_r^2$. Equation 10.101 retains the same form with dimensionless variables.

In the linear case, x and y have the same order of magnitude, and we set $u = U_0 + \epsilon u_1$, $v = \epsilon v_1$, $c^2 = c_0^2 + \epsilon (c^2)_1$ as we did in Sect. 10.3.1.

We then obtain $(U_0^2 - c_0^2)\, \partial u_1/\partial x - c_0^2\, \partial v_1/\partial y + \epsilon\, [(2u_0 u_1 - (c^2)_1)\, \partial u_1/\partial x$ $-(c^2)_1\, \partial v_1/\partial y + U_0\, v_1\, (\partial v_1/\partial x + \partial u_1/\partial y)] = 0.$

To the first order, we obviously obtain (10.93) from Sect. 10.3.1: $M_0^2\,\partial^2\varphi_1\partial x^2 - \Delta\varphi_1 = 0$.

This last result does not remain valid in the vicinity of the sonic flow, for which $M_0 = 1$. To study this configuration, we use the following coordinates and physical quantities:[10]

$$x = \bar{x},\ y = \epsilon^{n-1}\bar{y}\ u = U_0 + \epsilon^n u_1 + \epsilon^{2n} u_2 + \ldots,$$

$$v = \epsilon(v_1 + \epsilon^n v_2 + \ldots),\ c^2 = c_0^2 + \epsilon^n (c^2)_1 + \epsilon^{2n} (c^2)_2 + \ldots$$

Inserting these expressions in the basic equation (10.101), the following result is obtained:

$$\begin{aligned}
&[U_0^2 - c_0^2 + \epsilon^n(2U_0u_1 - (c^2)_1) \\
&+\epsilon^{2n}(u_1^2 + 2U_0u_2 - (c^2)_2)]\epsilon^n\,(\partial u_1/\partial\bar{x} + \epsilon^n\,\partial u_2/\partial\bar{x}) \\
&+(-c_0^2 - \epsilon^n(c^2)_1 + \epsilon^2 v_1^2 + 2\epsilon^{2+n}v_1v_2)(\epsilon^{2-n}\,\partial v_1/\partial\bar{y} + \epsilon^2\,\partial v_2/\partial\bar{y}) \quad (10.102)\\
&+(U_0 + \epsilon^n u_1 + \epsilon^{2n}u_2)\epsilon^2(v_1 + \epsilon^n v_2)\,[\partial u_1/\partial\bar{y} + \partial v_1/\partial\bar{x} \\
&+\epsilon^n\,(\partial u_2/\partial\bar{y} + \partial v_2/\partial\bar{x})] = 0.
\end{aligned}$$

Taking $U_0 = c_0$, we have

$$\epsilon^{2n}(2U_0u_1 - (c^2)_1)\frac{\partial u_1}{\partial\bar{x}} + O(\epsilon^{3n}) - \epsilon^{2-n}c_0^2\frac{\partial v_1}{\partial\bar{y}} + O(\epsilon^2) = 0. \qquad (10.103)$$

For $2n = 2 - n$ (i.e., $n = 2/3$), the following nontrivial result[11] is obtained:

$$[2U_0u_1 - (c^2)_1]\frac{\partial u_1}{\partial\bar{x}} - c_0^2\frac{\partial v_1}{\partial\bar{y}} = 0. \qquad (10.104)$$

For an ideal gas, equation $h + v^2/2 = h_0$ can be rewritten as $c^2/(\gamma-1) + v^2/2 = (\gamma+1)c_*^2/2(\gamma-1)$. It follows that $(c^2)_1 = (1-\gamma)U_0u_1$, so that $(\gamma+1)\,U_0u_1\,\partial u_1/\partial\bar{x} - c_0^2\,\partial v_1/\partial\bar{y} = 0$, and, considering that U_0 is the reference velocity,

$$(\gamma+1)u_1\frac{\partial u_1}{\partial\bar{x}} - c_0^2\frac{\partial v_1}{\partial\bar{y}} = 0. \qquad (10.105)$$

This result can be written as a function of the potential φ_1:

$$(\gamma+1)\varphi_{1,\bar{x}}\varphi_{1,\bar{x}\bar{x}} - \varphi_{1,\bar{y}\bar{y}} = 0. \qquad (10.106)$$

---

[10]In this case we set $\alpha = 0$ in (10.98).

[11]Other cases are presented in [217] and [218]. They lead to trivial solutions.

Fluid with a Single Near-Equilibrium Reaction

The basic equations for a steady flow of a compressible inviscid fluid with a single near-equilibrium reaction are as follows:

$$\begin{cases} d\rho/dt + \rho\boldsymbol{\nabla}\cdot\mathbf{v} = 0, \\ d(v^2/2)/dt + (1/\rho)\, dp/dt = 0, \\ \rho\, d\xi/dt = L\, A,\ T ds/dt - A\, d\xi/dt = 0, \end{cases} \tag{10.107}$$

where L is the phenomenological coefficient of (10.64). The reference quantities chosen are such that $l_r = v_r\tau_r$, $h_r = e_r = T_r s_r = p_r\vartheta_r = v_r^2 = A_r\xi_r$, $\xi_r = -(\partial\xi/\partial\vartheta)^0_{sA}$. We do not change the notation of the dimensionless parameters. The frozen velocity of sound is defined by

$$c_f^2 = (\frac{\partial p}{\partial\rho})_{s,\xi} = -\frac{1}{\rho^2}(\frac{\partial p}{\partial\vartheta})_{s,\xi} = \frac{1}{\rho^2}e_{\vartheta\vartheta}.$$

We have

$$-dp = e_{\vartheta\vartheta}\, d\vartheta + e_{\vartheta\xi}\, d\xi.$$

We deduce directly from this relation that

$$-\frac{1}{\rho}\frac{dp}{dt} = c_f^2\boldsymbol{\nabla}\cdot\mathbf{v} + \vartheta e_{\vartheta\xi}\, LA$$

and so we can write

$$\frac{d(v^2/2)}{dt} - c_f^2\boldsymbol{\nabla}\cdot\mathbf{v} = \vartheta e_{\vartheta\xi}\, LA. \tag{10.108}$$

The next steps involve differentiating the two sides of this equation with respect to time (we assume that the coefficient of A is constant to a first approximation) and expressing the time derivative of affinity as a function of the velocity vector. For the affinity, we have

$$A = -e_\xi,\ -dA = e_{\xi s}ds + e_{\xi\vartheta}d\vartheta + e_{\xi\xi}d\xi,$$

and since $Tds - Ad\xi = 0$,

$$-dA = (e_{\xi\xi} - \frac{A}{T}e_{\xi s})d\xi + e_{\xi\vartheta}d\vartheta$$

so

$$-\frac{dA}{dt} = (e_{\xi\xi} - \frac{A}{T}e_{\xi s})L\vartheta A + e_{\xi\vartheta}\frac{d\vartheta}{dt}.$$

Equation 10.108 gives the affinity and the continuity equation gives the time derivative of the volume as functions of velocity. We obtain

$$-\frac{dA}{dt} = \frac{e_{\xi\xi} - \frac{A}{T}e_{\xi s}}{\vartheta e_{\vartheta\xi}}\frac{d}{dt}(\frac{d(v^2/2)}{dt} - c_f^2\boldsymbol{\nabla}\cdot\mathbf{v}) + \vartheta e_{\vartheta\xi}\boldsymbol{\nabla}\cdot\mathbf{v}.$$

By differentiating the two sides of (10.108) with respect to time, we get

$$\frac{d}{dt}\left(\frac{d(v^2/2)}{dt} - c_f^2 \boldsymbol{\nabla}\cdot\mathbf{v}\right) + D\left(\frac{d(v^2/2)}{dt} - K\boldsymbol{\nabla}\cdot\mathbf{v}\right) = 0, \tag{10.109}$$

with $D = L\,\vartheta\,(e_{\xi\xi} - A\,e_{\xi s}/T)$ and $K = c_f^2 - \vartheta^2 e_{\vartheta\xi}^2/(e_{\xi\xi} - A\,e_{\xi s}/T)$.[12]

We will now derive the final transonic equation from these relations using the asymptotic expansion method [180, 218].

By assuming that the dimensionless terms $u_1, v_1, s_1, \vartheta_1, \xi_1$, like $\bar{x}$ and $\bar{y}$, are on the order of 1, we deduce the following result from (10.109):

$$\begin{aligned}&\tfrac{\partial}{\partial \bar{x}}[(M_{f0}^2 - 1)\tfrac{\partial u_1}{\partial \bar{x}} - \epsilon^{2-n}\tfrac{\partial v_1}{\partial \bar{y}} + \epsilon^n((M_{f0}^2 - 1)\tfrac{\partial u_2}{\partial \bar{x}} + \tfrac{2U_0 u_1 - (c_f^2)_1}{(c_f^2)_0}\tfrac{\partial u_1}{\partial \bar{x}}) + \ldots]\\ &+\epsilon^{(\delta-\alpha)n}\tfrac{D_0}{U_0}\tfrac{(c^2)_{e0}}{(c^2)_{f0}}[(M_{e0}^2 - 1)\tfrac{\partial u_1}{\partial \bar{x}} - \epsilon^{2-n}\tfrac{\partial v_1}{\partial \bar{y}} + \epsilon^n G + \ldots] = 0,\end{aligned} \tag{10.110}$$

with

$$G = (M_{e0}^2 - 1)\frac{\partial u_2}{\partial \bar{x}} + \frac{2U_0 u_1 - K_1}{(c^2)_{e0}}\frac{\partial u_1}{\partial \bar{x}} + \frac{D_1}{D_0}(M_{e0}^2 - 1)\frac{\partial u_1}{\partial \bar{x}}.$$

τ_p is the chemical time (at constant pressure), and $v_r = U_0$ is the reference speed. u and v are expressed as functions of the potential φ_1 for velocity disturbances (it is possible to introduce φ_1 since $\boldsymbol{\nabla}\times\mathbf{v}$ vanishes for the first order). Moreover, we have

$$s_1 = 0,\ \gamma_f = \left(\frac{\partial \ln p}{\partial \ln \rho}\right)_{s\xi} = \frac{\rho}{p}c_f^2,\ \gamma_e = \left(\frac{\partial \ln p}{\partial \ln \rho}\right)_{sA} = \frac{\rho}{p}c_e^2,$$

and we assume that the speeds of sound c_e and c_f are close, which makes it possible to write $c_{f0}^2/c_{e0}^2 - 1 = Z\epsilon^n$. If we choose to impose $n = 1$ and $\delta - \alpha = 0$, we obtain the linearized equations that we have already derived, and which do not enable us to study the transonic zone correctly. By making $n = 2/3$, we obtain satisfactory equations for the transonic zone with the borderline cases of frozen flow and equilibrium flow. For $\delta - \alpha = 0$ and with $\gamma = \gamma_{e0}$ we obtain[13]

$$\begin{cases}[(\gamma+1)\varphi_{1\bar{x}}\varphi_{1\bar{x}\bar{x}} - Z\varphi_{1\bar{x}\bar{x}} - \varphi_{1\bar{y}\bar{y}}]_{\bar{x}}\\ +(\gamma+1)\varphi_{1\bar{x}}\varphi_{1\bar{x}\bar{x}} - \varphi_{1\bar{y}\bar{y}} = 0,\\ \xi_{1\bar{x}} + \xi_1 + \varphi_{1\bar{x}} = 0,\\ \bar{x} = x,\ \bar{y} = \epsilon^{1/3}y,\ \vartheta_1 = u_1.\end{cases} \tag{10.111}$$

---

[12]Note that, for $A = 0$, these coefficients are simply $D = L\vartheta e_{\xi\xi} = \tau_r/\tau_{s\vartheta} = \tau_r/\tau_c = \theta$ and $K = c_f^2 - \vartheta^2 e_{\vartheta\xi}^2/e_{\xi\xi} = c_e^2$. We set $\theta = \epsilon^{\delta n}$ as in Sect. 7.2, and so, using (10.98), we obtain $\bar{x} = \epsilon^{(\alpha-\delta)n}x^+/l_c$, $\bar{y} = \epsilon^{(\alpha-\delta-1)n}y^+/l_c$.

[13]Note that other cases can be derived from (10.110) and are presented in [218]. These cases will not be studied here.

Solving the Transonic Equations

The near-equilibrium transonic monodimensional equations of Sect. 7.2 are easily solved. On the other hand, the flow is now two-dimensional, and the main task is to analytically solve the partial differential equations, which are nonlinear. We will try to find particular solutions.

In each case, we need to find a solution $\varphi_1(x, y)$ and $\xi_1(x, y)$ of the partial differential equations that satisfies certain boundary conditions, one of which is obtained by developing equation (10.108). Once the solution $\varphi_1(x, y)$ is known, it is possible to determine $\xi_1(x, y)$ by solving a linear first-order differential equation.

Only symmetrical solutions with respect to the axis Ox are applicable to a de Laval nozzle with an axis of symmetry. We will limit our study to these solutions of (10.111) and determine particular lines: the two sonic lines corresponding to the speeds of sound c_e and c_f, which we will call (S_e) and (S_f), respectively, and the throat line (H) (corresponding to the points where the tangent to the streamlines is parallel to the axis Ox and the characteristic lines).

The polynomial solution corresponding to a symmetrical flow with respect to the Ox axis is [217, 218]

$$\varphi_1 = \frac{m}{2}\bar{x}^2 + \left(\frac{\gamma+1}{2}m^2\bar{y}^2 + \alpha_0\right)\bar{x} + \frac{(\gamma+1)^2}{24}m^3\bar{y}^4 + \alpha_0\frac{\gamma+1}{2}m\bar{y}^2 + \beta_0, \quad (10.112)$$

where m, α_0, β_0 are constant coefficients,[14] and we have

$$\begin{cases} u_1 = m\bar{x} + (\gamma+1)m^2\bar{y}^2/2 + \alpha_0, \\ v_1 = (\gamma+1)m^2\bar{x}\bar{y} + (\gamma+1)^2)m^3\bar{y}^3/6 + \alpha_0(\gamma+1)m\bar{y}. \end{cases} \quad (10.113)$$

We distinguish two particular points on the Ox axis: the equilibrium sonic point E where $v = c_e$ (i.e., the point $x_E = -\alpha_0/m$), and the frozen sonic point F at fixed ξ_1 where $v = c_f$ (i.e., where $u_1 = Z/(\gamma+1)$ and $x_F = -(\alpha_0 - Z/(\gamma+1))/m$).[15] In this case, the condition (10.108) results in the equation

$$[(\gamma+1)\varphi_{1\bar{x}}\varphi_{1\bar{x}\bar{x}} - \varphi_{1\bar{y}\bar{y}}]_i = -Z\left(\frac{e_{\xi\xi 0}}{e_{\vartheta\xi 0}}\xi_1 + u_1\right)_i,$$

where the index i indicates an initial position. At $x = 0$, this condition becomes

$$m = \frac{e_{\xi\xi 0}}{e_{\vartheta\xi 0}}(\xi_1)_{\bar{x}=0} + \alpha_0.$$

---

[14] m and α_0 can be deduced from the values of the sonic points x_E and x_F that are defined shortly and determined by a monodimensional calculation, as mentioned in Sect. 7.2. β_0 is an arbitrary constant.

[15] $Z = \epsilon^{-n}(c_{f0}^2/c_{e0}^2 - 1)$ is a first-order coefficient that depends on the given value of $\epsilon \ll 1$ and the choice of n.

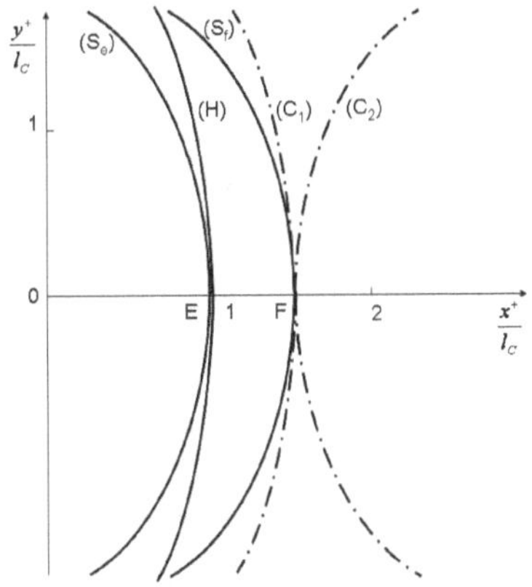

Fig. 10.9. Particular lines for a "nozzle solution"

The sonic lines are

$$\frac{\gamma+1}{2} m\bar{y}^2 + \bar{x} - \bar{x}_E = 0, \ (S_e)$$

$$\frac{\gamma+1}{2} m\bar{y}^2 + \bar{x} - \bar{x}_F = 0, \ (S_f)$$

The throat line has the equation

$$\frac{\gamma+1}{6} m\bar{y}^2 + \bar{x} - \bar{x}_E = 0, \ (H)$$

and passes through the point E.

Characteristic lines are obtained using the equation

$$[(\gamma+1)u_1 - Z](d\bar{y})^2 - (d\bar{x})^2 = 0, \tag{10.114}$$

where $\bar{x}$ and $\bar{y}$ are the coordinates of the current point on the curve. Those that cross the point F obey the following equations:

$$\frac{\gamma+1}{4} m\bar{y}^2 + \bar{x} - \bar{x}_F = 0, \ (C_1),$$

$$-\frac{\gamma+1}{2} m\bar{y}^2 + \bar{x} - \bar{x}_F = 0, \ (C_2).$$

(C_1) and (C_2) are the characteristics of the singular point and play a fundamental role in the flow analysis. The progress variable ξ_1 is given by the relation

$$\xi_1 = m - u_1, \tag{10.115}$$

with $e_{\xi\xi 0}/e_{\vartheta\xi 0} = 1$. These solutions are represented in the physical plane (x, y) by the curves of Fig. 10.9.

10.4 The Rankine–Hugoniot Relations

We observe two types of flame propagation in a mixture of fuel and oxidizer [124, 290]. When the propagation is slow (a few centimeters per second), we observe deflagration; on the other hand, fast propagation (some km/s) yields a detonation wave. For example, in a sufficiently long tube filled with a mixture of fuel and oxidizer, if we ignite the mixture at one end, a deflagration flame is generated that starts to travel at a constant speed. When this flame has covered a distance of about ten times the diameter of the tube, the flame accelerates and quickly changes into a detonation wave moving at constant speed. Here, we will only study planar waves traveling in the normal direction; in this case, the fluid motion can be regarded as steady and one-dimensional in the reference frame associated with the wave. We will thus eliminate the transitional zone from our study.

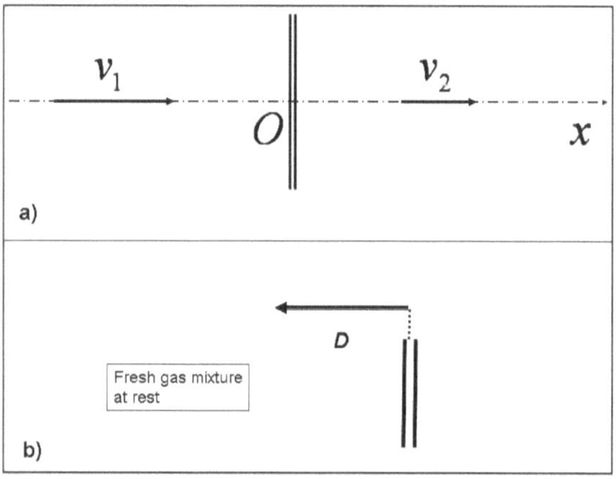

Fig. 10.10. System of axes: a, related to the wave; b, the wave travels into the fresh mixture

If we use a system of axes linked to the wave, the phenomenon can be considered to unfold as follows (Fig. 10.10a):

- Upstream of the wave ($x < 0$), the mixture of fresh gas moves at a constant velocity v_1
- With the crossing of the wave ($x = 0$), the fluid undergoes a discontinuity
- The mixture of burned gases downstream of the wave ($x > 0$) is moving at a uniform velocity of v_2.

If, on the other hand, we consider a system of axes linked to the motionless fresh gases, the wave would move into the fresh mixture (Fig. 10.10b).

The discontinuity equations for mass, momentum and energy, respectively (see Sect. 4.9), are as follows:

$$\rho_1 v_1 = \rho_2 v_2 = \dot{m}, \tag{10.116}$$

$$p_1 + \rho_1 v_1^2 = p_2 + \rho_2 v_2^2, \tag{10.117}$$

$$h_1 + v_1^2/2 = h_2 + v_2^2/2. \tag{10.118}$$

We need to add the equations of state for the gas mixture to these equations. We have two equations in the case of an ideal mixture of perfect gases:

$$\begin{cases} p_1/\rho_1 = n_1 R T_1, \; p_2/\rho_2 = n_2 R T_2, \\ h_1 = h_1(Y_{j1}, T_1), \; h_2 = h_2(Y_{j2}, T_2). \end{cases} \tag{10.119}$$

In state (1), the concentrations are those of the fresh gases, and the temperature is too low to start the chemical reactions. The mixture is chemically frozen. In state (2), we are dealing with the gas mixture in chemical equilibrium. The enthalpy h_2 thus refers to a different gas from gas (1). The difference in the enthalpies corresponds to the energy released by the reaction and to the enthalpy needed to heat the gas from T_1 to T_2. After simplifying, and assuming that the specific heat does not vary,[16] we get

$$h_2 - h_1 = -\Delta h + c_p(T_2 - T_1). \tag{10.120}$$

By eliminating v_1 and v_2 between the continuity equation and the momentum equation, we obtain

$$p_2 - p_1 = \rho_1 v_1^2 - \rho_2 v_2^2 = -\dot{m}^2\left(\frac{1}{\rho_2} - \frac{1}{\rho_1}\right) \tag{10.121}$$

or

$$(p_2 - p_1)/\left(\frac{1}{\rho_2} - \frac{1}{\rho_1}\right) = -\dot{m}^2. \tag{10.122}$$

Relation 10.122 corresponds to the "line of mass flow rate" of slope $(-\dot{m}^2)$ in the $(1/\rho, p)$ plane.

Proceeding in the same way with the energy equation, we get

$$h_2 - h_1 = -\frac{\dot{m}^2}{2}\left(\frac{1}{\rho_2^2} - \frac{1}{\rho_1^2}\right), \tag{10.123}$$

and by replacing $\dot{m}^2$ with its value (found in Eq. 10.122) we obtain

$$h_2 - h_1 = \frac{1}{2}\left(\frac{1}{\rho_2} + \frac{1}{\rho_1}\right)(p_2 - p_1). \tag{10.124}$$

[16] If we consider the crossing of a shock without combustion, we simply have $h_2 - h_1 = c_p(T_2 - T_1)$.

Relation 10.124 corresponds to the "adiabatic of detonation" in the $(1/\rho, p)$ plane. Relations 10.122 and 10.124 are known as the Rankine–Hugoniot relations.

If quantities ρ_1 and p_1 are given, ρ_2, p_2 are the unknown quantities of the problem.

We saw that $(h_2 - h_1)$ can be expressed as a function of a thermodynamic quantity (for example T). If the selected variables are p and ρ, (10.124) is represented by a curve that does not pass through the point $(1/\rho_1, p_1)$; indeed, if for example we make $T_2 = T_1$, we do not obtain $h_2 = h_1$, which contrasts with the case of a shock wave.

The solution is given by the intersection of the adiabatic defined by (10.124) and the straight line for the mass flow rate of slope $-\dot{m}^2$ defined by (10.122); see Fig. 10.11.

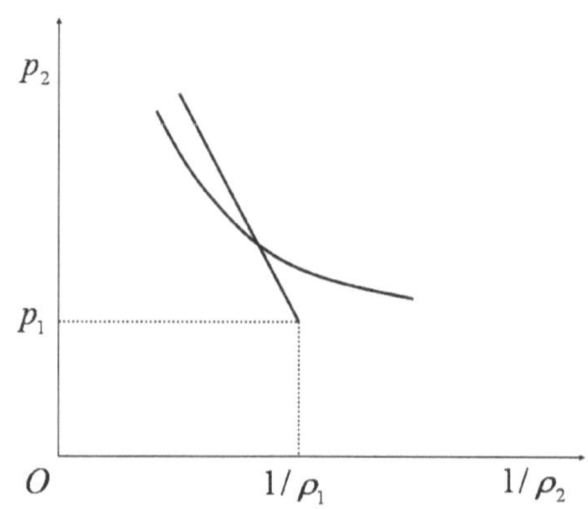

Fig. 10.11. Intersection of the adiabatic with the staight line for the mass flow rate

We will solve the problem for the simplified case already mentioned. We will also assume that the molar mass does not change during the crossing of the wave, so that

$$\frac{T_2}{T_1} = \frac{\rho_1}{\rho_2}\frac{p_2}{p_1}. \tag{10.125}$$

Let us now set

$$\rho_1/\rho_2 = \vartheta,\ p_2/p_1 = p,\ \tau = \Delta h/c_p T_1. \tag{10.126}$$

We then obtain, according to (10.122),

$$\frac{p-1}{\vartheta - 1} = -\frac{\dot{m}^2}{\rho_1 p_1}, \tag{10.127}$$

and according to (10.124),

$$p\vartheta - 1 - \tau = \frac{p_1}{2\rho_1 c_p T_1}(\vartheta + 1)(p - 1). \tag{10.128}$$

Taking into account the expression for the Mach number,

$$M_1 = v_1/c_1 = v_1/\sqrt{\gamma p_1/\rho_1} = \dot{m}/\sqrt{\gamma p_1 \rho_1}, \tag{10.129}$$

and the state law for perfect gases, we have

$$\frac{p-1}{\vartheta - 1} = -\gamma M_1^2, \tag{10.130}$$

$$p = \frac{(1-\gamma)\vartheta + 1 + \gamma + 2\gamma\tau}{(1+\gamma)\vartheta + 1 - \gamma}. \tag{10.131}$$

Equation 10.131 provides the adiabatic of Fig. 10.12.

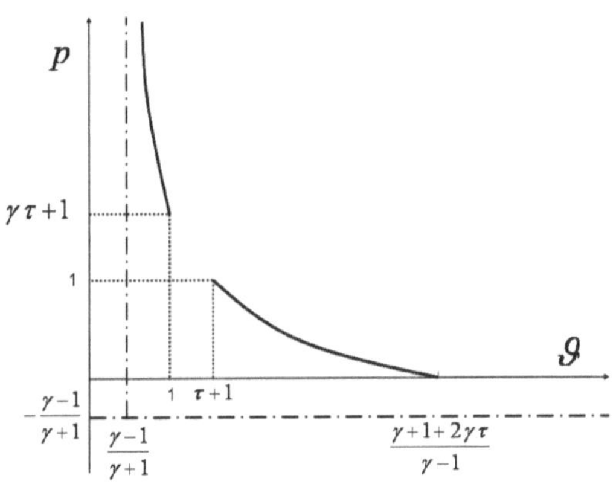

Fig. 10.12. The adiabatic curve: detonation and deflagration

The forbidden area corresponds to $M_1^2 < 0$.

The part of this curve where $\vartheta < 1$ corresponds to an increase in pressure—often a significant one. This is the detonation region. When $\tau = 0$ (i.e., in the absence of combustion), it simply reduces to a shock wave. In this case, by applying the second law of thermodynamics, we can show that the shock wave must be a compression wave. Thus, there cannot be a discontinuity wave in a one-component fluid with increasing density and decreasing pressure. This is not the case for combustion, where τ is positive.

The region corresponding to $p < 1$ is that of deflagration.[17]

[17] Generally speaking, the variation in pressure is so small here that it can be ignored. That is why we can assume a uniform pressure in our study of deflagration in Sect. 10.5.

Therefore, we have two quite distinct regions, the first corresponding to detonations and the other corresponding to deflagration waves. We can see that the discontinuity condition

$$[\mathcal{J}_S + \rho s(\mathbf{v} - \mathbf{W})]_-^+ \cdot \mathbf{N} = \dot{W}_{aS} \geq 0 \tag{10.132}$$

from Sect. 4.9, which gives

$$\rho_2 s_2 v_2 - \rho_1 s_1 v_1 = \dot{W}_{aS} \geq 0, \tag{10.133}$$

imposes more limitations on the domain in which these waves occur.

Let us initially define the two particular points where the straight line starting from the point $(1, 1)$ is tangential to the adiabatic. First of all, in the vicinity of these points, we can derive the equations

$$\begin{cases} dp = -\gamma M_1^2 \, d\vartheta, \\ [(1+\gamma)\vartheta + 1 - \gamma]dp + p(1+\gamma)d\vartheta = (1-\gamma)d\vartheta. \end{cases} \tag{10.134}$$

By eliminating dp, we have

$$-\gamma M_1^2[(1+\gamma)\vartheta + 1 - \gamma] + p(1+\gamma) - 1 + \gamma = 0. \tag{10.135}$$

Replacing M_1^2 with its value in terms of p and ϑ yields

$$p[(1+\gamma)\vartheta - \gamma] - \vartheta = 0. \tag{10.136}$$

Eliminating p gives the equation for the ϑ-coordinates of what are known as the Chapman–Jouguet points [45, 131]:

$$\vartheta^2 - 2(1+\tau)\vartheta + 1 + 2\gamma\tau/(1+\gamma) = 0. \tag{10.137}$$

We will now show that these *Chapman–Jouguet points* correspond to the extrema of entropy production.

The entropy s_2 is given by the intersection of the line drawn from the point $(1, 1)$ and the adiabatic; therefore s_2 depends on the ratio $(p-1)/(\vartheta-1)$, which is (as shown previously) proportional to the square of the Mach number M_1^2, where M_1 is the upstream Mach number.

Let us calculate the extrema of $(s_2 - s_1)$ (i.e., the extrema of the energy production through the wave). We have

$$\dot{W}_{aS}/\dot{m} = s_2 - s_1 = c_v \ln(p_2/p_1) - c_p \ln(\rho_2/\rho_1) = c_v \ln p + c_p \ln\vartheta. \tag{10.138}$$

The extremum corresponds to

$$d(s_2 - s_1)/c_v = d \ln p + \gamma d \ln\vartheta. \tag{10.139}$$

Let us replace p with its value

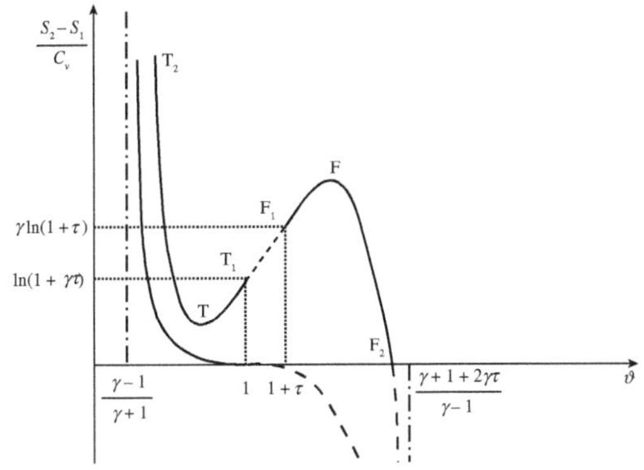

Fig. 10.13. Entropy production throughout the combustion wave and through the shock wave (curve corresponds to $\tau = 0$, with a horizontal tangent at the inflection point for $\vartheta = 1$)

$$\frac{1}{c_v}\frac{d(s_2 - s_1)}{d\vartheta} = \frac{1-\gamma}{(1-\gamma)\vartheta + 1 + \gamma + 2\gamma\tau} - \frac{1+\gamma}{(1+\gamma)\vartheta + 1 - \gamma} + \frac{\gamma}{\vartheta} \quad (10.140)$$

or

$$\vartheta^2 - 2(1+\tau)\vartheta + 1 + 2\gamma\tau/(1+\gamma) = 0. \quad (10.141)$$

Note that the extrema correspond to the Chapman–Jouguet points. In particular, the detonation corresponds to the minimum entropy production at the Chapman–Jouguet points.

We can state that the detonation wave thus obtained is stable.[18] The study of $(s_2 - s_1)/c_v$ as a function of ϑ gives the curve shown in Fig. 10.13.

When $\tau = 0$, we obtain the case of a shock wave that is characterized by a point of inflection with a horizontal tangent at $\vartheta = 1$.

Branch T_1T_2 of the curve of entropy production corresponds to detonation, while branch F_1F_2 corresponds to deflagration. The Chapman–Jouguet points are at points T and F. Only point T has any particular significance: it corresponds to Chapman–Jouguet detonation.[19]

This curve is shown on a (ϑ, p) plot in Fig. 10.14.

The speed of the detonation wave D is equal to the relative velocity of the fresh gases v_1. We can calculate this speed at point T in particular. We have

[18]However, the stability of the detonation wave generally depends on other factors too, such as the characteristics of the walls, and a thorough study is necessary to determine the zones of stability. We will not perform such a study here.

[19]The Chapman–Jouguet condition, which states that the detonation proceeds at a velocity such that the reacting gases only just attain sonic velocity (in the frame of the lead shock) when the reaction is finished, holds approximately in detonation waves.

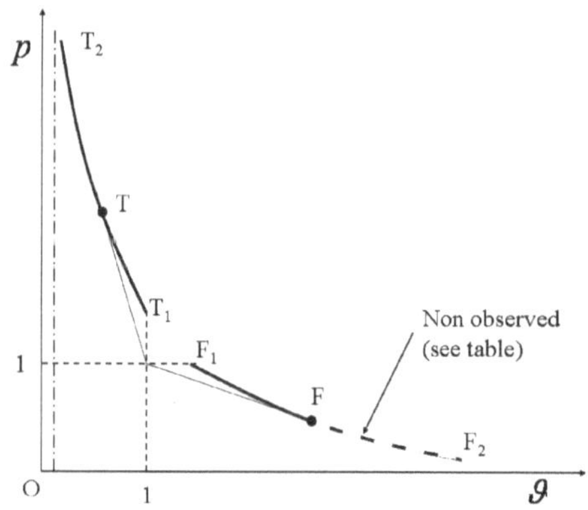

Fig. 10.14. Observed and unobserved waves

$$\vartheta = 1 + \tau - \sqrt{\tau^2 + 2\tau/(\gamma + 1)}. \qquad (10.142)$$

We replace ϑ with this value in the expression for p and use the equation

$$M_1^2 = -\frac{1}{\gamma}\frac{p-1}{\vartheta - 1} \qquad (10.143)$$

to calculate M_1, which allows us to deduce $v_1 = D$, since

$$v_1 = M_1 c_1. \qquad (10.144)$$

c_1 is the speed of sound in fresh gases, which is presumably known if they are at rest ($u_1 = 0 = v_1 - D$).

The downstream Mach number $M_2 = v_2/c_2$ relative to the wave is such that

$$M_2^2 = \frac{v_2^2}{c_2^2} = \frac{\rho_1^2 v_1^2/c_1^2}{\rho_2^2 c_2^2/c_1^2} = \frac{\rho_1^2 M_1^2}{\rho_2^2 (p_2/p_1)(\rho_1/\rho_2)} = \frac{\vartheta}{p} M_1^2. \qquad (10.145)$$

From this result, we can easily deduce the velocity v_2.

At Chapman–Jouguet point T, we obtain

$$M_{2T} = 1. \qquad (10.146)$$

A simple way to prove this is to express M_1^2 as a function of ϑ and p, as we did previously, and to use the following relation that is independent of τ and valid at the Chapman–Jouguet points:

$$p[(1+\gamma)\vartheta - \gamma] - \vartheta = 0. \qquad (10.147)$$

The fact that M_2 is equal to 1 in this case provides an element of stability for the detonation wave. Indeed, the acoustic waves cannot travel through the flow

Zone of the diagram	M$_2$	Type of wave	Conditions of observation	Remarks
T_2T	< 1	Strong detonation	Special	$M_1 > (M_1)_T$
T	$= 1$	Chapman–Jouget detonation	Usual	Wave travels into fresh gases at supersonic speed
TT_1	> 1	Weak detonation	Special	$M_1 > (M_1)_T$
F_1F	< 1	Weak deflagration	Usual	$M_1 < (M_1)_F < 1$ Wave travels into fresh gases at subsonic speed
F	$= 1$		Not observed	
FF_2	> 1		Not observed	

Table 10.1. Combustion waves

downstream to reach the detonation wave, and any deceleration of this wave will result in nonstationary phenomena. The summary presented in Table 10.1 provides some theoretical and experimental information on combustion waves.

10.5 Deflagration Waves

10.5.1 Steady Propagation of an Adiabatic Planar Flame

We now study the one-dimensional propagation of a planar premixed flame assuming that the Shvab–Zel'dovich approximation (see Sect. 7.4) holds [290].[20]

Taking into account these assumptions, we obtained the following single forms of the species balance and energy balance equations:

$$\begin{cases} \nabla \cdot (\rho \mathbf{v} \beta_j - \rho D \nabla \beta_j) = \dot{\zeta}, \\ \nabla \cdot (\rho \mathbf{v} \beta_T - \lambda / c_{p,\,f} \nabla \beta_T) = \dot{\zeta}, \end{cases} \tag{10.148}$$

where

$$\beta_j = Y_j / \nu_j \mathcal{M}_j, \ \beta_T = \sum_j Y_j \int_{T^0}^{T} c_{p,\,j} dT / \Delta H. \tag{10.149}$$

[20]Recall that the following assumptions are made in this approximation: there is steady flow in a suitably chosen reference frame; there is no thermodiffusion process ($D_T = 0$); external forces and viscosity are negligible; there is constant static pressure to a first approximation; Fourier's law of thermal conduction applies; Fick's law holds for diffusion and there is a single coefficient of diffusion for all species; the Lewis number is close to 1; a mixture of perfect gases is present; there is only one chemical reaction. For multireactive mixtures, see for example [142].

Let us now use this approximation to consider the plane of the deflagration flame in a reference frame associated with this wave. All of the parameters depend on just one variable, x (see Fig. 10.15).

As previously mentioned, a deflagration flame can be obtained experimentally in a long tube filled with a mixture of fuel and oxidizer that we ignite at one end. Thus, we can consider a tubular reactor supplied with fresh gas in such a way that the flame is fixed.[21]

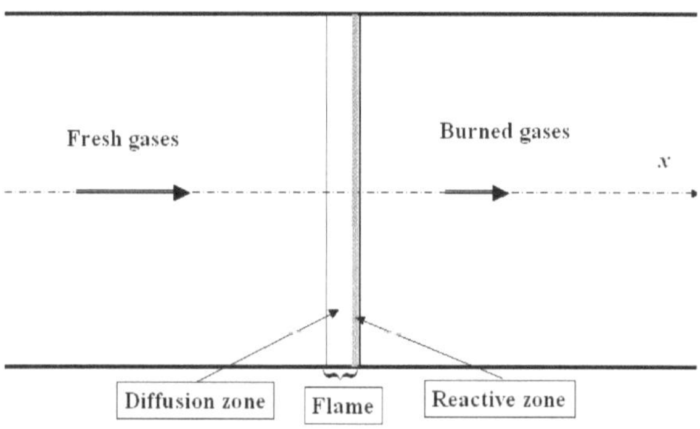

Fig. 10.15. Fresh gases and burned gases in the case of a thin flame

We denote unit mass flow rate by $\dot{m}$:

$$\dot{m} = \rho v, \tag{10.150}$$

and we make g the common value of the transfer coefficients:

$$g = \rho D = \lambda / c_{p,\,f}. \tag{10.151}$$

We now introduce the position variable η such that

$$d\eta / \eta = \dot{m}\, dx / g. \tag{10.152}$$

Let us consider the second-order chemical reaction

$$\mathrm{A} + \mathrm{B} \longrightarrow \mathrm{P}, \tag{10.153}$$

and, to simplify, we assume that before the reaction a mole of A is associated with a mole of B in a unit mass of the mixture:

[21] Note that deflagration (p =const.) is not the only possible mode (see Sect. 10.4); under certain conditions a very rapid detonation wave—a surface of discontinuity analogous to a shock wave—can occur.

$$\beta_A^0 = \beta_B^0 = -1,\ x = -\infty\ (\eta = 0), \tag{10.154}$$

where the right hand side is in moles per unit mass. The equations are then as follows:

$$\begin{cases} \eta^2\, d^2\beta_j/d\eta^2 = -g\dot{\zeta}/\dot{m}^2, \\ \eta^2\, d^2\beta_T/d\eta^2 = -g\dot{\zeta}/\dot{m}^2, \\ \dot{\zeta} = \rho^2 B T^s e^{-T_a/T}\beta_A\beta_B. \end{cases} \tag{10.155}$$

Let us eliminate $\dot{\zeta}$ between the equations in β_A and β_B; we then obtain a linear relation that gives $\beta_A - \beta_B$ as a function of η. The solution is assumed to have a finite value for $x = +\infty\ (\eta = \infty)$. We find that

$$\beta_A = \beta_B = -\beta. \tag{10.156}$$

Let us now eliminate $\dot{\zeta}$ between the β_A equation (or β_B) and the β_T one. This yields

$$d^2(\beta + \beta_T)/d\eta^2 = 0, \tag{10.157}$$

and a new linear relation is obtained:

$$\beta + \beta_T = a\eta + b. \tag{10.158}$$

We have

$$\begin{cases} \eta = 0:\ \beta = 1,\ \beta_T = 0 \implies b = 1, \\ \eta = \infty:\ \beta = 0,\ d\beta/d\eta = a. \end{cases} \tag{10.159}$$

At $\eta = \infty$, the reactions have completed and variations in β_T are only provided by heat exchange with the outside, which is characterized by a, the degree of adiabacity of the reactor.

Let us study the case of an adiabatic tube ($a = 0$). We thus find that

$$\beta + \beta_T = 1. \tag{10.160}$$

We must now solve the differential equation

$$\eta^2\, d^2\beta_T/d\eta^2 = -(g/\dot{m}^2)\rho^2 B T^s e^{-T_a/T}(1 - \beta_T)^2. \tag{10.161}$$

Let us denote as T_{ad} the temperature obtained at the end of the adiabatic combustion:

$$T_{ad} = T_0 + \sum_j \nu_j \mathcal{M}_j (q_f^0)_j / c_p. \tag{10.162}$$

We then have roughly

$$\beta_T \cong (T - T_0)/(T_{ad} - T_0). \tag{10.163}$$

By assuming that the perfect gas law holds for the mixture, we have

$$\rho = \bar{\mathcal{M}}p/RT, \tag{10.164}$$

where $\bar{\mathcal{M}}$ is the mean molar mass, and so we obtain

$$\eta^2\, d^2\beta_T/d\eta^2 = -\Lambda\,\omega(\beta_T) \tag{10.165}$$

with

$$\begin{cases} \Lambda = gT^s B\bar{\mathcal{M}}^2 p^2/\dot{m}^2 R^2 \cong \text{const.} \\ \omega(\beta_T) = [T_0 + \beta_T(T_{ad} - T_0)]^{-2}(1-\beta_T)^2 e^{-T_a/[T_0+\beta_T(T_{ad}-T_0)]}. \end{cases} \tag{10.166}$$

Numerical integration of this differential equation presents some difficulties. We obtain a suitable solution in the case of a thin reaction zone (Fig. 10.15) in $\eta = 1$ by considering a thickness equal to 2ϵ. In the vicinity of $\eta - 1$, we can use the approximation

$$\eta^2\, d^2\beta_T/d\eta^2 \cong d^2\beta_T/d\eta^2. \tag{10.167}$$

Multiplying the two sides by $(d\beta_T/d\eta)d\eta$ and integrating gives

$$[(d\beta_T/d\eta)^2/2]_{1-\epsilon}^{1+\epsilon} \cong -\Lambda \int_0^1 \omega(\beta_T)d\beta_T. \tag{10.168}$$

Λ is an eigenvalue of the problem that provides the mass flow rate of the steady regime.

At the boundaries $1+\epsilon$ and $1-\epsilon$, we have[22]

$$\begin{cases} \eta = 1+\epsilon: \ T = T_{ad},\ d\beta_T/d\eta = 0, \\ \eta = 1-\epsilon: \ d\beta_T/d\eta = 1. \end{cases} \tag{10.169}$$

Thus

$$\Lambda = (2\int_0^1 \omega(\beta_T)\, d\beta_T)^{-1}. \tag{10.170}$$

The integral is calculated through the use of approximations. For $\omega(\beta_T)$, we assume an expression of the form $(1-\beta_T)^{n-1}\beta_T^{p-1}$, so that

$$\int_0^1 \omega(\beta_T)\, d\beta_T = \Gamma(n)\Gamma(p)/\Gamma(n+p). \tag{10.171}$$

This approximation, which is valid for $T_a > 6T_{ad}$, was made by Rosen.

[22] $\eta < 1-\epsilon$ corresponds to a nonreactive zone with $d\beta_T/d\eta$ =const. The constant is equal to 1 since $\beta_T = 0$ for $\eta = 0$ and $\beta_T = 1$ for $\eta = 1$.

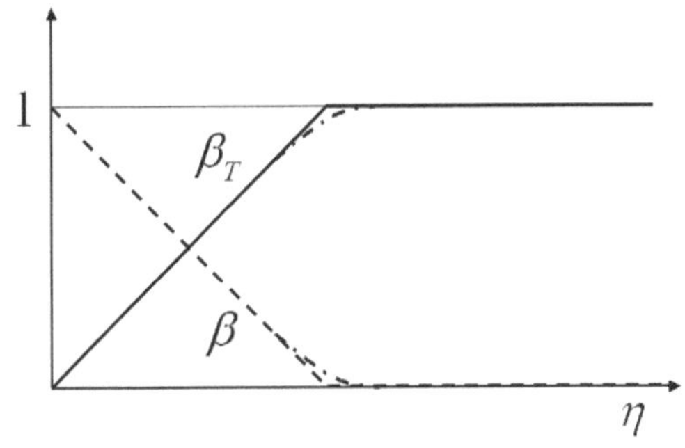

Fig. 10.16. Temperature and concentration profiles in a deflagration wave. *Dashed and continuous lines*: thin-flame assumption; *dash-dot line*: real curves

Figure 10.16 shows the profiles of β_T and β that are obtained for laminar planar premixed flames.

Using the value found for Λ and (10.166), we can deduce the mass flow rate

$$\dot{m} = (\bar{\mathcal{M}}p/R)\sqrt{gT^s B/\Lambda}$$

and thus the combustion velocity of the planar adiabatic laminar flame

$$s_L^0 = \dot{m}/\rho_u, \tag{10.172}$$

where ρ_u is the density of the unburnt mixture (the fresh gases of Fig. 10.15).

10.5.2 Curved Nonadiabatic Flames

For curved and/or nonadiabatic flames, the theory is more complex. Just as in Sect. 10.5.1, the flame structure (Fig. 10.17) consists of four regions (although the shape of the flame is not planar and we must characterize curvature effects in particular):

- The fresh gases
- The preheating zone or the diffusion zone of thickness ℓ_f, where convection, mass diffusion and thermal conduction are the most significant processes
- The reactive zone of thickness ℓ_δ, where diffusion, conduction and the chemical reaction are predominant, and which is represented by a jump[23]
- The burnt gases.

[23]For a reactive zone of thickness ℓ_δ, another small parameter $\beta = T_a/T_{ad}$ is introduced, where T_a and T_{ad} are respectively the activation temperature and the adiabatic combustion temperature. $1/\beta$ is the same order of magnitude as ℓ_δ/ℓ_f.

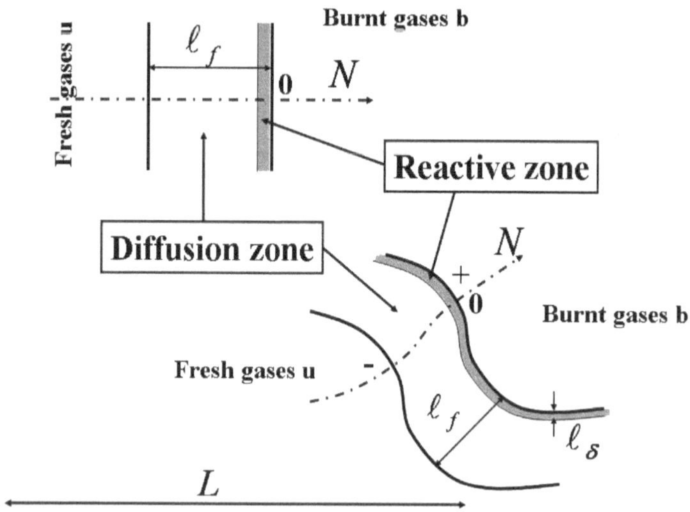

Fig. 10.17. The structures of planar and curved flames

A multiscale asymptotic expansion method with two small parameters ϵ and $1/\beta$ can be applied to the three-dimensional flow. Reference quantities are introduced: $\rho^- = \rho_u$ for density, T_{ad} for temperature, s_L^0 (the combustion velocity of the planar laminar adiabatic flame) for velocity, $\rho_u s_L$ for mass flow rate, and $p^- = p_u$ for pressure. The Lewis number Le is assumed to be different from 1, but an analysis by Clavin and Joulin [53] has shown that the difference must be on the order of $1/\beta$ (i.e., $Le = 1 + l/\beta$, where $l = O(1)$) to ensure analytical coherence. For the assumed reaction A $\longrightarrow$ B, which has a rate of order one, and at the scale ℓ_f, the reactive layer is replaced by a jump and the obtained dimensionless reaction rate per unit volume is [163, 219, 223]

$$W = \delta(N) \exp \beta((T^+ - 1), \tag{10.173}$$

where $\delta(N)$ is the Dirac delta function, and where the reduced temperature T_+ of the burnt gas just downstream of the reaction zone can be different from 1 (which is, by definition, the value for an adiabatic laminar planar flame; *note that the same symbols are used for dimensional and dimensionless parameters*). We have the following relation between the species and energy source terms and the dimensionless reaction rate per unit volume [219]:

$$W = -\epsilon \dot{W}_Y = \epsilon \dot{W}_T (1 - T^-). \tag{10.174}$$

In (10.174), the reduced source terms are $\dot{W}_Y$ for the concentration of A and $\dot{W}_T$ for the temperature. The reduced temperature of the fresh gas is

T_u. Equations 10.173 and 10.174 are valid for both incompressible [223] and compressible [219] flows. Equation 4.46 can be applied to total mass, mass of species A, temperature, and momentum. A corresponding equation can also be written for the same quantities for the fresh gases, without any flame.

In the following, we will consider excess quantities $\rho(f - f^-)$ instead of ρf.[24] The obtained reduced excess quantities are thus: $\rho(\vartheta - 1)$ for volume, $\rho(Y - 1)$ for mass fraction, $\rho(T - T^-)$ for temperature, and $\rho(\mathbf{v} - \mathbf{v}^-)$ for momentum. The space derivatives are decomposed into a normal part d/dN and a tangential part $(\mathbf{1} - \mathbf{N} \otimes \mathbf{N}) \cdot \boldsymbol{\nabla}$ using (11.18).

The material velocity $\mathbf{v}$, the composite velocity $\mathbf{V} = (\mathbf{1} - \mathbf{N} \otimes \mathbf{N}) \cdot \mathbf{v} + w\mathbf{N}$ as defined in (A.43) (see Sect. A.4 for the definition of w), and the relative normal velocity $u_\perp = (\mathbf{v} - \mathbf{V}) \cdot \mathbf{N}$ are used. We also use the variable n, where $dN = \epsilon dn$, for the normal coordinate instead of N.

With some assumptions about the orders of magnitude (the mean normal curvature $\boldsymbol{\nabla} \cdot \mathbf{N}$ of the surfaces—see Sect. A.3—and the time derivatives are assumed to be of order 1, ...), and using (10.174), we obtain the balance equations for excess quantities by subtracting between the balance equations written for the fluid mixture and the reference unburned gas:

$$\begin{cases} \partial(1-\rho)u_\perp/\partial n = \epsilon[d_{\mathbf{V}}(1-\rho)/dt + (1-\rho)\,\boldsymbol{\nabla}_{//}\cdot\mathbf{V}] \\ = \partial(u_\perp - u_\perp^-)/\partial n + \epsilon(\rho u_\perp - u_\perp^-)\boldsymbol{\nabla}\cdot\mathbf{N}, \\ \\ \partial[-(1/Le)\partial Y/\partial n + \rho(Y-1)u_\perp] + \epsilon[d_{\mathbf{V}}\rho(Y-1)/dt + \rho(Y-1) \\ \boldsymbol{\nabla}_{//}\cdot\mathbf{V}] + \epsilon[-(1/Le)\,\partial Y/\partial n + \rho(Y-1)u_\perp]\boldsymbol{\nabla}\cdot N = -W, \\ \\ \partial[-\partial T/\partial n + \rho(T-T^-)u_\perp] + \epsilon[d_{\mathbf{V}}\rho(T-T^-)/dt + \rho(T-T^-) \\ \boldsymbol{\nabla}_{//}\cdot\mathbf{V}] + \epsilon[-\partial T/\partial n + \rho(T-T^-)u_\perp]\boldsymbol{\nabla}\cdot\mathbf{N} = (1-T^-)W, \\ \\ \partial[(p-p^-)\mathbf{N} + \rho(\mathbf{v}-\mathbf{v}^-)u_\perp]/\partial n + \epsilon\,[d_{\mathbf{V}}\rho(\mathbf{v}-\mathbf{v}^-)/dt \\ +\rho(\mathbf{v}-\mathbf{v}^-)\boldsymbol{\nabla}_{//}\cdot\mathbf{V}] + \epsilon\,[(p-p^-)\mathbf{N} + \rho(\mathbf{v}-\mathbf{v}^-)u_\perp]\boldsymbol{\nabla}\cdot\mathbf{N} \\ = -\epsilon\,[(\rho-1)d_{\mathbf{V}}(\mathbf{v}^-)/dt + (\rho\mathbf{v}-\mathbf{v}^-)\cdot\boldsymbol{\nabla}\otimes\mathbf{v}^-]. \end{cases} \tag{10.175}$$

Note that, since the upstream flow is not taken to be uniform a priori, partial derivatives of the corresponding physical quantities are present in the equations. This was not the case in Sect. 10.5.1.

This system is analyzed [219] using the method of matched asymptotic expansions with two small parameters ϵ and $1/\beta$. The model is derived from that of Clavin and Joulin [53]. The expansions are performed to the first order of magnitude.

The results for zeroth order are the density, temperature and concentration profiles of species A. The momentum balance equation has a production term of order ϵ and gives

[24] This is necessary to ensure the convergence of integration across interfacial zones (see [94]).

$$\begin{cases} [p^{(0)}]_-^+ + \dot{m}^{(0)}[\mathbf{v}_\perp^{(0)}]_-^+ = 0 \\ [\mathbf{v}_{//}^{(0)}]_-^+ = 0 \end{cases} \quad (10.176)$$

to zeroth order. We observe that the velocity component parallel to the flame front is constant across the deflagration wave.

To the first order, after analyzing the expansions and after some rather involved calculations [219], we obtain the reduced temperature T^+ and the following dimensionless mass flow rate $\dot{m}^- = (\rho u_\perp)^-$, which exhibit the influence of the interfacial stretch $\nabla_{//} \cdot \mathbf{V}$ to zeroth order:

$$\dot{m}^- = 1 - \epsilon[\frac{\ln(1/T^-)}{(1-T^-)} - l\frac{T^-}{2}\int_0^{\frac{1-T^-}{T^-}} \frac{\ln(1+x)}{x}dx]\, \nabla_{//}\cdot \mathbf{V}_S^{(0)} + O(\epsilon^2) \quad (10.177)$$

where $\mathbf{V}_S$ is the mean velocity field (see Eq. 11.4). This is the main result of three-dimensional analysis. The detailed analysis, including the intermediate results, is not shown here. It can be found in the abovementioned papers. In Chap. 11 (Eq. 11.54), we give the expression for the combustion velocity s_L deduced from (10.177), as well as an alternative classical form of s_L in (11.55).

10.6 Structure of the Planar Detonation Wave

When the detonation wave is caused by a shock generated by an external source, the resulting heating of the gas causes the chemical reactions to start, which then continue downstream of the shock wave[25] The final point is on the adiabatic of detonation (Sect. 10.4). The final point is on the adiabatic of detonation (Sect. 10.4).

The wave is thus a layer of finite thickness along which the energy of reaction is gradually released. Throughout the process, the mass and momentum conservation equations are valid and the mass flow rate does not change. In other words, we remain on the line of constant mass flow rate (Michelson): $(p-1)/(\vartheta - 1) = -\gamma M_1^2$. In the Clapeyron diagram, the path is as follows [302] (Fig. 10.18):

- Point before the shock: I
- Shock: IN
- Combustion wave: line segment Na.

Point b cannot be reached in this manner, and only the points on branch T_2T in Fig. 10.14 correspond to this type of description.

[25]The structure of simple planar normal shock waves were studied in particular by Smolderen [262]. See also [94] for shock waves structure in a dusty gas.

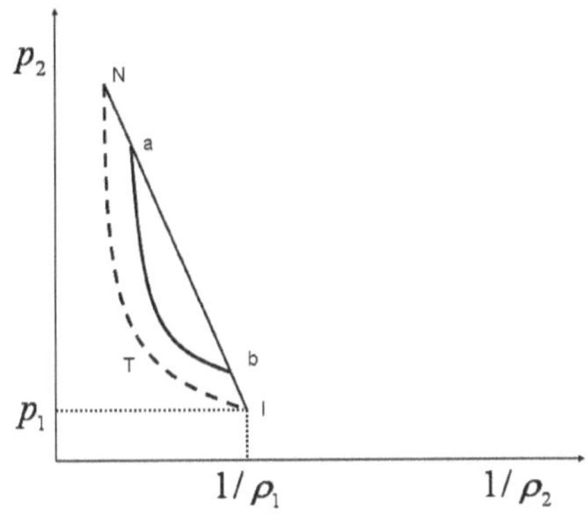

Fig. 10.18. Clapeyron diagram: detonation initiated by a shock

This result remains valid as long as the shock wave is the process that triggers combustion; the detonation wave does not necessarily have to be due to a shock wave, as was explained in Sect. 10.4, and the structure of a stable and steady detonation wave depends on all of the internal processes that occur inside the wave. Such processes include transfer phenomena.

Initially we regarded a combustion wave as a simple surface of discontinuity, which enabled us to determine the parameters on both sides knowing M_1. The speed of the wave was calculated for Chapman–Jouguet detonation, where $M_{1T} > 1$ and $M_{2T} = 1$.

In the same way as we studied the structure of the deflagration wave (see Sect. 10.5), we will now analyze the structure of a Chapman–Jouguet detonation wave. Despite the sharp gradients that arise in detonation waves, we will assume that the linear phenomenological relations that give transfer fluxes as functions of the gradients remain valid.

A detonation wave is not as difficult to analyze as the deflagration wave of Sect. 10.5: it is much faster and introduces significant pressure gradients; indeed, we cannot apply the Shvab–Zel'dovich approximation to it. In addition (and this is at least true of the most stable waves), its speed relative to the fresh gases is known—it is determined by the values of p and ϑ at the Chapman–Jouguet point T.

Our study of the structure of the detonation wave will thus proceed directly from the equations for steady one-dimensional flow.

In the reference frame associated with the wave, we have

$$d(\rho v)/dx = 0, \text{ and } \rho v = \dot{m}. \tag{10.178}$$

We will assume that the reaction is of the type A $\longrightarrow$ B to simplify matters. For such a reaction, and assuming that Fick's law holds, the species balance equation is $\frac{d}{dx}(\rho v Y_j - \rho D \frac{dY_j}{dx}) = \nu_j \mathcal{M}_j k\, C_A$.

Taking Y_B as the main unknown variable, we obtain (for ρD =const.)

$$\dot{m}\frac{dY_B}{dx} - \rho D\frac{d^2 Y_B}{dx^2} = \rho k(1 - Y_B) \text{ (species)} \tag{10.179}$$

We cannot ignore the pressure gradient in this case. The momentum equation is obtained after expressing the pressure tensor

$$\mathbf{P} = p\mathbf{1} - 2\mu\mathbf{D} = p\mathbf{1} - 2\mu[\frac{1}{2}(\boldsymbol{\nabla}\otimes\mathbf{v} + \widetilde{\boldsymbol{\nabla}\otimes\mathbf{v}}) - \frac{1}{3}\boldsymbol{\nabla}\cdot\mathbf{v}\,\mathbf{1}];$$

in other words, $\mathbf{P} = p\mathbf{1} - \frac{4}{3}\mu\frac{dv}{dx}$ here, and so we obtain

$$\dot{m}\frac{dv}{dx} + \frac{dp}{dx} - \frac{4}{3}\frac{d}{dx}(\mu\frac{dv}{dx}) = 0 \text{ (momentum)}. \tag{10.180}$$

The energy equation can then be written as follows:

$$\frac{d}{dx}\Big[\underbrace{\dot{m}(\sum_{j=1}^{N} Y_j h_j \quad \frac{p}{\rho}}_{\text{internal energy}} + \underbrace{\frac{v^2}{2}}_{\text{kinetic energy}}) - \underbrace{\lambda\frac{dT}{dx}}_{\text{conductive flux}}$$
$$+ \underbrace{\rho\sum_{j=1}^{N} Y_j h_j V_j}_{\text{heat flux due to mass diffusion}} + \underbrace{pv}_{\text{work associated with the normal pressure}}$$
$$- \underbrace{\frac{4}{3}\mu v\frac{dv}{dx}}_{\text{energy dissipated by the viscosity}} \Big] = 0.$$

The Schmidt and Prandtl numbers are assumed to be equal to 3/4. We then find that the preceding equation becomes

$$\frac{d}{dx}(\sum_{j=1}^{N}\rho_j v_j h_j) + \dot{m}\frac{d}{dx}(\frac{v^2}{2}) - g\frac{d^2}{dx^2}(c_p T + \frac{v^2}{2}) = 0,$$

where g is defined by (10.151) and is equal to $4\mu/3$, so

$$\frac{d}{dx}(\sum_{j=1}^{N}\rho_j v_j \int_{T_0}^{T} c_{p,\,j}) + \sum_{j=1}^{N}(q_f^0)_j\dot{W}_j + \dot{m}\frac{d}{dx}(\frac{v^2}{2}) - g\frac{d^2}{dx^2}(c_p T + \frac{v^2}{2}) = 0.$$

Thus, we have

$$\begin{cases} (d/dx)(\sum_{j=1}^{N}\rho_j v_j \int_{T_0}^{T} c_{p,\,j}) \cong \dot{m}\, d(c_p T)/dx, \\ \sum_{j=1}^{N}(q_f^0)_j\dot{W}_j = [(q_f^0)_B - (q_f^0)_A]\,\rho k(1 - Y_B) = -\Delta h\,\rho k(1 - Y_B), \end{cases} \tag{10.181}$$

where $\Delta h = \Delta H/\mathcal{M}$ is the enthalpy of reaction per unit mass of the mixture ($\Delta H = -\sum_j \nu_j \mathcal{M}_j (q_f^0)_j$ was defined in Sect. 2.4.1 and utilized in Sect. 7.4), and $\mathcal{M} = \mathcal{M}_A = \mathcal{M}_B$ is the molar mass of (both) species A and B.

Finally,

$$\dot{m}\frac{d}{dx}(c_pT + \frac{v^2}{2}) - g\frac{d^2}{dx^2}(c_pT + \frac{v^2}{2}) = \Delta h\, \rho k\, (1 - Y_B). \tag{10.182}$$

By setting

$$\Theta = (c_pT + v^2/2)/c_pT_1,\ \tau = \Delta h/c_pT_1, \tag{10.183}$$

where Θ is a reduced stagnation temperature, we obtain an energy equation similar to that found for diffusion:

$$\dot{m}d\Theta/dx - gd^2\Theta/dx^2 = \tau\rho k(1 - Y_B) \ \ \text{(energy)} \tag{10.184}$$

with the boundary conditions

$$\begin{cases} x \to -\infty: \ Y_B = Y_{B1} = 0, \\ x \to +\infty: \ Y_B = Y_{B2} = 1. \end{cases} \tag{10.185}$$

The similarity between Y_B and Θ/τ leads us to the linear relation

$$Y_B = (\Theta - \Theta_1)/(\Theta_2 - \Theta_1). \tag{10.186}$$

We now set

$$\begin{cases} d\xi = \dot{m}\, dx/g & \text{(reduced abscissa)}, \\ G = d\Theta/d\xi & \text{(gradient of reduced stagnation temperature)}, \\ U = v/c_1 & \text{(reduced velocity)}, \end{cases} \tag{10.187}$$

and obtain

$$d\Theta/d\xi - d^2\Theta/d\xi^2 = g\tau\rho k(1 - Y_B)/\dot{m}^2. \tag{10.188}$$

$d\Theta/d\xi$ can be replaced with G, $\rho = \dot{m}/c_1U$ and $k = Be^{-T_a/T}$ can be expressed as functions of Θ and U, and Y_B can be expressed as a function of Θ. A differential equation in G, Θ and U is then obtained for the **energy equation**

$$\frac{dG}{d\Theta} = 1 - \frac{g\tau}{G\dot{m}^2} f(\Theta, U), \tag{10.189}$$

where

$$f(\Theta, U) = \frac{B\dot{m}}{c_1U}\frac{\Theta_2 - \Theta}{\Theta_2 - \Theta_1} \exp\left[-\frac{T_a}{T_1}(\Theta - \frac{\gamma - 1}{2}U^2)^{-1}\right]. \tag{10.190}$$

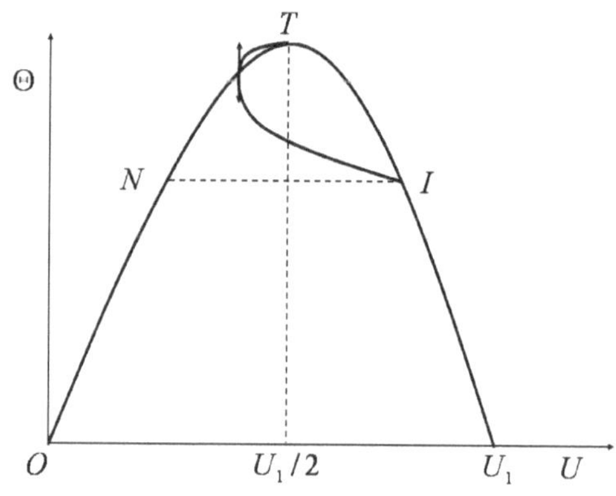

Fig. 10.19. Detonation in the Rayleigh diagram

The momentum equation (10.180) has a first integral and becomes

$$\dot{m}v + p - \frac{4}{3}\mu\frac{dv}{dx} = \dot{m}v_1 + p_1 \tag{10.191}$$

by assuming that motionless fresh gases are present at infinity upstream.

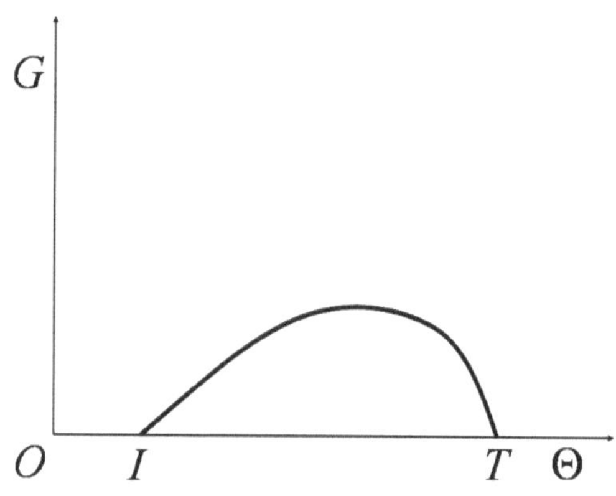

Fig. 10.20. Solution to the differential system

This equation becomes

$$\frac{dU}{d\xi} = \frac{p_1}{\dot{m}c_1}(\frac{p}{p_1} - 1) + U - U_1.$$

Using the relation

$$\frac{p}{p_1} = \frac{\rho}{\rho_1}\frac{T}{T_1} = \frac{U_1}{U}(\Theta - \frac{\gamma-1}{2}U^2)$$

leads to the following equation:

$$\frac{dU}{d\xi} = \frac{1}{U}(\frac{\Theta}{\gamma} + \frac{\gamma+1}{2\gamma}U^2) - \frac{1}{U_1}(\frac{\Theta_1}{\gamma} + \frac{\gamma+1}{2\gamma}U_1^2,) \tag{10.192}$$

and, by eliminating ξ, and because $dU/d\xi = G\,dU/d\Theta$, the **momentum equation** can be written in the form

$$dU/d\Theta = \frac{\Theta - g(U)}{\gamma U G}, \tag{10.193}$$

with

$$g(U) = \frac{U}{U_1}(\Theta_1 + \frac{\gamma+1}{2}U_1^2) - \frac{\gamma+1}{2}U^2. \tag{10.194}$$

We must therefore solve the system of two differential equations (10.189) and (10.193), where the variable is the reduced stagnation temperature characterized by Θ, and where the functions are the reduced velocity of the flow U and the reduced stagnation temperature gradient G.

The Rayleigh curve makes it possible to visualize the structure of the detonation wave (Fig. 10.19). It corresponds to the absence of a velocity gradient, which gives

$$dU/d\Theta = 0 \tag{10.195}$$

or

$$\Theta = g(U). \tag{10.196}$$

The initial and final points (infinity upstream and downstream) are located on the Rayleigh curve. The top of this curve corresponds to the Chapman–Jouguet point, and provides the final conditions. Indeed, at the top of the parabola, we have

$$\frac{d\,g(U)}{dU} = \frac{1}{U_1}(\Theta_1 + \frac{\gamma+1}{2}U_1^2) - (\gamma+1)U, \tag{10.197}$$

so

$$U = \frac{\Theta_1}{(\gamma+1)U_1} + \frac{U_1}{2}. \tag{10.198}$$

Moreover, at infinity downstream, we have

$$U_2 = \frac{v_2}{c_1} = \frac{v_2}{c_2}\frac{c_2}{c_1} = M_2\sqrt{\frac{T_2}{T_1}} = M_2\sqrt{\Theta_2 - \frac{\gamma-1}{2}U_2^2} \tag{10.199}$$

or

$$(1 - \frac{\gamma - 1}{2} M_2^2)\, U_2^2 = M_2^2\, \Theta_2 = M_2^2\, g(U_2). \tag{10.200}$$

With $M_2 = 1$ we thus obtain

$$U_2 = \frac{\Theta_1}{(\gamma + 1) U_1} + \frac{U_1}{2}. \tag{10.201}$$

This calculation shows that the top of the parabola corresponds well to the Chapman–Jouguet point.[26]

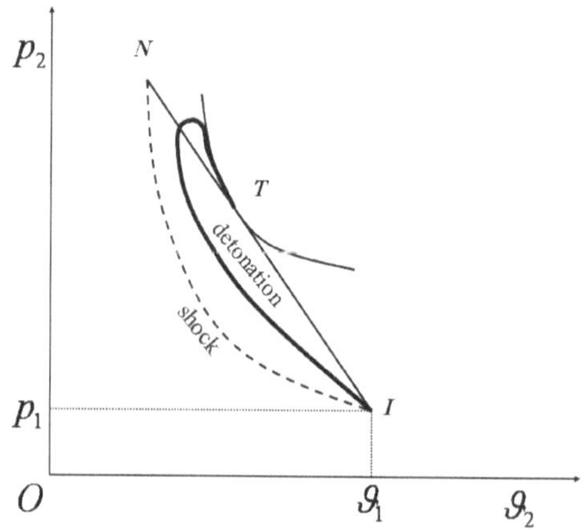

Fig. 10.21. Situation in the Clapeyron diagram

The detonation wave travels from I to T, following a curve which crosses the Rayleigh curve at a point with a vertical tangent, since the gradients $dU/d\Theta$ vanish along this curve. In an extreme case, the path followed would be INT, and the horizontal part would correspond to Θ =const. (i.e., to a shock without a chemical reaction), and path NT to the chemical reaction.

We can solve the differential system and determine the evolution of G as a function of Θ, just like that of U as a function of Θ; the curve shown in Fig. 10.20 is obtained. Figure 10.21 shows the situation in the Clapeyron diagram.

[26] In the Clapeyron plane, the Rayleigh curve corresponds to the equation $p/p_1 - 1 = \gamma U_1^2(1 - \rho_1/\rho)$, or $(p/p_1 - 1)/(\vartheta/\vartheta_1 - 1) = -\gamma M_1^2$. This is the straight line that passes by the initial point ($p/p_1 = \vartheta/\vartheta_1 = 1$). To perform this calculation we must simply replace U with $M_1 \vartheta/\vartheta_1$ and Θ with $\vartheta p/\vartheta_1 p_1 + ((\gamma - 1)/2) M_1^2 \vartheta^2/\vartheta_1^2$ in (10.196).

10.7 Spherical Waves

10.7.1 Case of Small Movements

Let us assume, as in Sect. 10.2.2, that the reference configuration is chemical equilibrium in a one-reaction mixture. The motion is governed by the equation in φ_1, which, in the case of spherical symmetry, becomes

$$\tau_{v0}\frac{\partial}{\partial t}[\frac{\partial^2\varphi_1}{\partial t^2} - \frac{c_{f0}^2}{r^2}\frac{\partial}{\partial r}(r^2\frac{\partial\varphi_1}{\partial r})] + \frac{\partial^2\varphi_1}{\partial t^2} - \frac{c_{e0}^2}{r^2}\frac{\partial}{\partial r}(r^2\frac{\partial\varphi_1}{\partial r}) = 0. \quad (10.202)$$

By simply setting

$$\varphi_1 = \psi_1/r, \quad (10.203)$$

we obtain a planar wave equation in the potential ψ_1

$$\tau_{v0}\frac{\partial}{\partial t}(\frac{\partial^2\psi_1}{\partial t^2} - c_{f0}^2\frac{\partial^2\psi_1}{\partial r^2}) + \frac{\partial^2\psi_1}{\partial t^2} - c_{e0}^2\frac{\partial^2\psi_1}{\partial r^2} = 0. \quad (10.204)$$

The monochromatic solutions are of the form

$$\varphi_1 = \frac{1}{r}\,\hat{\varphi}\,\exp\left[-\omega\gamma(\omega)r\right]\,\exp\left[i\omega(\frac{r}{c(\omega)} - t)\right], \quad (10.205)$$

where $\hat{\varphi}$ is a constant.

When c_e and c_f are close to each other, we have

$$\begin{cases} \omega\gamma(\omega) = \frac{c_{f0}-c_{e0}}{c_{e0}^2}\frac{\omega^2\tau_{v0}}{1+\omega^2\tau_{v0}^2} \\ c^2(\omega) = \frac{c_{e0}^2+\omega^2\tau_{v0}^2c_{f0}^2}{1+\omega^2\tau_{v0}^2}, \end{cases} \quad (10.206)$$

where $\gamma(\omega)$ and $c(\omega)$ are the attenuation and the speed of the wave, respectively. In addition to chemical damping, characterized by $\omega\gamma(\omega)$, there is geometrical damping that behaves as $1/r$. The density disturbance ρ_1 will take the same form as φ_1:[27]

$$\rho_1 = \frac{1}{r}\,\hat{\rho}\,\exp\left(-\omega\gamma r\right)\,\exp\left[i\omega(\frac{r}{c} - t)\right]. \quad (10.207)$$

Let us apply this result to the following experimental situation (Fig. 10.22, after [286]). Consider a container with two compartments, one on top of the other. The shock wave generated by a bullet moving at supersonic speed V_0 in the top compartment meets a series of small, equidistant holes located along a straight line $x'x$ in the wall separating the two compartments. The passage

[27] Wegener et al. [286] note that, for a medium with a number of nonequilibrium processes, the wavefront will still decay as $r^{-(n-1)/2}\exp(-r/\lambda)$ with $n = 1, 2, 3$ for plane, cylindrical, and spherical waves, respectively.

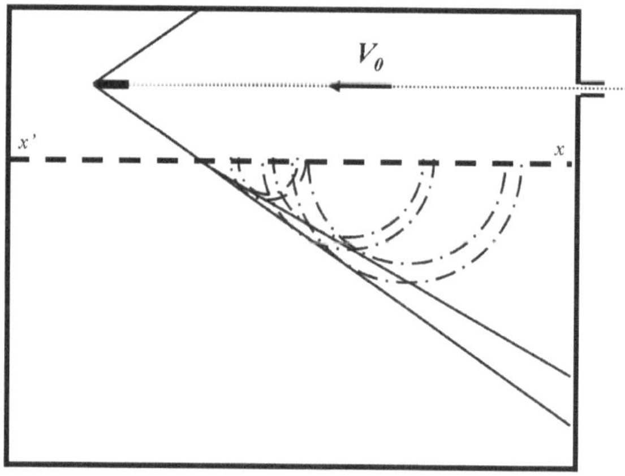

Fig. 10.22. Schematic of the Wegener, Chu and Klikoff experiment

of the shock wave above each small hole generates a wave with spherical symmetry in the lower container, which was initially in chemical equilibrium [286].

The spherical wave has an upper limit on its speed of c_{f0}, which is damped as $(1/r)e^{-iK_i r}$, where

$$K_i = \lim(\gamma\omega)_{\omega\to\infty} = (c_{f0} - c_{e0})/\tau_{v0}c_{e0}^2. \tag{10.208}$$

It also has a lower limit on its speed $c_{eo} < c_{f0}$, which is damped as $1/r$.

The hull of the upper limit on the speed is a half-cone with a vertex half-angle of α_f such that $\sin\alpha_f = c_{f0}/V_0$. The hull of the lower limit on the speed is a half-cone with the same top and the same axis, but with a half-angle of α_e such that $\sin\alpha_e = c_{e0}/V_0$ (Fig. 10.22).

The schlieren photographic technique highlights the geometrical and chemical damping of these waves and makes it possible to experimentally deduce the chemical time τ_{v0}. A good way to do this is to observe the damping along the external cone and to compare it with what occurs in a nonreactive medium.

10.7.2 Spherical Flames

Spherical diffusion flames are discussed in Chap. 12 in relation to droplet combustion. Spherical premixed flames can be ignited in a mixture of fuel and oxidizer by a spark. Formulae for the propagation of such a flame have been reported by several authors, and they avoid the complications associated with thermodiffusive instabilities [51, 202, 206]. We will only present a result from recent work on spherical deflagration wave instability here.

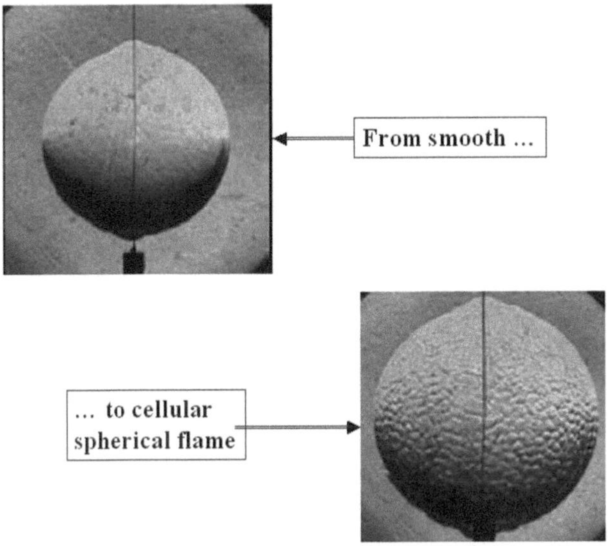

Fig. 10.23. Sequence of schlieren images for typical flame propagation, showing the transition from a smooth (t = 1375 μs, R = 1.95 cm) to a cellular (t = 1750 μs, R = 2.47 cm) spherical premixed flame (G. Jomaas, private communication, 2009; reprinted with permission from [128])

The linear stability of an unsteady spherical flame was studied analytically by Bechtold and Matalon [11], and in a recent paper Jomaas, Law and Bechtold [128] successfully compared theoretical results to experimental ones. The basic (corrected) formula for the instability threshold is:

$$Pe_c = R_c/\ell_f = Pe_1(\sigma) + Ze(Le - 1)Pe_2(\sigma), \tag{10.209}$$

where Pe_c is the critical Péclet number for transition to cellularity, R_c is the flame radius at transition, ℓ_f is the laminar flame thickness, $Ze = E_a(T_{ad} - T_u)/RT_{ad}^2$ is the Zel'dovich number, E_a is the overall activation energy, T_{ad} is the adiabatic flame temperature, and σ is the ratio of the density of the burned mixture to that of the unburned mixture. Pe_1 provides the contribution from hydrodynamic instability, and Pe_2 the additional influence of diffusional-thermal instability (see Sect. 8.2.1). Cellular instability (Fig. 10.23, after Jomaas et al. [128]) can appear during the growth of the flame radius, and the study shows the influence of each source of instability when a near-equidiffusive fuel (acetylene) or nonequidiffusive fuels (hydrogen and propane) are used.

10.7.3 Blast Waves

Let us consider a local explosion of energy E_0 in a motionless atmosphere characterized by ρ_1 and p_1 (Fig. 10.25). The wave is assumed to be very strong, so that the ratio p_2/p_1 of the pressures of the downstream and upstream flows is infinite [159, 251, 302]. For a shock wave, this corresponds to $M_1 \to \infty$. We will assume an ideal gas (with constant γ).

The classical balance relations at discontinuity are valid on both sides of the spherical shock wave obtained.

Fig. 10.24. Yakov Borisovich Zel'dovich, Soviet physicist (1914–1987) (image from the website Wikipedia)

Using Hugoniot's relation (10.31)

$$\frac{v_1^2}{2} + \frac{c_1^2}{\gamma - 1} = \frac{\gamma + 1}{2(\gamma - 1)} c^{*2}$$

and Prandtl's relation (10.34) $v_1 v_2 = c^{*2}$, and taking conservation of the mass flow rate into account, we get

$$\frac{v_2}{v_1} = \frac{\rho_1}{\rho_2} = \frac{2}{\gamma + 1} M_1^{-2} + \frac{\gamma - 1}{\gamma + 1}; \tag{10.210}$$

i.e., for a strong wave,

$$\frac{\rho_1}{\rho_2} = \frac{\gamma - 1}{\gamma + 1} = \frac{v_2}{v_1}. \tag{10.211}$$

The absolute velocity downstream of the spherical shock wave is thus

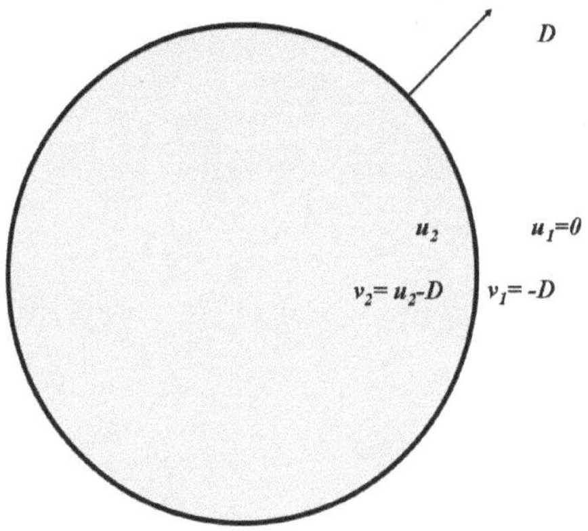

Fig. 10.25. Blast wave

$$u_2 = \frac{2}{\gamma + 1} D. \tag{10.212}$$

We also have (Hugoniot's relation)

$$\frac{v_2^2}{2} + \frac{c_2^2}{\gamma - 1} = \frac{v_1^2}{2} + \frac{c_1^2}{\gamma - 1}, \tag{10.213}$$

which, after dividing by c_1^2, gives

$$\frac{c_2^2}{c_1^2} = \frac{T_2}{T_1} = 1 + \frac{2(\gamma - 1)}{(\gamma + 1)^2} \frac{1}{M_1^2} (M_1^2 - 1), (\gamma M_1^2 + 1) \tag{10.214}$$

and, for very large Mach numbers,

$$\frac{T_2}{T_1} \cong \frac{2\gamma(\gamma - 1)}{(\gamma + 1)^2} M_1^2. \tag{10.215}$$

Finally,

$$\frac{p_2}{p_1} = \frac{\rho_2}{\rho_1} \frac{T_2}{T_1}, \tag{10.216}$$

which gives

$$\frac{p_2}{p_1} \cong \frac{2\gamma}{\gamma + 1} M_1^2 \tag{10.217}$$

for large M_1, or

$$p_2 \cong 2\rho_1 v_1^2/(\gamma + 1) = 2\rho_1 D^2/(\gamma + 1). \tag{10.218}$$

Let us seek self-similar solutions, with the only parameters being E_0, ρ_1, r and t (since p_1, which is much lower than p_2, can be ignored).

The single Π ratio is

$$\Pi_\rho = \rho_1 r^5 t^{-2}/E_0, \tag{10.219}$$

and the problem thus depends on the single dimensionless variable

$$\xi = r(\rho_1/E_0 t^2)^{1/5}. \tag{10.220}$$

We denote the value of ξ for the wavefront as ξ_0, and its distance from the center will be (for constant ξ_0)

$$R(t) = \xi_0 (E_0/\rho_1)^{1/5}\, t^{2/5}. \tag{10.221}$$

The speed D of the wave will be equal to

$$D = dR(t)/dt = 2/5\, \xi_0 (E_0/\rho_1 t^{-3})^{1/5} = 2R/5t \tag{10.222}$$

or

$$D = 2/5\, \xi_0^{2/5} (E_0/\rho_1)^{1/2} R^{-3/2}. \tag{10.223}$$

Therefore, just downstream of the wave (inside a sphere of radius $R(t)$) we have

$$\begin{cases} u_2(t) = \frac{4}{5(\gamma+1)}\, \xi_0 \left(\frac{E_0}{\rho_1}\right)^{1/5} t^{-3/5}, \\ p_2(t) = \frac{8\rho_1}{25(\gamma+1)}\, \xi_0^2 \left(\frac{E_0}{\rho_1}\right)^{2/5} t^{-6/5}, \\ \rho_2 = \frac{\gamma+1}{\gamma-1}\, \rho_1. \end{cases} \tag{10.224}$$

These relations provide the variations in u_2, p_2 and ρ_2 just downstream of the wave as functions of time.

We can also express these quantities as follows (taking into account Eq. 10.222):

$$\begin{cases} u_2(t) = \frac{4}{5(\gamma+1)}\, \frac{R}{t}, \\ p_2(t) = \frac{8\rho_1}{25(\gamma+1)}\, \frac{R^2}{t^2}, \\ \rho_2 = \frac{\gamma+1}{\gamma-1}\, \rho_1 = \text{const.} \end{cases} \tag{10.225}$$

The self-similar solutions will be such that the ratios u/u_2, p/p_2 and ρ/ρ_2 are functions of ξ only. Let us then write (as done by Landau)

$$\begin{cases} u(t) = \frac{4}{5(\gamma+1)}\, \frac{r}{t}\, u', \\ p(t) = \frac{8\rho_1}{25(\gamma+1)}\, \frac{r^2}{t^2}\, p', \\ \rho = \frac{\gamma+1}{\gamma-1}\, \rho_1\, \rho'. \end{cases} \tag{10.226}$$

For self-similar solutions, the new parameters u', p' and ρ' are indeed functions of the single variable ξ, since we have

$$r(t)/R(t) = \xi/\xi_0. \tag{10.227}$$

For $\xi = \xi_0$ (i.e., just downstream of the wave), we of course obtain $u' = p' = \rho' = 1$. Taking into account the spherical symmetry of this case and the fact that we have an inviscid fluid in adiabatic evolution, the classical mass and entropy balance equations that are valid inside the spherical wave give

$$\frac{\partial \rho}{\partial t} + \frac{1}{r^2}\frac{\partial(r^2 \rho u)}{\partial r} = 0, \tag{10.228}$$

$$\frac{\partial(p/\rho^\gamma)}{\partial t} + u\frac{\partial(p/\rho^\gamma)}{\partial r} = 0. \tag{10.229}$$

We then obtain

$$(u' - \frac{\gamma+1}{2})\frac{d\ln\rho'}{d\ln\xi} + \frac{du'}{d\ln\xi} = -3u' \ \text{(continuity)} \tag{10.230}$$

$$\frac{d\ln(p'/\rho'^\gamma)}{d\ln\xi} = \frac{5(\gamma+1) - 4u'}{2u' - \gamma - 1} \ \text{(entropy)} \tag{10.231}$$

with the new parameters. The entropy equation is still

$$(u' - \frac{\gamma+1}{2})(\frac{d\ln(p'/\rho'^\gamma)}{d\ln\xi} + 5) = 3u'. \tag{10.232}$$

Combination with the continuity equation yields

$$(u' - \frac{\gamma+1}{2})(\frac{d\ln(p'/\rho'^{\gamma-1})}{d\ln\xi} + 5) + \frac{du'}{d\ln\xi} = 0. \tag{10.233}$$

This relation is easily integrated and gives

$$\frac{p'}{\rho'^{\gamma-1}}(2u' - \gamma - 1) = -(\gamma-1)(\frac{\xi_0}{\xi})^5. \tag{10.234}$$

The energy equation (or the one for momentum) must be added to the preceding relations. However, it is more useful to obtain another first integral directly. To do this, we note that the conservation of total energy inside the sphere of radius $R(t)$ (corresponding to ξ_0) and the similarity of the problem imply that energy is preserved inside any sphere of radius $r(t)$ corresponding to ξ =const.

Consider r, the radius of a sphere of parameter ξ at time t; at time $t+dt$, this radius becomes

$$r(t+dt) \cong r + (dr/dt)dt, \tag{10.235}$$

where dr/dt will be considered at constant ξ

$$dr/dt = 2r/5t. \tag{10.236}$$

Investigating the energy balance between spheres of radii r and $r(t+dt)$ leads us to conclude that the total energy contained in this volume, $4\pi r^2 \rho(e + u^2/2)(dr/dt)dt$, is derived solely from the flow that crosses a sphere of radius r : $4\pi r^2\, \rho u\, (h + u^2/2)\, dt$. This leads to the relation

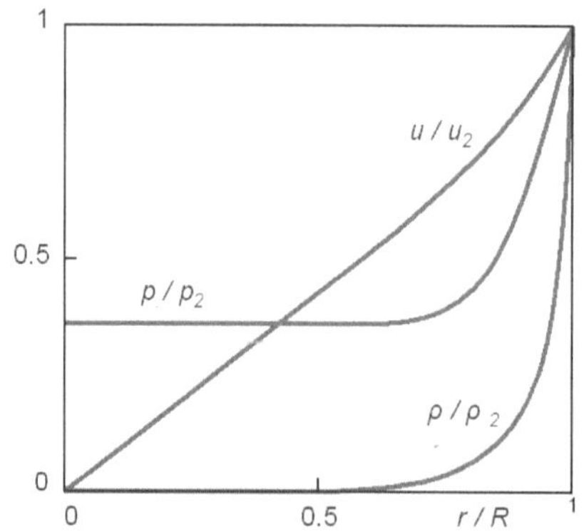

Fig. 10.26. Blast wave: evolutions of the reduced parameters (redrawn after [147])

$$u(h + u^2/2) = (e + u^2/2)\; 2r/5t. \tag{10.237}$$

Since $e = p/(\gamma - 1)\rho$ and $h = \gamma p/(\gamma - 1)\rho$, we have

$$\frac{p}{\rho} = \frac{\gamma - 1}{2} u^2 \frac{2r/5t - u}{\gamma u - 2r/5t}, \tag{10.238}$$

and by shifting to the parameters u', p', ρ', we get

$$\frac{p'}{\rho'} = u'^2 \frac{\gamma + 1 - 2u'}{2\gamma u' - \gamma - 1}. \tag{10.239}$$

This adiabaticity relation does not depend on ξ.

Relations 10.230, 10.234 and 10.239 allow us to solve this problem completely, taking into account the boundary conditions. Equations 10.234 and 10.239 permit us to express ρ' and p' according to u' and ξ. By introducing the value obtained for ρ' into (10.230), we can obtain a differential equation in u' and $du'/d\ln\xi$ that can easily be integrated.

Thus, knowing u', ρ' and p' as functions of ξ and ξ_0, we can then determine ξ_0 by writing the conservation of total energy in a sphere of radius $R(t)$:

$$\int_0^{R(t)} \rho(p/(\gamma-1)\rho + u^2/2)\, 4\pi r^2 dr = E_0.$$

This integral is expressed as a function of ξ_0. Calculations carried out for ambient air($\gamma = 7/5$) lead to the value $\xi_0 = 1.033$.

Landau and Lifschitz [147] (see also [219]) give the evolutions of u/u_2, ρ/ρ_2 and p/p_2, obtained numerically, as functions of r/R (equal to ξ/ξ_0; see Fig. 10.26).

11

Interface Phenomena

This chapter is devoted to interfaces of various kinds. In addition to interfaces that separate two immiscible pure fluids at mechanical and thermodynamic equilibrium, we consider moving surfaces with surface tension and internal viscosity. Heat and mass transfer can also occur through and along interfaces, and this can induce evaporation and condensation processes, chemical and thermal Marangoni effects, etc. Chemical reactions may also take place. Various miscibility scenarios can be considered too.

Some general information on interfaces is given in Sect. 11.1, which presents several types of interfaces along with an analysis of scales. Some thermodynamic aspects are also discussed.

The basic equation that characterizes the balance for a quantity F is established in Sect. 11.2 with the aid of specified assumptions. Choosing appropriate coordinates and an appropriate reference frame enables us to integrate—as far as we can—the three-dimensional equations for the interfacial layer in order to obtain the two-dimensional equations for material surfaces.

Simple interface balance laws are obtained in Sect. 11.3 for surface variables that obey the Gibbs relation.

The problem of closure relations is studied in Sect. 11.4. In some cases, constitutive relations are deduced directly from linear irreversible thermodynamics for two-dimensional interfaces. In other cases, it is necessary to perform integration through the interfacial layer to obtain closure relations.

In Sect. 11.5 we use the method of virtual power to obtain the momentum equation when the interface shows a certain degree of resistance to the strain, particularly when bending moments are applied to it.

Finally, in Sect. 11.6 we provide some background on second-gradient theory, which in particular allows us to link surface tension to a bulk coefficient called the capillarity coefficient.

R. Prud'homme, *Flows of Reactive Fluids*,
DOI 10.1007/978-0-8176-4659-2_11,

11.1 General Information About Interfaces

11.1.1 Interfaces and Interfacial Layers

Interfaces are material media that are similar to surfaces at the macroscopic scale. However, interfaces are not simply geometric surfaces. At a finer scale, down to the microscopic scale, the interface becomes an "interfacial layer"—a three-dimensional medium where the variations of some properties in the direction normal to the interface are definitely more significant than those in the direction tangential to it [94]. Thus, an "interface" and an "interfacial layer" differ in terms of scale length; the thickness of an interfacial layer is always small compared to a macroscopic reference length. Figure 11.1 shows the evolution of a quantity f across an interfacial layer and across a corresponding interface. For example, this quantity could be the density in a system consisting of a liquid and a gas separated by an interface, a solid plate, or a shell which has one of its surfaces in contact with the liquid and the other in contact with the gas. In such cases there are clear physical discontinuities at the macroscopic scale. However, fluid boundary layers or combustion waves can also be considered interfaces (and we will do so here), because parameters such as velocity, temperature or concentration vary strongly over a short distances across them.

Interface modeling is very often a complex problem. In theory, it involves reducing what occurs inside the thickness of the "interfacial layer" to a description of surface quantities and jumps in bulk properties. It is easy to see that such an approach can lead to errors, either because we lose too much information by doing so or because describing the suface properties of the medium is even more complicated than describing its bulk properties. Therefore, we will largely limit ourselves here to situations where it is known that this approach is appropriate, although we will seek to study cases that are as generally applicable as possible; i.e., we will assume moving and deformable interfaces. We will also extend the method to some other less familiar cases. The main task will be to establish the balance equations and their closure relations.

Some types of interface are presented in Sect. 11.1.2. Surface geometry and surface motions are described in Sects. A.3 and A.2 of Appendix A. Thermodynamic definitions are given in Sect. 11.1.3.

11.1.2 Types of Interfaces

In the field of physics, we can distinguish between three types of interface:

- The interface is the surface that separates two media with different natures or physical states. The most traditional of these is an interface that separates phases. One example is the surface of a moving liquid with surface tension.

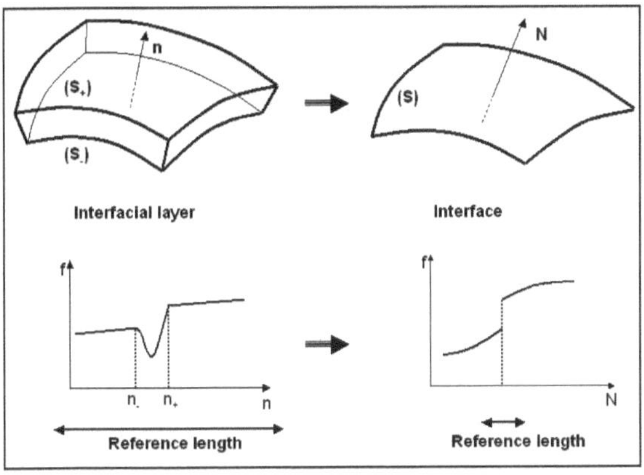

Fig. 11.1. From an interfacial layer to an interface

- The interface is a film (flexible or not) of matter that is in contact with fluids or solids on both of its surfaces. A membrane separating two fluids is a typical example. A deformable solid shell is another example (see Sect. 11.5).
- The interface is an interfacial zone of a single material medium that exhibits a strong gradient. This is the case for boundary layers of various types, contact surfaces, combustion waves and shock waves. For waves, a large amount of matter continuously crosses the interfacial zone so the transverse flow is important, while the transverse flow is weak in other cases (diffusion layers).

In all cases, we can consider exchanges of mass, momentum, heat and energy in the form of work inside the interface and between the interface and the bulk.

We will take a special interest in situations where the interfacial layer can be regarded as a continuous medium; i.e., where its analysis does not require shifting to the molecular scale. We will thus describe the medium at an intermediate scale, called sometimes a "mesoscopic" scale, which is sufficiently small but is still larger than the intermolecular distance (for gases, larger than the mean free path), using the classical equations of continuous media. The balance laws for the interface will be obtained by integrating these equations along a normal coordinate through the interfacial layer. In all cases, we will assume that the geometry of the interfacial layer is described by layers of surfaces (S) between limiting surfaces (S_-) and (S_+), at which the properties hardly vary (Fig. 11.2).

This integration will be facilitated by employing certain assumptions relating to the internal velocity field. It is generally more convenient to introduce a velocity field $\mathbf{V}$ that does not vary very much as the interfacial layer is crossed.

Let us refer to the jump equations to determine the most suitable velocity field. The momentum equation (4.99) given in Sect. 4.9 leads to two quite distinct situations. In the first, the normal flow is zero (this is called a contact surface). Here, the pressure does not vary and the normal component of the relative velocity of the surface is zero. In the second, the normal flow is nonzero (this is called a shock wave in the case of a simple compressible fluid). Here, the tangential velocity does not vary as the surface is crossed, but the speed in the normal direction does, and its variation is dependent on the pressure jump.[1]

The following velocity fields are thus possible for the interfacial layers:

1. For boundary layers (Chap. 9) and shear layers:

$$\mathbf{V} = \mathbf{v}_{\perp} \tag{11.1}$$

2. For shock waves, premixed flames, and evaporating surfaces (see Fig. 11.2 and Chap. 10):[2]

$$\mathbf{V} = \mathbf{v}_{//} + \mathbf{w}. \tag{11.2}$$

11.1.3 Thermodynamic Definitions

The thermostatic relations between internal energy, entropy, density and chemical concentrations obtained in Chap. 2 for three-dimensional media are valid for the bulk materials in contact with the two sides of the interfacial layer. They are also valid inside the interfacial layer for some types of interfaces.

---

[1]If the surfaces (S) follow the motion of the material (the local barycentric motion for a fluid mixture), we can consider the simplest case, where the velocity is constant as the interfacial layer is crossed. When modeling the interfacial layer, each point P on surface (S) is then characterized by a single velocity $\mathbf{V} = \mathbf{v}$ that only depends on the position of P on the surface.

[2]In this case where the interface is crossed by the fluid, in addition to the material velocity field $\mathbf{v}$ (the barycentric velocity for a mixture), it will be necessary to consider the velocity field $\mathbf{V}$ at point P. $\mathbf{V}$ is a continuous field with a component normal to S that is the displacement velocity of S in the normal direction (i.e., $\mathbf{w}$; see Sect. A.2; the component normal to (S_+) is the displacement velocity of (S_+) in the normal direction, i.e., $\mathbf{w}_+$; and the component normal to (S_-) at any point on (S_-) is the velocity of (S_-) in the normal direction, $\mathbf{w}_-$). Moreover, the tangential component of $\mathbf{V}$ is equal to that of $\mathbf{v}$. If we move from (S_-) to (S_+) through a continuous family of nonsecant surfaces (S) , each of these surfaces will have a local normal velocity $\mathbf{w}$.

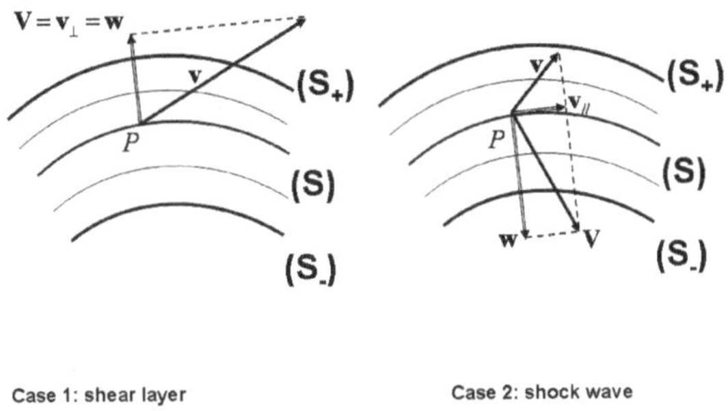

Fig. 11.2. Definition of the velocity vector inside an interfacial layer (2D representation)

They are not sufficient, however, to describe capillary interfacial layers. In this case, we need other thermostatic relations where the internal energy and the other thermodynamic potentials depend upon the density gradient too (see Sect. 11.6).

If f is the quantity F per unit mass and ρ is the density of the mixture, we define the mass per unit area of the surface (see Sect. A.2) as

$$\rho_a = \int_{C_3} \rho \, dN = \int_{N_-}^{N_+} \rho \, dN = \epsilon \int_{n_-}^{n_+} \rho \, dn, \qquad (11.3)$$

where $dN = \epsilon \, dn$ with $\epsilon = \delta/L \ll 1$; L is the hydrodynamic length and δ is a reference length that characterizes the variation of the internal physical process across the interface. The quantity n designates a measured length at the scale of the thickness δ of the interfacial layer, whereas N is the same length at the hydrodynamic scale.

We define the mean surface quantity per unit mass f_S at any point of the interface (Figs. 11.1, A.2):

$$\rho_a f_S = \int_{N_-}^{N_+} \rho f dN = \epsilon \int_{n_-}^{n_+} \rho f dn. \qquad (11.4)$$

Mathematically, a change of scale is very important because it can transform finite values (at a large scale) into infinite ones (at smaller scales). In particular, the convergence properties of integrals can be modified. In some cases it may be suitable to introduce excess quantities. If the interface is located on average at $n = 0$, and f_- and f_+ are the values of f in the bulk media on each side of the interface, then the excess quantity defined by

$$\rho_a \, \Delta f_S = \epsilon \int_{-\infty}^{0} (\rho f - \rho^- f^-) \, dn + \epsilon \int_{0}^{\infty} (\rho f - \rho^+ f^+) \, dn \tag{11.5}$$

has a finite value in general [94]. Excess quantities will not be considered in the following section, but it is necessary to consider them when studying stretched premixed flames such as those in Sect. 11.4.2 [94], for instance.

Thermostatic relations between the internal energy e_S, the entropy s_S, the density ρ_a and the chemical concentrations Y_{jS} are not evident a priori. They result from the integration of thermostatic relations through the three-dimensional interfacial layer. In some cases (see Sect. 11.3), simplified relations analogous to those for the bulk can be assumed (see Eqs. 11.28).

For interfacial media with internal capillarity, we must integrate equations of the type (11.89) that are valid for such media.

11.2 General Form of Interface Balance Laws

The aim of the concept of an interface is to reduce what occurs in a interfacial layer to a description relating to a material surface [65, 85, 95, 102, 259]. This requires the introduction of surface state parameters as well as flux densities and production rates. These new quantities obey constitutive relations that need to be determined in each case. Very often, this means that it is necessary to study the bulk of the interfacial layer, which explains why we previously proceeded to transform the bulk equations.

11.2.1 Appropriate Form of Bulk Equations

The equations of a continuous medium (single-component or mixed) can be written in the integral form

$$\frac{d_{\mathbf{V}}}{dt} \int_{\mathcal{V}} \rho f \, d\mathcal{V} + \int_{\partial \mathcal{V}} [\boldsymbol{\mathcal{J}}_F + \rho f (\mathbf{v} - \mathbf{V})] \cdot \mathbf{n} dS = \int_{\mathcal{V}} \dot{W}_F \, dV, \tag{11.6}$$

where $\mathbf{v}(\mathbf{x}, t)$ is the material velocity field of the continuous medium, and where $\mathbf{V}(\mathbf{x}, t)$ is an arbitrary velocity field that is continuous and differentiable. The derivative $d_{\mathbf{V}}/dt$ means that we follow the motion velocity $\mathbf{V}$ of the closed volume $(\mathcal{V})$. $\mathbf{n}$ is the normal external to the closed surface $(\partial \mathcal{V})$ surrounding the related volume $(\mathcal{V})$. As in Chap. 4, f is the quantity F per unit mass, $\boldsymbol{\mathcal{J}}_F$ is the flux density of quantity F, and $\dot{W}_F$ is the production rate of F. In local form, this equation becomes

$$\frac{d_{\mathbf{V}}(\rho f)}{dt} + \rho f \boldsymbol{\nabla} \cdot \mathbf{V} + \boldsymbol{\nabla} \cdot [\boldsymbol{\mathcal{J}}_F + \rho f (\mathbf{v} - \mathbf{V})] = \dot{W}_F. \tag{11.7}$$

Let us choose to describe the interfacial layer in an orthogonal curvilinear reference frame such as that defined by Eq. A.6, with the surfaces (S) corresponding to x_3 =const. If the velocity at point P on (S) is $\mathbf{v}$, its projection on the plane that is tangent to (S) at P is written

$$\mathbf{v}_{//} = (\mathbf{1} - \mathbf{N} \otimes \mathbf{N}) \cdot \mathbf{v}, \tag{11.8}$$

where $\mathbf{N}$ is the unit normal to (S) at P, oriented in a direction chosen in advance.

Moreover, the normal component of $\mathbf{v}$ is

$$v_{\perp} = \mathbf{N} \cdot \mathbf{v}. \tag{11.9}$$

If w is the speed of displacement of surface (S) in the normal direction, we can choose the field $\mathbf{V}(\mathbf{x}, t)$ such that

$$\mathbf{V} \cdot \mathbf{N} = \mathbf{w} \cdot \mathbf{N} \Rightarrow V_{\perp} = w. \tag{11.10}$$

The relative velocity is

$$\mathbf{u} = \mathbf{v} - \mathbf{V} \tag{11.11}$$

and its normal component becomes

$$u_{\perp} = v_{\perp} - w. \tag{11.12}$$

Also let us note that

$$\nabla \cdot \mathbf{V} = \nabla_{//} \cdot \mathbf{V} + \nabla_{\perp} \cdot \mathbf{V}, \tag{11.13}$$

with

$$\nabla_{\perp} \cdot \mathbf{V} = \frac{1}{h_3}(w_{,3} + V_1 \frac{h_{3,1}}{h_1} + V_2 \frac{h_{3,2}}{h_2}). \tag{11.14}$$

Since the interface is defined physically, let us choose to describe the interfacial layer as *a set of parallel surfaces* (S). This supposes that the order of magnitude of the interfacial thickness does not vary in space. It is a restrictive assumption. The set of surfaces (S) become deformed in time, but these surfaces remain parallel. h_3 then depends only on x_3, and the relations (A.63) are valid. If we also assume that *the velocity* $\mathbf{w}$ *is uniform along a common normal* $\mathbf{N}$, we find that

$$\nabla_{\perp} \cdot \mathbf{V} = 0, \; \nabla \cdot \mathbf{V} = \nabla_{//} \cdot \mathbf{V}. \tag{11.15}$$

The divergence of vector $\mathbf{V}$ represents the stretching rate for each surface (S), the points of which have the velocity $\mathbf{V}$. To ensure that this stretching rate is realistic, it is sufficient to choose $\mathbf{V}$ in such a manner that $\mathbf{V}_{//}$ is equal to $\mathbf{v}_{//}$.

Let us now consider an unspecified field of vectors $\boldsymbol{\varphi}(\mathbf{x}, t)$. We have

$$\boldsymbol{\nabla} \cdot \boldsymbol{\varphi} = \boldsymbol{\nabla} \cdot \boldsymbol{\varphi}_{//} + \varphi_{\perp} \boldsymbol{\nabla} \cdot \mathbf{N} + \frac{\partial \varphi_{\perp}}{\partial N} \tag{11.16}$$

according to (A.21).

However, according to the assumptions given in Sect. A.4,

$$\boldsymbol{\nabla}_{\perp} \cdot \boldsymbol{\varphi}_{//} = 0, \tag{11.17}$$

so that we finally obtain

$$\boldsymbol{\nabla} \cdot \boldsymbol{\varphi} = \boldsymbol{\nabla}_{//} \cdot \boldsymbol{\varphi}_{//} + \varphi_{\perp} \boldsymbol{\nabla} \cdot \mathbf{N} + \frac{\partial \varphi_{\perp}}{\partial N}. \tag{11.18}$$

Equation 11.7 then becomes

$$\begin{aligned} &d_{\mathbf{V}} \rho f / dt + \rho f \boldsymbol{\nabla}_{//} \cdot \mathbf{V} + \boldsymbol{\nabla}_{//} \cdot [\mathcal{J}_{F//} + \rho f u_{//}] \\ &+ (\mathcal{J}_{F\perp} + \rho f u_{F\perp}) \boldsymbol{\nabla} \cdot \mathbf{N} + (\partial / \partial N)(\mathcal{J}_{F\perp} + \rho f u_{F\perp}) = \dot{W}_F. \end{aligned} \tag{11.19}$$

Thus, by setting

$$\mathbf{J}_{\mathbf{V}F} = \mathcal{J}_F + \rho f \mathbf{u}, \tag{11.20}$$

we obtain the following final relation:

$$\frac{d_{\mathbf{V}} \rho f}{dt} + \rho f \boldsymbol{\nabla}_{//} \cdot \mathbf{V} + \boldsymbol{\nabla}_{//} \cdot \mathbf{J}_{\mathbf{V}F} + \partial J_{\mathbf{V}F\perp} / \partial N = \dot{W}_F, \tag{11.21}$$

where

$$\boldsymbol{\nabla}_{//} \cdot \mathbf{J}_{\mathbf{V}F} = \boldsymbol{\nabla}_{//} \cdot \mathbf{J}_{\mathbf{V}F//} + J_{\mathbf{V}F\perp} \boldsymbol{\nabla} \cdot \mathbf{N}. \tag{11.22}$$

We need to make assumptions in order to solve the balance equations inside the interfacial layer. The principal assumption relates to the orders of magnitude of the gradient components: for some quantities (depending on the problem), the order of magnitude of the normal gradient (i.e., gradient in the normal direction) is higher than that of the tangential gradient. Thus, for a medium with constant density, the normal density gradient is zero. In the same way, for interfaces of the wave type, the normal gradient of the tangential velocity is small. For a flame for example, the pressure gradient is negligible, whereas the temperature and reactant concentration gradients are high. For a capillary surface, the pressure gradient is high, as is the density gradient.[3] Pressure and temperature gradients are high for a shock wave or a detonation wave. These examples illustrate the wide range of situations that

[3] An interfacial layer is a capillary medium; see Sect. 11.6

are possible, and we cannot hope to solve the most general problem without addressing physical considerations a priori.

In addition, we defined the interfacial layer as a set of parallel surfaces (S), which enabled us to obtain (11.21). However, an interface is not always a layer that is delimited by two parallel surfaces (S_-) and (S_+). It follows that by dictating that the surfaces are parallel, we will see some normal gradients that vary in order of magnitude from one point to another. Nevertheless, we will limit ourselves here to cases where this is not the case.

Certain terms that are not normal gradients can be of a similar order of magnitude to them. This is the case for certain production rates (chemical or volume for example), and for the stretching rate of a planar flame in front of a stagnation plane (see Sect. 7.7).

The radius of curvature of the interface will always be assumed to be very large compared to the interfacial thickness. This is one of the conditions that makes it possible to consider the medium (the interfacial layer) as a surface. Zones with a small radius of curvature require special study, and this will not be performed here (see [206] for how to solve this complex problem for premixed flames).

In spite of these difficulties, we will see that a number of interfaces are accurately described by surface balance equations.[4]

11.2.2 General Balance Equation of an Interface

Equation 11.21 is easily integrated into the crossing of the interfacial layer if the velocity $\mathbf{V}$ and the differential operator $\boldsymbol{\nabla}_{//}$ are conservative along normal curves; i.e., they do not change if x_3 changes between $x_{3-} = N_-$ and $x_{3+} = N_+$ for a given x_1, x_2, t (see Sect. A.3). We can set[5]

$$\mathbf{V} = \mathbf{V}_S, \; \boldsymbol{\nabla}_{//} = \boldsymbol{\nabla}_S. \tag{11.23}$$

Let us also set (see Eq. 11.20 for the definition of $\mathbf{J}_{\mathbf{V}F}$)

$$\begin{cases} \rho_a f_S = \int_{N_-}^{N_+} \rho f \, dN, \\ \mathbf{J}_{\mathbf{V}Fa} = \int_{N_-}^{N_+} \mathbf{J}_{\mathbf{V}F} \, dN, \\ \dot{W}_{Fa} = \int_{N_-}^{N_+} \dot{W}_F \, dN. \end{cases} \tag{11.24}$$

[4]Finally, note that, whenever the interface exhibits a certain degree of resistance to pleating, it is necessary to establish momentum balance equations that relate to not only resultants of forces but also to torque moments. The preferable approach in this case is to use the virtual power method. This is presented in Sect. 11.5.

[5]Note that integrating the bulk equations without this latter restrictive hypothesis leads to the same form (11.25) of balance equation for the property F of an interface [94]. The result is obtained as an approximation, taking into account the thinness of the interfacial layer in particular. Also note that, in the cited work, a material velocity vector $\mathbf{v}_S$ is introduced for the interface such that $\rho_a \mathbf{v}_S = \int_{N_-}^{N_+} \rho \mathbf{v} dN$.

We obtain, with the aid of the previous assumptions (in particular the fact that the thickness $N_+ - N_-$ is constant and uniform[6]),

$$\frac{d_S\left(\rho_a f_S\right)}{dt} + \rho_a f_S\, \boldsymbol{\nabla}_S \cdot \mathbf{V}_S + \boldsymbol{\nabla}_S \cdot \mathbf{J}_{\mathbf{V}Fa} + \left[\mathbf{J}_{\mathbf{V}F\perp}\right]_-^+ = \dot{W}_{Fa}. \qquad (11.25)$$

Here, the quantity $\rho_a f_S$ is the surface density of F (the quantity F per unit area), $\mathbf{J}_{\mathbf{V}Fa}$ is the vector density of flux of F per unit length, and $\dot{W}_{Fa}$ is the surface production rate of F. The jump $[\mathbf{J}_{\mathbf{V}F\perp}]_-^+$ is also equal to

$$\left[\mathcal{J}_{F\perp} + \rho f\, \mathbf{u}_\perp\right]_-^+ = \left[(\boldsymbol{\mathcal{J}}_F + \rho f(\mathbf{v} - \mathbf{V})) \cdot \mathbf{N}\right]_-^+. \qquad (11.26)$$

If, instead of deducing (11.25) from the bulk equation (11.21), we postulate the existence of interfacial quantities, we can also obtain (11.25) directly. We initially perform an integral balance of the property F on the portion of surface (Σ) bounded by the curve (C) (Fig. 11.3). We then deduce the local equation with the aid of some mathematical transformations (see [10, 217]).

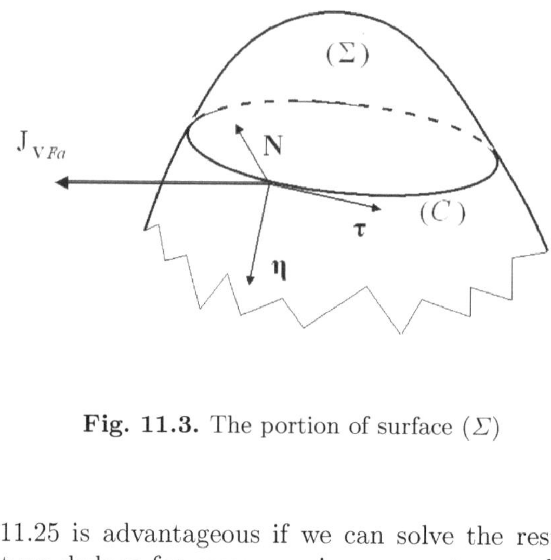

Fig. 11.3. The portion of surface (Σ)

Equation 11.25 is advantageous if we can solve the resulting system of equations that we deduce for mass, species, momentum and energy directly, using initial and boundary conditions.

This implies that—with the aid of an appropriate choice for the velocity $\mathbf{V}$ that obeys (11.10)—we know constitutive relations for the interface. In some cases, these constitutive relations are obtained experimentally (for example

[6] In particular, in this cases, if we set $d_{\mathbf{V}_S}/dt = d_S/dt$ to simplify the notation, we have $d_S\left(\int_{N_-}^{N_+} \rho f dN\right)/dt = \int_{N_-}^{N_+} \left[d_{\mathbf{V}}(\rho f)/dt\right] dN$ because $d_{\mathbf{V}}(dN)/dt$ vanishes.

surface tension); in other cases, they result from the asymptotic analysis of the system deduced from (11.21) (e.g., premixed flames; see Sect. 10.5.2, etc.).

The equations for the bulk in contact with the interface will have also to be determined. In certain cases, the jump $[J_{VF\perp}]$ also obeys a constitutive relation that can be determined a priori (adsorption–desorption). We will now provide some examples to illustrate these remarks.

11.3 Interface Balance Laws When Surface Variables Obey Classical Thermodynamic Relations

Generally, the thermodynamic states of a two-dimensional interface depend not only on the thermodynamic variables of the surface but also on those of the volumes in contact with it [65]. This is what we find when we integrate the bulk variables along the normal coordinate N, as in the first relation of (11.24). For the interfaces considered in this section, the surface variables obey thermodynamic laws that are similar to those obeyed by the bulk variables [260]. *This assumes that these interfaces behave, at equilibrium, like autonomous systems.* This is the case for some real interfaces, and it is important to know how to study them. We initially write down the laws of state by regarding the interface variables as being autonomous with respect to the bulk variables. The complementary constitutive relations are then obtained from the expression for the surface entropy production rate.

11.3.1 Classical Thermodynamic Relations

Thus, the laws of state obey[7]

$$\begin{cases} \rho s = (1/T)\, \rho\, e + p/T - \sum_j (g_j/T)\, \rho_j, \\ d\rho s = (1/T)\, d(\rho\, e) - \sum_j (g_j/T)\, d\rho_j \end{cases} \tag{11.27}$$

in the bulk volumes in contact with the interface, and

$$\begin{cases} \rho_a s_S = (1/T)\, \rho_a e_S + \sigma/T_S - \sum_j (g_{jS}/T_S)\, \rho_{ja}, \\ d\rho_a s_S = (1/T)\, d\rho_a e_S - \sum_j (g_{jS}/T_S)\, d\rho_{ja} \end{cases} \tag{11.28}$$

at the interface.[8]

[7] ρf is the density of F; i.e., the value of F per unit volume. $\rho_a f_S$ is the surface density of F as defined by (11.24).

[8] Thermodynamic relations were given in Sect. 2.6, but for molar quantities, denoted $\bar{F}_a$. The two types of relations obtained are similar, but (11.28) is better suited to the following study of interface balance laws.

11.3.2 Interface Balance Laws for Species, Mass, Momentum and Energy

If $\mathbf{V}_S = \mathbf{v}_S$, the surface material velocity vector, the balance equations are

$$\begin{cases} d_S\rho_a/dt + \rho_a \boldsymbol{\nabla}_S \cdot \mathbf{v}_S + [\rho u_\perp]^+_- = 0 \quad \text{(surface mass)}, \\ d_S(\rho_a Y_{jS})/dt + \rho_a Y_{jS} \boldsymbol{\nabla}_S \cdot \mathbf{v}_S + [\mathcal{J}_{j\perp} + \rho_j u_\perp]^+_- + \boldsymbol{\nabla}_S \cdot \boldsymbol{\mathcal{J}}_{ja} \\ = \dot{W}_{ja} \quad \text{(surface mass of the species)}, \\ d_S(\rho_a \mathbf{v}_S)/dt + \rho_a \mathbf{v}_S \boldsymbol{\nabla}_S \cdot \mathbf{v}_S + [\mathbf{P} \cdot \mathbf{N} + \rho \mathbf{v} u_\perp]^+_- - \boldsymbol{\nabla}_S \cdot \boldsymbol{\sigma} \\ = \sum_j (\rho_a Y_{jS}) \mathbf{f}_{jS} \quad \text{(surface momentum)}, \\ d_S(\rho_a e_S + \rho_a k_S)/dt + (\rho_a e_S + \rho_a k_S) \boldsymbol{\nabla}_S \cdot \mathbf{v}_S \\ +[q_\perp + \mathbf{v} \cdot \mathbf{P} \cdot \mathbf{N} + \rho(e+k)u_\perp]^+_- + \boldsymbol{\nabla}_S \cdot (\mathbf{q}_a - \mathbf{v}_S \cdot \boldsymbol{\sigma}) \\ = \sum_j (\rho_a Y_{jS}) \mathbf{v}_{jS} \cdot \mathbf{f}_{jS} \quad \text{(total surface energy)}, \end{cases} \tag{11.29}$$

with

$$\begin{cases} \rho_a k_S = \rho_a v_S^2/2, \\ d_S\,(\rho_a f_S)/dt = d_{\mathbf{v}_S}\,(\rho_a f_S)/dt = \partial(\rho_a f_S)/\partial t + \mathbf{v}_S \cdot \boldsymbol{\nabla}(\rho_a f_S). \end{cases} \tag{11.30}$$

Multiplying the momentum equation by $\mathbf{v}_a$ yields the kinetic energy balance

$$\begin{cases} d_S(\rho_a k_S)/dt + \rho_a k_S \boldsymbol{\nabla}_S \cdot \mathbf{v}_S + [\mathbf{v} \cdot \mathbf{P} \cdot \mathbf{N} \\ +\rho_K u_\perp]^+_- - \boldsymbol{\nabla}_{//} \cdot (\mathbf{v}_S \cdot \boldsymbol{\sigma}) = \sum_j \rho_{ja} \mathbf{v}_S \cdot \mathbf{f}_{jS} \\ -\boldsymbol{\sigma}\text{:}\, \boldsymbol{\nabla}_{//} \otimes \mathbf{v}_S + [\frac{1}{2}\rho(\mathbf{v} - \mathbf{v}_S)^2 u_\perp + (\mathbf{v} - \mathbf{v}_S) \cdot \mathbf{P} \cdot \mathbf{N}]^+_-. \end{cases} \tag{11.31}$$

Subtracting from the energy equation leads to

$$\begin{cases} d_S(\rho_a e_S)/dt + \rho_a e_S \boldsymbol{\nabla}_S \cdot \mathbf{v}_S + [q_\perp + \rho e u_\perp]^+_- \\ +\boldsymbol{\nabla}_S \cdot \mathbf{q}_a = \sum_j \boldsymbol{\mathcal{J}}_{ja} \cdot \mathbf{f}_{jS} \\ +\boldsymbol{\sigma}\text{:}\, \boldsymbol{\nabla}_S \otimes \mathbf{v}_S - [\frac{1}{2}\rho(\mathbf{v} - \mathbf{v}_S)^2 u_\perp + (\mathbf{v} - \mathbf{v}_S) \cdot \mathbf{P} \cdot \mathbf{N}]^+_-. \end{cases} \tag{11.32}$$

11.3.3 Interfacial Entropy Production

Taking into account the state relations, we can write

$$\begin{cases} d_a(\rho_a s_S)/dt = (1/T_S)\, d_a(\rho_a e_S)/dt - \sum_j (g_{jS}/T_S)\, d_S \rho_{ja}/dt, \\ \boldsymbol{\mathcal{J}}_{Sa} = (1/T_S)\, \mathbf{q}_a - \sum_j (g_{jS}/T_S)\, \boldsymbol{\mathcal{J}}_{ja}, \\ \boldsymbol{\mathcal{J}}_S = (1/T)\, \mathbf{q} - \sum_j (g_j/T)\, \boldsymbol{\mathcal{J}}_j. \end{cases} \tag{11.33}$$

The entropy balance is

$$\frac{d_S\,(\rho_a s_S)}{dt} + \rho_a s_S \boldsymbol{\nabla}_S \cdot \mathbf{v}_S + [\mathcal{J}_{S\perp} + \rho_S u_\perp]_-^+ + \boldsymbol{\nabla}_S \cdot \boldsymbol{\mathcal{J}}_{Sa} = \dot{W}_{Sa}. \tag{11.34}$$

Finally, using all of the preceding relations and eliminating the entropic variables yields the surface entropy production rate:

$$\begin{aligned} \dot{W}_{Sa} &= (1/T_S)\, \boldsymbol{\sigma}_v\colon \boldsymbol{\nabla}_S \otimes \mathbf{v}_S + \mathbf{q}_a \cdot \boldsymbol{\nabla}_S (1/T_S) \\ &- \textstyle\sum_j \boldsymbol{\mathcal{J}}_{ja} \cdot [\boldsymbol{\nabla}_a (g_{jS}/T_S) - (\mathbf{f}_{jS}/T_S)] - \sum_j (g_{jS}/T_S)\, \dot{W}_{ja} \\ &+[(q_\perp + \rho h u_\perp)(1/T - 1/T_S) \\ &- \textstyle\sum_j (\mathcal{J}_{j\perp} + \rho_j u_\perp)(g_j/T - g_{jS}/T_S) \\ &-(1/T_S)\, [(\mathbf{v} - \mathbf{v}_S) \cdot \boldsymbol{\Pi} \cdot \mathbf{N} - (1/2)\, \rho(\mathbf{v} - \mathbf{v}_S)^2 u_\perp]_-^+, \end{aligned} \tag{11.35}$$

where the enthalpy density $\rho h = \rho e + p$, the viscous pressure tensor $\boldsymbol{\Pi} = \mathbf{P} - p\mathbf{1}$, and the viscous surface stress tensor $\boldsymbol{\sigma}_v = \boldsymbol{\sigma} - \sigma(\mathbf{1} - \mathbf{N} \otimes \mathbf{N})$.

The first four terms of the right hand side represent the respective effects of the surface viscosity, thermal transfer, species diffusion, and surface chemical reactions[16].

The jump term $[\ ]_-^+$ relates, as in Sect. 4.9, to exchanges between the interface and the volumes in contact with it (successively: energy, mass, momentum, and kinetic energy exchanges).

The last term, which is third degree in the relative velocity, is generally negligible. It must be neglected in a theory that does not take into account terms with the diffusion velocity squared. The term $(\mathbf{v} - \mathbf{v}_a)$ is like a diffusion velocity of the fluid relative to the interface, and so the square of this term must vanish in the entropy production expression.

11.4 Constitutive Relations of Interfaces

11.4.1 Constitutive Relations Deduced Directly from Linear Irreversible Thermodynamics for Two-Dimensional Interfaces

Generalities

Generalized fluxes and forces are apparent in the expression for $\dot{W}_{Sa}$ given in Sect. 11.3 (Eq. 11.35). These must be rearranged to highlight the phenomena that are actually independent.

$\boldsymbol{\sigma}_v$ is a symmetric tensor such that the contracted product of it with $\mathbf{N}$ (on the left or on the right) vanishes. The velocity gradient tensor can be replaced with that for the deformation rates (or the strain rate tensor)

$$\mathbf{D}_S = \frac{1}{2}[(\boldsymbol{\nabla}_S \otimes \mathbf{v}_S) \cdot \mathbf{1}_{//} + \mathbf{1}_{//} \cdot (\widetilde{\boldsymbol{\nabla}_S \otimes \mathbf{v}_S})]. \tag{11.36}$$

We will separate the terms in the g_{jS}/T_S gradient that depend on temperature from those that do not depend on temperature. The terms that do depend on temperature will be gathered together with the second term of $\dot{W}_{Sa}$ in $\boldsymbol{\nabla}_S(1/T_S)$ to highlight thermal conduction. Such a rearrangement is similar to that performed for the entropy production in bulk (see Sect. 7.3). For the jump term, we will also gather the factor of $1/T - 1/T_S$ in $g_j/T - g_{jS}/T_S$ together with the term of thermal transfer.

At equilibrium, the generalized fluxes and forces vanish. The generalized forces are $\mathbf{D}_S/T_S$, $\boldsymbol{\nabla}_S(1/T_S)$, $(\boldsymbol{\nabla}_S(g_{jS}/T_S) - \mathbf{f}_{jS}/T_S)$, surface chemical affinities, $(1/T - 1/T_S)$, $(g_j/T - g_{jS}/T_S)$, and $-(1/T_S)(\mathbf{v} - \mathbf{v}_S)$.

The constitutive relations are obtained (except in the case of chemical kinetics) by writing linear relations between fluxes and forces which assume that the forces are of low intensity.

The simplified theory provided here, where interfaces are assumed to be autonomous, applies rather well to capillary surfaces, to adsorption–desorption phenomena, to some membranes, to elastic and viscoelastic films, to evaporating surfaces, etc. To solve it, the laws of state must be known in each case, as well as the constitutive relations and their transfer coefficients.

Note that some parts of a given interfacial medium can be regarded as simplified interfaces while others cannot. This is the case for an evaporating layer with diffusion: the surface of the liquid is often considered a simplified interface, but we cannot consider the diffusion layer above it in this way.

Constitutive Relations for Momentum

Surface Stresses

Let us consider a liquid surface in contact with a gas that neither evaporates nor condenses, so the only exchanges that occur involve momentum. The surface mass and the surface momentum are negligible, so the terms in $\rho_a f_S$ disappear. The relative normal velocity $u_\perp = v_\perp - w$ vanishes and the surface stress tensor $\boldsymbol{\sigma}$ is such that

$$\boldsymbol{\sigma} \cdot \mathbf{N} = \mathbf{0}, \tag{11.37}$$

where the direction of the normal $\mathbf{N}$ is unchanged from one side of the interface to the other. We then obtain the following for the momentum:

$$[\mathbf{P} \cdot \mathbf{N}]_-^+ - \boldsymbol{\nabla}_S \cdot \boldsymbol{\sigma} = \mathbf{0}. \tag{11.38}$$

The pressure tensor $\mathbf{P}$ incorporates the thermodynamic pressure and the viscous tension. The tensor $\boldsymbol{\sigma}$ includes the thermodynamic tension and possibly a surface viscosity (see later).

In the equilibrium case, $\mathbf{P} = p\mathbf{1}$ and $\boldsymbol{\sigma} = \sigma(\mathbf{1} - \mathbf{N} \otimes \mathbf{N}) = \sigma\,\mathbf{1}_{//}$, so that

$$[p]_-^+ \,\mathbf{N} - \boldsymbol{\nabla}_S \sigma + \sigma \boldsymbol{\nabla} \cdot \mathbf{N}\,\mathbf{N} = \mathbf{0}. \tag{11.39}$$

By projecting onto the tangent plane we can see that $\boldsymbol{\nabla}_S \sigma$ is equal to zero (so σ =const. at rest), and by projecting onto the normal plane we observe that

$$[p]_-^+ + \sigma \boldsymbol{\nabla}_S \cdot \mathbf{N} = 0. \tag{11.40}$$

This is the *Laplace relation.* This relation was obtained through thermostatic considerations in Sect. 2.6 and used to study a vapor bubble inside a liquid under zero-gravity conditions. Equation 11.38 was used with $\boldsymbol{\sigma} = \sigma(\mathbf{1}-\mathbf{N}\otimes\mathbf{N})$ and a temperature-dependent surface tension in Sect. 7.9, in the study of the Marangoni effect and the Bénard–Marangoni instability.

Surface Viscosities

In the presence of surface viscosities, we have $\boldsymbol{\sigma}_v = \boldsymbol{\sigma}_{v//} = \boldsymbol{\sigma} - \sigma\,\mathbf{1}_{//}$ for the viscous part of the interfacial stress tensor, which is not equal to zero. We assume (just as Slattery did) that $\boldsymbol{\sigma}_v$ is a linear function of the gradient $\boldsymbol{\nabla}_S \otimes \mathbf{v}_S$ (Boussinesq fluid surface; see [260]). Requiring objectivity and isotropy yields

$$\boldsymbol{\sigma}_{v//} = [(\kappa_a - \epsilon_a)\boldsymbol{\nabla}_S \cdot \mathbf{v}_S]\,\mathbf{1}_{//} + 2\epsilon_a \mathbf{D}_S, \tag{11.41}$$

where $\mathbf{1}_{//}$ corresponds to the point considered on the interface, κ_a is the surface dilatational viscosity, ϵ_a is the surface shear viscosity, and $\mathbf{D}_S$ is a surface strain tensor defined by (11.36). This tensor is such that $\mathbf{1}_{//}\cdot\mathbf{D}_S\cdot\mathbf{1}_{//} = \mathbf{D}_S$. The surface viscosity coefficients are functions of interface state variables. In a more sophisticated theory, we can imagine that surface viscosities are functions of the tensorial invariants of $\mathbf{D}_S$; i.e., $\boldsymbol{\nabla}_S \cdot \mathbf{v}_S$ and $\mathrm{tr}\,(\mathbf{D}_S \cdot \mathbf{D}_S)$ [260].

The surface viscosity term for the entropy production, which is present in (11.41) and (11.36), is

$$\begin{cases} \boldsymbol{\sigma}_v\mathbf{:}\ \boldsymbol{\nabla}_S \otimes \mathbf{v}_S = [(\kappa_a - \epsilon_a)\boldsymbol{\nabla}_S \cdot \mathbf{v}_S]\mathbf{1}_{//} + 2\epsilon_a \mathbf{D}_S \mathbf{:} \boldsymbol{\nabla}_S \otimes \mathbf{v}_S \\ = (\kappa_a - \epsilon_a)(\boldsymbol{\nabla}_S \cdot \mathbf{v}_S)^2 + 2\epsilon_a \mathbf{D}_S\mathbf{:}\mathbf{D}_S. \end{cases} \tag{11.42}$$

It is positive or zero for the entire $\mathbf{v}_S$ field. By writing $\mathbf{D}_S = \frac{1}{2}(\boldsymbol{\nabla}_S \cdot \mathbf{v}_S)\mathbf{1}_{//} + \boldsymbol{\Delta}_S$, where $\boldsymbol{\Delta}_S$ is a 2D tensor with zero trace, we obtain

$$\kappa_a \geq 0,\ \ \epsilon_a \geq 0.$$

Relation 11.41 is an example of a constitutive relation that includes surface viscosities. The viscosity coefficients are functions, in particular, of the stretch $\nabla_S \cdot \mathbf{v}_S$.

Scriven and Sternling [250] studied the Marangoni instability in a more complex situation than Pearson [197] did (see Sect. 7.9), taking into account surface deformations and interface viscosity coefficients. Here, the unit vector $\mathbf{N}$ is normal to the surface and oriented from the liquid phase to the gas phase, and the interface boundary conditions at $z = f(x, y, t)$ are more complex than in (7.127):

$$\begin{cases} \nabla \cdot \mathbf{N} = \partial f/\partial t, \ \mathbf{N} \cdot \nabla T + \Lambda(T - T_\infty) = 0, \\ [\mathbf{P}]_-^+ - \nabla_S \cdot \sigma(\mathbf{1} - \mathbf{N} \otimes \mathbf{N}) - \nabla_S \cdot \boldsymbol{\sigma}_v = 0. \end{cases} \tag{11.43}$$

In addition to the Nusselt and Marangoni numbers, other dimensionless numbers are also used: the surface viscosity number $V_i = (\kappa_a + \epsilon_a)/\mu\, d$, and the capillary (or crispation) number $Cr = \mu\kappa/\sigma^*\delta$ (see Table 5.3).

Neutral stability curves obey the relation

$$Ma = \frac{8\alpha(\alpha\cosh\alpha + Nu\,\sinh\alpha)[\alpha - \sinh\alpha\cosh\alpha + (V_i/2)\,\alpha\,(\alpha^2 - \sinh^2\alpha)]}{\alpha^3\cosh\alpha - \sinh^3\alpha - 8\,Cr\alpha^3\,\sinh}. \tag{11.44}$$

This result can be compared to Pearson's; see (7.132). The neutral stability curves of Scriven and Sternling [250], which represent neutral stability with respect to stationary disturbances, differ from those of Pearson. For example, in contrast to what is observed for a planar surface, there is no longer a critical Marangoni number in the case of an isothermal bottom, whatever the Nusselt number, except at the limit of $Cr = 0$ (an example is given in Fig. 11.4). If $Vi = 0$ and $Cr = 0$, the planar surface case of Fig. 7.24 is obtained again. The same conclusion (i.e., that there is no critical Marangoni number for $Cr \neq 0$) remains valid in all cases with an isothermal or isolated bottom.

Surface Reactions, Adsorption–Desorption, and Evaporation

A succinct discussion of surface reactions, adsorption-desorption and evaporation phenomena is now presented.

Surface Chemical Reactions

Many types of surface chemical reactions are possible. In [94], coupling involving the catalytic surface reaction $A_2 \longrightarrow 2A$ is considered. The fluid in the bulk, which consists of A_2 and A, is in contact with a catalytic planar interface. The reaction is not athermal and results in the release of heat, and thus heat transfer. Viscosity is ignored. The medium under the interface is assumed to be impermeable, adiabatic and at rest ($\mathbf{v} = \mathbf{0}$). At the interface

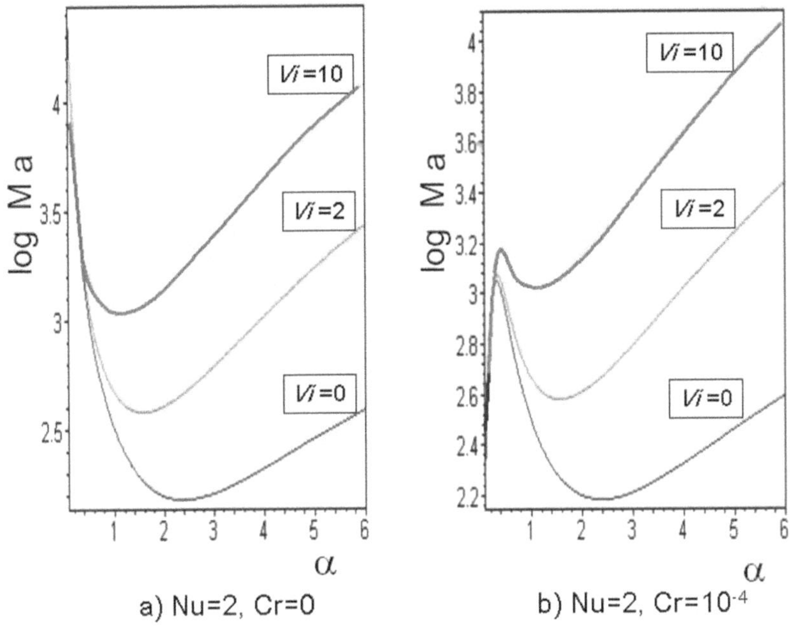

Fig. 11.4. Neutral stability curves obtained for a free surface with $Nu = 2$ and an isothermal bottom: **a** in the case of a planar surface ($Cr = 0$); **b** in the case of a free surface ($Cr = 10^{-4}$)

the temperature is T_S and $\mathbf{v}_S = \mathbf{V}_S = \mathbf{0}$. In this case, the general expression (11.35) for the production at the interface, ignoring surface mass diffusion and terms involving the square of the velocity, involves five generalized forces: the affinity of the chemical reaction A_S, the jump in the inverse temperature between the fluid and the surface $(1/T - 1/T_S)$, the jumps $(g_S^A/T_S - g^A/T)$ (this term relating to the interaction between the bulk and the surface corresponds to the adsorption–desorption phenomena presented in a moment) and $(g_S^{A_2}/T_S - g^{A_2}/T)$, and the surface gradient $\boldsymbol{\nabla}_S(1/T_S)$. The corresponding generalized fluxes are the reaction rate, the surface heat flux, and the normal components of the heat flux, the diffusion flux of A and the diffusion flux of A_2 from the fluid.

When it is possible to use linear relations (i.e., near equilibrium), the flux–force relations are deduced by developing the matrix $\mathbf{L}$ of phenomenological coefficients. The transport coefficients satisfy the symmetry relations of Onsager ($L_{\alpha\beta} = L_{\beta\alpha}$) and are such that $\mathbf{L}$ is positive.

Thus, we show the dependence of the reaction rate on not only the chemical affinity A_S but also on the temperature jump and the surface chemical potential jump. Coupling between the chemical reaction and the heat transfer along the surface is still forbidden. These conclusions are quite general and should apply to any surface reaction.

Adsorption–Desorption

Adsorption is a process in which atoms or molecules move from a bulk phase onto a solid or liquid surface.[9] In the reverse process, called desorption, atomic and molecular species leave the surface and enter the surrounding gas or vacuum.

The surface quantity here is the number of moles j adsorbed per unit area. On a solid surface at rest in the absence of surface diffusion, $\mathbf{V} = \mathbf{0}$ and

$$\partial n_{ja}/\partial t + [J_{j\perp}]_{-}^{+} = 0, \tag{11.45}$$

with $[J_{j\perp}]_{-}^{+} = J_{j\perp}^{+}$ for exchanges between the interface and the region above it. The adsorption flow $J_{\perp}^{+}$ is generally provided by a adsorption kinetics law.

If A is the gaseous species, S is a vacant site on the surface, and AS is the adsorbed species, we have the following adsorption reaction (physisorption in this case):

$$\mathrm{A} + \mathrm{S} \overset{k_{ad.}}{\rightarrow} \mathrm{AS}$$

where $k_{ad.}$ is the specific adsorption rate, which can generally be written in the Arrhenius form:

$$k_{ad.} = A_{ad.} \exp\left(-E_{ad.}/RT\right),$$

where $A_{ad.}$ =const. and $E_{ad.}$ is the activation energy of adsorption. If C_A is the number of moles of A per unit volume, C_{Sa} is the number of moles of active sites S per unit surface area, and C_{Aa} is the number of moles of occupied sites AS, the flux of the adsorbed species is then

$$J_{A\perp\, ad.}^{+} = -\mathcal{M}_A k_{ad.}\, C_A\, C_{Sa}. \tag{11.46}$$

For the reverse reaction (desorption), AS $\overset{k_{de.}}{\rightarrow}$ A + S, we obtain $J_{A\perp\, de.}^{+} = \mathcal{M}_A k_{de.}\, C_{Aa}$, where $k_{de.} = A_{de.} \exp\left(-E_{de.}/RT\right)$.

For an adsorption–desorption process, we have $J_{A\perp}^{+} = \mathcal{M}_A(k_{de.}\, C_{Aa} - k_{ad.}\, C_A\, C_{Sa})$.

At adsorption–desorption equilibrium, we obtain $J_{A\perp\, de.}^{+} \cong -J_{A\perp\, ad.}^{+}$, so $C_A\, C_{Sa}/C_{Aa} = k_{de.}/k_{ad.} = K_{eq}$, where K_{eq} is the equilibrium constant.

[9]Weak interactions analogous to those that occur between molecules in liquids give rise to what is called physical adsorption or physisorption. Strong interactions ($> 40kJ/mol$) similar to those found between atoms within a molecule give rise to chemical adsorption or chemisorption. The adsorbed molecule remains intact in physisorption, but in chemisorption the molecule can be broken into fragments on the surface, in which case the process is called dissociative chemisorption.

In the general case we must consider reactions of the type

$$\sum_{j=1}^{N} \nu_j' A_j + \sum_{j=N+1}^{M} \nu_j' S_j \rightleftharpoons \sum_{j=1}^{N} \nu_j'' A_j + \sum_{j=N+1}^{M} \nu_j'' S_j.$$

Evaporation–Condensation

Evaporation–condensation problems are also considered in [94]. Some of these problems (in the case of evaporation–condensation at equilibrium) are addressed in Sect. 7.5, and in Sects. 12.4.3 and 12.4.4.

When the interface is not too far from equilibrium, it is possible to deduce linear constitutive relations from the linear thermodynamics of irreversible processes. Far from equilibrium, nonlinear rates of evaporation must be considered. A kinetic theory of evaporation–condensation has been developed using several approaches [170, 182, 296].

A Hertz–Knudsen law is generally written to express the evaporation mass flow rate of a liquid under reduced pressure in the case of a pure substance. The mass flux through a vapor–liquid interface is determined via the kinetic theory and the Boltzmann equation, using an equilibrium distribution function corresponding to local pressure and temperature. In front of the interface, a Knudsen layer with a typical thickness of approximately one mean free path separates the vapor–liquid interface and the bulk. In the classical Hertz–Knudsen theory, it is assumed that all vapor molecules interacting with the interface have the distribution of the bulk vapor, while those leaving the interface have a distribution associated with the interface. In reality, these two different molecular streams interact in the Knudsen layer, which alters the distributions of both and consequently modifies the evaporation and condensation mass flow rates. The following generalized Hertz–Knudsen equation is thus obtained [20]:

$$\dot{m} = (\frac{M}{2\pi R})^{1/2}(\alpha_v \frac{p_{sat}}{(T_S)^{1/2}} - \alpha_c \frac{p^+}{(T^+)^{1/2}}), \tag{11.47}$$

where α_v is the evaporation coefficient, α_c is the condensation coefficient, M is the molecular weight of the liquid, R is the universal gas constant, p_{sat} is the equilibrium vapor pressure at the surface temperature T_S, which is equal to the temperature T^- of the liquid at the evaporation interface level, while p^+ and T^+ are the pressure and temperature of the gas phase above the liquid.

For a mixture, the evaporation rate is a function of both the surface temperature and the composition. Bose and Palmer [25], who studied vapor recoil (a phenomenon that can lead to Hickman instability, see Sect. 7.5), used a modified form of the classical Hertz–Knudsen equation for the case where there are two species. This modified equation has a single evaporation–condensation coefficient α and assumes that there is no interaction between species A and B:

$$\begin{cases} \dot{m}_A = \alpha \left(\frac{M_A}{2\pi R}\right)^{1/2} \left(\frac{Y_A^- p^\bullet_{sat,A}}{(T_S)^{1/2}} - \frac{Y_A^+ p^+}{(T^+)^{1/2}}\right), \\ \\ \dot{m}_B = \alpha \left(\frac{M_B}{2\pi R}\right)^{1/2} \left(\frac{Y_B^- p^\bullet_{sat,B}}{(T_S)^{1/2}} - \frac{Y_B^+ p^+}{(T^+)^{1/2}}\right). \end{cases} \tag{11.48}$$

Here, $\dot{m}_A$ and $\dot{m}_A$ are the mass evaporative fluxes of species A and B, M_A and M_B are the molecular weights of the respective species, $p^\bullet_{sat,A}$ and $p^\bullet_{sat,B}$ are the pure-component vapor pressures of these species at the surface temperature T_S, and Y_A and Y_B are the mass fractions of the species.

11.4.2 Constitutive Relations for Interfaces Deduced by Applying Irreversible Thermodynamics to Three-Dimensional Interfacial Layers

In contrast to the case shown in Sect. 11.4.1, we will deduce contitutive relations by integrating the bulk equations across the interfacial layer, thus following the method described in Sect. 11.2.2. Whereas most of the matter flow crosses the interface when waves are considered (flames, shock waves or combustion waves), most of the flow occurs parallel to the interface for boundary layers and shear layers.

Premixed Flames

One-dimensional deflagration waves with high activation energy chemical reactions (thin flames) were presented in Sect. 10.5. The main result of this theory is the determination of the laminar mass flow rate and subsequently the combustion velocity.

These results are modified in the case of curved and unsteady two-dimensional flames. For deflagration waves with weak stretching rates and low curvatures, the parallel velocity vector of the fluid $\mathbf{v}_{//}$ is preserved at the crossing of the interface to a first approximation. Therefore, we will take the velocity defined by (11.2) as the velocity $\mathbf{V}$: $\mathbf{V} = \mathbf{v}_{//} + \mathbf{w}$.

The quantity $\boldsymbol{\nabla}_S \cdot \mathbf{V}_S$ is then the stretching rate of the flame (see Sect. A.4).

We assume that

$$\mathcal{J}_{\mathbf{V}Fa} = \mathbf{0} \text{ and } J_{\mathbf{V}F\perp} = \mathcal{J}_{\mathbf{V}F\perp} + \rho\, f\, u_\perp, \tag{11.49}$$

where $\mathbf{u} = \mathbf{v} - \mathbf{V}_S$.

Moreover, to avoid problems with integral divergence (occurring with f_S), we will use the excess quantity Δf_S, such that

$$\rho_a \Delta f_S = \epsilon \int_{-\infty}^{0} \rho(f - f^-),\, dn$$

where f^- is any specific quantity of the unburned fresh gases.[10] We then have the following for the general interface balance equation of quantity F:

$$\frac{d_S\,(\rho_a \Delta f_S)}{dt} + \rho_a \Delta f_S\, \nabla_S \cdot \mathbf{V}_S + [\mathcal{J}_{\mathbf{V}F\perp} + \rho\,(f - f^-)\,u_\perp]_-^+ = \dot{W}_{Fa}, \quad (11.50)$$

where $u_\perp$ is the relative laminar flame velocity.

Constitutive relations can be found for curved premixed flames. This problem was studied via the formalism of interface theory in [94, 219, 223], and has been the subject of numerous research works in the field of combustion (see [51, 53, 135, 163, 171, 258]).

The three-dimensional flame structure consists of four regions, as described in Sects. 10.5 and 10.5.2:

- The fresh gases
- The preheating or diffusion zone of thickness ℓ_f, where convection, mass diffusion and thermal conduction are the most significant processes
- The reactive zone of thickness ℓ_δ, where diffusion, conduction and chemical reaction are predominant, and which is represented by a jump
- The burnt gases.

The second and third regions constitute the interfacial layer.

As specified at the beginning of this section, here we consider the flame to be an interface with properties that are deduced from the previous three-dimensional theory [53] by integrating across the interfacial region.

From the detailed three-dimensional analysis summarized in Sect. 10.5.2, and for the single chemical reaction A $\longrightarrow$ B,[11] we can deduce (see [219]) the following interfacial source terms (which are included among the constitutive relations for the interface) for volume, mass fraction of A and temperature, respectively, in dimensionless variables:

$$\begin{cases} \dot{W}_{\mathcal{V}a} = [v_\perp], \\ \dot{W}_{Ya} = -\exp\left[\beta(T^+ - 1)/2\right], \\ \dot{W}_{Ta} = (1 - T^-)\exp\left[\beta(T^+ - 1)/2\right]. \end{cases} \quad (11.51)$$

[10] Here $\epsilon = \ell_f/L$ where $\ell_f = \kappa/u_\perp$ (the thickness of the flame deduced from the form of the energy equation applied to the diffusion zone), and where L is the hydrodynamic length. Moreover, $dN = \epsilon dn$ ($n = 0$ at the flame front where the reaction occurs).

[11] This is a first-order chemical reaction, and the reactive layer is replaced at the scale δ with a jump. The dimensionless reaction rate per unit volume obtained is (see [163, 219, 223]) $\dot{W}_R = \delta(N)\exp[\beta(T^+ - 1)/2]$, where $\delta(N)$ is the Dirac delta function, and where the reduced temperature of the burnt gas T^+ can differ from unity (which is, by definition, the value for an adiabatic laminar planar flame; note that the same symbols are used for dimensional and dimensionless parameters).

The following interfacial equations of state (which are also included among the interfacial constitutive relations) are also derived from this analysis:

$$\begin{cases} \rho_a \Delta T_S = T^- \rho_a \Delta \vartheta_S = \epsilon T^- \ln(1/T^-)/\dot{m}, \\ \rho_a \Delta Y_S = -[\epsilon T^-/\dot{m}(1-T^-)] \, [\ln(1/T^-) \\ -(Le-1)[T^-/(1-T^-)] \int_0^{(1-T^-)/T^-} x^{Le-2} \ln(1+x) dx]. \end{cases} \tag{11.52}$$

If we apply the series expansion $\varphi = \varphi^{(0]} + \epsilon\varphi^{(1)} + \epsilon^2\varphi^{(2)} + O(\epsilon^3)$ to any parameter of (11.51), (11.50) becomes [219] the approximate balance equation for a flame considered to be an interface.

Finally, asymptotic analysis yields (10.177) of Chap. 10:

$$\begin{aligned} \dot{m}^- &= 1 - \epsilon\,[\ln(1/T^-)/(1-T^-) - l\,(T^-/2) \\ &\int_0^{(1-T^-)/T^-} \ln(1+x)/x\,dx]\, \nabla_{//} \cdot \mathbf{V}_S^{(0)} + O(\epsilon^2). \end{aligned} \tag{11.53}$$

This result shows that the mass flow rate differs from the mass flow rate of the adiabatic plane laminar flame by a factor that is proportional to the stretch $\nabla_{//} \cdot \mathbf{V}_S^{(0)}$, and that the Lewis number plays a very important role.

Since $\dot{m}^- = u_\perp^-/u^-$, we obtain the following for the combustion velocity $u_\perp^- = v_\perp^- - V_{S\perp}$ (denoted s_L below):

$$\frac{s_L}{s_L^0} \cong 1 - \epsilon\,[\frac{\ln(1/T_u)}{(1-T_u)} - \beta\frac{T_u}{2}\int_0^{\frac{1-T_u}{T_u}} \frac{\ln(1+x)}{x} dx]\, \nabla_{//} \cdot \mathbf{V}_S. \tag{11.54}$$

In this formula, which is valid for a stretched flame with a high activation energy and a Lewis number that is close to unity, s_L^0 is the planar adiabatic laminar combustion speed,[12] ϵ is the diffusion flame thickness divided by a reference hydrodynamic length, T_u is the reduced temperature of the unburnt gases (i.e., the temperature of the unburnt gases divided by T_{ad}), $\beta = T_a/T_{ad} \gg 1$, and $\nabla_{//} \cdot \mathbf{V}_S$ is the surface strain rate.

Equation 11.54 can be written in an alternative form [54, 53, 173, 202]:

$$s_L/s_L^0 \cong 1 - \mathcal{L}(\kappa + S/s_L^0), \tag{11.55}$$

where the term $\nabla_{//} \cdot \mathbf{V}_S$ of (11.54) has been decomposed into two terms: κ and S/s_L^0, where κ is the mean curvature of the flame surface, $S = -\mathbf{N} \cdot \nabla \otimes \mathbf{v} \cdot \mathbf{N}$ is the strain rate of the flow in the flame region, and $\mathcal{L}$ the Markstein length with respect to the unburnt mixture, as defined by

[12] This combustion speed must be calculated for the actual chemical reaction A $\longrightarrow$ B, in contrast to the case in Sect. 10.5, where the chemical reaction was A + B $\longrightarrow$ P. The determination of s_L^0 is not very different, as shown for example in [94].

$$\frac{\mathcal{L}}{\ell_f} = \frac{1}{\gamma} \ln \frac{1}{1-\gamma} + \frac{Ze(Le-1)}{2} \frac{1-\gamma}{\gamma} \int_0^{\frac{\gamma}{1-\gamma}} \frac{\ln(1+x)}{x} dx, \tag{11.56}$$

where ℓ_f is the flame thickness, $\gamma = (\rho_u - \rho_b)/\rho_u$, and the Zel'dovich number $Ze = E_a(T_b - T_u)/RT_b^2$.

Moreover, note that the quantities $\rho_a \Delta f_S$ are always proportional to the jump conditions $[f]_-^+$ and to the inverse of the combustion unit mass flow rate $\dot{m}$, with an appropriate coefficient. We then have

$$\rho_a \Delta f_S = L\epsilon \frac{c_p}{\lambda} \bar{D} \frac{[f]_-^+}{\dot{m}}, \tag{11.57}$$

where L is a proportionality coefficient and $\bar{D}$ is a diffusion coefficient (ρD or λ/c_p). In the cases considered, the coefficient L is assumed to be constant.

Using the appropriate constitutive relations and the reaction rate given in system (11.51), this two-dimensional interface theory yields the same results as the classical three-dimensional theory [219] when applied to a premixed stretched flame with a high activation energy.

Boundary Layers

The discussion of boundary layers presented below does not actually produce any new results beyond those given in Chap. 9; we include it here to show how boundary layers can be interpreted with respect to interfaces.

Balance Equation for Boundary Layers

The velocity $\mathbf{V}$ as defined by (11.2) is not now conservative throughout the interface, and thus we find ourselves in the situation corresponding to case 1 of Fig. 11.2. For boundary layers above a rigid blunt body, we choose the local velocity of the surface of this blunt body $\mathbf{v}^-$ (see Fig. 11.5) as $\mathbf{V}$; this velocity is zero in the reference frame associated with the body.

Here, we consider the excess quantity $\rho_a \Delta f_S = \epsilon \int_0^\infty \rho(f - f^+) dn$ (see Eq. 11.5). The specific quantity $f^+(\mathbf{x}, t)$ will refer to the external flow (i.e., at the (+) side of the interface).

Therefore, in the reference frame associated with the obstacle ($\mathbf{V} = \mathbf{v}^-$), (11.25) becomes

$$d_{\mathbf{V}}(\rho_a \Delta f_S)/dt + \boldsymbol{\nabla}_S \cdot \mathbf{J}_{\mathbf{V}Fa} + [\mathbf{J}_{\mathbf{V}F\perp}]_-^+ = \dot{W}_{Fa}. \tag{11.58}$$

If the normal convective terms are negligible, $[\mathbf{J}_{\mathbf{V}F\perp}]_-^+$ reduces to $[\boldsymbol{\mathcal{J}}_{\mathbf{V}F\perp}]_-^+$, and the convective part of $\mathbf{J}_{\mathbf{v}Fa}$ reduces to

$$\int_{N_-}^{N_+} \rho_F \mathbf{v}_{//} \, dN. \tag{11.59}$$

We now provide some examples to illustrate this potential description of boundary layers in terms of interfaces.

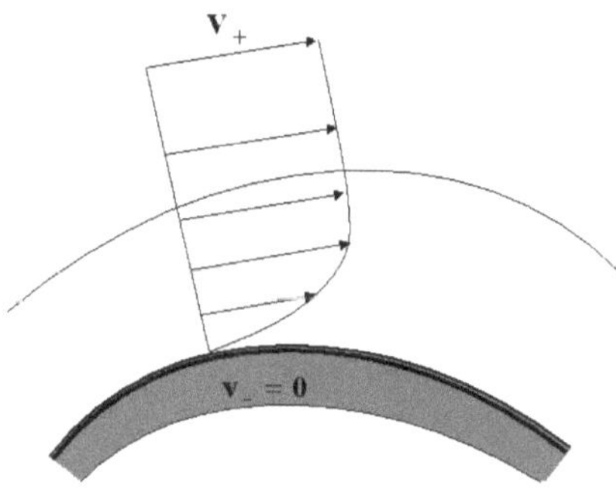

Fig. 11.5. Velocity profile in a boundary layer for $\mathbf{v}_- = \mathbf{0}$

Steady Boundary Layer Above a Flat Plate

Let us assume an incompressible fluid and planar two-dimensional flow with constant pressure. The x-component of the velocity at infinity is U_∞, and the plate occurs at $y = 0$ and $x > 0$. We have

$$\frac{\partial u}{\partial x} + \frac{\partial v}{\partial y} = 0, \tag{11.60}$$

$$\frac{\partial \rho u(u - U_\infty)}{\partial x} + \frac{\partial \rho v(u - U_\infty)}{\partial y} - \mu \frac{\partial^2 u}{\partial y^2} = 0. \tag{11.61}$$

The self-similar solution is well known (see Sect. 9.3). Recall that if we write $\eta = y\sqrt{U_\infty/2\nu x}$, $\nu = \mu/\rho$, we find that

$$u = U_\infty f'(\eta), \; v = (\nu U_\infty/2x)^{1/2}(\eta f'(\eta) - f(\eta)), \tag{11.62}$$

where f obeys the Blasius equation $f'' + f\ f'' = 0$, with the boundary conditions $f(0) = f'(0) = 0$, $f'(\infty) = 1$.

Integrating (11.61) with respect to y gives the equation sought:

$$\boldsymbol{\nabla}_S \cdot \mathbf{J}_{ua} + [\mathbf{J}_{u\perp}]_-^+ = 0, \tag{11.63}$$

where, for excess longitudinal convective flow and friction stress at the wall,[13]

[13]The friction stress τ_+ is zero.

$$\begin{cases} \nabla_S \cdot \mathbf{J}_{ua} = \partial J_{ua}/\partial x, \\ J_{ua} = \int_0^\infty \rho u(u - U_\infty)\, dy, \\ [\mathbf{J}_{u\perp}]_-^+ = [\tau]_-^+ = -\tau_-. \end{cases} \tag{11.64}$$

The flux J_{ua} can be written

$$J_{ua} = -\mu \sqrt{\frac{2xU_\infty}{\nu}} U_\infty f''(0), \tag{11.65}$$

where $f''(0) = 0.4696$.

Relation (11.65) can be interpreted via dimensional considerations.

Indeed, if ρ, U_∞ and x are selected as basic quantities, two dimensionless ratios (see Sect. 5.1.2) appear: $\Pi_\mu = \mu/\rho U_\infty x = Re_x$ and $\Pi_J = J_{ua}/\rho U_\infty^2 x = -K_u \psi(\Pi_\mu)$.

Therefore, we find that

$$J_{ua} = -K_u \rho U_\infty x [u]_-^+ \psi(\mu/\rho U_\infty x). \tag{11.66}$$

Result (11.65) shows that the function ψ must be a square root. The dimensionless phenomenological coefficient K_u is then equal to $K_u = \sqrt{2} f''(0) \cong 0.665$. K_u will not be constant in other boundary layers. Finally, we can use relations of this type to express any flux $\mathbf{J}_{Fa}$ as a function of the jump $[f]_-^+$ with the dimensionless coefficient K_F.

Unsteady Boundary Layer

Another classical phenomenon is that represented by a medium and a flat plate that are both initially at rest. At time $t = 0$, and for $t > 0$, the plate moves at a constant velocity V along Ox.

The momentum equation for the interface is then

$$\frac{\partial(\rho_a \Delta u_S)}{\partial t} - [\tau]_-^+ = 0. \tag{11.67}$$

Classical calculation makes it possible to determine that

$$\rho_a \Delta u_S = \int_{-\infty}^{0} \rho u\, dy = \sqrt{4\nu t/\pi}\, \rho V.$$

We can see that $\rho_a \Delta u_S$ is proportional to the momentum jump $[\rho u]_-^+ = \rho V$. Moreover, dimensional considerations allow us to envisage the form of $(\rho_a \Delta u_S)$. Since time t and viscosity ν are the basic parameters, the momentum transfer thickness (the thickness of the boundary layer) is $d_u = \sqrt{\nu t}$.

By introducing the dimensionless phenomenological coefficient L_u, we find that

$$\rho_a \Delta u_S = L_u \rho [u]_-^+ d_u = -L_u \sqrt{\nu t} \rho V, \tag{11.68}$$

which gives $L_u = -\sqrt{4/\pi}$ here.

Boundary Layer with Diffusion

Boundary layers with diffusion were analyzed in particular in Sects. 9.4.1 and 9.6.2. Here, for flow over a flat plate in steady mode, the equation for the interface is

$$\frac{\partial J_{Ya}}{\partial x} + [\mathcal{J}_{Y\perp}]_-^+ = \dot{W}_{Ya}, \tag{11.69}$$

where $J_{Ya} = \int_0^\infty \rho u(Y - Y^+)dy$.

Dimensional analysis enables us to write a formula that is similar to (11.66): $J_{Ya} = -K_Y \rho U_\infty x[Y]_-^+ \psi(\rho U_\infty x/\mu,\ \mu/\rho D)$. The function ψ is generally a product of powers, so that

$$J_{Ya} = -K_Y \rho U_\infty x[Y]_-^+ Re_x^a\, Sc^b. \tag{11.70}$$

In laminar flow, a is equal to $(-1/2)$, and the power b varies according to whether the Schmidt number is small or large compared to 1.

For example, if the Schmidt number is small, the boundary layer of diffusion is very thick compared to the viscous layer, so that velocity can be regarded as being constant in the diffusion zone. b is then $(-1/2)$.

Thus, if $Y^+ = 0$ and $Y^- = 1$, we obtain

$$J_{Ya} = -K_Y \rho U_\infty \sqrt{Dx/U_\infty}. \tag{11.71}$$

Exact calculations give the same result, with $K_Y = \sqrt{4/\pi}$.

If the Schmidt number is large, the power b becomes equal to $(-2/3)$. These results are linked to those already obtained for the exchange coefficient and the Sherwood number. By calculating $\partial J_{Ya}/\partial x$, we immediately obtain the expression for the Sherwood number, $Sh = (K_Y/2)Re_x^{1/2} Sc^{b+1}$, which is valid for laminar boundary layers. This result gives the physical significance of coefficient K_Y.

11.5 Interfaces with Resistance to Wrinkling

The study of interfaces that have a certain degree of stiffness, particularly in response to bending, is linked to shell and plate theory [136, 189, 272].

In order to study this type of interface, we can consider the case where the velocity field is provided by kinematic torsors [14][14]

[14] In the preface of the English version of his book, J.-M. Berthelot noted that: "The development of this textbook is based on a generalized use of the concept of 'torseur' (in French). We think that this concept is not really used in the English textbooks. We will call this concept as 'torsor.' In the textbook, the English formulation was thus transposed from the French formulation [15]." P. Germain utilizes the term "distributeur" instead of "torseur cinématique" [99].

$$\{\mathcal{C}\} = \left\{ \begin{array}{c} \boldsymbol{\omega} \\ \mathbf{v}_P \end{array} \right\}. \tag{11.72}$$

When the velocity field is given by the kinematic torsor field $\mathcal{C}$, the velocity of a point M (such that $\mathbf{PM}$ is collinear with the normal $\mathbf{N}$ to (S) at point P) in the interfacial layer is given by

$$\mathbf{v}_M = \mathbf{v}_P + \boldsymbol{\omega} \times \mathbf{PM}. \tag{11.73}$$

The motion of the segment $P'P''$ (Fig. 11.6) is locally the motion of a rigid body, and $\boldsymbol{\omega}$ and $\mathbf{v}_P$ are solely functions of the space coordinates of P in (S) and of time.

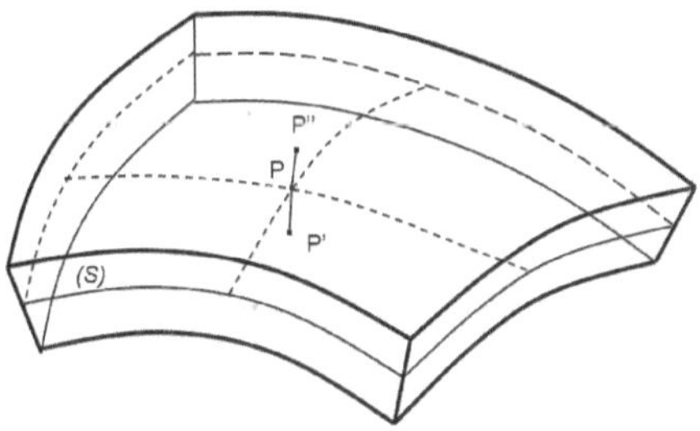

Fig. 11.6. Interfacial layer in the case of a kinematic torsor

We then have

$$d\mathbf{v}_M = d\mathbf{v}_P + \boldsymbol{\omega} \times d\mathbf{PM} + d\boldsymbol{\omega} \wedge \mathbf{PM}. \tag{11.74}$$

This expression will allow us to determine $\boldsymbol{\nabla} \otimes \mathbf{v}_M$; its symmetrical part is the tensor of the strain rates of the medium. We can that $d\mathbf{v}_M$ corresponds to the torsor

$$\{d\mathcal{C}\} = \left\{ \begin{array}{c} d\boldsymbol{\omega} \\ d\mathbf{v}_P + d\mathbf{P} \times \boldsymbol{\omega} \end{array} \right\}, \tag{11.75}$$

Note that the product of a kinematic torsor $\{\mathcal{C}\}$ with the torsor of forces $[\mathcal{F}]$ applied to a rigid body gives the power developed by these forces during the motion: $\mathcal{P} = \{\mathcal{C}\} \cdot [\mathcal{F}]$.

We will utilize fields of torsors (because the torsors considered depend on the position on the surface) and Germain's formalism here.

where $d\mathbf{P}$ is such that $\mathbf{N} \cdot d\mathbf{P} = 0$.

As $\boldsymbol{\omega}$ is tangent at P to (S), we introduce the vector $\mathbf{l}$ such that

$$\mathbf{l} = \boldsymbol{\omega} \times \mathbf{N}, \ \boldsymbol{\omega} = \mathbf{N} \times \mathbf{l} \tag{11.76}$$

to simplify calculations. Thus,

$$d\mathbf{P} \times \boldsymbol{\omega} = (d\mathbf{P} \cdot \mathbf{l})\mathbf{N} = d\mathbf{P} \cdot (\mathbf{l} \otimes \mathbf{N}). \tag{11.77}$$

The strain rate torsor will therefore be defined as

$$\{\mathcal{D}_{ij}\} = \left\{ \begin{matrix} \mathcal{S}(\omega_{j,i}) \\ \mathcal{S}(v_{Pj,i} + l_j N_j) \end{matrix} \right\} \tag{11.78}$$

($\mathcal{S}$: symmetrical part). However, since $\boldsymbol{\omega}$ and $\mathbf{v}_P$ are only defined on the surface, we must consider

$$\{\mathcal{D}_{ij}\} = \left\{ \begin{matrix} \mathcal{S}[(\delta_{i,j} - N_i N_k)\omega_{j,k}] \\ \mathcal{S}[(\delta_{i,j} - N_i N_k)v_{Pj,k} + l_j N_j] \end{matrix} \right\}. \tag{11.79}$$

As we did previously, we indicate the tangential gradient on the surface (S) by the symbol $\boldsymbol{\nabla}_{//}$. Thus, the the strain rate torsor can be written

$$\{\mathcal{D}\} = \left\{ \begin{matrix} \mathcal{S}(\boldsymbol{\nabla}_{//} \otimes \boldsymbol{\omega}) \\ \mathcal{S}(\boldsymbol{\nabla}_{//} \otimes \mathbf{v} + \mathbf{l} \otimes \mathbf{N}) \end{matrix} \right\} = \left\{ \begin{matrix} \boldsymbol{\chi} \\ \boldsymbol{\eta} \end{matrix} \right\}. \tag{11.80}$$

To obtain the momentum equations, it is convenient to use the *method of virtual power*, which gives results that are applicable to deformable solid interfaces. The virtual power principle is expressed in the following terms. In a Galilean referential and for an absolute time, the virtual power of the acceleration quantities $\mathcal{A}^*$ of any part $\mathcal{V}$ of a material system $\mathcal{V}_0$ ($\mathcal{V} \subset \mathcal{V}_0$) is equal to the sum of the virtual powers $\mathcal{P}^*_{(i)}$ and $\mathcal{P}^*_{(e)}$ of the internal and external forces for any virtual motion defined on $\mathcal{V}$ and belonging to $\mathcal{V}^*$.[15] Moreover, the virtual power of the internal forces is an objective quantity; i.e., it is zero for all rigid virtual motion [97, 98, 99]:

$$\mathcal{P}^*_{(i)} + \mathcal{P}^*_{(e)} = \mathcal{A}^*. \tag{11.81}$$

$\mathcal{P}^*_{(i)}$ is obtained by multiplying D^* by a field of stress torsors with the reduction elements $\boldsymbol{\gamma}$ and $\boldsymbol{\mu}$, which are both symmetrical tensors, so that

$$\mathcal{P}^*_{(i)} = - \int_{\Sigma} (\boldsymbol{\gamma} \colon \boldsymbol{\eta}^* + \boldsymbol{\mu} \colon \boldsymbol{\chi}^*) d\Sigma. \tag{11.82}$$

[15] The elements of $\mathcal{V}^*$ are vector fields $\mathbf{v}^*$ defined on $\mathcal{V}_0$. The space $\mathcal{V}^*$ is taken to be linear and normal. In addition, all of the fields $\mathcal{V}^*$ associated with the rigid motions defined on $\mathcal{V}_0$ are in $\mathcal{V}^*$. The different expressions $\mathcal{A}^*$, $\mathcal{P}^*_{(i)}$ and $\mathcal{P}^*_{(e)}$ are linear forms defined on $\mathcal{V}^*$. Depending on the space $\mathcal{V}^*$ chosen and on the expressions for $\mathcal{A}^*$, $\mathcal{P}^*_{(i)}$ and $\mathcal{P}^*_{(e)}$ selected, we obtain different theories for the continuous medium [97].

This expression is transformed to give the virtual velocities $\mathbf{v}^*$ and $\boldsymbol{\omega}^*$. Using the Stokes–Ostrogradsky theorem, we finally obtain [219]

$$\begin{cases} \mathcal{P}^*_{(i)} = -\int_\Sigma [\mathbf{v}^* \cdot (\boldsymbol{\nabla}_{//} \cdot \boldsymbol{\sigma} + \boldsymbol{\nabla}_{//} \cdot (\mathbf{N} \otimes \mathbf{g})) \\ +\boldsymbol{\omega}^* \cdot (\boldsymbol{\nabla}_{//} \cdot \mathbf{M} + \boldsymbol{\nabla}_{//} \cdot (\mathbf{N} \otimes \mathbf{m}) - \mathbf{N} \times \mathbf{g})] d\Sigma \\ -\oint_c [\mathbf{v} * \cdot \boldsymbol{\sigma} + \mathbf{g}\mathbf{v}^* \cdot \mathbf{N} + \boldsymbol{\omega}^* \cdot \mathbf{M}] \cdot \boldsymbol{\eta} ds. \end{cases} \tag{11.83}$$

The virtual power of the external forces is written in a similar form:

$$\mathcal{P}^*_{(e)} = \int_\Sigma (\mathbf{p} \cdot \mathbf{v}^* + \boldsymbol{\varphi} \cdot \boldsymbol{\omega}^*) d\Sigma + \oint_c (\mathbf{T} \cdot \mathbf{v}^* + \boldsymbol{\phi} \cdot \boldsymbol{\omega}^*) ds. \tag{11.84}$$

The virtual power of the acceleration quantities is

$$\mathcal{A}^* = \int_\Sigma \rho_a \frac{d\mathbf{v}_a}{dt} \cdot \mathbf{v}^* \, d\Sigma. \tag{11.85}$$

Thus, the equations of motion [99] are derived directly from the virtual power principle:

$$\begin{cases} \boldsymbol{\nabla}_{//} \cdot \boldsymbol{\sigma} + \boldsymbol{\nabla}_{//} \cdot (\mathbf{N} \otimes \mathbf{g}) + \mathbf{p} = \rho_a \, d\mathbf{v}_a / dt, \\ \boldsymbol{\nabla}_{//} \cdot \mathbf{M} + \boldsymbol{\nabla}_{//} \cdot (\mathbf{N} \otimes \mathbf{m}) - \mathbf{N} \times \mathbf{g} + \boldsymbol{\varphi} = \mathbf{0}, \\ (\boldsymbol{\sigma} + \mathbf{N} \otimes \mathbf{g}) \cdot \boldsymbol{\eta} = \mathbf{T}, \\ \mathbf{M} \cdot \boldsymbol{\eta} = \boldsymbol{\phi}. \end{cases} \tag{11.86}$$

Here, $\mathbf{p}$ is the surface density of forces; $\boldsymbol{\varphi}$ is the surface density of torques, $\boldsymbol{\varphi} \cdot \mathbf{N} = 0$; $\mathbf{T}$ is the density of forces per unit length; and $\boldsymbol{\phi}$ is the density of torques per unit length, $\boldsymbol{\phi} \cdot \mathbf{N} = 0$. The tensor $\boldsymbol{\sigma}$ is the tensor of tensions (or the membrane stress tensor). The vector $\mathbf{g}$ is the shear-forces vector. $\mathbf{M}$ and $\mathbf{m}$ characterize the bending moments.

We could have separated the normal part $v^*_\perp \mathbf{N}$ from the tangential part $\mathbf{v}^*_{//}$ of the virtual velocity vector. This would have allowed us to distinguish the normal constraints from the tangential constraints. We can also perform this decomposition starting from the equations of motion.

We will not proceed any further with the description of this theory. Thin shells have been studied in many papers with varying degrees of specialization, several of which have been cited in this section. The preceding equations also apply to plates (in this case, it is sufficient to make $\mathbf{N} = \mathbf{e}_3$; $\boldsymbol{\nabla}_{//} = \mathbf{e}_\alpha \partial / \partial x_\alpha$, $\alpha = 1, 2$).

We can also treat the case of cylindrical shells around an axis $\mathbf{e}_3$, where there are forces normal to the axis. In the planar case, we again obtain the equations for planar curvilinear media.

All of the moments vanish for flexible interfaces (membranes). We find that $\mathbf{N} \times \mathbf{g} = \mathbf{0}$; i.e., that $\mathbf{g}$ is null, just like $T_\perp$. Obviously, there is no shearing action.

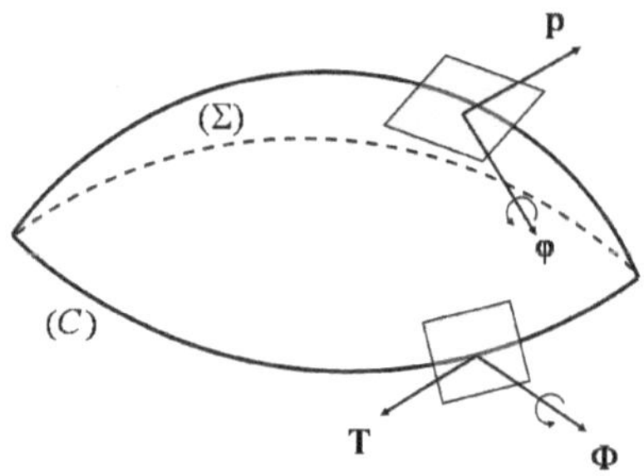

Fig. 11.7. Representation of the forces acting on the surface

Fig. 11.8. Paul Germain, professor at the Pierre and Marie Curie University in Paris (1920–2009) (M.-H. Debost, private communication, 2009; see also [174])

The equations become

$$\rho_a\, d_S\mathbf{v}_S/dt - \boldsymbol{\nabla}_{//}\cdot\boldsymbol{\sigma} + [\mathbf{P} + \rho(\mathbf{v}-\mathbf{v}_S)\otimes(\mathbf{v}-\mathbf{v}_S)]_-^+\cdot\mathbf{N} = \rho_a\mathbf{f}_S, \quad (11.87)$$

where $\mathbf{f}_S$ is the surface force (due to gravity for example). Equation 11.87 can be deduced directly from (11.29) when it is applied to a single-component medium.

In this section, we focused on the momentum equations of interfaces that are resistant to wrinkling. To study such interfaces in more depth, we need to establish the balance equations for mass, species, energy and entropy, just as we did for flexible interfaces. We also need the constitutive relations and laws of state. One well-known case in this regard is that of elastic shells, where linear relations are obtained between strains and stresses. However, much more work is needed for the other constitutive relations. Some of the relations that were established for flexible interfaces can be used to this end.

11.6 Concepts of Second-Gradient Theory

Thermodynamic Relations

Let us now consider the case of 3D interfacial fluid layers with internal capillarity, and assume a one-component fluid for simplicity. For a volume $\mathcal{V}$ with a small thickness $\Delta\xi$ around a coordinate surface, the internal energy E of a stratified layer is a first-order homogeneous function of the quantities $\mathcal{S}$, $\mathcal{V}$, ξ, the mass m and a complementary extensive variable $\mathcal{S}$ (which is homogeneous to area but is not equal to the area of the part of S_3 present in the considered volume). We then have $\mathcal{S} = \int_{\mathcal{V}} (1/\vartheta)|d\vartheta/d\xi|\, dV$, and the internal energy per unit mass becomes $e = e(s, \vartheta, |d\vartheta/d\xi|)$. We can write

$$\begin{cases} e = T\, s - p\,\vartheta + \Lambda\,|d\vartheta/d\xi| + g, \\ de = T\, ds - p\, d\vartheta + \Lambda\, d|d\vartheta/d\xi|, \\ 0 = s\, dT - \vartheta\, dp + |d\vartheta/d\xi|\, d\Lambda + dg. \end{cases} \tag{11.88}$$

More usually, we would write [94, 256]

$$e = e(s, \rho, |\boldsymbol{\nabla}\rho)|^2,\; de = T\, ds - p\, d(\frac{1}{\rho}) + \lambda(\boldsymbol{\nabla}\rho)\cdot d(\boldsymbol{\nabla}\rho), \tag{11.89}$$

where λ is the capillary coefficient.

3D Closure Relations

In simple cases, it is possible to relate the surface tension σ to the capillarity coefficient λ [256] by writing

$$\sigma = -\int_{N_-}^{N_+} p^*(\xi)\, d\xi \cong \int_{N_-}^{N_+} \lambda|\boldsymbol{\nabla}\rho)|^2\, d\xi, \tag{11.90}$$

where p^* is the effective pressure defined by

$$p^* = p - \rho\boldsymbol{\nabla}\cdot(\lambda\boldsymbol{\nabla}\rho). \tag{11.91}$$

This description can be extended to fluid mixtures with internal capillarity and mass diffusion [94]. In this case, the effective pressure becomes

$$p^* = p - \sum_{j=1}^{N} \sum_{i=1}^{N} \rho_i \nabla \cdot (\lambda_{ji} \nabla \rho_j). \tag{11.92}$$

A fluid can also exhibit capillarity near a wall. This is considered in some wetting problems [208, 257].

11.7 Conclusions Regarding Interface Equations

The utilization of interface equations often leads to discontinuities, and it is often difficult to solve such problems numerically. Therefore, some authors have even tried to obtain continuous equations for discontinuous situations. Phase field models have been described for the solidification of alloys. The interface is considered a transition region where averaged local quantities weighted by the liquid and solid volume fractions are introduced. An enthalpy method has been used at the macroscopic scale [79], whereas balance equations are deduced by minimizing the free energy in functional analysis at a smaller scale [205]. Jamet et al. [123] use a second-gradient method with an artificially thickened interface, and Jamet and Petitjeans [122] apply phase field models to diffusion interfaces. Research into phase field methods for fluid interfaces would certainly prove useful.

12

Multiphase Flow Concepts

Multiphase flows can be gas flows with solid and/or liquid particles, or liquid flows with vapor bubbles, solid and/or liquid particles [138]. Exchanges between phases, chemical reactions, and particle disintegration and agglomeration can occur within them. The example of a gas-droplet multiphase flow[1] illustrates their complexity. In this chapter, we will attempt to obtain a complete set of equations, if possible, with closure relations that allow the analytical or numerical solution of such a flow.

Section 12.1 deals with the generation of a two-phase flow, especially in the case of droplets. The atomization of a liquid is needed in combustion engines for example. We do not study the problem of dust generation here,

[1] As an example of this complexity, consider (as Young does [295]) a multiphase medium consisting of a neutral carrying gas (mass m_G), steam (mass m_V), and liquid drops of the same composition as the steam and that are of mass class i (each mass class m_i consists of presumably spherical drops of radius r_i). A control volume $\mathcal{V}_m$ contains a mass m_m of the multiphase medium, and so $m_G+m_V+\Sigma_i m_i = m_m$. In this control volume, the respective volumes of the gas phase (we assume an ideal gas mixture) and the particles of class i are $\mathcal{V}$ and $\mathcal{V}_i$, and we have $\Sigma \mathcal{V}_i + \mathcal{V} = \mathcal{V}_m$. Let us now define $g = mG/m_m$ and $y_i = m_i/(m_V + \Sigma_i m_i)$ and $y = \Sigma_i y_i$. The following specific densities are then introduced: $\rho_{Gs} = m_G/\mathcal{V}$, $\rho_{Vs} = m_V/\mathcal{V}$, $\rho_{is} = m_i/\mathcal{V}_i$, $\rho_s = (m_G + m_V)/\mathcal{V} = \rho_{Gs} + \rho_{Vs}$, $\rho_m = m_m/\mathcal{V}_m$.

1. We can easily show that $g + (1-g)(1-y) + (1-g)y = 1$, according to the basic relation $m_G + m_V + \Sigma_i m_i = m_m$.
2. We can show that $1/\rho_m = (g + (1-g)(1-y))/\rho_s + [\Sigma_i(1-g)y_i]/\rho_{is}$, according to the basic relation $\Sigma_i \mathcal{V}_i + \mathcal{V} = \mathcal{V}_m$.
3. The mass fractions in the gas are $Y_G =, Y_V =, Y_G + Y_V = 1$. The partial densities are $\rho_G = \epsilon\rho_{Gs}$, $\rho_V = \epsilon\rho_{Vs}$, $\rho_i = \epsilon_i\rho_{is}$, $\rho = \rho_G + \rho_V = \epsilon\rho_s$. Note that if the particles are all composed of a single species with a constant specific density, then $\epsilon_i = \epsilon_p$ for any i and we have $\rho_i = \epsilon_p\rho_{is}$, $\rho_i = (1-\epsilon)\rho_{is}$, $\rho_p = (1-\epsilon)\rho_{ps}$.

ϵ is expressed as a function of the preceding quantities: $\epsilon = (g + (1-g)(1-y))\rho_m/\rho_{Gs}$. We also have $\epsilon = (g + (1-g)(1-y))\rho_m/\rho_s$.

R. Prud'homme, *Flows of Reactive Fluids*,
DOI 10.1007/978-0-8176-4659-2_12, © Springer Science+Business Media, LLC 2010

but it is important to recognize that it is a significant challenge to generate a homogeneous suspension of solid particles [30].

In Sect. 12.2 we will deal with simplified macroscopic balance equations for two-phase flows using the method described in Chap. 4. We assume that the medium can be considered a continuum. The only exchanges considered are those of momentum and energy, and all of the particles are the same size. The forms of the constitutive relations are deduced from the principles of irreversible thermodynamics (Chap. 3). The simplified equations obtained are then applied to two standard problems that can be solved without using a computer: the study of small-perturbation propagation in the vicinity of a motionless reference state (i.e., linearized sound propagation), and the motion of a vortex in a dilute suspension.

In Sect. 12.3 we investigate the balance equations of a flow with evaporating droplets. The spray considered is always assimilated into a continuum. These equations are applicable to the study of spray flame propagation with a minimal model for instance.

Problems addressed at the particle scale are presented in Sect. 12.4. The description of fluid–particle exchange is a problem in itself, and particle-scale (mesoscopic scale) modeling is needed to determine (for example) friction effects, the regression rate of a droplet during combustion, etc. Four examples of such problems are then studied at the particle scale. The first two relate to an inert particle exchanging momentum and energy. The third problem is that of the steady combustion of a fuel droplet in a combustive atmosphere that is at rest and has uniform properties at infinity; i.e., far from the droplet (the volume of the atmosphere is assumed to be at rest, uniform, and infinitely large). The theoretical results obtained are interpreted and compared with those obtained experimentally. Finally, the fourth problem is the transient vaporization of a droplet.

In each case, the respective exchange coefficients that specify the constitutive relations are determined. We calculate the regression coefficient for evaporating droplets, and we determine the cut-off frequency that delimits the area of instability for transient vaporization.

The results obtained at the particle scale must then be inserted as closure relations into the balance equations for the multiphase mixture obtained in Sect. 12.2, or into more complex balance equations.[2]

Of course, the types of equations obtained in this chapter and the methods used have their limits. As long as the flow regime is quite continuous and regular, similar to the laminar flow of a homogeneous fluid, the balance equations obtained will be valid if we make certain assumptions (in particular that the

[2]Various methods are used to establish the balance equations starting from the properties at the particle scale. We will mention those that start from a probabilistic balance equation similar to that of Chap. 4 and apply this equation to the various quantities [10, 140]. Readers interested in this problem should refer to the works of Faeth [81] (on sprays with evaporation and combustion) and Yang et al. [294] (on propellant vaporization and combustion).

volume fraction of particles is low). However, under external constraints such as mass flow rate variations, significant agglomerations can appear, as can fluid pockets. Medium stirring can become very strong, reducing the ability of the equations established to accurately reflect the situation. In addition, this chapter does not treat cases where the particle size or concentration is large. It is worth noting that many of the problems associated with this field are still unsolved, although research into them is underway. This means that our ambitions are limited. However, the cases treated in this chapter highlight their complexity and show which cases are easy to solve.

12.1 Formation of a Two-Phase Flow: Droplet Generation

Fogs and sprays are created from liquids for domestic and industrial use. The process of forming droplets is termed "atomization." One of the most interesting uses of sprays occurs in combustion. The engines of ground, sea, air or space vehicles often utilize the burning of droplets. These droplets are generated using injectors, the characteristics of which vary according to the conditions.

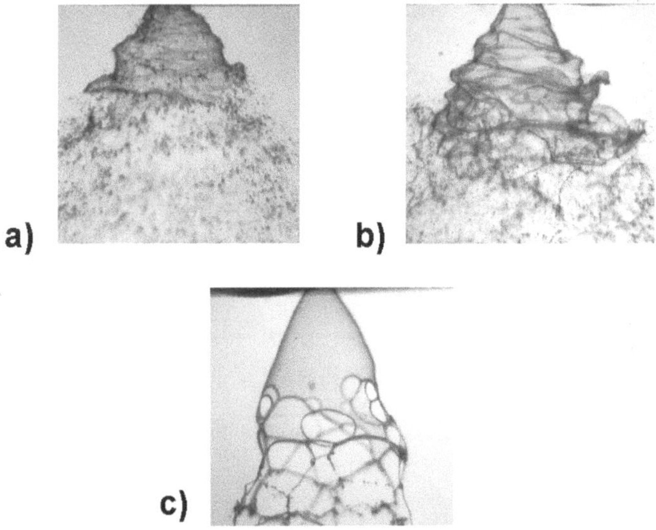

Fig. 12.1. Influence of the liquid viscosity on the distortion of a swirling conical sheet (photos from CORIA). The viscosity of the liquid increases from **a** to **c**. (C. Dumouchel, private communication, 2009; see also [74])

Examples are given in Figs. 12.1, 12.5 and 12.13. In Fig. 12.1, a swirling, conical thin liquid sheet injected into an environment at rest is destabilized to produce a droplet cloud.[3] A review of some droplet generation problems has recently been published [221].

The process of primary atomization is shown in Fig. 12.2 for the case of a liquid planar sheet that is subjected to an external flow. When the drops formed initially are shattered, the process is called secondary atomization.

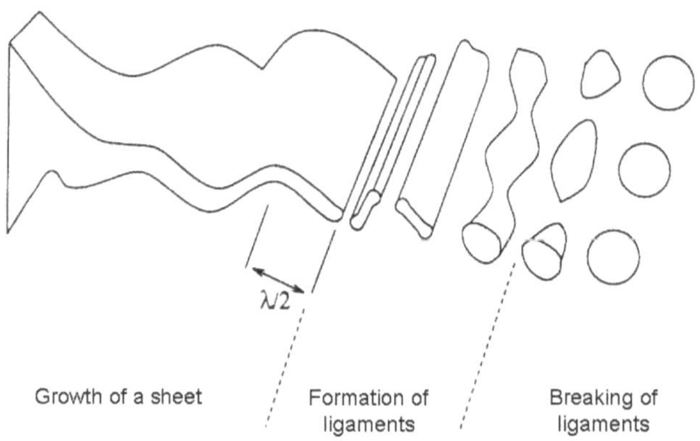

Fig. 12.2. Successive mechanisms leading to droplet formation

To study the destabilization of the liquid sheet before it breaks up, we can carry out a linearized study. We consider the case of a layer of viscous liquid (L) of thickness h (planar in the reference state) in the presence of a gas flow (G), which occurs on both sides of the liquid layer and parallel to it with the same relative velocity U (see Fig. 12.3).

In the reference frame of the liquid layer, the small disturbance of the flow parameter f can be written in the form

[3] If we compare Fig. 12.1a and b, it is clear that increasing the viscosity of the liquid does not modify the processes of distortion and atomization, but it does increase their characteristic length scales. Figure 12.1c shows that the behavior exhibited at very high values of viscosity is totally different: atomization can occur even when there is no perturbation to deform the interface; the mean liquid flow ensures that the liquid–gas interface increases until surface tension forces oppose this increase by rearranging the flow into ligaments and drops.

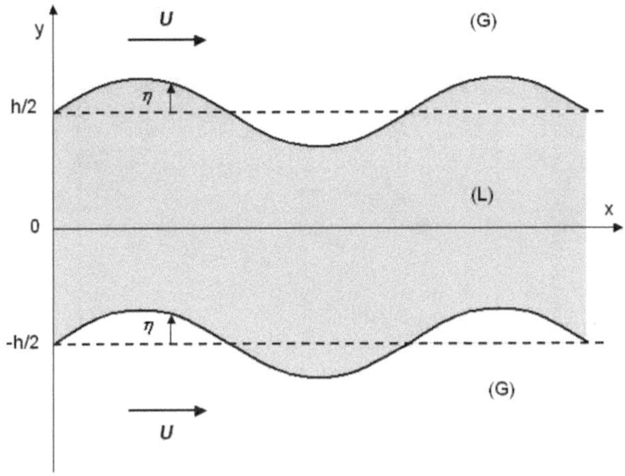

Fig. 12.3. Sinuous undulations $y = \pm h/2 + \eta(x,\, t)$ of the planar liquid sheet

$$f = \hat{f}(y)\, \exp i(kx - \omega t),$$

and the dispersion equation that links the wavenumber k to the pulsation ω [58, 56] can be written

$$\tanh(kh/2)(\Omega{+}2)^2{-}4\,\tanh(mh/2)\,\sqrt{\Omega+1}{+}\rho(\Omega{+}i\,Re)^2{+}Oh^{-2} = 0, \quad (12.1)$$

where $\Omega = -i\,\omega\,\rho_L/\mu_L\,k^2$, $\rho = \rho_G/\rho_L$, $m^2 = k^2\,(\Omega + 1)$, and in particular the Reynolds number $Re = \rho_L\,U/\mu_L\,k$ and the Ohnesorge number $Oh = \mu_L\sqrt{k/\rho_L\,\sigma}$. We have $Oh^{-2} = Re^2/We$, with $We = \rho_L\,U^2/\sigma\,k$ (see Table 5.3).

Knowing that $\omega = \omega_r + i\omega_i$, we then trace the curves that give the growth rate ω_i as a function of k (an example is given in Fig. 12.4 for $h = 3$ mm, $U = 10$ m s^{-1}, $\rho = 1.29 \times 10^{-3}$ kg m^{-3}, $\sigma = 0.075$ N m^{-1}). The effective wavenumber will be that which corresponds to the maximum growth rate.

The linearized method applies to many other liquid layer configurations in the presence of gas flow (see in particular [94, 145, 220]). The disruption of a liquid sheet by sinuous waves or dilation waves has been studied by Dombrowski and Hooper [70] and more recently by [114] for the case of a viscous liquid sheet in a high-speed viscous gas.

The shattering of a liquid ligament by the Rayleigh mode[4] has been described in particular by Raynal [232].

[4]The Rayleigh instability involves deforming and then breaking a liquid cylinder into spherical droplets under the effects of surface tension.

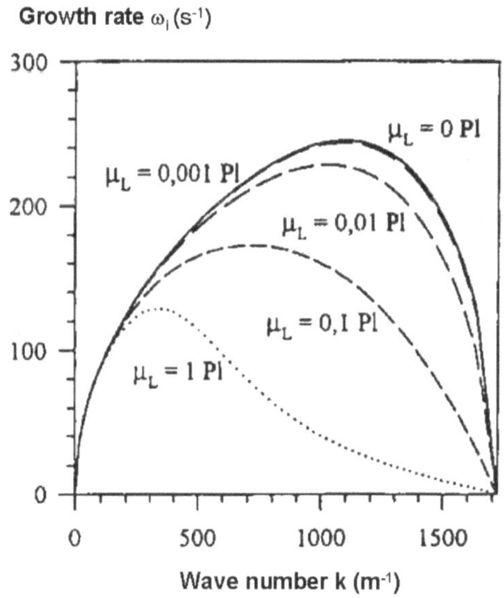

Fig. 12.4. Dispersion diagrams for several values of viscosity coefficient (C. Dumouchel, private communication; see also [58])

In the coaxial atomizer depicted in Fig. 12.5, a liquid (such as the liquid oxygen, "LOx," in the cryogenic engine of an Ariane rocket) is injected through a tube into a cylindrical flow of gas (gaseous hydrogen in this case) at high speed.

The resulting atomization process is complex, three-dimensional, and involves the formation of filaments (or ligaments, or fingering), as shown in Fig. 12.6.

As one can imagine, these processes are not easy to model theoretically or numerically. Transport processes, interface phenomena such as capillary tension, and the various instability phenomena characterized by dimensionless numbers (such as the Weber number $We = \rho_G(U_L - U_G)^2\delta_G/\sigma$, where the subscripts L and G denote "liquid" and "gas," respectively, U is the speed, δ_G is the boundary layer thickness, and σ is the surface tension) must be taken into account.

There are several methods that can be used to experimentally study droplet formation from ligaments. We will mention in this regard the recent study by Antkowiak et al. [3], who carried out and analyzed "Pokrovski's experience" of the generation of a liquid column due to the impact of a tube containing a liquid with a concave free surface (resulting from the meniscus, or a bubble) on a horizontal solid surface under the influence of gravity. This column fragments from the top (Fig. 12.7).

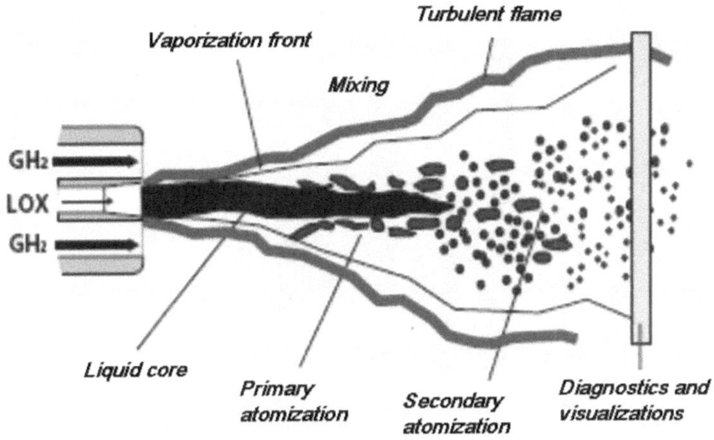

Fig. 12.5. The physical phenomena involved in the flame from a cryotechnic coaxial injector (extracted from the ONERA website)

Fig. 12.6. Formation of digitations by the liquid jet of a coaxial injector (after Marmotant and Villermaux 2004; reprinted, with permission, from [172])

Fig. 12.7. Liquid column fragmenting from the top (A. Antkowiak, private communication, 2009; reprinted with permission from [3])

Several methods are used to obtain the *distribution of the droplet diameters* in a spray.

In numerical methods such as VOF (Volume-of-Fluid), the space divided into domains, each of them containing possibly any liquid gas interface. This method is developed in particular by Zaleski [297].

The method of maximum entropy is efficient to predict the distribution of the diameter of droplets in the spray of a thin liquid layer in a quiet atmosphere. We use the Shannon entropy with a set of constraints written on the basis of classical conservation laws, namely, conservation of mass, energy, and momentum. In a variant, these constraints are replaced by a theoretical mean diameter, calculated from the wavelength of the most amplified disturbance obtained by linearized analysis. Various cases of linear analysis of surface disturbances can be envisaged, inviscid or viscous. Extending the method to the case of thick layers in the presence of fast flow is complex but a method of resolution can be sketched.

The distribution function in size and velocity of spray droplets can be predicted using a theoretical formulation based on the principle of maximum entropy [247, 125]. The Shannon entropy for the subsystems of possible states $1, 2, ..., n$ with probabilities $p_1, p_2, ..., p_n$, is written:

$$S(p_1, p_2, ..., p_n) = -k \sum_{i=1}^{n} p_i \ln p_i$$

In the continuous formulation and when only the reduced diameter occurs, the entropy becomes:

$$S = -k \int_0^{\infty} f(d) \ln f(d) d\,d$$

In classical models, the associated distribution function — discrete [50] or continuous [50], [153], [253] — is determined taking into account the constraints of speed, flow momentum and kinetic and surface energy [220]. This formalism requires the definition of a control volume [50], [153], [246] and experimental information such as average diameters measured in drops. Distributions are calculated and in fairly good agreement with experimental results for injectors producing liquid sheets flat, cylindrical or conical [280].

Cousin et al. [59] have developed a method giving the distribution of droplet size from a formalism based on the maximization of entropy in which the required information is derived mainly from theoretical considerations.

12.2 Simplified Model of a Flow with Particles

Here, we will assume that only one type of rigid isothermal particle is present, and that there are no changes in mass or adsorption or desorption reactions. The fluid is an inert gas comparable to a nonviscous fluid, except very close to the particles, where it exerts a frictional force $\mathbf{F}$, and to which it transfers an amount Q of heat per unit time and unit volume [140].

Apart from these exchange terms, each component behaves like a fluid. The gas and the (pseudo) fluid of particles have respective velocities of $\mathbf{v}_g$ and $\mathbf{v}_p$. The densities are defined per unit volume of the mixture. The particles do not interact with each other and so there is no interparticle pressure.

12.2.1 Variables That Characterize Two-Phase Flow

At the macroscopic scale, each "mixture particle" [5] is a composite system. It includes two components: a fluid phase (gaseous or liquid) and a condensed phase (which can be liquid or solid)—a system of very small particles comparable to a fluid. In this section we consider a fluid phase consisting of a gas, and any quantities referring to this gas will be have a subscript of g. However, the general balance equations that we will establish will also be valid for a liquid fluid phase so long as the state law is not specified. The condensed phase will be indicated by the subscript p.

We initially define specific quantities for each "fluid:"

$$\rho_{ps} = \frac{m_p}{\mathcal{V}_p} = \frac{\mu_p}{(4/3\,\pi R^3)}. \tag{12.2}$$

[5] Not to be confused with the condensed particles contained in the two-phase flow.

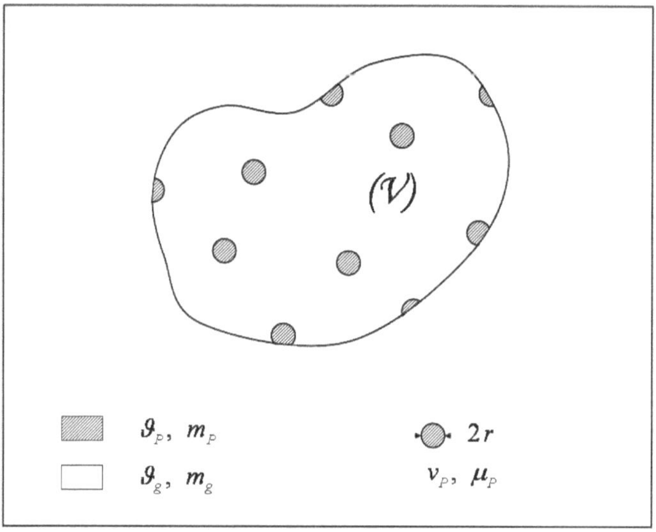

Fig. 12.8. Control volume of a mixture

Here, μ_p is the mass of a condensed particle and R is its radius if it is spherical,[6] m_p is the total mass of the condensed species in a small control volume, and $\mathcal{V}_p$ is the volume of the condensed species (Fig. 12.8). Similarly, for the fluid phase we have

$$\rho_{gs} = \frac{m_g}{\mathcal{V}_g}. \tag{12.3}$$

We will denote the total volume of mixture particles as $\mathcal{V} = \mathcal{V}_g + \mathcal{V}_p$. Thus, the partial densities are

$$\rho_p = \frac{m_p}{\mathcal{V}} = \epsilon_p \rho_{ps} = (1-\epsilon)\rho_{ps}, \ \rho_g = \frac{m_g}{\mathcal{V}} = \epsilon_g \rho_{gs} = \epsilon \rho_{gs}, \tag{12.4}$$

and the average density of the mixture is $\rho = \rho_g + \rho_p$.

The porosity of the mixture is then $\epsilon_g = \epsilon = \mathcal{V}_g/\mathcal{V} = \rho_g/\rho_{gs}$. We also define $\epsilon_p = \mathcal{V}_p/\mathcal{V} = 1-\epsilon$, and assume in what follows that ϵ is close to unity ($\epsilon_p \ll 1$).

On average, in a mixture particle, the condensed phase will have an average velocity $\mathbf{v}_p$ and the uncondensed phase will have a velocity $\mathbf{v}_g$.

The thermodynamic state of the condensed phase will be defined only by the temperature T_p, whereas that of the fluid phase will be defined by T_g and ρ_{gs}. Therefore, the internal energies per unit of mass (note that ρ_{ps} is assumed constant) are

$$e_{ps} = e_{ps}(T_p), \ de_{ps} = T_p \, ds_{ps} = c_c \, dT_p, \tag{12.5}$$

[6]In this simplified model, the particles are assumed to be spherical and have a single diameter and physical state.

where c_c is the specific heat of the condensed phase per unit mass, and (with $\rho_{gs} \cong \rho_g$ because $\epsilon \cong 1$)

$$e_{gs} = e_{ps}(T_g, \rho_g), \; de_{gs} = T_g \, ds_{gs} - p \, d(1/\rho_g). \tag{12.6}$$

If a perfect gas is considered, the fundamental energy law is deduced from the following laws of state:

$$p = \rho_g r_g T_g, \; de_g = c_v dT_g. \tag{12.7}$$

There are two types of exchanges between the condensed phase and the fluid phase when the condensed particles are rigid and do not evaporate: exchanges of frictional force $\mathbf{F}$ and heat Q from the fluid to the particles per unit volume of mixture. Since each component has its own velocity, we can define the material derivatives

$$d_p/dt = \partial/\partial t + \mathbf{v}_p \cdot \boldsymbol{\nabla}, \quad d_g/dt = \partial/\partial t + \mathbf{v}_g \cdot \boldsymbol{\nabla}. \tag{12.8}$$

12.2.2 Balance Equations

Particle Balance

For the pseudo-fluid of particles, if there are no interactions between the condensed particles the specific pressure does not intervene. The temperature is the average temperature of the condensed particles, which are assumed to have uniform temperatures. Locally, each particle has the same temperature.

By following the motion of a connected domain $(\mathcal{V}_p)$ in this fluid of particles, we obtain

$$\frac{d_p}{dt} \int_{\mathcal{V}_p} \rho_p d\mathcal{V} = 0, \tag{12.9}$$

which leads to (as in Chap. 4)

$$\frac{d_p \rho_p}{dt} + \rho_p \boldsymbol{\nabla} \cdot \mathbf{v}_p = 0 \;\; \text{(mass)}. \tag{12.10}$$

The momentum balance will be

$$\frac{d_p}{dt} \int_{\mathcal{V}_p} \rho_p \mathbf{v}_p d\mathcal{V} = \int_{\mathcal{V}_p} \boldsymbol{\mathcal{F}} d\mathcal{V}. \tag{12.11}$$

Taking the mass balance into account, we deduce that

$$\rho_p \frac{d_p \mathbf{v}_p}{dt} = \boldsymbol{\mathcal{F}} \;\; \text{(momentum)}. \tag{12.12}$$

For energy, it is necessary to take into account the heat flux from the gas to the condensed species and the power of the force $\boldsymbol{\mathcal{F}}$:

$$\frac{d_p}{dt} \int_{\mathcal{V}_p} \rho_p (e_{ps} + v_p^2/2) d\mathcal{V} = \int_{\mathcal{V}_p} (\mathcal{Q} + \boldsymbol{\mathcal{F}} \cdot \mathbf{v}_p) d\mathcal{V}. \tag{12.13}$$

We can then deduce the local equation:

$$\rho_p \frac{d_p(e_{ps} + v_p^2/2)}{dt} = \mathcal{Q} + \mathcal{F} \cdot \mathbf{v}_p \quad \text{(energy)}. \tag{12.14}$$

By taking the scalar product of the two sides of the momentum equation by $\mathbf{v}_p$ and by subtracting this equation term-by-term from the energy equation, we obtain

$$\rho_p \frac{d_p e_{ps}}{dt} = \mathcal{Q} \quad \text{(internal energy)}. \tag{12.15}$$

Note that $\mathcal{F}$ and $\mathcal{Q}$, which were introduced as exchanges between the gaseous and condensed species, should in fact be regarded as production terms for the mixture.

Gas-Phase Balance

Establishing the balance equations for the gas involves following the motion of a domain $(\mathcal{V}_g)$ of the gas. The equations obtained are then, locally,

$$\frac{d_g \rho_g}{dt} + \rho_g \boldsymbol{\nabla} \cdot \mathbf{v}_g = 0 \quad \text{(mass)}, \tag{12.16}$$

$$\rho_g \frac{d_g \mathbf{v}_g}{dt} + \boldsymbol{\nabla} p = -\mathcal{F} \quad \text{(momentum)}, \tag{12.17}$$

$$\rho_g \frac{d_g(e_{gs} + v_g^2/2)}{dt} + \boldsymbol{\nabla} \cdot (p\mathbf{v}_g) = -\mathcal{Q} - \mathcal{F} \cdot \mathbf{v}_g \quad \text{(energy)}, \tag{12.18}$$

$$\rho_g \frac{d_g e_{gs}}{dt} + p \boldsymbol{\nabla} \cdot \mathbf{v}_g = -\mathcal{Q} - \mathcal{F} \cdot (\mathbf{v}_p - \mathbf{v}_g) \quad \text{(internal energy)}, \tag{12.19}$$

since the heat fluxes are equal in size but opposite in sign, and the point at which the force is applied moves with the condensed phase.

Entropy Balance and Phenomenological Relations

The expression for the entropy production rate of the mixture is important for determining the constitutive relations that provide the expressions for $\mathcal{F}$ and $\mathcal{Q}$.

Since there are no irreversible processes in either the particles or the gas, the only irreversibility is associated with exchanges between the gas and the particles. Therefore,

$$\rho_p \frac{d_p s_{ps}}{dt} + \rho_g \frac{d_g s_{gs}}{dt} = \dot{W}_S \geq 0. \tag{12.20}$$

The thermodynamic laws lead to

$$\begin{cases} ds_{ps} = \frac{1}{T_p} de_{ps}, \ de_{ps} = c_c dT_p, \\ ds_{gs} = \frac{1}{T_g} de_{gs} + \frac{p}{T_g} d(\frac{1}{\rho_g}), \ de_{gs} = c_v dT_g, \\ \rho_{gs} \cong \rho_g. \end{cases} \tag{12.21}$$

By using these Gibbs relations and the balance equations, we thus obtain

$$\dot{W}_S = (\frac{1}{T_p} - \frac{1}{T_g})\mathcal{Q} + \frac{1}{T_g}\boldsymbol{\mathcal{F}} \cdot (\mathbf{v}_g - \mathbf{v}_p). \tag{12.22}$$

The principles of the thermodynamics of irreversible processes (see Sect. 3.2) allow us to consider the following linear laws when the generalized forces are not too large:

$$\mathcal{Q} = L_1\,(\frac{1}{T_p} - \frac{1}{T_g}), \ \boldsymbol{\mathcal{F}} = L_2\,\frac{1}{T_g}(\mathbf{v}_g - \mathbf{v}_p), \tag{12.23}$$

where L_1 and L_2 are coefficients of proportionality that are known as "phenomenological coefficients." The internal energy and momentum balances of the (pseudo) fluid of particles then give us

$$\begin{cases} \rho_p c_c d_p T_p/dt = (L_1/T_p T_g)(T_g - T_p), \\ \rho_p d_p \mathbf{v}_p/dt = (L_2/T_g)(\mathbf{v}_g - \mathbf{v}_p). \end{cases} \tag{12.24}$$

Two characteristic times τ_T and τ_v can be defined:

$$\tau_T = \rho_p c_c T_g T_p / L_1, \ \tau_v = \rho_g T_g / L_2, \tag{12.25}$$

which are the thermal and friction relaxation times, respectively. Finally, we have

$$\begin{cases} d_p T_p/dt = (T_g - T_p)/\tau_T, \\ d_p \mathbf{v}_p/dt = (\mathbf{v}_g - \mathbf{v}_p)/\tau_v. \end{cases} \tag{12.26}$$

Note that the ratio τ_T/τ_v is similar to a Prandtl number.

The balance equations can therefore be written in their final forms, which leads to the system

$$\begin{cases} d_p \rho_p/dt + \rho_p \boldsymbol{\nabla} \cdot \mathbf{v}_p = 0, \\ d_g \rho_g/dt + \rho_g \boldsymbol{\nabla} \cdot \mathbf{v}_g = 0, \\ \rho_p\, d_p \mathbf{v}_p/dt + \rho_g\, d_g \mathbf{v}_g/dt + \boldsymbol{\nabla} p = \mathbf{0}, \\ \rho_p\, d_p(e_{ps} + v_p^2/2)/dt + \rho_g\, d_g(e_{gs} + v_g^2/2)/dt + \boldsymbol{\nabla} \cdot (p\mathbf{v}_g) = 0, \\ d_p \mathbf{v}_p/dt = (\mathbf{v}_g - \mathbf{v}_p)/\tau_v, \\ d_p T_p/dt = (T_g - T_p)/\tau_T. \end{cases} \tag{12.27}$$

The entropy production rate becomes

$$\dot{W}_S = \frac{\rho_p c_c}{\tau_T T_g T_p}(T_g - T_p)^2 + \frac{\rho_p}{\tau_v T_g}(\mathbf{v}_g - \mathbf{v}_p)^2 \geq 0. \tag{12.28}$$

We will now consider two borderline cases.

At gas–particle equilibrium, the generalized forces cancel, which means that $T_g = T_p$ and $\mathbf{v}_g = \mathbf{v}_p$. The particles, which are very small, are carried by the gas and have the same temperature as it. The balance equations for the mixture are identical to those of a nonviscous fluid of velocity $\mathbf{v} = \mathbf{v}_g = \mathbf{v}_p$, of density $\rho = \rho_g + \rho_p$, of pressure p, and of temperature $T = T_g = T_p$. The laws of state obtained differ from those of the gas.

We deduce from the equations for the particles and the gas that (since $\mathbf{v} = \mathbf{v}_g = \mathbf{v}_p$)

$$d\rho_p/\rho_p = d\rho_g/\rho_g$$

for each particle of the mixture. It follows that

$$\rho_p = \kappa\rho_g.$$

If the coefficient κ is uniform for all particles and for an ideal gas, the classical relations

$$p = \rho\frac{r_g}{1+\kappa}T = \rho\bar{r}T, \; e = \frac{e_g + \kappa e_p}{1+\kappa} = \bar{c_v}T$$

are obtained.

These results show that, in this case, the mixture behaves like an ideal gas of specific heat

$$\bar{c_v} = (c_v + \kappa c_c)/(1+\kappa)$$

and of molar mass

$$\bar{\mathcal{M}} = (1+\kappa)\mathcal{M}_g.$$

Another situation that leads to $\dot{W}_S = 0$ occurs when the generalized fluxes $\mathcal{Q}$ and $\boldsymbol{\mathcal{F}}$ vanish, which in turn leads, according to (12.12) and (12.15), to T_p =const. and $\mathbf{v}_p$ =const. The particle motion is not influenced by that of the gas; exchanges with the gas are, to some extent, *frozen*. This is the case for very large particles.

These two extreme cases aside, it is necessary to take all of the equations of the problem into account, and the relaxation times τ_T and τ_v intervene.

12.2.3 Application to the Study of Small Disturbances

The reference configuration is that of gas–particle equilibrium at rest. It is thus characterized by

$$\mathbf{v}_{g0} = \mathbf{v}_{p0} = \mathbf{0}, \; T_0, \; \rho_{p0} = \kappa\rho_{g0}, \; p_0 = \rho_{g0} r_g T_0$$

(the ideal gas is already assumed). The subscript "$_0$" refers to average quantities in this linearized theory. Disturbances are indicated by the subscript "$_1$," leading to [140]

$$\mathbf{v}_g = \mathbf{v}_{g1},\ \mathbf{v}_p = \mathbf{v}_{p1},\ T_g = T_0 + T_{g1},\ T_p = T_0 + T_{p1},$$

$$\rho_g = \rho_{g0} + \rho_{g1},\ \rho_p = \kappa\rho_{g0} + \rho_{p1},\ p = p_0 + p_1.$$

The equations that describe the small motions are thus

$$\partial\rho_{g1}/dt + \rho_{g0}\nabla\cdot\mathbf{v}_{g1} = 0, \tag{12.29}$$

$$\partial\rho_{p1}/dt + \kappa\rho_{g0}\nabla\cdot\mathbf{v}_{p1} = 0, \tag{12.30}$$

$$\rho_{g0}\frac{\partial\mathbf{v}_{g1}}{dt} + \kappa\rho_{g0}\frac{\partial\mathbf{v}_{p1}}{dt} + \nabla p_1 = \mathbf{0}, \tag{12.31}$$

$$\rho_{g0}\,\partial e_{gs1}/dt + \kappa\rho_{g0}\,\partial e_{p1}/dt + p_0\nabla\cdot\mathbf{v}_{g1} = 0, \tag{12.32}$$

$$\partial\mathbf{v}_{p1}/dt = (\mathbf{v}_{g1} - \mathbf{v}_{p1})/\tau_v, \tag{12.33}$$

$$\partial T_{p1}/dt = (T_{g1} - T_{p1})/\tau_T, \tag{12.34}$$

$$p_1 = r_g\,T_0\,\rho_{g1} + r_g\,\rho_{g0}\,T_{g1}, \tag{12.35}$$

$$e_{gs1} = c_v\,T_{g1} = c_p/\gamma\,T_{g1}, \tag{12.36}$$

$$e_{ps1} = c_c\,T_{p1} = \beta\,c_p\,T_{p1},\ \beta = c_c/c_p. \tag{12.37}$$

Equation 12.30 is the only one that contains ρ_{p1}. Equation 12.32 can be rewritten with the aid of (12.36) and (12.37) as

$$\frac{\partial T_{g1}}{\partial t} + \kappa\beta\gamma\,\frac{\partial T_{p1}}{\partial t} + (\gamma - 1)\,T_0\nabla\cdot\mathbf{v}_{g1} = 0. \tag{12.38}$$

Combining this with (12.29), we obtain the first integral

$$T_{g1} + \kappa\beta\gamma\,T_{p1} + \frac{(\gamma - 1)\,T_0}{\rho_{g0}}\rho_{g1} = 0. \tag{12.39}$$

Then, combining this last equation with (12.35) yields

$$p_1 = \gamma\,r_g\,T_0\,\rho_{g1} - r_g\,\rho_{g0}\,\kappa\beta\gamma\,T_{p1}. \tag{12.40}$$

The entropy production rate given in Sect. 12.2.2 is equal to zero except at second order. It follows that

$$\kappa \frac{\partial s_{p1}}{\partial t} + \frac{\partial s_{g1}}{\partial t} = 0 \tag{12.41}$$

or

$$\kappa s_{p1} + s_{g1} = 0. \tag{12.42}$$

Since

$$\begin{cases} s_g = c_v \ln T_g - r_g \ln \rho_g + \text{const.}, \\ s_p = c_c \ln T_p + \text{const.}, \end{cases} \tag{12.43}$$

we have

$$\kappa\, s_{p1} + s_{g1} = \kappa\beta\, c_p\, T_{p1}/T_0 + c_v\, T_{g1}/T_0 - r_g\, \rho_{g1}/\rho_0 = 0, \tag{12.44}$$

which again gives the first integral (12.39) and nothing more than this.

Equations 12.34 and 12.39 give us

$$\frac{\partial T_{p1}}{\partial t} = \frac{1}{\tau_T}(-\kappa\beta\gamma\, T_{p1} + \frac{(\gamma-1)T_0}{\rho_{g0}}\rho_{g1} - T_{p1}). \tag{12.45}$$

The system to be solved is thus

$$\begin{cases} \partial\rho_{g1}/dt + \rho_{g0}\,\boldsymbol{\nabla}\cdot\mathbf{v}_{g1} = 0, \\ \rho_{g0}\,\partial\mathbf{v}_{g1}/dt + \kappa\rho_{g0}\,\partial\mathbf{v}_{p1}/dt + \gamma\, r_g\, T_0 \boldsymbol{\nabla}\rho_{g1} - r_g\,\rho_{g0}\,\kappa\beta\gamma\,\boldsymbol{\nabla} T_{p1} = \mathbf{0}, \\ \partial\mathbf{v}_{p1}/dt = (\mathbf{v}_{g1} - \mathbf{v}_{p1})/\tau_v, \\ \partial T_{p1}/\partial t = (1/\tau_T)\,[(\gamma-1)T_0\,\rho_{g1}/\rho_{g0} - (1+\kappa\beta\gamma)T_{p1}.] \end{cases} \tag{12.46}$$

Let us eliminate ρ_{g1} and $\mathbf{v}_{g1}$. We are left with

$$\begin{cases} (\partial/\partial t)\,[\tau_T \partial T_{p1}/\partial t + (1+\kappa\beta\gamma)T_{p1}] \\ +(\gamma-1)T_0\boldsymbol{\nabla}\cdot(\tau_v\,\partial\mathbf{v}_{p1}/dt + \mathbf{v}_{p1}) = 0, \\ (\partial/\partial t)\,[\tau_v\,\partial\mathbf{v}_{p1}/dt + (1+\kappa)\mathbf{v}_{p1}] \\ +\gamma\, r_g/(\gamma-1)\,\boldsymbol{\nabla}[\tau_T \partial T_{p1}/\partial t + (1+\kappa\beta)T_{p1}] = \mathbf{0}. \end{cases} \tag{12.47}$$

Let us now write, for a one-dimensional wave,

$$T_{p1} = \hat{T}_p\, e^{i(\omega t - \mathbf{K}\cdot\mathbf{x})}, \quad \mathbf{v}_{p1} = \hat{\mathbf{v}}_p\, e^{i(\omega t - \mathbf{K}\cdot\mathbf{x})}, \tag{12.48}$$

where ω is the wave pulsation and $\mathbf{K}$ the wavenumber vector. The system has a single solution if the following complex equation is verified:

$$\frac{K^2}{\omega^2}c_0^2 = \frac{(1+\kappa\beta\gamma+i\omega\tau_T)(1+\kappa+i\omega\tau_v)}{(1+\kappa\beta+i\omega\tau_T)(1+i\omega\tau_v)}, \tag{12.49}$$

where the square of the speed of sound in gas

$$c_0^2 = \gamma r_g T_0. \tag{12.50}$$

The complex number K is such that

$$K/\omega = 1/c(\omega) + i\gamma(\omega), \tag{12.51}$$

where c is the speed of the wave and $\omega\gamma(\omega)$ is the degree of damping per unit length (see Sect. 10.2.2).

When the frequency is close to zero, we obtain the real value

$$\frac{c_0^2}{c_e^2} = \frac{(1+\kappa\beta\gamma)(1+\kappa)}{(1+\kappa\beta)}. \tag{12.52}$$

The disturbance is sufficiently slow so that the mixture has time to return to equilibrium. The speed c is the equilibrium speed c_e.

When ω tends towards infinity, we obtain the real value

$$c = c_0. \tag{12.53}$$

The disturbance is so fast that the particles do not have enough time to react. This is the frozen case, and the speed of sound is that of the gas.

The shape of the curve $c(\omega)$ depends on the respective orders of magnitude of τ_T and τ_v. Note that $c_0 > c_e$ because $\gamma > 1$.

For very small and light particles and for finite values of the frequency, the speed of sound will be close to c_e. For large particles, which are more difficult to move, it will be closer to c_0.

Figure 12.9 gives the results for particular coefficient values when the Prandtl number is equal to 2/3, which corresponds to $\tau_T = \beta\tau_v$.

We have just studied the propagation of sound in a dilute suspension using simplified balance equations. In particular, we assumed a linear law for the expression of the force exerted by the fluid on the particles. We will see in Sect. 12.4.1 that other terms should also be taken into account. The propagation of sound can be studied while accounting for these other terms [69] (see Eq. 12.138).

12.2.4 Application to the Study of a Vortex in a Dilute Suspension

Basic Equations for a Given Gaseous Velocity Field

In contrast to the preceding case, here we will assume that the particles in dilute suspension do not act on the carrying fluid (which has a given field velocity $\mathbf{v}_g$). We will not take heat exchange into account; thus, the only

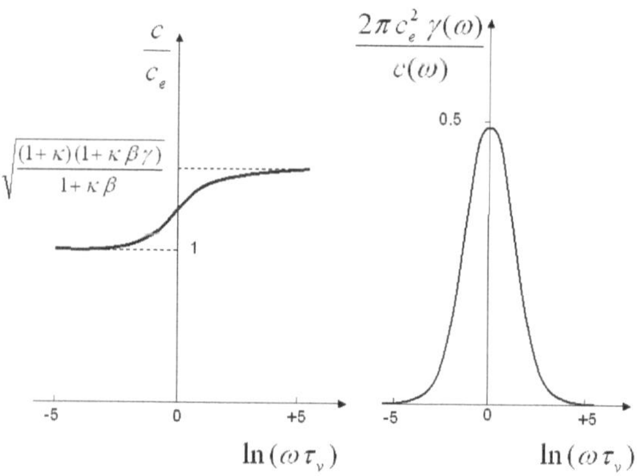

Fig. 12.9. Dispersion and absorption curves for a dilute suspension (P. Kuentzmann, private communication, 2009; see also [140])

exchanges considered are those of momentum. We only take into account the Stokes force (12.128), so the equation to solve is as follows:

$$\frac{d_p \mathbf{v}_p}{dt} = \frac{\mathbf{v}_g - \mathbf{v}_p}{\tau_v}, \tag{12.54}$$

where $\mathbf{v_p}$ is the particle velocity, $d_p\mathbf{v_p}/dt$ is the particle acceleration, and τ_v is the particle/gas relaxation time, which is assumed to be constant.

The issue of the influence of other forces was investigated by several authors in different situations such as the propagation of small periodic perturbations [69], and the effect of vortex motion [273, 274] with or without heat or mass transfer. The influence of gravity has been investigated, over the last few years in particular, by Ganan-Calvo and Lasheras [92]; Tio et al. [273], [274]; and Lasheras and Tio [148]. These authors have found several stable and unstable structures, closed orbits and attractors for particles whose specific densities are generally larger than the density of the fluid.

Marcu et al. [169] have studied the effect of gravity on particles that move inside a Burgers vortex. Burgers vortex-like structures appear to be important in turbulence (see Chap. 8). In the case of zero gravity, there is a critical value of the Stokes number below which the particle is driven by the flow towards the center of the vortex. Gravity effects lead to more complex situations and one or three equilibrium points appear away from the center depending on the Stokes and Froude numbers. In the vicinity of these equilibrium points, changes in stability are observed as functions of terminal velocity and the strain parameter. Nodes, foci, closed trajectories, and saddle points are obtained.

The particle trajectories will be determined, in the two-dimensional case, as the motions in the (x, y), the (r, t) and (θ, t), the (r, v_{pr}) and $(r, v_{p\theta})$, or the (r, θ) planes, preferably after having made the equation terms dimensionless.

The analysis of particle motion in a force field is nothing new for those who study mechanics or particle physics, since they are continually confronted with this type of problem. The present section focuses on vortex movements and solutions of (12.54) with a view to applying the results when studying suspensions.

The problems examined at the particle continuum scale are multiple (we consider this a macroscopic scale as opposed to the microscopic scale of the particle and the molecular scale internal to the particle or gas).

First, the velocity field will be investigated. This can be defined in Eulerian coordinates or Lagrangian coordinates. In the case of suspensions with uniform concentrations at $t = 0$, some very simple and characteristic situations are of great interest in order to understand the phenomenology of the motion. Generally, Eulerian coordinates can then be used to clearly define what is understood by a rotationally invariant solution near the center of the vortex and by a steady or unsteady solution. These results have to be converted to Lagrangian coordinates to take advantage of the knowledge gained from the analysis of the motion of an isolated particle. In addition, questions will be asked concerning the continuity of the solutions and the physical assumptions made in relation to the continuum approach.

In cylindrical (or two-dimensional polar) Eulerian coordinates, it is always assumed that the radial and angular components of the velocity vector depend only on the radial coordinate r and the time (the velocity field is invariant upon rotation at time t):

$$v_{pr} = v_{pr}(r, t), \; v_{p\theta} = v_{p\theta}(r, t). \tag{12.55}$$

In Lagrangian coordinates, due to the rotational invariance of the initial positions, we can write[7]

$$r = r(a, \tau), \; a = r(a, 0), \; t = \tau, dr = \frac{\partial r}{\partial a} da + \frac{\partial r}{\partial \tau} d\tau, dt = d\tau;$$

i.e., by inversion,

$$da = \frac{dr - (\partial r/\partial \tau) d\tau}{\partial r/\partial a}.$$

The Eulerian partial derivatives $\partial v_r/\partial t$, $\partial v_\theta/\partial t$ are then given in terms of Lagrangian coordinates, yielding

[7] We will retain the notation r instead of r_p for the position of the particle when there is no ambiguity.

$$\frac{\partial v_\theta}{\partial t} = \frac{r}{\partial r/\partial a}(\frac{\partial \dot{\theta}}{\partial \tau}\frac{\partial r}{\partial a} - \frac{\partial \dot{\theta}}{\partial a}\frac{\partial r}{\partial \tau}), \tag{12.56}$$

$$\frac{\partial v_r}{\partial t} = \frac{1}{\partial r/\partial a}(\frac{\partial \dot{r}}{\partial \tau}\frac{\partial r}{\partial a} - \frac{\partial \dot{r}}{\partial a}\frac{\partial r}{\partial \tau}). \tag{12.57}$$

The terms on the right hand side are zero for a steady velocity field, leading to the double condition

$$\frac{\partial \dot{\theta}/\partial \tau}{\partial \dot{\theta}/\partial a} = \frac{\partial \dot{r}/\partial \tau}{\partial r/\partial a} = b(a), \tag{12.58}$$

where $b(a)$ is the initial radial velocity as a function of a.

Condition (12.58) must be verified for any rotationally invariant steady flow. It should be noted that r depends only on a and τ, but that the polar angle θ depends on the initial polar angle too:

$$\theta = \theta(a,\, \alpha,\, \tau) = \theta(a,\, \alpha,\, 0),$$

whereas θ depends only on a and τ. This means that

$$\theta = \alpha + \theta_0(a,\, \tau),\;\; \theta_0(a,\, 0) = 0. \tag{12.59}$$

Condition (12.59) is always satisfied for any rotationally invariant flow, regardless of whether it is steady or not.

The particle concentration ρ_p (defined as the mass of condensed phase per unit volume of the mixture) for a rotationally invariant flow depends only on r and t in Eulerian variables, and therefore on a and τ in Lagrangian variables. In Eulerian variables, it obeys the equation

$$\frac{\partial(r\rho_p)}{\partial t} + \frac{\partial(v_{pr} r \rho_p)}{\partial r} = 0, \tag{12.60}$$

and in Lagrangian variables

$$\frac{\partial}{\partial \tau}(r\rho_p\frac{\partial r}{\partial a}) = 0,$$

yielding

$$\rho_p(a,\tau) = \frac{a\rho_p(a,0)}{r(\partial r/\partial a)}. \tag{12.61}$$

It should be noted that a steady velocity field—which thus obeys the double equation (12.58) if it is rotationally invariant—can give an unsteady density distribution, as demonstrated by the following examples.

Vortex with Concentrated Vorticity

The two-dimensional fluid flow is irrotational beyond a singularity located at one point. The circulation Γ is constant on any closed curve surrounding the center of the vortex. The gas velocity vector components in polar coordinates are therefore

$$v_{gr} = 0, \; v_{g\theta} = \Gamma/2\pi r. \tag{12.62}$$

This velocity field verifies the momentum and continuum equations for constant-density fluid flows (ρ_g is constant). From a practical standpoint, this type of vortex is the idealization of the flow created by an infinitely thin rod rotating at high velocity around itself in a fluid at rest at infinity. If the rod is removed, the vortex becomes free, and—assuming that its center remains steady—the vortex is damped by viscosity (see Sect. 8.2.2) according to

$$v_{gr} = 0, \; v_\theta = \frac{\Gamma}{2\pi r}(1 - e^{-\frac{r^2}{4\nu t}}), \tag{12.63}$$

where $\nu = \mu/\rho$ is the kinematic viscosity. The vorticity is no longer concentrated but extends to a viscous core with a radius that increases as $\sqrt{\nu t}$.

Analysis of Particle Motion

The reference time t_{ref} is the relaxation time τ_v involved in the particle momentum equation. It is the relaxation time needed for the particle to reach the same velocity as the fluid. The given circulation Γ is then used to define reference length and velocity such that

$$l_{ref} = r_v = \sqrt{\Gamma \tau_v/2\pi}, \; v_{ref} = \sqrt{\Gamma/2\pi\tau_v}. \tag{12.64}$$

The chosen reference length r_v is the distance from the center of the vortex at which the fluid travels $1/2\pi$ radians in a time τ_v, and the corresponding angular velocity component is then the reference velocity. Reference time and length decrease with decreasing particle density and radius and with increasing viscosity. Reference length decreases with Γ. When made dimensionless in this way, and considering (12.62), (12.54) becomes

$$\frac{d_p \mathbf{v}_p}{dt} = \frac{1}{r}\mathbf{e}_\theta - \mathbf{v}_p. \tag{12.65}$$

For simplicity, the same symbols are used in (12.65) and the following equations for dimensionless parameters. If θ and r are the dimensionless angular and radial coordinates of the particle, we then have (along θ and r, respectively)

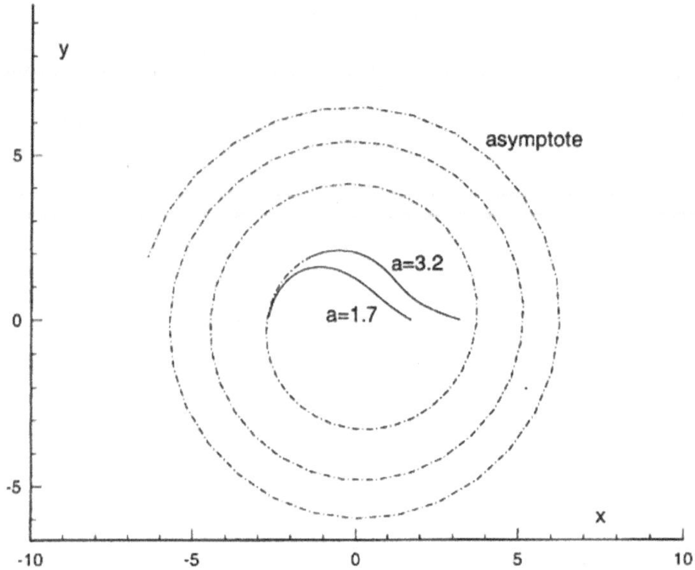

Fig. 12.10. Example of path determined for $a^2\beta = 2$, $b = 0$ [224]. Reprinted with the permission of the *International Journal of Multiphase Flow*

$$\begin{cases} (1/r)\,d(r^2\dot{\theta})/dt = 1/r - r\dot{\theta}, \\ \ddot{r} = r\dot{\theta}^2 = -\dot{r}, \end{cases} \tag{12.66}$$

where $\dot{r}$, $\ddot{r}$, $\dot{\theta}$ are the time derivatives of the position variables of the particle. The solution to this system of equations depends on four integration constants that are the values at time $t = 0$:

$$r(0) = a,\ \theta(0) = \alpha,\ \dot{r}(0) = b,\ \dot{\theta}(0) = \beta. \tag{12.67}$$

First integration of (12.66) and substitution yield [263]

$$\begin{cases} r^2\dot{\theta} = 1 + (a^2\beta - 1)e^{-t}, \\ \ddot{r} + \dot{r} = [1 + (a^2\beta - 1)e^{-t}]^2/r^3. \end{cases} \tag{12.68}$$

Solving (12.68) with (12.67) gives the particle trajectories and the particle position as a function of time.[8]

[8] We have not made use here of a characteristic hydrodynamic length that is independent of the characteristics of the vortex and particle. If such a hydrodynamic length L were to be introduced, there would be no characteristic dimensionless number, and a characteristic time $T = 2\pi r L^2/\Gamma$ could be defined that corresponds to the time taken for the fluid to travel $1/2\pi$ radians of the circle of radius L. A

The asymptotic behavior of this system (12.68) was studied in [224], and the results obtained compared favorably with the numerical results. An example is presented in Fig. 12.10, which shows two trajectories that, after initally moving away from each other, tend to converge in a spiral that is a common asymptote.

Before examining a particle continuum, we will recall the physical assumptions involved. For an individual particle in the vortex:

- The particle size must be small compared with the distance to the center of the vortex.
- The Reynolds number of the particle must be sufficiently small to verify Stokes's theory.
- The conditions must be satisfied for the other drag terms to be negligible compared with the Stokes terms. In particular, this means that the gas density must be very small compared with the particle density and that the accelerations must not be too large.

For the continuum (two-phase medium), the assumptions regarding dilute suspensions are assumed to hold. In particular, the volume occupied by the particles must be negligible compared with the volume of gas. This assumption does not hold for high particle concentrations, ρ_p. We will see that there are cases where the particles tend to concentrate in certain regions. If the interparticle distance becomes too small, the theory of suspensions without particle interactions will no longer be valid; indeed, if this distance becomes very small and high gradients are present, even the continuum hypothesis may no longer be valid, in which case shocks can occur.

Analysis of the Motion of the Particle Continuum

The above analysis naturally leads us to consider the particle continuum that comprises the suspension in terms of Lagrangian coordinates. In effect, the above equations and results are valid, but the question of continuity arises and discontinuity waves may occur. We will then examine this description using Eulerian coordinates, assuming a uniform particle size.

Stokes number St could then be introduced by comparing the two times τ_v and T: $St = \tau_v/T = r_v^2/L^2$.

The advantage of this approach is that the variations in τ_v and r_v with the parameters of the particle and vortex are directly determined by the Stokes number. However, the advantage of the approach we actually used is the simplicity of the equations obtained. The usual discussion concerning the value of St is replaced with a discussion of the orders of the dimensionless variables r and t (small values of St mean that times that are over τ_v and radii that are larger than r_v are considered; conversely, high values of St correspond to small times and distances from the vortex center). As we will see later, the immediate consequence is that if $r \gg l_{ref}$, the particle will respond quickly to the changing flow direction and the radial inertial drift outwards will be weaker. If $r \ll l_{ref}$, the inertia has a strong impact on the particle's trajectory.

The particle continuum obeys (12.54), which takes the form of (12.65) for a gas vortex with concentrated vorticity and yields (12.66) in Lagrangian coordinates, with

$$\begin{cases} r = r(a,\tau),\ \dot{r} = dr/d\tau,\ a = a(r,0),\ b = \dot{r}(a,0), \\ \theta = \alpha + \theta_0(a,\tau),\ \dot{\theta} = \partial\theta/\partial\tau, \\ 0 = \theta_0(a,0),\ \beta = \dot{\theta}(a,0), \end{cases} \tag{12.69}$$

assuming that the solutions are rotationally invariant, as in (12.59). However, considering a continuum also means that we must specify the functions

$$b = b(a),\ \beta = \beta(a). \tag{12.70}$$

The forms of these functions determine whether or not the particle flow is a continuum and the steady or unsteady nature of certain solutions.

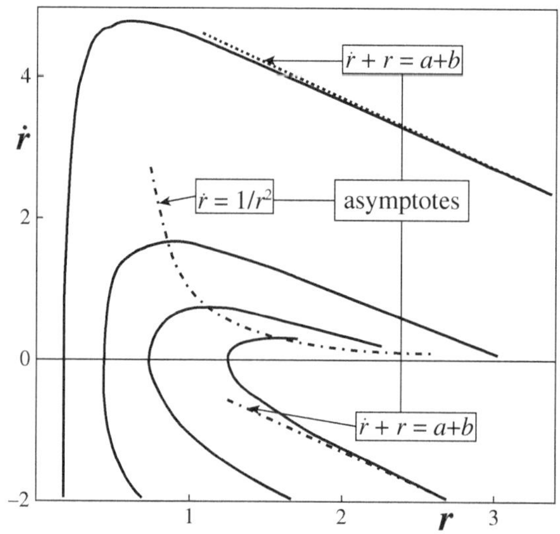

Fig. 12.11. A single system of curves for $K = 1$ (redrawn from [224]; reprinted with the permission of the *International Journal of Multiphase Flow*)

Instances of the general case where the product $a^2\beta$ is considered constant and equal to K are very interesting; (12.68) then becomes

$$\begin{cases} r^2\dot{\theta} = 1 + (K-1)e^{-t}, \\ \ddot{r} + \dot{r} = [1 + (K-1)e^{-t}]^2/r^3. \end{cases} \tag{12.71}$$

For $K = 1$, the particles have the same angular component of velocity as the fluid, but their radial velocity differs from the fluid due to the centrifugal effect. We get

$$\begin{cases} r\dot{\theta} = 0 \Rightarrow v_{pr} = 1/r, \\ \ddot{r} + \dot{r} = 1/r^3. \end{cases} \tag{12.72}$$

We can change variables and use r instead of t for given values of a and b; i.e., for a given trajectory. Setting $\dot{r} = v(a, b, r)$ yields the following result for the second part of (12.71):

$$v\frac{\partial v}{\partial r} + v = \frac{1}{r^3}[1 + (K-1)\exp(-\int_a^r \frac{dr}{v})]. \tag{12.73}$$

This equation is simplified considerably when $K = 1$. Only one system of curves is obtained. For a given trajectory, different initial conditions will lead to the same trajectory, as demonstrated in Fig. 12.11 for $b = 0$ and $b = -2$ after computation using (12.72).

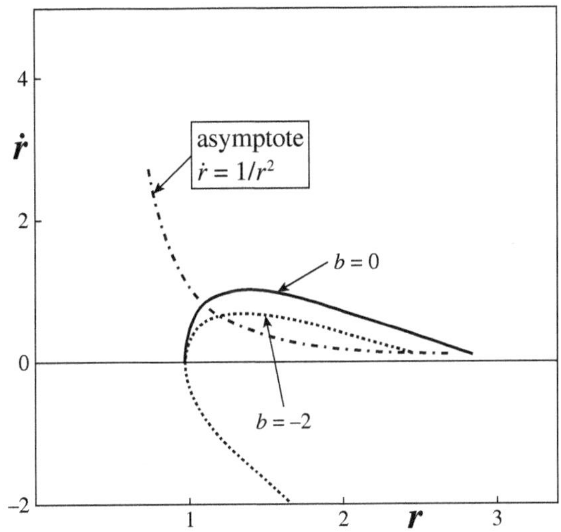

Fig. 12.12. Influence of b on the path for $K = 2$ (redrawn from [224]; reprinted with the permission of the *International Journal of Multiphase Flow*)

In contrast to the $K = 1$ case, the $K = 2$ case (for instance) shows that the trajectories obtained with $b = -2$ after computation are different from those plotted with $b = 0$. If a starting point of $b = -2$, $a = 1.7$ is chosen, the result from this new initial point is a new trajectory (Fig. 12.12). In addition, the locus of the maxima of v_{pr} is not unique.

12.3 Flow with Evaporating Droplets

After injecting them into an atmosphere or combustion chamber, the droplets form a two-phase flow with the surrounding gaseous mixture. This flow contains droplets of diverse sizes, and these sizes vary over time if vaporization and combustion occur, as shown in Fig. 12.13).

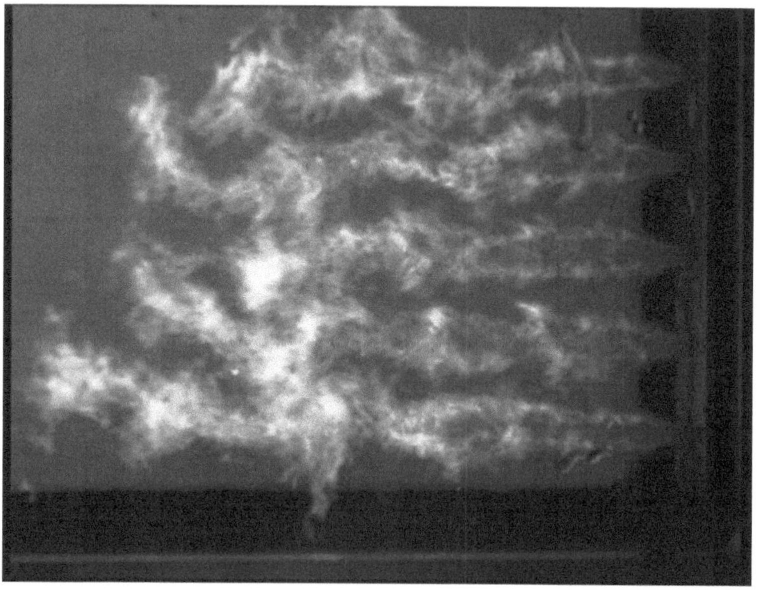

Fig. 12.13. Formation and combustion of a spray of liquid oxygen droplets in gaseous hydrogen (L. Vingert, private communication, 2005)

The balance equations obtained are more complex than the previous ones obtained with no mass source terms, and it is of course much more complicated to solve them analytically and numerically than those described in the previous section [55, 72, 188, 263, 290, 295]. These difficulties increase if there are interdroplet interactions and the flow is turbulent.

12.3.1 Flow Variables

We will assume that the two-phase medium is a suspension of very thin droplets. For a monodisperse suspension (droplets with the same radius r), we can represent the spatial repartition of the droplet and the fluid schematically, as shown in Fig. 12.8, and the variables that characterize the flow are the same as in Sect. 12.2.1.

In the case of a polydisperse suspension, it is usual to consider particle size classes. If we follow (in the Lagrangian sense) the motion of the particles of the mixture (i.e., of the suspension), the size associated with a class of droplets will be defined for instance in the injection zone, and will evolve further under the effect of vaporization, but will stay in the same class until they completely disappear.

We generally consider classes of particles, and we introduce the local particle number of each class. Each class (m) satisfies a conservation equation. The particles of a given class are such that, for example, at the initial time $t = 0$ (at the injection point or at the inlet of a pipe), their radii are between $r^{(m)}$ and $r^{(m)} + \delta r^{(m)}$. It is assumed that, for $t \geq 0$, the radii of the particles decrease under the effect of vaporization, but that they stay in their class during their motion and do not disappear. If germination occurs or droplets disappear, we must introduce specific terms. We will assume here that there is only one class of particles, and that the droplets do not germinate, disappear, coalesce or break up.

12.3.2 Balance Equations for the Flow

Droplet Balance

The method of obtaining balance equations is in this case similar to described in Sect. 12.2.2. In addition, we must take the mass source term due to particle vaporization into account (one-component liquid droplets), write the balance equation for the number of particles, and to consider that the gas phase is this time a gaseous mixture with transfer fluxes and possible chemical reactions. Let us consider the number n of particles per unit volume. The particle number $n(\mathbf{x}, t)$ is assumed to obey the integral conservation equation

$$\frac{d_p}{dt} \int_{\mathcal{V}_p} n d\mathcal{V} = 0. \tag{12.74}$$

The time derivative d_p/dt means (as described in Sect. 12.2.1) that the volume $\mathcal{V}_p$ follows the particle motion everywhere. It can be deduced that

$$\int_{\mathcal{V}_p} \frac{d_p}{dt}(n d\mathcal{V}) = \int_{\mathcal{V}_p} \frac{d_p n}{dt} d\mathcal{V} + \int_{\mathcal{V}_p} n \boldsymbol{\nabla} \cdot \mathbf{v}_p \, d\mathcal{V} = 0.$$

Therefore,

$$\frac{\partial n}{\partial t} + \boldsymbol{\nabla} \cdot (n \mathbf{v_p}) = 0. \tag{12.75}$$

To obtain the mass balance we write

$$\rho_p = n\, M, \quad \frac{d_p}{dt} \int_{\mathcal{V}_p} \rho_p d\mathcal{V} = \frac{d_p}{dt} \int_{\mathcal{V}_p} n\, M d\mathcal{V} = \int_{\mathcal{V}_p} n \frac{dM}{dt} d\mathcal{V},$$

where M is the droplet mass. $dM/dt = -\dot{M}$, and we deduce the local mass balance equation

$$\frac{d_p\rho_p}{dt} + \rho_p \nabla \cdot \mathbf{v}_p = \dot{W}_p \text{ (mass)}, \tag{12.76}$$

where $\dot{M}$ is the vaporization rate of each droplet and $\dot{W}_p = -n\dot{M}$ is the droplet production rate[9] per unit volume of the mixture due to vaporization. For momentum, we know that $M d_p \mathbf{v}_p/dt = \mathbf{F}$. Thus,

$$\frac{d_p}{dt}\int_{\mathcal{V}_p} \rho_p \mathbf{v}_p d\mathcal{V} = \int_{\mathcal{V}_p} \dot{W}_p \mathbf{v}_p d\mathcal{V} + \int_{\mathcal{V}_p} \rho_p \frac{d_p \mathbf{v}_p}{dt} d\mathcal{V},$$

or

$$\partial(\rho_p \mathbf{v}_p)/dt + \nabla \cdot (\rho_p \mathbf{v}_p \otimes \mathbf{v}_p) = \dot{W}_p \mathbf{v}_p + \boldsymbol{\mathcal{F}}.$$

Rearranging via (12.76), we find

$$\rho_p \frac{d_p \mathbf{v}_p}{dt} = \boldsymbol{\mathcal{F}} \text{ (momentum)}, \tag{12.77}$$

which is identical to (12.12), which was found without vaporization. Note that this equation can be deduced directly by multiplying the two sides of the equation $M d_p \mathbf{v}_p/dt = \mathbf{F}$ by n. However, in both methods, considering a droplet continuum implies that the time derivative d/dt becomes a material derivative d_p/dt.

For internal energy we have

$$\frac{d_p}{dt}\int_{\mathcal{V}_p} \rho_p \bar{e}_p d\mathcal{V} = \int_{\mathcal{V}_p} \dot{W}_p \bar{e}_p d\mathcal{V} + \int_{\mathcal{V}_p} \rho_p \frac{d_p \bar{e}_p}{dt} d\mathcal{V}.$$

The local form of this equation is deduced as

$$\frac{d_p(\rho_p \bar{e}_p)}{dt} + \rho_p \bar{e}_p \nabla \cdot \mathbf{v}_p = \dot{W}_p \bar{e}_p + \mathcal{Q} + \dot{W}_p (h_S - \bar{h}_p).$$

Combining with (12.76), we obtain

$$\rho_p\, d_p \bar{e}_p/dt = \mathcal{Q} + \dot{W}_p (h_S - \bar{h}_p) \text{ (internal energy)}. \tag{12.78}$$

Here, the subscript S means "at the droplet surface," and the bar indicates an average quantity for the liquid droplet. It is only when the droplet temperature is uniform that the enthalpy term is equal to the the latent heat l at the droplet temperature T_p.

The kinetic energy equation is deduced from momentum balance (12.12):

$$\rho_p \frac{d_p k_p}{dt} = \boldsymbol{\mathcal{F}} \cdot \mathbf{v}_p, \tag{12.79}$$

and the total energy equation for the particles of the spray becomes

$$\rho_p \frac{d_p(\bar{e}_p + k_p)}{dt} = \mathcal{Q} + \dot{W}_p (h_S - \bar{h}_p) + \boldsymbol{\mathcal{F}} \cdot \mathbf{v}_p \text{ (energy)}. \tag{12.80}$$

---

[9]This quantity is negative when vaporization prevails; it is the opposite of the gas-phase production rate.

Gas-Phase Balance

Here, we first write the equations for the mixture (the spray), and then we establish the equations for the gaseous phase by subtracting them from those for the particles. The method used to establish the equations is the same as that employed previously, and takes into account that there is conservation of total mass, momentum (if there are no external volume forces), and total energy in the mixture. The gas itself is a mixture of reacting species and is a viscous fluid. We only give the results here.

For the gas phase we obtain

$$\frac{d_g \rho_g}{dt} + \rho_g \nabla \cdot \mathbf{v}_g = -\dot{W}_p \quad \text{(mass)}, \tag{12.81}$$

$$\rho_g \frac{d_g \mathbf{v}_g}{dt} + \nabla p = -\boldsymbol{\mathcal{F}} - \dot{W}_p[\mathbf{v}_p - \mathbf{v}_g] \quad \text{(momentum)}, \tag{12.82}$$

$$\begin{aligned} &\rho_g d_g (e_{gs} + v_g^2/2)/dt + \nabla \cdot (\mathbf{q} + \mathbf{v}_g \cdot \mathbf{P}) = \\ &-\dot{W}_p (h_S - \tfrac{p}{\rho_{ps}} + k_p - e_g - k_g) - \mathcal{Q} - \boldsymbol{\mathcal{F}} \cdot \mathbf{v}_p \quad \text{(energy)}, \end{aligned} \tag{12.83}$$

$$\begin{aligned} &\rho_g d_g e_{gs}/dt + \nabla \cdot \mathbf{q} = -\nabla \otimes \mathbf{v}_g : \mathbf{P} - \dot{W}_p (h_S - p/\rho_{ps} + k_p - e_g \\ &-(\mathbf{v}_p - \mathbf{v}_g)^2/2) - \mathcal{Q} - \boldsymbol{\mathcal{F}} \cdot (\mathbf{v}_p - \mathbf{v}_g) \quad \text{(internal energy)}, \end{aligned} \tag{12.84}$$

since the heat flows are equal in size and of opposite signs, and the point of application of the force moves with the condensed phase.

The chemical species balance must be

$$\begin{cases} \rho_g d_g Y_v/dt + \nabla \cdot \boldsymbol{\mathcal{J}}_{Dv} = \dot{W}_v - (1 - Y_v)\dot{W}_p \quad \text{(vapor of droplets)}, \\ \rho_g d_g Y_j/dt + \nabla \cdot \boldsymbol{\mathcal{J}}_{Dj} = \dot{W}_j + Y_j \dot{W}_p \quad \text{(other species)}, \end{cases} \tag{12.85}$$

where the subscript v designates the gaseous species that forms the vapor of the one-component droplet, and j (for $j = 2, \dots N$) corresponds to the other species.

Entropy Balance Law for the Spray and Phenomenological Relations

Here, the entropy balance equation has additional terms compared to (12.20). It can be written

$$\rho_p \frac{d_p \bar{s}_p}{dt} + \rho_g \frac{d_g s_g}{dt} - \dot{W}_p (s_g - \bar{s}_p) + \nabla \cdot \frac{\mathbf{q} - \sum_j g_j \boldsymbol{\mathcal{J}}_{Dj}}{T_g} = \dot{W}_S \geq 0. \tag{12.86}$$

Just as we did for the inert particle case of Sect. 12.2.2, we use the basic thermodynamic laws for the gas (which is a chemical mixture this time) and the particles. The final expression for the entropy production rate is as follows:

$$
\begin{aligned}
&\dot{W}_S = \mathcal{Q}(1/T_p - 1/T_g) + \boldsymbol{\mathcal{F}} \cdot (\mathbf{v}_g - \mathbf{v}_p)/T_g + \mathbf{q} \cdot \boldsymbol{\nabla}(1/T_g) \\
&- \textstyle\sum_j \boldsymbol{\mathcal{J}}_{Dj} \cdot \boldsymbol{\nabla}(g_j/T_g) - (1/T_g)\boldsymbol{\Pi} : \boldsymbol{\nabla} \otimes \mathbf{v_g} - \textstyle\sum_j (g_j/T_g)\dot{W}_j \\
&- \dot{W}_p[(h_{vS} - \bar{h}_p)/T_p - (1/T_g)(h_{vS} - p_g/\rho_{ps} - e_g + (\mathbf{v}_p - \mathbf{v}_g)^2/2 \\
&- p_g/\rho_g - \textstyle\sum_j g_j Y_j - g_v + T_g(s_g - \bar{s}_p))],
\end{aligned}
\tag{12.87}
$$

where T_p is the local temperature of the pseudo-fluid of droplets and $\boldsymbol{\Pi} = \mathbf{P} - p\mathbf{1}$. We again find the usual terms for reacting gaseous mixtures and for suspensions with no mass exchange, but we also find a term in $(-\dot{W}_p)$. This term is

$$
-\dot{W}_p \, [(\frac{1}{T_g} - \frac{1}{T_p})h_{vS} + (\frac{\bar{g}_p}{T_p} - \frac{g_v}{T_g}) - \frac{p_g}{T_g \rho_{ps}} + \frac{1}{T_g}\frac{(\mathbf{v}_p - \mathbf{v}_g)^2}{2}].
$$

Neglecting the relative kinetic energy and the term in $1/\rho_{ps}$, we obtain the following for the gas–droplet exchange term in the entropy production rate:

$$
\dot{W}_S^{g-p} = (\frac{1}{T_p} - \frac{1}{T_g})(\mathcal{Q} + \dot{W}_p h_{vS}) + \boldsymbol{\mathcal{F}} \cdot \frac{\mathbf{v}_g - \mathbf{v}_p}{T_g} - \dot{W}_p(\frac{\bar{g}_p}{T_p} - \frac{g_v}{T_g}). \tag{12.88}
$$

The present phenomena have three origins:

- The temperature difference
- The chemical potential difference
- The velocity difference.

Between the generalized fluxes $\mathcal{Q} + \dot{W}_p h_{vS}$, $\boldsymbol{\mathcal{F}}$, $\dot{W}_p$ and the conjugate generalized forces in the last relation, we can write—if the generalized fluxes and forces are sufficently small—linear phenomenological relations that follow the laws of TIP (see Sect. 3.2). Couplings can appear with other terms in (12.87). If there are no couplings, we have:

$$
\begin{cases}
\mathcal{Q} + \dot{W}_p h_{vS} = L_1 \, (1/T_p - 1/T_g), \; \boldsymbol{\mathcal{F}} = (L_2/T_g)(\mathbf{v}_g - \mathbf{v}_p), \\
\dot{W}_p = L_3(g_v/T_g - \bar{g}_p/T_p),
\end{cases}
\tag{12.89}
$$

where L_1, L_2 and L_3 are phenomenological coefficients. These coefficients can be deduced from analyzing the problem at the particle scale (which is the subject of the following section). Very often vapor and liquid are in equilibrium at the droplet surface, and the droplet mass flow rate is given directly by a formula such as (12.173).

To discern a possible liquid–vapor disequilibrium at the droplet surface, it is possible to rearrange the formula giving the entropy production rate.[10] Finally, (12.88) becomes

$$\dot{W}_S^{g-p} = (1/T_p - 1/T_g)\bar{\mathcal{Q}} + \boldsymbol{\mathcal{F}} \cdot (\mathbf{v}_g - \mathbf{v}_p)/T_g - \dot{W}_p(g_{pS} - g_{vS})/T_p,$$

where $\bar{\mathcal{Q}} = \mathcal{Q} + \dot{W}_p(h_{vS} - \bar{h}_v)$. New phenomenological relations can be written in which vaporization disequilibrium appears through the relation

$$\dot{W}_p = L_3' \frac{g_{vS} - g_{pS}}{T_p}.$$

12.3.3 Application to the Study of Spray Flame Propagation

Modeling a Spray of Small Droplets

Let us now consider a spray flame propagating through a fresh mixture consisting of small fuel droplets, fuel vapor, and air. We will first study the steady 1-D solution to the problem, and we will then summarize the theory of Nicoli, Haldenwang and Suard [184] for analyzing the linear stability of the flame obtained.

The configuration is similar to that of Sect. 10.5, except that this time the fresh mixture contains small fuel droplets that form a dilute suspension. We consider a single reaction with a high activation energy. This leads to a thin premixed flame. Droplet vaporization is assumed to occur inside the preheating zone, and the reaction occurs in the gaseous phase. This configuration is shown in Fig. 12.14 for the 1-D spray flame. The following assumptions are made:

- The combustion and spray scales are well separated, so $\ell_d \ll \ell_f$, where ℓ_d is the inter-droplet distance and ℓ_f the reactive-diffusive length. It can be proven that this condition is equivalent to $Da \ll 1$, where Da is the Damköhler parameter (equal to the ratio of a characteristic vaporization time τ_{vap} to the chemical time τ_{chem}). Indeed,

$$D_a = \tau_{vap}/\tau_{chem} = (\ell_d/\ell_f)^2.$$

[10] The difference between the ratios of chemical potential to temperature (i.e., $\frac{g_p}{T_p} - \frac{g_v}{T_g}$) in (12.88) can be decomposed into three terms. The first term corresponds to droplet internal exchanges. It vanishes if droplet quantities such as temperature are uniform, which we will assume hereafter. In the opposite case, we must take into account internal droplet relaxation. The second term relates to vaporization–condensation at the droplet surface. It is equal to zero at evaporation equilibrium. The last term is related to exchanges between the droplet surface and the bulk of the gaseous phase. Part of this term can be regrouped with $\mathcal{Q}$.

In other words, we have $g_{vS}/T_S - g_v/T_g \cong \bar{h}_v(1/T_S - 1/T_g)$, where $\bar{h}_v$ is the average local enthalpy per unit mass of the vapor between T_S, at the droplet surface, and T_g in the bulk.

- The medium is thus considered to be a continuum consisting of three species: air, gaseous fuel, and liquid fuel. The diffusion of liquid droplets in this continuum is ignored.
- We will assume that the latent heat vanishes here.
- We will also make the following assumptions for the gas phase: there is no thermal diffusion process ($D_T = 0$); external forces and viscosity are negligible; we have a constant, static pressure to a first approximation; the Fourier law holds for the thermal conduction; Fick's law holds for the diffusion, and there is a single coefficient of diffusion for all species; the Lewis number is equal to 1; there is only one chemical reaction; a mixture of perfect gases is present.

Scales related to the gaseous premixed flame of Zel'dovich and Kamenetskii (see [301]), hereafter termed the "ZFK flame," are used. We introduce $\theta(x, y, t) = (T - T_u)/(T_b - T_u)$ for the temperature,[11] and fuel mass fractions Y_l and Y_g divided by the initial overall fuel mass fraction $(Y_{lu} + Y_{gu})$. The time and length units are, respectively, $(D_{th})_b/U^2_{ZFK}$ and $(D_{th})_b/U_{ZFK}$, with

$$U_{ZFK} = \sqrt{\frac{2Le\lambda\rho_b^2 Y_{O_2} B}{C_p\rho_u^2 Ze^2}} \exp\left(-\frac{E_a}{2RT_b}\right).$$

We finally obtain the balance laws in a nondimensional form:

$$\begin{cases} \partial Y_l/\partial t + \mathbf{v}\cdot\boldsymbol{\nabla} Y_l = -\dot{W}_{vap}, \\ \partial Y_g/\partial t + \mathbf{v}\cdot\boldsymbol{\nabla} Y_g = \Delta Y_g + \dot{W}_{vap} - \dot{W}_{chem}, \\ \partial\theta/\partial t + \mathbf{v}\cdot\boldsymbol{\nabla}\theta = \Delta\theta + \dot{W}_{chem}, \end{cases} \tag{12.90}$$

with

$$\begin{cases} \dot{W}_{chem} = (Ze^2/2)\, Y_g \exp[-Ze\,(1-\theta)], \\ \dot{W}_{vap} = (Y_l/Da)\, H(\theta - \theta_v), \end{cases} \tag{12.91}$$

where H is the Heaviside function.

Steady 1-D Analysis of a Spray Flame

This analysis proceeds in a relatively similar fashion to that in Sect. 10.5.1, but with a transition spray gas. An asymptotic method will be applied, based on the four zones shown in Fig. 12.14. The results for each zone are given below.

Zone I $= (-\infty < x \le x_v)$, where x_v is such that $\theta(x_v) = \theta_v$ (i.e., this is where the vaporization starts):

[11] θ is similar to c in (8.110) or β_T in (10.163).

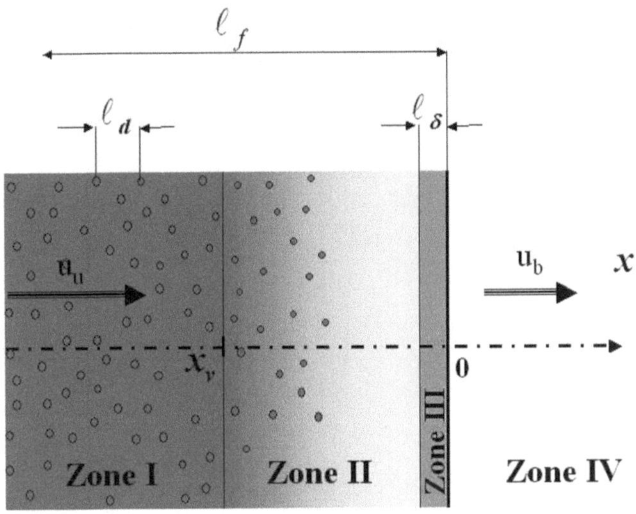

Fig. 12.14. 1-D spray flame configuration of Nicoli et al. [184]. Zone I, with x_v, is where the vaporization starts. Zone II is where the vaporization is assumed to occur. Zone III is the the asymptotically thin reaction zone. Zone IV is where the burnt gas is in equilibrium

$$\begin{cases} \theta = \exp(U_{SF}\, x) < \theta_v, \; Y_l = Y_{lu} = 1, \\ Y_g = Y_{gv} \exp(U_{SF}\, x). \end{cases} \tag{12.92}$$

Zone II = ($x_v \leq 0^-$):

$$\begin{cases} \theta = \exp(U_{SF}\, x) > \theta_v, \; Y_l = \exp\,(x - x_v)/Da, \\ Y_g = 1 + [-1 + \exp\,(x_v/Da)/(1 + 1/DaU_{SF})] \exp(U_{SF}\, x) \\ - \exp\,[(x - x_v)/Da]/(1 + 1/DaU_{SF}), \end{cases} \tag{12.93}$$

$$\begin{cases} \theta = \exp(U_{SF}\, x) < \theta_v, \; Y_l = Y_{lu} = 1, \\ Y_g = Y_{gv} \exp(U_{SF}\, x), \end{cases} \tag{12.94}$$

where U_{SF} is the spray flame speed.
Zone III = ($0^- \leq 0^+$) (see Eq. 10.173 in Sect. 10.5.2):

$$\begin{cases} \theta^- = \theta^+ = 1, \; (Y_l)^- = (Y_l)^+ = 0, \; (Y_g)^- = (Y_g)^+ = 0, \\ d\theta/dx|_- = 1, \; d\theta/dx|_+ = 0, \; dY_g/dx|_- = -1, dY_g/dx| + - = 0. \end{cases} \tag{12.95}$$

Zone IV = ($0^+ < x < \infty$) (where $\theta = 1$).

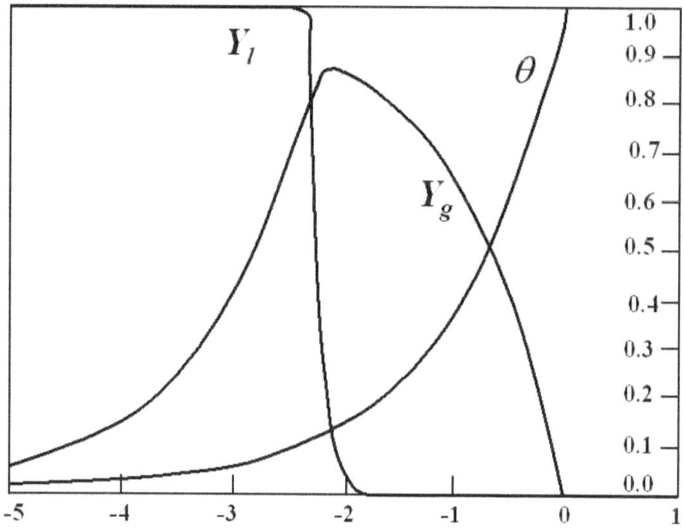

Fig. 12.15. Steady spray flame structure with $Da = 0.1$, $\theta_v = 0.1$ (P. Haldenwang, private communication, extracted from [184])

When $\theta_v^{1/Da} \ll 1$, which is obtained for $Da < 1$ and $0 < \theta_v < 1$ and ensures the quasi-complete vaporization of the droplets upstream of the reaction zone, the spray flame speed is found to be identical to the ZFK gaseous flame speed. The results are shown in Fig. 12.15. Numerical confirmation of the latter result can be found in [184, 268].

Linear Stability Analysis

Equations are rewritten in the perturbed frame, which is moving at the speed of the perturbed flame:

$$\begin{cases} \xi = [x - x_{SF}(y,\, t)]\, U_{SF}/D_{th}, \\ \eta = y\, U_{SF}/D_{th}, \\ \tau = t\, U_{SF}^2/D_{th}. \end{cases} \tag{12.96}$$

The flow parameters are written in the form $f = \bar{f} + \epsilon\, \hat{f}(\xi)\, \exp(\omega\tau + ik\eta)$, where ϵ is the (infinitely small) dimensionless amplitude of front corrugation. An asymptotic analysis was performed by Nicoli, Haldenwang and Suard [184], who studied the stability of the spray flame in the 1-D and 2-D cases. Threshold data are then obtained for pulsating flat spray flames ($k = 0$) and for pulsating wrinkled spray flames ($k \neq 0$). The analytical method used is that employed for matched asymptotic expansions. Four domains are considered, as in the preceding study of steady planar flames.

We obtain a dispersion relation that takes the following simplified form for $\theta_v^{1/Da} \ll 1$:

$$(1-\frac{2r^-}{Ze})\frac{[\omega + r^- + Da\,(\omega^2 - k^2)]\,\exp\,(-2r^-\xi_v)}{(k\,Da)^2 - Da - (\omega Da + 1)^2} + \frac{2r^-}{Ze}(r^- - r^+) = 0, \quad (12.97)$$

where $r^{\pm} = (1/2)\,[1 \pm \sqrt{1 + 4(\omega + k^2)}]$.

The theoretical and numerical investigations allowed the authors to interpret their results physically. The advanced instability mechanism complies with the following sequence: (a) assume that the flame accelerates for some reason; (b) the preheating zone then shortens; (c) vaporization consequently occurs closer to the reaction zone; (d) gaseous fuel enhancement therefore occurs in the reaction zone, and; (e) the flame speeds up because the heat release is enhanced. Step (e) is the feedback stage that is essential for instability.

In a recent paper [185], the authors compare the stabilities of the flames obtained with gas fuel and spray fuel in lean mixtures. Their conclusion is as follows:

"Our stability analysis supplies the stability diagram in the plane $\{Le, \delta\}$ for various Ze values[12] and shows that, for all Le, the plane front becomes unstable for high liquid loading. At large or moderate Lewis number, we show that the presence of droplets substantially diminishes the onset threshold of the oscillatory instability, making the appearance of oscillatory propagation easier. Oscillations can even occur for $Le < 1$ when sufficient spray substitution is operated. The pulsation frequency occurring in this regime is a tunable function of δ. At low Lewis number, the substitution of a spray for a gas leads to a more complex situation in which two branches can coexist: the first one still corresponds to the pulsating regime, while the other is related to the diffusive–thermal cellular instability."

12.4 Problems at the Particle Scale

In Sect. 12.2 we deduced the expressions for the force $\mathcal{F}$ and heat flow $\mathcal{Q}$ acting on the particles per unit volume of the mixture from phenomenological laws. Two characteristic times (τ_T and τ_v) appeared, although we were not able to specify their values. We can obtain these values by studying the problem at the scale of the condensed particle (known as the mesoscopic scale). This theory provides expressions that are generally corrected by empirical formulae.

If we ignore the simplifying assumptions of Sect. 12.2 to some extent, we are confronted with the presence of many phenomena that will be evoked thereafter. One of these is the variation in particle mass due to evaporation or

[12] δ is the spray liquid loading: the ratio of liquid fuel mass to overall fuel mass.

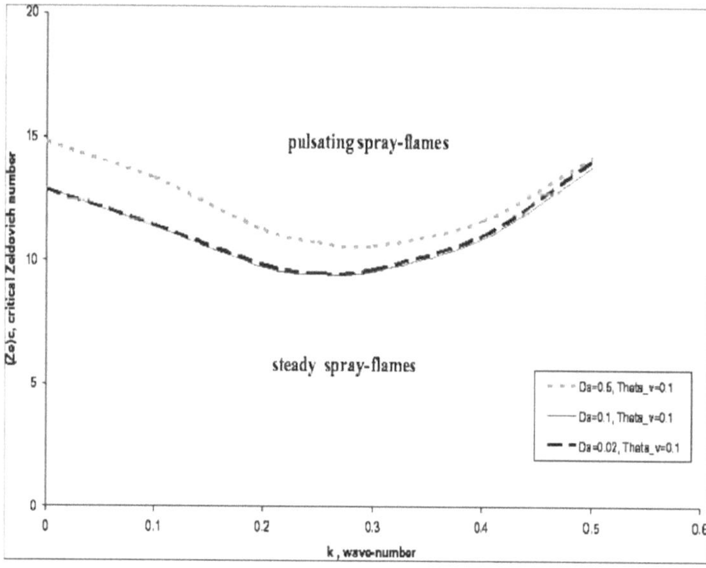

Fig. 12.16. Neutral curve of the oscillatory instability of a spray flame for various Damkhöler numbers (P. Haldenwang, private communication, 2009; see also [184])

condensation. We will study the case of a drop in the presence of a diffusion flame. In each of these problems the particle will be assumed to be spherical.

12.4.1 Force Exerted by a Fluid on a Spherical Particle

The fluid will initially be assumed to be incompressible, and several situations will be studied before we arrive at a realistic formula [90, 96].

Case of an Incompressible Perfect Fluid

A sphere located in a steady flow of a nonviscous fluid moving with an irrotational velocity at infinite U_∞ is only subjected to the Archimedes force. The motion of the fluid does not induce any other force. If R is the radius of the sphere, the force is thus directed vertically upwards and has a value of

$$F = \frac{4}{3}\pi R^3 \rho_{gs}\, g. \tag{12.98}$$

Induced Inertial Force

The flow can be described using the stream function ψ or the velocity potential ϕ, which are both harmonics. Variables x, y (orthonormal coordinates) and r

are defined in Fig. 12.17. This solution is valid only in the case of cylindrical symmetry (where the axis of symmetry is Ox). The stream function ψ obeys the equation

$$\frac{\partial^2 \psi}{\partial x^2} + \frac{\partial^2 \psi}{\partial y^2} - \frac{1}{y}\frac{\partial \psi}{\partial y} = 0. \tag{12.99}$$

A doublet of intensity $K = -2\pi U_\infty R^3$ yields

$$\psi = \frac{K}{4\pi}\frac{\sin^2\theta}{r} = \frac{K}{4\pi}\frac{y^2}{r^3}, \tag{12.100}$$

and a uniform flow gives

$$\psi = U_\infty \frac{r^2 \sin^2\theta}{r} = U_\infty \frac{y^2}{2}. \tag{12.101}$$

By superposing these two equations we get

$$\phi = U_\infty x - \frac{K}{4\pi}\frac{x}{r^3}, \quad \psi = U_\infty \frac{y^2}{2} + \frac{K}{4\pi}\frac{y^2}{r^3}. \tag{12.102}$$

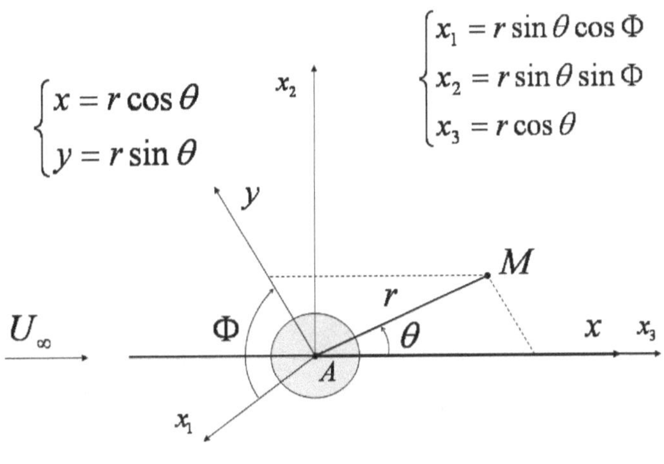

Fig. 12.17. Cylindrical symmetry configuration. The Cartesian frame (x_1, x_2, $x_3 = x$) is attached to the center (A) of the sphere. Fluid motion does not depend on Φ

If, instead of being motionless in a uniform flow at infinity,[13] the sphere exhibits uniform motion with a translational velocity of $\mathbf{V}_A$ in a motionless atmosphere at infinity, the result is the same in a reference frame associated with the sphere. If $\mathbf{V}_A$ is not constant, the field of the flow is determined at any time in a similar way, but the intensity of the doublet is variable over time, resulting in the existence of a force arising from the sphere's acceleration. That force, which is proportional to the acceleration γ_A, utilizes an induced mass. To obtain its value, it is necessary to shift from the reference frame, in which the motion of the sphere is observed, to the reference frame related to the sphere (Fig. 12.18).

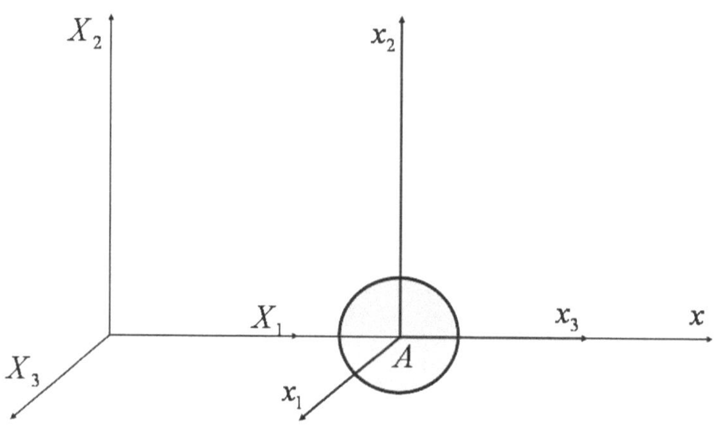

Fig. 12.18. System of axes related to the sphere

We then have

$$\phi(X_1, X_2, X_3, t) = \varphi(x_1, x_2, x_3, t). \tag{12.103}$$

We note that, at any point on the sphere's surface, the relative normal velocity of the flow is zero; i.e.,

$$(\mathbf{v} - \mathbf{V}_A) \cdot \mathbf{n} = 0 \tag{12.104}$$

[13]The symmetry of the pressure field acting on the sphere's surface leads to the flow exerting a net force of zero on the sphere. However, this zero drag directly contradicts observations of substantial drag on bodies moving relative to fluids. This peculiar result is known as "d'Alembert's paradox."

or

$$\partial\varphi/\partial r = V_A \cos\theta. \tag{12.105}$$

The potential

$$\varphi = -\frac{R^3}{2r^2} V_A \cos\theta \tag{12.106}$$

obeys this condition. At any point M, and for any vector $\mathbf{AM}$, we thus obtain

$$\phi = R^3 \frac{\mathbf{V}_A \cdot \mathbf{AM}}{2\, AM^3}. \tag{12.107}$$

Let us note that

$$\partial\mathbf{AM}/\partial t = -\mathbf{V}_A. \tag{12.108}$$

The application of the unsteady Bernoulli equation makes it possible to calculate the forces exerted on the sphere. Only the term $\partial\phi/\partial t$ gives a nonzero result, in addition to the Archimedes force that has already been evoked. We have

$$\frac{\partial\phi}{\partial t} = -\frac{R^2}{2}\left(\frac{\gamma \cdot \mathbf{AM}}{r^3} - \frac{\mathbf{V}_A^2}{r^3} + 3\frac{\mathbf{V}_A \cdot \mathbf{AM}}{r^5}\right). \tag{12.109}$$

The last two terms disappear from the expression for the force due to symmetry, and integration over the sphere's surface (Σ) gives

$$\mathbf{F}' = -\frac{\rho}{2R}\int_\Sigma (\gamma \cdot \mathbf{AM})\mathbf{AM} ds = -\frac{2}{3}\pi\rho R^3 \gamma_A = -m'\gamma_A. \tag{12.110}$$

The induced mass m' is equal to half the mass the sphere would have if it was filled with the external fluid. The induced inertial force $\mathbf{F}'$ also exists in the presence of a viscous fluid, but with a different expression.

Case of an Incompressible Viscous Fluid

Archimedes Force

The particles of a two-phase flow are swept along by the carrier fluid, mainly under the influence of viscous friction forces [107]. In most cases, the Reynolds number (calculated using the relative gas–particle velocity and the particle radius or diameter) is very low. In the vicinity of the particle, the inertia of the fluid is weak compared to viscous forces, so nonlinear terms of the momentum equation are neglected. Thus, to a first approximation, the system that describes the steady flow is

$$\begin{cases} \nabla p = \mu \Delta \mathbf{v}, \\ \nabla \cdot \mathbf{v} = 0. \end{cases} \tag{12.111}$$

These equations are at the heart of Stokes' theory, which enables the friction force to be calculated accurately. On the other hand, the flow obtained is not satisfactory, since the streamlines remain very disturbed far from the sphere and there is no wake, which contradicts experimental observations. To obtain a better description far from the sphere, we must use Oseen's theory, which takes into account the convection induced by the velocity at infinity: $\mathbf{U}_\infty$. We can then write the system

$$\begin{cases} \rho \mathbf{U}_\infty \cdot \nabla \otimes \mathbf{v} + \nabla p = \mu \Delta \mathbf{v}, \\ \nabla \cdot \mathbf{v} = 0, \end{cases} \tag{12.112}$$

which amounts to linearizing the inertial terms.

Let us first establish the complete system of equations for the general case, without ignoring the nonlinear terms and for axisymmetric flow. The variables used are those of the preceding section. In order to eliminate the pressure, we consider the curls of the two sides of the momentum equation, which then becomes

$$\frac{\partial \boldsymbol{\omega}}{\partial t} + \nabla \times (\boldsymbol{\omega} \times \mathbf{v}) = \nu \Delta \boldsymbol{\omega}, \tag{12.113}$$

where $\boldsymbol{\omega} = 1/2\ \nabla \times \mathbf{v}$, $\nu = \mu/\rho$. The continuity equation becomes

$$\frac{\partial (yu)}{\partial x} + \frac{\partial (yv)}{\partial y} = 0, \tag{12.114}$$

which gives the stream function ψ such that

$$u = \frac{1}{y}\frac{\partial \psi}{\partial y}, \quad v = -\frac{1}{y}\frac{\partial \psi}{\partial x}. \tag{12.115}$$

We show that $\boldsymbol{\omega}$ is collinear to the $\mathbf{k}$ vector of the standard basis $\{\mathbf{i}, \mathbf{j}\ \mathbf{k}\}$ of the three-dimensional Cartesian coordinate system (see Fig. 12.19) and that its algebraic value along Oz is

$$\omega = -\frac{1}{2y}\left(\frac{\partial^2 \psi}{\partial x^2} + \frac{\partial^2 \psi}{\partial y^2} - \frac{1}{y}\frac{\partial \psi}{\partial y}\right) \tag{12.116}$$

The equation in $\boldsymbol{\omega}$ then becomes

$$\frac{\partial \omega}{\partial t} + u\frac{\partial \omega}{\partial x} + v\frac{\partial \omega}{\partial y} - \frac{\omega v}{y} = \frac{\nu}{y}\left[\frac{\partial^2 (\omega y)}{\partial x^2} + \frac{\partial^2 (\omega y)}{\partial y^2} - \frac{1}{y}\frac{\partial (\omega y)}{\partial y}\right]. \tag{12.117}$$

The last two equations (12.116) and (12.117) enable us to study the flow. Let us now introduce dimensionless quantities, which are obtained by dividing

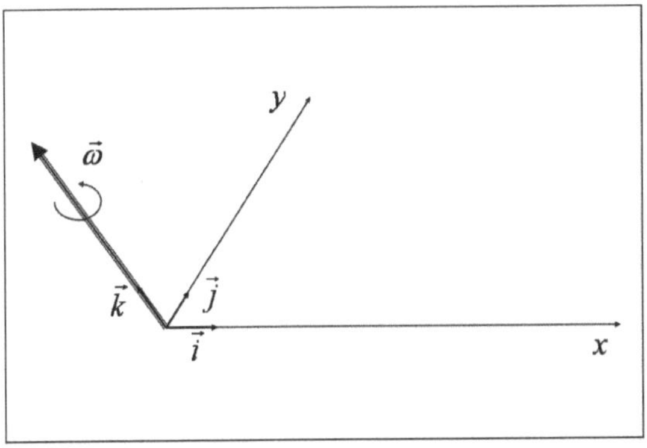

Fig. 12.19. The swirl ω in space

the time, the velocity, the length, the stream function, and the swirl by the reference quantities R/U_∞, U_∞, R, $U_\infty R^2$, and U_∞/R, respectively.

Without changing the notation, the previous equations give the following dimensionless relations:

$$\frac{\partial^2\psi}{\partial x^2} + \frac{\partial^2\psi}{\partial y^2} - \frac{1}{y}\frac{\partial\psi}{\partial y} = -2\omega y, \tag{12.118}$$

$$\frac{\partial^2(\omega y)}{\partial x^2} + \frac{\partial^2(\omega y)}{\partial y^2} - \frac{1}{y}\frac{\partial(\omega y)}{\partial y} = Re\, y\,(\frac{\partial\omega}{\partial t} + u\frac{\partial\omega}{\partial x} + v\frac{\partial\omega}{\partial y} - \frac{\omega v}{y}), \tag{12.119}$$

where $Re = U_\infty R/\rho$ is the Reynolds number.

For *Stokes' theory*, the convective terms of the right hand side of (12.119) vanish ($Re \to 0$), and it becomes

$$\frac{\partial^2(\omega y)}{\partial x^2} + \frac{\partial^2(\omega y)}{\partial y^2} - \frac{1}{y}\frac{\partial(\omega y)}{\partial y} = 0. \tag{12.120}$$

This is the same equation as (12.99), which is obeyed by the stream function of an irrotational, incompressible, nonviscous fluid: ψ is replaced simply by ωy. We therefore know a particular solution: the doublet of the form

$$\omega y = \frac{K\sin^2\theta}{2r}. \tag{12.121}$$

Using this solution, (12.119) gives

$$\frac{\partial^2\psi}{\partial r^2} + \frac{1}{r^2}\frac{\partial^2\psi}{\partial\theta^2} - \frac{\cos\theta}{r^2\sin^2\theta}\frac{\partial\psi}{\partial\theta} = -\frac{K\sin^2\theta}{r}. \tag{12.122}$$

By writing

$$\psi = f(r)\sin^2\theta, \tag{12.123}$$

we obtain

$$f"(r) - \frac{2}{r^2}f(r) = -\frac{K}{r}. \tag{12.124}$$

We immediately obtain the solution. By considering the boundary conditions at the level of the surface of the sphere where $\mathbf{v} = \mathbf{0}$, we see that

$$\psi(1,\theta) = 0, \quad \frac{\partial\psi}{\partial r}(1,\theta) = 0, \tag{12.125}$$

and by accounting for the conditions at infinity, where $\psi = y^2/2$, we obtain $K = -3/2$ for the doublet intensity and the solution

$$\psi = (r^2 - \frac{3r}{2} + \frac{1}{2r})\frac{\sin^2\theta}{2}. \tag{12.126}$$

The terms in r^2 and $1/2r$ are identical to those of an irrotational flow of a perfect fluid. The solution is the superposition of a uniform flow and a doublet. The term $(-3r/4\,\sin^2\theta)$ is sometimes called a stokeslet.

Since the velocity field is known, we can calculate the forces acting on the sphere by writing, for example [96], that their power is equal to the power dissipated by the viscosity:

$$U_\infty F = 4\mu\int_{\mathcal{V}}\omega^2\,d\mathcal{V}. \tag{12.127}$$

We obtain Stokes' force (in dimensional variables)

$$F = 6\pi\mu U_\infty R. \tag{12.128}$$

The corresponding force per unit mass is then

$$\frac{\mathcal{F}}{\rho_p} = \frac{F}{4/3\,\pi R^3\rho_{ps}} = \frac{9}{2}\frac{\mu U_\infty}{R^2\rho_{ps}}, \tag{12.129}$$

or, if we replace $\mathbf{U}_\infty$ by $\mathbf{v}_g - \mathbf{v}_p$,

$$\boldsymbol{\mathcal{F}} = \rho_p\frac{\mathbf{v}_g - \mathbf{v}_p}{\tau_v}, \tag{12.130}$$

where

$$\tau_v = 2R^2\rho_{ps}/9\mu. \tag{12.131}$$

This is the expression for the friction relaxation time given by Stokes' theory.

In *Oseen's theory*, the dimensionless system to be solved becomes

$$\begin{cases} \partial^2\psi/\partial x^2 + \partial^2\psi/\partial y^2 - (1/y)\,\partial\psi/\partial y = -2\omega y, \\ \partial^2(\omega y)/\partial x^2 + \partial^2(\omega\, y)/\partial y^2 - (1/y)\,\partial(\omega y)/\partial y = Re\, y\, u\, \partial\omega/\partial x. \end{cases} \tag{12.132}$$

Oseen's solution for a sphere is as follows:

$$\psi = (r^2 + \frac{1}{2r})\frac{\sin^2\theta}{2} - \frac{3}{2}\frac{1+\cos\theta}{Re}[1 - e^{-\frac{Re\, r}{2}(1-\cos\theta)}]. \tag{12.133}$$

We obtain terms for uniform flow and the doublet, as well as a final term that is sometimes called an oseenlet.

The resulting force is more difficult to calculate. We will therefore only quote the following expressions [107, 215]:

$$F = 6\pi\mu U_\infty R\,(1 + \frac{3}{8}Re - \frac{19}{320}Re^2 + \frac{71}{2560}Re^3 - \frac{30179}{2150400}Re^4 + \ldots) \tag{12.134}$$

and

$$F = 6\pi\mu U_\infty R(1 + \frac{3}{8}Re + \frac{9}{40}Re^2 \ln Re + O(Re^2)). \tag{12.135}$$

We will not comment on these results here.

It is common to use *empirical formulae* that include a corrective coefficient as well as the Stokes relaxation time $(\tau_v)_{St}$:

$$(\tau_v)_{emp} = \frac{1}{\mathcal{C}_v}(\tau_v)_{St}. \tag{12.136}$$

For example, Carlson and Hoglund [41] give $\mathcal{C}_v$ as a function of the Mach number and the Reynolds number based on the diameter of the sphere, $Re_D = 2Re$:

$$\mathcal{C}_v = \frac{(1 + 0.15Re^{0.687})(1 + \exp(0.427M^{-4.63} - 3Re_D^{-0.88}))}{1 + MRe_D(3.82 + 1.28\exp(-1.25Re_D M^{-1}))}. \tag{12.137}$$

There are other empirical formulae too.

Accelerating the particles modifies the force. The following is obtained for condensed particles in a gas using Stokes' theory (see [55, 57, 117, 175, 271]):

$$\begin{aligned} F &= 6\pi\mu U_\infty R(\mathbf{v}_g - \mathbf{v}_p) + \tfrac{2}{3}\pi R^3\rho_{gs}(d_p\mathbf{v}_g/dt - d_p\mathbf{v}_p/dt) \\ &-\tfrac{4}{3}\pi R^3\boldsymbol{\nabla}p + 6R^2\sqrt{\pi\rho_{gs}\mu}\int_t^0(d_p\mathbf{v}_g/dt' \\ &-d_p\mathbf{v}_p/dt')/\sqrt{t-t'}\;dt' + \tfrac{4}{3}\pi\rho_{ps}\mathbf{g}. \end{aligned} \tag{12.138}$$

Note that in this case we generally have (except at the critical point) $\rho_{gs} \ll \rho_{ps}$, and the nonstationary additive terms (Basset–Boussinesq forces) are negligible [69].

12.4.2 Heat Exchange

The convective terms are neglected in Stokes' theory. If we attempt to study thermal transfer in the same way, we end up determining the heat flux for a moving spherical particle (Fig. 12.20).

In the case of a sphere at rest with a temperature T_0 that is plunged into a motionless fluid atmosphere of temperature T_∞ a long distance from the particle, the only phenomenon present is thermal conduction in the fluid. The equation for this problem reduces to

$$\nabla^2 T = 0; \tag{12.139}$$

i.e., in spherical coordinates,

$$\frac{\partial}{\partial r}(r^2 \frac{\partial T}{\partial r}) = 0. \tag{12.140}$$

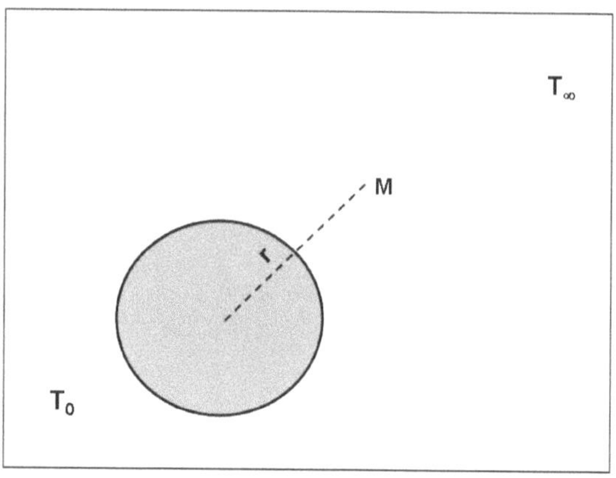

Fig. 12.20. Thermal exchange of a sphere

For a sphere of unit radius, the solution is

$$T = T_\infty + \frac{T_0 - T_\infty}{r}. \tag{12.141}$$

The heat flow through a sphere of unspecified radius r is constant and equal to

$$Q = -4\pi r^2 \lambda \frac{\partial T}{\partial r} = 4\pi\lambda(T_0 - T_\infty). \tag{12.142}$$

This refers to the heat flow from the inside towards the outside of the sphere.

In the case of a moving sphere, a surface exchange coefficient α per unit area is introduced, and the Nusselt number Nu is defined by

$$Nu = 2R\alpha/\kappa, \tag{12.143}$$

where $2R$ is the diameter of the sphere, $\kappa = \lambda/\rho c_p$ is the thermal diffusivity, and α is the heat exchange coefficient per unit area. In the limiting case of a sphere at rest, we obtain the classical result

$$Nu = 2. \tag{12.144}$$

If we take into account the fluid flow, dimensional analysis (Sect. 9.6.2) gives us

$$\Pi_\kappa = \kappa/\nu,\ \Pi_\alpha = \alpha/(2R)^{u-v}U_\infty^v \nu^{1-v}, \tag{12.145}$$

where

$$u = 2v - 1, \tag{12.146}$$

and, according to the Vashi–Buckingham theorem,

$$\Pi_\alpha = \psi(\Pi_\kappa). \tag{12.147}$$

To account for the fact that $Nu = 2$ must be true for $U_\infty = 0$, it is necessary to use $\alpha - 2\kappa/D$ instead of α in the definition of Π_α. With a power law for ψ, we then obtain

$$Nu = 2 + ARe^v Pr^{1-w}. \tag{12.148}$$

In laminar flow $v = 1/2$, and for solid spheres $1 - w = 1/3$.

The formula of Carlson and Hoglund, which is more precise, takes into account the Mach number and gives

$$Nu = 2\frac{1 + 0.230Re^{0.55}}{1 + 6.84MRe^{-1}(1 + 0.230Re^{0.55})} = 2\mathcal{C}_T. \tag{12.149}$$

Other formulae are given by Clift et al. [55] for Nu, as well as for Sh, which is similar to Nu but applies to mass diffusion ($Sh = 2R\alpha_C/\rho D$; see Sect. 5.4). It is also worth referring to [1, 151] if vaporization is present. Knowing the exchange coefficient per unit area, we can calculate the heat flux per unit volume of particle:

$$\mathcal{Q} = \frac{3\alpha}{R}(T_g - T_p),\ \alpha = \frac{\kappa}{R}\mathcal{C}_T. \tag{12.150}$$

According to (12.2.2), we then have

$$\mathcal{Q} = \frac{\rho_{ps} c_c}{\tau_T} (T_g - T_p). \tag{12.151}$$

We thus deduce the expression for τ_T:

$$\tau_T = \frac{R^2 \rho_{ps} c_c}{3 \kappa C_T}. \tag{12.152}$$

12.4.3 Steady Combustion of a Fuel Drop in a Combustive Atmosphere at Rest

Let us now consider a spherical, homogeneous fuel drop that yields a vapor H in an atmosphere of an oxidizer O [106, 264, 290]. Combustion takes place according to the one-step reaction

$$\mathrm{H} + \mathrm{O} \longrightarrow \mathrm{P}$$

The stoichiometric coefficients are assume to be equal to one for simplification purposes, but the following theory does not change if we use other stoichiometric coefficients. We assume a spherical symmetry. The liquid fuel is termed LH and its vapor is termed H. The concentration of the combustion product P around the droplet depends on the distance from the droplet at which the species mix with the combustion products.

We consider two situations:

- The chemical reaction and the subsequent flame occur around a cloud of droplets (we will assume that the droplets do not interact with each other). We will then have to study the evaporation of a drop.
- Combustion occurs around each droplet.

Quasi-steady evolution is assumed in each situation (i.e., the flow is steady in the gaseous phase, where the radius of the droplet is assumed to be constant compared to the timescale for diffusion). However, droplet evolution is unsteady at a longer timescale, and an important aim of our calculations is to determine the law for the droplet diameter as a function of time.

Evaporation of a Droplet

Here, the flame occurs around thousands of droplets. Each droplet evaporates, forming a fog.

The d^2 Law for a One-Component Liquid Droplet in its Vapor

Let us initially treat the general case of a drop at uniform and constant temperature and density that is present in its vapor in quasi-steady mode. This corresponds to the case where the mixing of the species with the combustion products occurs very far from the droplet. The surroundings of the droplets consist purely of fuel vapor. We assume that the physical coefficients of the gas (thermal conductivity λ and specific heat c_p) are functions of the temperature. The pressure and the temperature at infinity are given constants. The quasi-steady equations are reduced to mass conservation and energy conservation:[14]

$$\dot{M} = 4\pi r^2 \dot{m},\ \dot{m} = \rho u,\ \dot{M} c_p \frac{dT}{dr} - 4\pi \frac{d}{dr}(\lambda r^2 \frac{dT}{dr}) = 0. \tag{12.153}$$

At the liquid–vapor interface $r = r_s$, we have

$$T = T_s,\ 4\pi r_s^2 \lambda(T_s)(\frac{dT}{dr})_{r=r_s} = \dot{M}\, l(T_s), \tag{12.154}$$

and, at infinity,

$$T = T_\infty. \tag{12.155}$$

Upon performing the first integration of (12.153), we deduce that

$$\dot{M} \int_{T_s}^{T} c_p(T) dT = 4\pi[\lambda(T) r^2 \frac{dT}{dr} - \lambda(T_l) r_s^2 (\frac{dT}{dr})_{r=r_s}]. \tag{12.156}$$

Taking into account (12.154), the latter relation is

$$\frac{dr}{r^2} = \frac{4\pi\lambda}{\dot{M}} \frac{dT}{\int_{T_s}^{T} c_p(T) dT + l(T_s)}, \tag{12.157}$$

which gives, after integration,

$$\frac{1}{r_s(t)} - \frac{1}{r} = \frac{4\pi}{\dot{M}(t)} \int_{T_s}^{T} \frac{\lambda(\theta) d\theta}{\int_{T_s}^{\theta} c_p(\theta) d\theta + l(T_s)}. \tag{12.158}$$

Applying (12.158) at infinity and using (12.155) yields

$$\dot{M}(t) = 4\pi r_s(t) \int_{T_s}^{T_\infty} \frac{\lambda(\theta) d\theta}{\int_{T_s}^{\theta} c_p(\theta) d\theta + l(T_s)}. \tag{12.159}$$

As

$$\dot{M}(t) = -dM/dt = -4\pi r_s^2 \rho_L dr_s/dt, \tag{12.160}$$

we can deduce the d^2 law:

$$d^2 = d_0^2 - Kt, \tag{12.161}$$

where

$$d = 2r_s,\ K = \frac{8}{\rho_L} \int_{T_s}^{T_\infty} \frac{\lambda(\theta) d\theta}{\int_{T_s}^{\theta} c_p(\theta) d + l(T_s)}.$$

[14]The pressure is constant, as assumed in the Shvab–Zel'dovich approximation (Sect. 7.4).

Case of a One-Component Liquid Droplet in a Gaseous Mixture with Constant Physical Coefficients

This is the case when the species mix with the combustion products around the droplet or a group of droplets but the flame occurs far from the droplet(s). Thus, the vaporization of the droplet of pure fuel occurs in a mixture of gaseous fuel H and combustion products P.

The equations for quasi-steady gas flow are as follows:

$$\dot{M} = 4\pi r^2 \dot{m}, \ \dot{m} = \rho u, \tag{12.162}$$

$$\dot{M}\frac{dY_j}{dr} - 4\pi\rho D r \frac{d^2(rY_j)}{dr^2} = 0, \tag{12.163}$$

$$\dot{M}c_p\frac{dT}{dr} - 4\pi\lambda r \frac{d^2(rT)}{dr^2} = 0, \tag{12.164}$$

with the following boundary conditions:

- At infinity ($r = \infty$):

$$Y_j = Y_{j\infty}, \ T = T_\infty \tag{12.165}$$

- At the droplet surface ($r = r_s$):

$$Y_{js}, \ 4\pi\rho D r_s^2 (\frac{dY_j}{dr})_{r=r_s} = -\dot{M}(Y_{jL} - Y_{js}) \tag{12.166}$$

$$T_s, \ 4\pi\frac{\lambda}{c_p} r_s^2 (\frac{dT}{dr})_{r=r_s} = -\dot{M}\frac{l}{c_p}. \tag{12.167}$$

Therefore, the solution to the diffusion equation is

$$Y_j = a_j + b_j e^{-\xi_M}, \ \xi_M = \dot{M}/4\pi\rho D r. \tag{12.168}$$

Two boundary conditions are sufficient to determine the coefficients a_j and b_j. As we have seen, three conditions need to be satisfied: one condition at infinity ($r = \infty$, $\xi_M = 0$) and two at the interface (r_s, ξ_{Ms}). We set

$$B_M = \frac{Y_{js} - Y_{j\infty}}{Y_{jL} - Y_{js}}. \tag{12.169}$$

Note that the Spalding parameter $B_M = (Y_{Hs} - Y_{H\infty})/(1 - Y_{Hs})$ does not depends on j. This yields $\xi_{Ms} = \ln(1 + B_M)$, and thus

$$\dot{M} = 4\pi\rho D\, r_s \ln(1 + B_M), \tag{12.170}$$

from which we deduce the d^2 law

$$d^2 = d_0^2 - Kt, \tag{12.171}$$

where $d = 2r_s$, $\dot{M} = -dM/dt = -4\pi\rho r_s^2 dr_s/dt$.

On the other hand, the solution to the conduction equation is

$$T = a_T + b_T e^{-\xi_T},\ \xi_T = \dot{M}c_p/4\pi\lambda r. \tag{12.172}$$

Again, three conditions need to be satisfied to determine the two temperature constants, and we get

$$\dot{M} = 4\pi\frac{\lambda}{c_p}r_s \ln(1 + B_T), \tag{12.173}$$

where $B_T = c_p(T_\infty - T_s)/l$ this time. The parameters B_T and B_M depend on the boundary conditions, not all of which are known. Thus, it is necessary to determine the temperature and the concentration at the surface. To do this, two relations are needed.

The Spalding parameter for heat exchange B_T is related to the Spalding parameter for mass exchange B_M by the equation

$$\dot{M} = 4\pi\frac{\lambda}{c_p}r_s \ln(1 + B_T) = 4\pi\rho D r_s \ln(1 + B_M). \tag{12.174}$$

So B_T is a function of B_M, which depends on the mass fraction of gaseous fuel at the droplet surface (if the Lewis number Le is equal to unity, we have $B_T = B_M$). This gives

$$\frac{\mathcal{M}_H Y_{P\infty} X_{Hs}}{\mathcal{M}_P X_{P\infty}(1 - X_{Hs})} = \frac{c_p(T - T_s)}{l}.$$

The mass fraction Y_{Hs} is related to the temperature T_s by the equilibrium relation $\mu_{HL} = \mu_H$. When μ_{HL}, the molar chemical potential of the liquid fuel is a function of T only. For an ideal gas mixture, this leads to $pX_{Hs} = p_{sat}(T_s)$. According to Lefebvre and Chin [49, 154], the equilibrium law is then

$$p_{sat}(T_s) = \exp[a - b/(T_s - c)],$$

where a, b and c are constant coefficients.

We can then relate Y_{Hs} to the surface molar fraction X_{Hs}, and the surface mass fraction Y_{Hs} is found to be a function of the surface temperature T_s and the total pressure p, which is assumed to be uniform and constant: $Y_{Hs} = f(T_s, p)$. Thus, B_T is function of T_s only.

The latent heat itself is a function of the temperature at the surface of the droplet. Indeed, from the Clapeyron relation, we have (per mole of pure substance):

$$L = \frac{RT^2}{p}(\frac{dp}{dT})_{sat}.$$

This relation, which is applied here to a unit mass of H (see Sect. 7.5), becomes

$$l = bRT_s^2/(T_s - c)^2.$$

Consequently, we obtain the surface temperature and concentrations and thus the value of the regression coefficient K of the droplet:

$$K = 8(\lambda/c_p)\ln[1 + c_p(T_\infty - T_s)/l].$$

Combustion of a Droplet with a Flame at a Finite Distance from the Droplet

We now study the case of vaporization where a diffusion flame is at a finite distance from the droplet. This diffusion flame can be maintained in a quasi-steady mode. We will assume spherical symmetry. The drop consists of the single pure substance H. This problem is similar to the Emmons problem of Chap. 9, but here the configuration is spherical and there is no forced convection. The energy provided by the reaction at the flame level diffuses towards the droplet surface, thus causing surface evaporation and the release of gaseous fuel (i.e., gaseous H). The pressure is assumed to be constant and the temperature of the droplet is assumed to be uniform and constant (this temperature is that of saturating vapor, T_s). At infinity the temperature is T_∞ and the mass fraction of the oxidizer is $Y_{O\infty}$. Here, in contrast to the Emmons problem, the momentum equation is not taken into account.

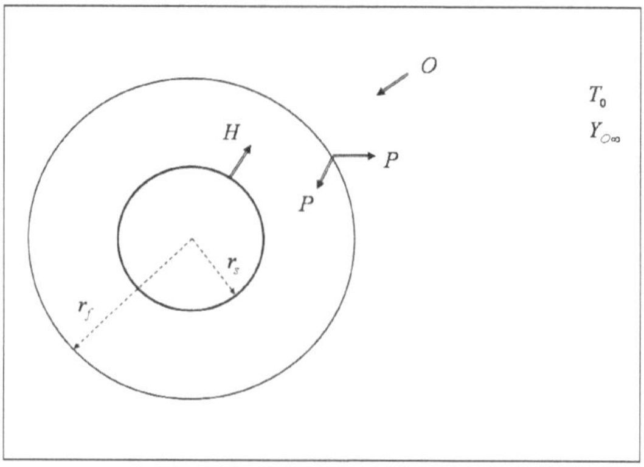

Fig. 12.21. Combustion of a droplet with a flame at a finite distance from it

Figure 12.21 illustrates the particular case where combustion takes place inside a concentric spherical flame. This is the case treated below, with simplifying assumptions [290]. If the density of the condensed phase is ρ_L and the radius of the drop is r_s, the mass flow of the gas through a sphere of radius r_s is given by (12.160). If we assume that $\dot{M}$ is related to r_s, we can deduce the evolution of the radius of the droplet as a function of time by simple integration.

Experimental observations and the theory provided below show that there is a linear relation between the mass flow rate $\dot{M}$ and the droplet diameter $d = 2r_s$.

We can then deduce, as was done in Sect. 12.4.3, that the square of the droplet diameter is a linear function of time (Eq. 12.161). This linear decrease in r_s^2 over time can also be observed in experiments.

For the theoretical analysis, we use the Shvab–Zel'dovich approximation presented in Sect. 7.4.

Recall that the two following equations were obtained:

$$\boldsymbol{\nabla} \cdot (\rho \mathbf{v} \beta_j - \rho D \boldsymbol{\nabla} \beta_j) = \dot{\zeta},$$

$$\boldsymbol{\nabla} \cdot (\rho \mathbf{v} \beta_T - \lambda / c_{p\,f} \boldsymbol{\nabla} \beta_T) = \dot{\zeta},$$

where β_j is a concentration variable and β_T is a temperature variable: $\beta_j = Y_j / \nu_j \mathcal{M}_j$, $\beta_T = \sum_j Y_j \int_{T^0}^{T} c_{p,\,j} dT / \Delta H$.

We will assume that the Lewis number is equal to 1, so $\lambda / c_{pf} = \rho D = g$, where g is assumed to be constant.

The boundary conditions (surface of the drop) are similar to those of the Emmons problem. The following system is thus obtained:

$$\begin{cases} 4\pi r^2 \rho v = \dot{M} \cong \text{const.}, \\ d\beta_j/dr - d/dr[(4\pi r^2 g/\dot{M}) d\beta_j/dr] = 4\pi r^2 \dot{\zeta}/\dot{M}, \\ d\beta_T/dr - d/dr[(4\pi r^2 g/\dot{M}) d\beta_T/dr] = 4\pi r^2 \dot{\zeta}/\dot{M}, \\ -g(d\beta_j/dr)_{r=r_s} = -(\rho v)_l (\beta_{js} - \beta_{jL}), \\ -g(d\beta_T/dr)_{r=r_s} = -(\rho v)_l \, l/\Delta H. \end{cases} \tag{12.175}$$

Let us write

$$\xi = \dot{M}/4\pi g r. \tag{12.176}$$

For the quantity $\beta_T - \beta_O$, we obtain

$$d^2(\beta_T - \beta_O)/d\xi^2 + d(\beta_T - \beta_O)/d\xi = 0. \tag{12.177}$$

Integration leads to

$$\beta_T - \beta_O = A_T + B_T \, e^{-\xi}. \tag{12.178}$$

At the limit $r = \infty$, $\xi = 0$, we have[15]

$$-\beta_{O\infty} = Y_{O\infty}/\mathcal{M}_O = A_T + B_T. \tag{12.179}$$

At $r = r_s$, the boundary condition is

$$(d(\beta_T - \beta_O)/d\xi)_{r=r_s} = l/\Delta H \tag{12.180}$$

or

$$-B_T e^{-\xi_s} = l/\Delta H. \tag{12.181}$$

At $r = r_s$, we also have

$$\beta_{Ts} = A_T + B_T\, e^{-\xi_s} \tag{12.182}$$

since $\beta_{Os} = 0$; the oxidizer is completely consumed at the level of the flame. The temperature T (and thus β_T) is known. We can therefore deduce the values

$$\begin{cases} A_T = \beta_{Ts} + l/\Delta H, \\ B_T = -(\beta_{O\infty} + \beta_{Ts} + l/\Delta H), \\ (l/\Delta H)e^{\xi_s} = \beta_{O\infty} + \beta_{Ts} + l/\Delta H, \end{cases} \tag{12.183}$$

which lead to

$$\dot{M} = 4\pi g r_s \ln((\beta_{O\infty} + \beta_{Ts})\Delta H/l + 1). \tag{12.184}$$

The mass flow rate is then proportional to the droplet radius, and the d^2 law

$$d^2 = d_0^2 - Kt \tag{12.185}$$

can be deduced, where

$$K = \frac{8g}{\rho_L} \ln[(\beta_{O\infty} + \beta_{Ts})\Delta H/l + 1]. \tag{12.186}$$

The variable $\beta_T - \beta_O$ evolves according to the law

$$\beta_T - \beta_O = \beta_{Ts} + l/\Delta H - (\beta_{O\infty} + \beta_{Ts} + l/\Delta H)e^{-\xi}. \tag{12.187}$$

Studying $\beta_H - \beta_O$ will give us the position of the flame. We will assume that the flame is thin and that combustion is stoichiometric: $\beta_H = \beta_O$.

We find of course that

$$\beta_H - \beta_O = A + Be^{-\xi}. \tag{12.188}$$

[15] Note that B_T is an integration constant here that should not be confused with the Spalding parameter for heat exchange.

The constants A and B are obtained by writing the conditions at infinity and at the droplet surface level:

$$\begin{cases} -\beta_{O\infty} = A + B, \\ \beta_{Hs} = A + Be^{-\xi_s}, \\ d(\beta_H/d\xi)_s + \beta_{Hs} + 1/\mathcal{M}_H = 0. \end{cases} \tag{12.189}$$

From these three relations we deduce that

$$A = -1/\mathcal{M}_H, \; B = 1/\mathcal{M}_H - \beta_{O\infty}, \tag{12.190}$$

and the value of the mass fraction Y_{Hs} at the surface of the drop is deduced from

$$\beta_{Hs} = -1/\mathcal{M}_H + (1/\mathcal{M}_H - \beta_{O\infty})\, e^{-\xi_s}. \tag{12.191}$$

The location of the spherical thin flame is given by

$$\beta_H - \beta_O = 0 = -1/\mathcal{M}_H + (1/\mathcal{M}_H - \beta_{O\infty})\, e^{-\xi_f}. \tag{12.192}$$

We then obtain

$$\frac{r_f}{r_s} = \frac{\ln[1 + (\beta_{O\infty} + \beta_{Ts})\Delta H/l]}{\ln(1 - \mathcal{M}_H \beta_{O\infty})}. \tag{12.193}$$

The concentration and temperature profiles are obtained as follows.

For $r_s < r < r_f$, $\beta_O = 0$ and $\dot{\zeta} = 0$, so

$$\begin{cases} \beta_H = A + Be^{-\xi}, \\ \beta_T = A_T + B_T e^{-\xi}, \end{cases} \tag{12.194}$$

(the four coefficients were determined previously).

When $r_f < r$, we always have $\dot{\zeta} = 0$ as well as $\beta_H = 0$. Thus,

$$\begin{cases} \beta_O = -A - Be^{-\xi}, \\ \beta_T = A_T - A + (B_T - B)e^{-\xi}. \end{cases} \tag{12.195}$$

We can see that, with the assumption of a reactive thin flame and an infinitely fast reaction, the evolutions are controlled by the processes of convection and diffusion. The chemical reaction only plays a role through the energy it provides. It would not be the same with a premixed flame obtained with a droplet of mixture, for example, because the combustion speed would then also depend on the chemical kinetics, as we saw in the study of deflagration waves (Chap. 10).

The obtained theoretical profiles with respect to $e^{-\xi}$ ($\xi = \dot{M}/4\,\pi\, g\, r$) are shown in Fig. 12.22.

The results of this theory were compared with those of experiments carried out earlier by Kumagai [141] (see also [219]) under satisfactory microgravity

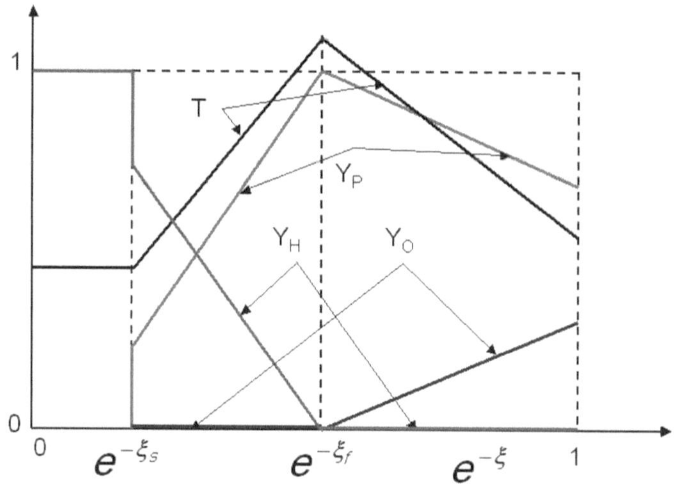

Fig. 12.22. Temperature and mass fraction profiles of a burning droplet

conditions (gravity must be close to zero to ensure that spherical symmetry is respected at least).

There is good agreement with the linear law for the regression of the radius of the drop squared as a function of time according to (12.185) and (12.186). However, this is not the case for the ratio r_f/r_s of the radius of the flame to that of the droplet: (12.193) indicates that the r_f/r_s ratio is constant, whereas experimental results show that this ratio is not.

More elaborate calculations that take into account the unsteady characteristics of the process and the variations in the transfer coefficients as functions of temperature [178, 248] lead to results that are in better agreement with the experimental results. Note that the values of r_s and r_f/r_s have a very significant influence on the performance of a rocket motor in which the flow includes condensed drops during combustion. The evolution of the droplet radius as a function of time yields the combustion time and the length of the combustion chamber. The ratio r_f/r_s indicates the possible interactions between the drops undergoing combustion in the flow (see Borghi and Lacas [24, 68]).

The preceding theories only apply in the absence of interactions. The effects of coupling between the flow and the individual process that the drop undergoes during combustion are much more difficult to evaluate.

Droplet Combustion with Condensation of the Burned Products

When dealing with a drop of liquid oxygen burning in an atmosphere of hydrogen (which is the case in a cryogenic rocket engine), we can use the same treatment as employed in the preceding section for a drop surrounded by a flame and the chemical reaction H + O $\rightarrow$ P (or, more realistically, $\nu_H \mathrm{H} + \nu_O \mathrm{O} \rightarrow \nu_P \mathrm{P}$). This time, in this simplified model with a single chemical reaction, the oxidant O is gaseous oxygen (O_2), the reducer H is gaseous hydrogen H_2, and the combustion products can be summarized as water vapor (H_2O). Figure 12.22 remains valid provided we replace H with O, and vice versa.

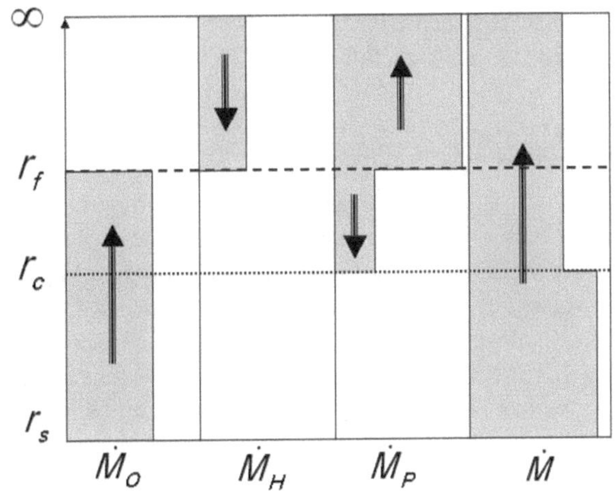

Fig. 12.23. Mass flow rates for a scenario where a flame surrounds a droplet of O and where the burned product P condenses

In reality, this scenario presents several issues. Apart from the fact that the Lewis number is different from 1, the conditions are often transcritical (see Sect. 12.4.5). On the other hand, water vapor turns into ice near the liquid oxygen droplet, which has a temperature of around 90 K. The latter problem has been studied [144]. The assumptions of simplified analytical theory (we assume in particular that the ice occupies negligible volume and is thus a mass sink for the gaseous species) are illustrated in Fig. 12.23 in relation to the fluxes of the gaseous species. A more complete theory where we take the diffusion into account, and where the condensed species is considered a gaseous species, confirms the fact that water has a concentration of zero in the extreme vicinity of the liquid oxygen droplet, and therefore can not dissolve in it [144].

12.4.4 Transient Vaporization of a Droplet

In liquid propellant engines, combustion instabilities occur due to coupling between the combustion process and the acoustics of the chamber. The phenomenon of evaporation is a possible mechanism for combustion instability. Thus, let us now study the dynamic response of a vaporizing droplet in the field of an acoustic disturbance, by assuming local liquid–vapor equilibrium at the droplet surface [225].

According to Delplanque and Sirignano [67, 231], "the criterion of Rayleigh stipulates that a small pressure disturbance could grow if the considered process adds energy in phase (or with a small variation of phase) with the pressure." To quantify this criterion in studies of droplet evaporation, a response factor has been defined as follows:

$$N = \int_{V,t} q'(V,t)p'(V,t)dt\,dV / \int_{V,t} (p'(V,t))^2 dt\,dV, \tag{12.196}$$

where $p' = (p - \bar{p})/\bar{p}$ is the relative pressure disturbance and $q' = (q - \bar{q})/\bar{q}$ is the relative disturbance of the heat or mass flow rate. We consider a drop located in a velocity node in a cavity. It is easily shown that, for sinusoidal oscillations with the same period, the response factor is $N = |\hat{q}| \cos\theta / |\hat{p}|$, where $|\hat{q}|$, $|\hat{p}|$ are the moduli and θ is the difference in phase between q' and p'. The quantities N and θ are functions of the oscillation frequency.

If $N > 0$, evaporation has a destabilizing effect, whereas it has a stabilizing effect for $N < 0$. The separation between the two domains corresponds to the cut-off frequency.

The Heidmann and Wieber Model

Here, we consider that a stationary drop (such as that presented by Heidmann [115]) represents the average situation in a combustion chamber.

This drop is spherical and has a radius r_s. It is continuously supplied in liquid form by injectors (a hypodermic syringe would simulate this well) in a flow equal to the flow corresponding to the stationary mode of reference (Fig. 12.24).

We limit ourselves to the linearized theory and consider the small relative disturbances of the unspecified parameter f to be sinusoidal functions of time (just as we did above for p and q) such that $f' = (f - \bar{f})/\bar{f}$, $f' = \hat{f}e^{i\omega t}$.

We then proceed to solve this problem as follows: we initially establish the perturbation equations for the system; we then write the equations in $\hat{f}$; and finally we deduce the response factor and dephasing:

$$N = \Re(\hat{\dot{M}}/\hat{p}),\ \theta = \arg(\hat{\dot{M}}/\hat{p}).$$

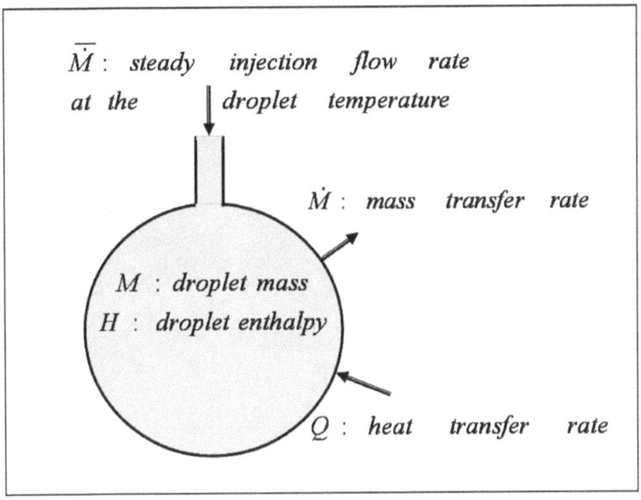

Fig. 12.24. A vaporizing drop continuously supplied by a steady flow rate

The model of Heidmann and Wieber [116] makes the assumption that the liquid phase has an infinite thermal conductivity in the steady and unsteady regimes (i.e., there is a uniform temperature inside the drop). These assumptions can influence the response of the drop.

In order to evaluate the response factor and the cut-off frequency more accurately, it is possible to extend the model of Heidmann by taking the thermal evolution inside the droplet in the unsteady regime into account.[16] We will, however, retain the assumption of an average droplet in the established regime. With regard to the gas phase, we use the approach of Spalding [264], which assumes a quasi-steady mode.

Linearized Equations of the Gas Phase

We assume that the vapor phase is an ideal mixture of perfect gases in stationary evolution, and that, the gaseous mixture behaves as an ideal gas in isentropic evolution far from the droplet (note that the subscript "$_\infty$" will be replaced by the subscript "$_c$" for the combustion chamber here). The drop is at rest in an atmosphere that is also at rest an infinite distance from the droplet. According to Sect. 12.4.3, we therefore obtain the following system of equations for the gas:

[16] As well as the motion of the liquid inside the droplet if there is a velocity disequilibrium.

$$\begin{cases} M = 4/3\,\rho_L r_s^3,\ \dot{M} = 4\pi\rho D r_s \ln(1+B_M), \\ B_M = (Y_{Hs} - Y_{Hc})/(1 - Y_{Hs}), \\ Y_H = \mathcal{M}_H X_H/(\mathcal{M}_H X_H + \mathcal{M}_P X_P), \\ p_C X_{Hs} = \exp\left[(a-b)/(T_s - c)\right], \\ dM/dt = -\dot{M} + \overline{\dot{M}}, \end{cases} \tag{12.197}$$

$$\begin{cases} B_T = c_p(T_\infty - T_s)/(l + Q_L/\dot{M}),\ T_c/p_c^{(\gamma-1)/\gamma} = \text{const.}, \\ 4\pi\lambda/c_p\, r_s \ln(1+B_T) = 4\pi\rho D r_s \ln(1+B_M), \\ Le = 1 \Rightarrow B_M = B_T,\ l = b\,RT_s^2/[\mathcal{M}_H(T_s - c)^2]. \end{cases} \tag{12.198}$$

In the first equation of system (12.198), the quantity Q_L is the heat flux accumulated by the droplet, which in turn causes its temperature to change. The other part of the heat flux arising from the gas phase, $\dot{M}l$, is used to vaporize the droplet.

Now the two systems are linearized. For system (12.197) we have

$$\begin{cases} r_s' = M'/3,\ \dot{M}' = M'/3 + \frac{\bar{B}_M}{(1+\bar{B}_M)\ln(1+\bar{B}_M)} B_M', \\ B_M' = \frac{Y_{Hs}Y_{Pc}}{(\bar{Y}_{Hs}-\bar{Y}_{Hc})(1-\bar{Y}_{Hs})} \frac{\mathcal{M}_H}{\mathcal{M}_H X_{Hs} + \mathcal{M}_P X_{Ps}} X_{Hs}',\ X_{Hs}' = \bar{b}T_s' - p_c', \\ \bar{\tau}_v \frac{dM'}{dt} = -\dot{M}', \end{cases} \tag{12.199}$$

where $\bar{b} = b\bar{T}_s/(\bar{T}_s - c)^2$, $\bar{\tau}_v = \bar{M}/\bar{\dot{M}} = \rho_L c_p \bar{r}_s^2/3\lambda \ln(1+\bar{B}_T)$. Finally, system (12.199) gives the following equation after eliminating the intermediate quantity:

$$\frac{d\dot{M}'}{dt} + \frac{\dot{M}'}{3\bar{\tau}_v} = \alpha(\bar{b}\frac{dT_s'}{dt} - \frac{dp_c'}{dt}), \tag{12.200}$$

where

$$\alpha = \frac{\bar{B}_M}{(1+\bar{B}_M)\ln(1+\bar{B}_M)} \frac{\bar{Y}_{Pc}\bar{Y}_{Hs}}{Y_{Ps}(\bar{Y}_{Hs} - \bar{Y}_{Hc})} \frac{\mathcal{M}_H}{\mathcal{M}_H X_{Hs} + \mathcal{M}_P X_{Ps}}.$$

If we now introduce the form $f' = \hat{f}\exp(i\omega t)$ for each disturbance f', (12.200) gives

$$\hat{\dot{M}} = \alpha\frac{iu}{1+iu}(\bar{b}\hat{T}_s - \hat{p}_c), \tag{12.201}$$

where the dimensionless reduced frequency $u = 3\omega\bar{\tau}_v$.

For system (12.198), we deduce that

$$\begin{cases} \Delta Q_L = Q_L - \bar{Q}_L = -\bar{\dot{M}}\bar{l}(\frac{\bar{T}_s}{\bar{T}_c-\bar{T}_s}T'_s - \frac{\bar{T}_c}{\bar{T}_c-\bar{T}_s}T'_c + B'_T + l'), \\ B'_T = B'_M,\ l' = -\frac{2c}{\bar{T}_s-c}T'_s. \end{cases} \tag{12.202}$$

System (12.202) gives the equation

$$\Delta Q_L = \dot{\bar{M}}\bar{l}(\bar{a}p'_c - \bar{\mu}T'_s), \tag{12.203}$$

where

$$\begin{cases} \bar{a} = \frac{\bar{T}_c}{\bar{T}_c-\bar{T}_s}\frac{\gamma-1}{\gamma} + \varphi,\ \bar{\mu} = \frac{\bar{T}_s}{\bar{T}_c-\bar{T}_s} - \frac{2c}{\bar{T}_s-c} + \bar{b}\varphi, \\ \varphi = \frac{\bar{Y}_{Pc}\bar{Y}_{Hs}}{\bar{Y}_{Ps}(\bar{Y}_{Hs}-\bar{Y}_{Hc})}\frac{\mathcal{M}_H}{\mathcal{M}_H\bar{X}_{Hs}+\mathcal{M}_P\bar{X}_{Ps}}. \end{cases} \tag{12.204}$$

Writing $\Delta Q_L = \Delta\hat{Q}_L \exp(i\omega t)$, we then deduce from (12.203) that

$$\Delta\hat{Q}_L = \dot{\bar{M}}\bar{l}(\bar{a}\hat{p}_c - \bar{\mu}\hat{T}_s). \tag{12.205}$$

Linearized Equations of the Liquid Phase

To solve this problem, we must also obtain equations for the liquid phase. These equations and the resulting expression for ΔQ_L depend on the assumptions made regarding thermal exchange inside the droplet. In the theory of Heidmann and Wieber, the droplet temperature is uniform. We thus have

$$Mc_L\frac{dT_s}{dt} = Q_L \tag{12.206}$$

and the linearized equation

$$\Delta Q_L = \bar{M}c_L\bar{T}_s\frac{dT'_s}{dt}. \tag{12.207}$$

We deduce from (12.207) that

$$\Delta\hat{Q}_L = \bar{M}c_L\bar{T}_s i\omega\hat{T}_s. \tag{12.208}$$

Transfer Function of the Heidmann and Wieber Model

The elimination of $\Delta\hat{Q}_L$ between the equations (12.205) and (12.208) affords the following equation in $\hat{T}_s$ and $\hat{p}_c$:

$$(\bar{\lambda}\, iu/3 + \bar{\mu})\hat{T}_s = \bar{a}\hat{p}_c, \tag{12.209}$$

where $\bar{\lambda} = c_L\bar{T}_s/\bar{l}$.

In order to study the stability of the drop during the course of evaporation, we must eliminate $\hat{T}_s$ between (12.201) and (12.209). This leads to the following expression for the transfer function Z_0:

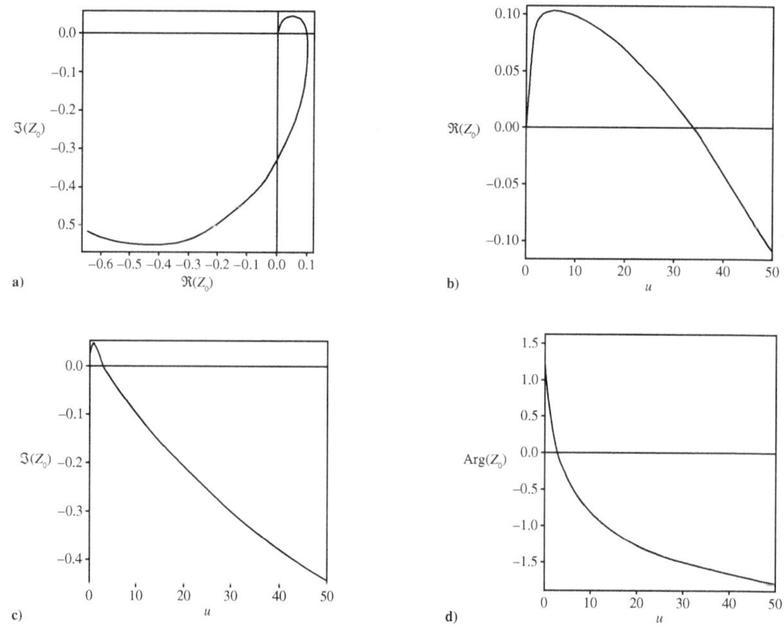

Fig. 12.25. The transfer function Z_0 for a droplet with a uniform temperature in a gaseous mixture for arbitarily chosen coefficient values: $A = 10$, $B = 100$. **a** Z_0 in the complex frame for u = 0–150; **b** $N/\alpha = \Re(Z_0)$; **c** the imaginary part of Z_0, $\Im(Z_0)$, as a function of the reduced frequency u; **d** dephasing (in radians) as a function of the reduced frequency u

$$Z_0 = \frac{\hat{\dot{M}}}{\alpha \hat{p}_c} = \frac{iu}{1+iu}\frac{A-iu}{B+iu}, \tag{12.210}$$

which can be used to deduce the response factor, the cut-off frequency and the dephasing (see Fig. 12.25 and [226] for more realistic cases). In (12.210), A and B are constant coefficients:[17]

$$\begin{cases} A = \frac{3(\bar{a}\bar{b}-\bar{\mu})}{\lambda} = \frac{3\bar{l}}{c_L \bar{T}_s}\left(\frac{\bar{b}\bar{T}_c(\gamma-1/\gamma)-\bar{T}_s}{\bar{T}_c-\bar{T}_s} + \frac{2c}{\bar{T}_s-c}\right), \\ B = \frac{3\bar{\mu}}{\lambda} = \frac{3\bar{l}}{c_L \bar{T}_s}\left(\frac{\bar{T}_s}{\bar{T}_c-\bar{T}_s} - \frac{2c}{\bar{T}_s-c}\right. \\ \left. +\bar{b}\frac{\bar{Y}_{Pc}\bar{Y}_{Hs}}{\bar{Y}_{Ps}(\bar{Y}_{Hs}-\bar{Y}_{Hc})}\frac{\mathcal{M}_H}{\mathcal{M}_H\bar{X}_{Hs}+\mathcal{M}_P\bar{X}_{Ps}}\right). \end{cases} \tag{12.211}$$

[17] According to the Heidmann interpretation, the term $(A - iu)$ is related to the heating dynamics of the droplet, even when the term $(B+iu)$ is rather characteristic of interactions between the chamber pressure and the droplet temperature.

Model with Nonuniform Temperature Inside the Droplet

Equations of the Gas Phase

We assume quasi-steady evolution in the gas phase. Thus, the perturbation equations are the same as (12.201) and (12.205) in Sect. 12.4.4.

Equations of the Liquid Phase

The Heidmann droplet of Fig. 12.24 can be fed in various ways with fuel at an average temperature of $\bar{T}_s$ and an average mass flow rate of $\bar{\dot{M}}$. The total mass balance for the droplet is $dM/dt = \bar{\dot{M}} - \dot{M}$. In steady mode, $\dot{M} = \bar{\dot{M}}$, $dM/dt = 0$, $M = \bar{M}$. If the thermal conductivity of the drop is infinite, the drop has a uniform temperature that is equal to its surface temperature T_s, whatever the process used to feed the drop with fuel.

This is not the case if the drop has a finite thermal conductivity. Two characteristic times should be taken into account: a residence time $\tau_v = \bar{M}/\bar{\dot{M}}$ in the reactor (the drop), and a time for transfer by thermal diffusion $\bar{\tau}_T = \bar{r}_s^2/\kappa_L$, where the thermal diffusivity of the liquid is $\kappa_L = \lambda_L/\rho_L c_L$. We can estimate that the conduction mode will dominate if $\bar{\tau}_T \ll \bar{\tau}_v$. The case where the droplet has infinite thermal conductivity returns in this configuration with uniform temperature (Sect. 12.4.4). On the other hand, for $\bar{\tau}_T \gg \bar{\tau}_v$, the convection of fuel will dominate. In extreme cases, a thermally insulating drop will have a uniform temperature of $\bar{T}_s$. The two thermal transfer modes will coexist if $\bar{\tau}_T \approx \bar{\tau}_v$.[18]

The equations of thermal transfer for the liquid phase are (for a sufficiently small θ^{-1})

$$\begin{cases} \partial T_l/\partial t - (\kappa_l/r)\partial^2(rT_l)/\partial r^2 = 0, \ (\partial T_l/\partial r)_{r=0} = 0, \\ 4\pi r_s^2 \lambda_l (\partial T_l/\partial r)_{r=r_s} = Q_L, \ T_l(r_s,t) = T_s. \end{cases} \tag{12.212}$$

In the case of small disturbances, these equations are

[18] If we wanted to consider the mass supplied to the drop in more detail, we could introduce a local rate of mass supply, such as $\dot{\omega}$, without taking thermal dilation into account. This gives, in spherical symmetry, $\partial(r^2\rho_l v_l)/\partial r = r^2\dot{\omega}$, with the condition $4\pi\int_0^{\bar{r}_s}\dot{\omega}r^2\,dr = 4\pi\bar{r}_s^2\rho_l v_l = \bar{\dot{M}}$. A simple theoretical case is that of a point source placed at the center of the droplet. Taking into account the induced motion, the equations for the drop are as follows: $\bar{\dot{M}} = 4\pi r^2\rho_l v_l$ =const., $\partial T_l/\partial t + v_l\partial T_l/\partial r - \kappa_l/r\,\partial^2(rT_l)/\partial r^2 = 0$, $4\pi r_s^2\lambda_l(\partial T_l/\partial r)_{r=r_s} = Q_L = Q - \dot{M}l$, $T_l(r_s,t) = T_s$, $T_l/(0,t) = T_0$, $v_l(0,t) = \infty$. Therefore, at any point in the liquid, $v_l = \bar{\dot{M}}/4\pi r^2\,\rho_l$, so that $r\partial T_l/\partial t + \bar{\dot{M}}/4\pi r\,\rho_l\partial T_l/\partial r - \kappa_l\,\partial^2(rT_l)/\partial r^2 = 0$. Writing $\theta = 9\lambda_L\bar{\tau}_v/\rho_L c_L\bar{r}_s^2 = \bar{\tau}_v/\bar{\tau}_T$, we can see that $\bar{\dot{M}}/4\pi r\,\rho_l = (\kappa_l/(3\theta)(r_s/r)$. Thus, the convective term is negligible for sufficiently high values of θ.

$$\begin{cases} \partial T_l'/\partial t - (\kappa_l/r)\partial^2(rT_l')/\partial r^2 = 0, \ (\partial T_l'/\partial r)_{r=0} = 0, \\ 4\pi r_s^2 \lambda_l \bar{T}_s (\partial T_l'/\partial r)_{r=r_s} = \Delta Q_L, \ T_l'(r_s,t) = T_s'. \end{cases} \tag{12.213}$$

We seek solutions of the type $T_l'(r,t) = \hat{T}_l(r)e^{i\omega t}$. We must take into account the fact that $r = r_s$ means $r = \bar{r}_s(1+r')$. In addition, the solutions will have to obey the condition that the minimum occurs at the center of the sphere. We then obtain

$$i\omega(r\hat{T}_l) - \kappa_L\,\partial^2(r\hat{T}_l)/\partial r^2 = 0. \tag{12.214}$$

The general solution of the equation above is $\hat{T}_l = C^+\,e^{s_0 r}/r + C^-\,e^{-s_0 r}/r$. Feeding this solution into the equation above yields the characteristic equation $s^2 = i\omega/\kappa_L$, which has solutions of $s = \pm s_0$, $s_0 = (1+i)\sqrt{\omega/2\kappa_L}$. To ensure that the general solution remains finite in the center, $C^+ + C^- = 0$ must hold, giving $r\hat{T}_l = C(e^{s_0 r} + e^{-s_0 r})$.

In the vicinity of the center of the sphere, $\hat{T}_l = \frac{C}{r}(e^{s_0 r} + e^{-s_0 r}) = 2s_0 r(1 + \frac{s_0^2}{6}r^2 + o(r^2))$, $\frac{dT_l}{dr} = 2s_0 C(\frac{s_0^2}{3} + o(r))$, and the condition that the fradient is zero at the center is automatically obeyed.

On the surface of the droplet, we can consider $r = \bar{r}_s$ except for second-order terms. Indeed, the temperature disturbance develops as follows: $T_l'(r_s,t) = T_l'(\bar{r}_s(1+r'),t) = T_l'(\bar{r}_s,t) + (\partial T_l'/\partial r)_{r=\bar{r}_s}\bar{r}_s r_s' + \ldots$.

We assume that T_l' and its successive partial derivatives with respect to r are on the same order as r_s'. Ignoring the second order, $T_l'(r_s,t) \cong T_l'(\bar{r}_s,t)$. We therefore find that $\hat{T}_l(\bar{r}_s,t) = \hat{T}_s$.

The condition on the temperature gives $\bar{r}_s\hat{T}_s = C(e^{s_0\bar{r}_s} + e^{-s_0\bar{r}_s})$, which provides the constant for the problem, $C = \bar{r}_s\hat{T}_s/(e^{s_0\bar{r}_s} + e^{-s_0\bar{r}_s})$, and gives the following temperature profile:

$$r\hat{T}_l = \hat{T}_s\frac{\bar{r}_s}{r}\frac{e^{s_0 r} + e^{-s_0 r}}{e^{s_0\bar{r}_s} + e^{-s_0\bar{r}_s}}. \tag{12.215}$$

Let us now examine the condition on the heat flux. We easily find that

$$\Delta\hat{Q}_L = -4\pi\bar{r}_s\lambda_L\bar{T}_s E(u)\hat{T}_s, \tag{12.216}$$

where

$$\begin{cases} E(u) = 1 - s_0 r_s \coth(s_0 r_s), \ \text{with } s_0 r_s = (1+i)\sqrt{3u/2\theta}, \\ \theta = 9\lambda_L\bar{\tau}_v/\rho_L c_L\bar{r}_s^2 = \frac{\bar{\tau}_v}{\bar{\tau}_T}. \end{cases} \tag{12.217}$$

The time $\bar{\tau}_T$ is almost equal to the characteristic thermal diffusion time inside the droplet.

The three fundamental equations of the problem are thus (12.201), (12.205) and (12.216).

After eliminating $\Delta\hat{Q}_L$ and $\hat{T}_s$ between these equations, we deduce the expression for the new transfer function Z:

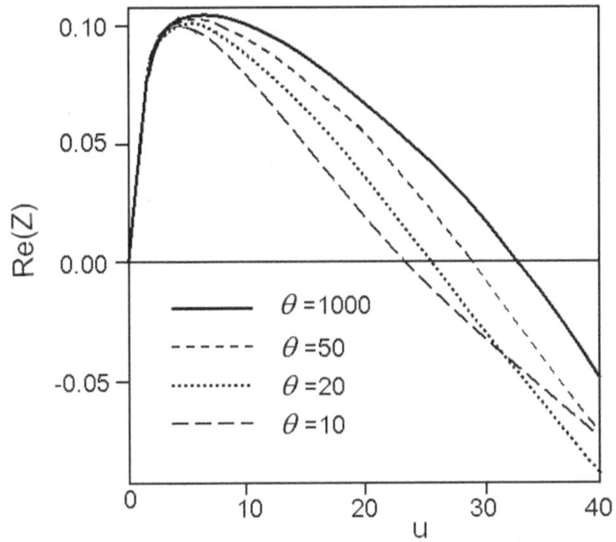

Fig. 12.26. The reduced response factor N/α as a function of the reduced frequency u for increasing values of θ and for arbitarily chosen values of the coefficients $A = 10,\ B = 100$. The reduced cut-off frequencies u_c are obtained for $N/\alpha = \Re(Z) = 0$

$$Z = \frac{\hat{\dot{M}}}{\alpha \hat{p}_c} = \frac{iu}{1+iu}\frac{A+\theta E(u)}{B-\theta E(u)}, \tag{12.218}$$

where A and B are defined by (12.211).

Note that $Z(0) = Z_0(0) = 0$. Also, for $u \to \infty$ or for high values of θ, $\theta E \approx -iu$ and $Z \approx Z_0$. We can see that the cut-off frequency and then the instability domain are smaller in the model with a temperature gradient inside the droplet than in the model with no temperature gradient (see Fig. 12.26 and [226] for more realistic cases).

12.4.5 Other Cases of Droplet Vaporization and Combustion

If we are interested in obtaining not only analytical solutions to these problems relating to droplet evaporation and combustion but numerical solutions too (in order to calculate real flows), we cannot use some of the simplifying assumptions that we employed previously. Important assumptions of the preceding section can then be examined, and we must then perform a study to determine a system of equations that will lead to the solution.

We assumed for example that there was no translational motion of the droplets relative to the fluid. In practical situations, in particular in the vicinity of the injectors of an internal combustion engine, there is actually a relative velocity between the droplets and carrying gas. This induces relative motion and friction and modifies heat exchange. The liquid phase itself exhibits internal motions [127]. Readers interested in this particular topic can consult many works on it [1, 48, 60].[19]

The combustion process depends on the chemical composition of the injected propellants, the injector design, and the injection conditions. The initial diameter distribution of the droplets is an important parameter of this problem. Interactions between droplets play a role [24, 68, 228, 284], and cannot be ignored for dense clouds. Droplets are often considered to be spherical, but this assumption may not be valid, in particular for cases with low gas–liquid surface tension in the presence of a nonzero relative velocity. Pope et al. [209] studied numerically transient combustion regimes for moving droplets and suspended droplets; for a moving droplet, the Reynolds number decreases over time (due to both the relative velocity and reduction in droplet size) but the Damköhler number increases with time; for a suspended droplet, both the Reynolds number and the Damköhler number decrease over time due to the reduction in droplet size.

In cryogenic engines, the pressure is supercritical and the droplets are injected at a subcritical temperature into a high-temperature combustion chamber. Thus, transcritical conditions are encountered. This case can also arise with diesel engines.

These circumstances have led to several recent studies on, for instance, the expansion of a droplet at the vapor–liquid critical point [4, 212]) (the expansion of a droplet of critical fluid is presented in Sect. A.5.2), or the vaporization of liquid droplets at supercritical conditions [47, 66, 67, 112, 143, 187, 293, 294].

The subjects of reactivity and phase change are of fundamental interest, and there are a vast range of potential applications of supercritical fluids, from the space launcher industry to organic waste processing.

[19] *The Hill vortex.* One example of an analytical model for the internal motions of the liquid phase is the Hill vortex. If the radius of the droplet is r_s, the stream function of the Hill vortex is defined, in spherical coordinates in the case of cylindrical symmetry about the Oz axis, by

$$\psi = U_s(r^2 \sin^2\theta/2)(1 - r^2/r_s^2), \; r \le r_s, \; \psi = U_s(r^2 \sin^2\theta/2)(1 - r^3/r_s^3), \; r \ge r_s,$$

where U_s is the velocity at the droplet surface. Writing $d\psi = -rU_\theta \sin\theta \, dr + r^2 U_r \sin\theta \, d\theta$, the velocity components of the liquid are then

$$U_r = U_s(1 - r^2/r_s^2)\cos\theta, \; U_\theta = -U_s(1 - 2r^2/r_s^2)\sin\theta.$$

The stream surfaces are toroidal, and have Oz as their axis of symmetry.

A

Appendix

A.1 Tensor Notation

In oriented three-dimensional Cartesian space, only orthonormal frames of the same orientation are considered. Dyadic notation is used systematically. The components of tensors of order 1 (i.e., vectors), 2, 3, ... are denoted by the indices $\alpha, \beta, \gamma, ..., or$i, j, k, ... that take the values 1, 2, 3, ... The Einstein convention is used for the summation.

Tensors

The following notation is adopted for tensors and their components:

- Vector: $\mathbf{V}$, $(\mathbf{V})_\alpha = V_\alpha$
- Second-order tensor: $\mathbf{P}$, $(\mathbf{P})_{\alpha\beta} = P_{\alpha\beta}$
- Second-order unit tensor: $\mathbf{1}$, $(\mathbf{1})_{\alpha\beta} = \delta_{\alpha\beta}$, where $\delta_{\alpha\beta}$ is the Kronecker symbol such that $\delta_{\alpha\beta} = 0$ for $\alpha \neq \beta$, and $\delta_{\alpha\beta} = 1$ for $\alpha = \beta$
- Third-order tensor: $\mathbf{B}$, $(\mathbf{B})_{\alpha\beta\gamma} = B_{\alpha\beta\gamma}$, and so on.

Tensor Products

The tensor product is denoted by the symbol $\otimes$.[1] For example,

$$(\mathbf{V} \otimes \mathbf{P})_{\alpha\beta\gamma} = V_\alpha P_{\beta\gamma},\ (\mathbf{P} \otimes \mathbf{V})_{\alpha\beta\gamma} = P_{\alpha\beta} V_\gamma,\ (\mathbf{P} \otimes \mathbf{P}')_{\alpha\beta\gamma\delta} = P_{\alpha\beta} P'_{\gamma\delta}.$$

Contracted Products

A dot ("·") is used for contracted products of two tensors where the summation is made on two neighboring indices. For example,

$$\mathbf{V} \cdot \mathbf{V}' = \sum_\alpha V_\alpha V'_\alpha = V_\alpha V'_\alpha,\ (\mathbf{V} \cdot \mathbf{P})_\beta = V_\alpha P_{\alpha\beta},$$

[1]This symbol is sometimes omitted when ambiguous interpretation is not possible.

R. Prud'homme, *Flows of Reactive Fluids*,
DOI 10.1007/978-0-8176-4659-2, © Springer Science+Business Media, LLC 2010

$$(\mathbf{P}\cdot\mathbf{V})_\alpha = P_{\alpha\beta}V_\beta,\ (\mathbf{P}\cdot\mathbf{P}')_{\alpha\delta} = P_{\alpha\beta}P'_{\beta\delta}.$$

In particular, we have $\mathbf{1}\cdot\mathbf{P} = \mathbf{P}\cdot\mathbf{1} = \mathbf{P}$.

In the same way, two dots (":") are used for double-contracted tensor products. We perform a summation after equalizing the nearest indices (not summed beforehand) of the once-contracted product. Then

$$\mathbf{P} : \mathbf{P}' = P_{\alpha\beta}P'_{\beta\alpha} = \sum_{\alpha\beta} P_{\alpha\beta}P'_{\beta\alpha} = \sum_{\alpha}(\mathbf{P}\cdot\mathbf{P}')_{\alpha\alpha}.$$

In matrix language, the contracted product of two second-order tensors is equivalent to the product of the corresponding matrices, and the double-contracted product is equal to the trace of this product. In particular,

$$\mathbf{P} : \mathbf{1} = \mathbf{1} : \mathbf{P} = \mathrm{tr}\mathbf{P} = P_{\alpha\alpha}.$$

The triple-contracted product of two third-order tensors gives $\mathbf{B}\vdots\mathbf{B}' = B_{\alpha\beta\gamma}B'_{\gamma\beta\alpha}$, and so on.

These are classical rules, and the corresponding notation is unambiguous. This is not the case for tensor derivatives. The following rules are used in this book:

- Gradient of a scalar a: $(\boldsymbol{\nabla} a)_\alpha = \partial a/\partial x_\alpha = a_{,\alpha}$.
- Gradient of a vector $\mathbf{V}$: $(\boldsymbol{\nabla}\otimes\mathbf{V})_{\alpha\beta} = \partial V_\beta/\partial x_\alpha = V_{\beta,\alpha}$.
- If we consider $\boldsymbol{\nabla}$ to be a vector, $(\boldsymbol{\nabla})_\alpha = \partial x_\alpha$, then we can simply apply the rules of tensor products, for example $(\boldsymbol{\nabla}\otimes\mathbf{V})_{\alpha\beta} = (\boldsymbol{\nabla})_\alpha(\mathbf{V})_\beta = (\partial/\partial x_\alpha)V_\beta = V_{\beta,\alpha}$. The same is true of the other operations, as we can see below.
- Divergence of a vector $\mathbf{V}$: we have $\boldsymbol{\nabla}\cdot\mathbf{V} = \partial V_\alpha/\partial x_\alpha = V_{\alpha,\alpha}$.
- Divergence of a second-order tensor $\mathbf{P}$: we have[2] $(\boldsymbol{\nabla}\cdot\mathbf{P})_\gamma = \partial P_{\alpha\gamma}/\partial x_\alpha = P_{\alpha\gamma,\alpha}$, which is obtained by contracting $(\boldsymbol{\nabla}\cdot\mathbf{P})_{\alpha\beta\gamma}$ on the two indices α and β.

  For the expansion of $\boldsymbol{\nabla}\cdot(\mathbf{V}\cdot\mathbf{P})$, we get $(\mathbf{V}\cdot\mathbf{P})_\beta = V_\alpha P_{\alpha\beta}$, $\boldsymbol{\nabla}\cdot(\mathbf{V}\cdot\mathbf{P}) = \partial(V_\alpha P_{\alpha\beta})/\partial x_\alpha = (\partial(V_\alpha)/\partial x_\beta)P_{\alpha\beta} + V_\alpha(\partial P_{\alpha\beta}/\partial x_\beta)$, and so

$$\boldsymbol{\nabla}\cdot(\mathbf{V}\cdot\mathbf{P}) = \boldsymbol{\nabla}\otimes\mathbf{V} : \mathbf{P} + \mathbf{V}\cdot(\boldsymbol{\nabla}\cdot\mathbf{P}),$$

  with $(\widetilde{\mathbf{P}})_{\alpha\beta} = (\mathbf{P})_{\beta\alpha}$.

[2]Note that different notations are used for second-order tensor divergence. Some authors write $(\boldsymbol{\nabla}\cdot\mathbf{P})_\alpha = \partial P_{\alpha\gamma}/\partial x_\gamma$. A simple transposition of the tensor $\mathbf{P}$ makes the distinction between the notations. Both results are similar when $\mathbf{P}$ is a symmetric tensor.

A.2 Motion and Field of Deformation of an Interfacial Layer

At a small scale, an interfacial layer is a particular three-dimensional medium that is delimited by two surfaces (S_-) and (S_+), one of which can be sent to infinity if the scale of analysis is very small.

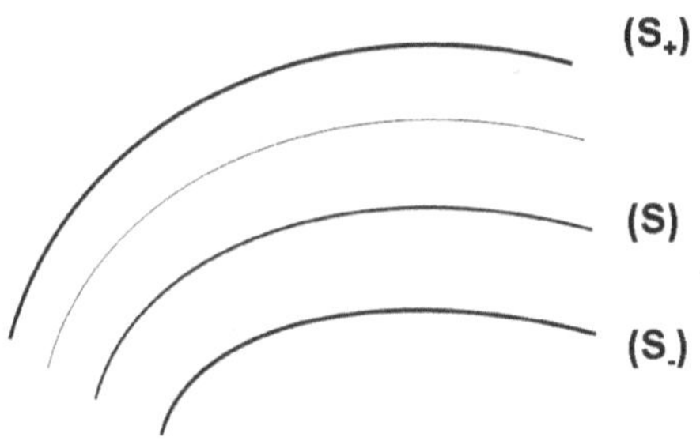

Fig. A.1. The family of surfaces for an interfacial layer (2D representation)

We assume that (S_-) and (S_+) belong to a family of nonsecant surfaces (S) with forms that vary in a continuous way (Fig. A.1). The forms of this family of surfaces depend on time and, at a given moment, can be described by an orthogonal curvilinear frame of reference.

We will examine the situation where the family consists of parallel surfaces (S) (see Eq. A.61). This is aside from the choice that will be made in Sect. 11.2 to establish the balance equations. We are initially interested in an element (S) of this family for which we will give an implicit equation and then a description in orthogonal curvilinear coordinates. The various differentiation operations for tensors of order 0 to 2 with respect to space variables and time will be expressed in these coordinates.[3]

All of the results established here will be used to establish the balance equations by the classical method and the method of virtual power.

Geometry and Kinematics of a Surface in Intrinsic Coordinates

Consider a surface (S) that is defined in implicit form by the equation

[3]The corresponding field of deformations is established in Sect. 11.5 for the case of the velocities of a field of kinematic torsors.

$$\mathcal{F}(\mathbf{x},\, t) = 0. \tag{A.1}$$

The unit normal $\mathbf{N}$ at a point $\mathbf{x}$ on this surface at time t is

$$\mathbf{N} = \pm \boldsymbol{\nabla}\mathcal{F} / \left|\boldsymbol{\nabla}\mathcal{F}\right| . \tag{A.2}$$

Here, we have assumed that $\mathcal{F}$ has all the necessary properties of continuity and differentiability. The orientation of the normal can be chosen arbitrarily. The orientation of the normal can be changed by using the function $-\mathcal{F}$ to define the same surface.

The relations (A.1) and (A.2) define the unit normal at a given point, and by choosing the sign (+),

$$\mathbf{N} = \boldsymbol{\nabla}\mathcal{F} / \left|\boldsymbol{\nabla}\mathcal{F}\right| \tag{A.3}$$

defines a vector field that covers the whole defined space of $\mathcal{F}(\mathbf{x},\, t)$.

According to (A.1), we have, while following the motion of surface (S),

$$\boldsymbol{\nabla}\mathcal{F} \cdot d\mathbf{x} + \frac{\partial \mathcal{F}}{\partial t} dt = 0, \tag{A.4}$$

so that the normal velocity of the surface becomes

$$w = -\frac{\partial \mathcal{F}}{\partial t} / \left|\boldsymbol{\nabla}\mathcal{F}\right| , \tag{A.5}$$

which depends on the orientation of $\mathbf{N}$. However, the normal velocity vector $\mathbf{w} = w\mathbf{N}$ does not depend on this orientation.

$\mathbf{w}(\mathbf{x},\, t)$ defines a field of vectors and $w(\mathbf{x},\, t)$ a field of scalars in the defined field of $\mathcal{F}(\mathbf{x},\, t)$.

A.3 Geometry of Interfaces and Interfacial Layers in Curvilinear Coordinates

Orthogonal Curvilinear Coordinates

Equation A.1 does not make it possible to specify the motion of a particular point on surface (S) other than by its normal speed. If we want to characterize this motion (in order to study the motion of the continuous medium inside an interfacial layer), it is necessary to obtain the velocity field at any point. Orthogonal curvilinear coordinates provide a convenient description of the motion and characteristics of surfaces. While the expressions obtained are sometimes complex, it is often instructive to use orthogonal curvilinear coordinates to carry out some investigations, even if we must then return then to using more condensed formulae.

Denoting Cartesian coordinates by x, y and z and orthogonal curvilinear coordinates by x_1, x_2 and x_3, we can then write the following coordinate change formulae:

$$\begin{cases} x = x(x_1,\, x_2,\, x_3,\, t), \\ y = y(x_1,\, x_2,\, x_3,\, t), \\ z = z(x_1,\, x_2,\, x_3,\, t). \end{cases} \tag{A.6}$$

We will denote the vector of components $x_{,j}$, $y_{,j}$, $z_{,j}$ (where $_{,j}$ means $\partial/\partial x_j$) as $\mathbf{h}_j$, and assume that none of the vectors $\mathbf{h}_j$ ($j = 1,\, 2,\, 3$) are zero vectors. The curvilinear coordinates are orthogonal if the trihedron $\{\mathbf{h}_1,\, \mathbf{h}_2,\, \mathbf{h}_3\}$ is orthogonal. We will assume that this is the case by definition. At each point in space, we can define the basis $\{\mathbf{e}_1,\, \mathbf{e}_2,\, \mathbf{e}_3\}$ such that

$$\mathbf{h}_j = h_j\, \mathbf{e}_j \tag{A.7}$$

(see Fig. A.2). The differentiation formulae at a fixed time t can then be written in the matrical form

$$\left\| \begin{matrix} dx \\ dy \\ dz \end{matrix} \right\| = \left\| \, h_1\mathbf{e}_1 \; h_2\mathbf{e}_2 \; h_3\mathbf{e}_3 \, \right\| \left\| \begin{matrix} dx_1 \\ dx_2 \\ dx_3 \end{matrix} \right\| . \tag{A.8}$$

If we set

$$dX_j = h_j\, dx_j, \tag{A.9}$$

the correspondence between (dX_1, dX_2, dX_3) and (dx, dy, dz) is obtained by a change of Cartesian coordinates.

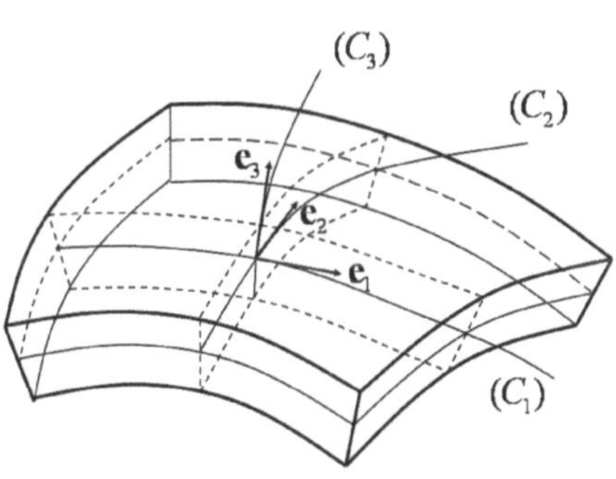

Fig. A.2. Reference basis for orthogonal curvilinear coordinates at the current point

At any time t, the curves (C_1), (C_2), (C_3) that pass through the point P are defined by fixed pairs of variables x_j. Thus, (C_1) is defined by its fixed values of x_2 and x_3. A surface can be defined by keeping just one of the variables x_j constant, as illustrated by system (A.6).

We will make x_3 constant in order to define surface (S). Therefore, the normal $\mathbf{N}$ to (S) at the point P is simply the vector $\mathbf{e}_3$.[4]

Expressions for vector or tensor curvilinear partial derivatives are needed for practical applications. Exterior calculus enables them to be easily obtained, and will be used in some of the demonstrations that follow. For example, a differential volume element can be written as

$$dx \wedge dy \wedge dz = h_1 h_2 h_3 \, dx_1 \wedge dx_2 \wedge dx_3. \tag{A.10}$$

Gradient of a Scalar

Let us now consider a scalar function $a(x_1, x_2, x_3, t)$. The gradient of a is deduced using the differential form $\omega = a$:

$$\boldsymbol{\nabla} a = \frac{a_{,1}}{h_1}\mathbf{e}_1 + \frac{a_{,2}}{h_2}\mathbf{e}_2 + \frac{a_{,3}}{h_3}.\mathbf{e}_3 \tag{A.11}$$

The parallel gradient is[5]

$$\boldsymbol{\nabla}_{//} a = \frac{a_{,1}}{h_1}\mathbf{e}_1 + \frac{a_{,2}}{h_2}\mathbf{e}_2, \tag{A.12}$$

and the gradient of the normal to (S) becomes

$$\boldsymbol{\nabla}_{\perp} a = \frac{a_{,3}}{h_3}\mathbf{e}_3. \tag{A.13}$$

Gradient and Divergence of a Vector Field

Suppose that we have a vector $\mathbf{V} = V_x\mathbf{i} + V_y\mathbf{j} + V_z\mathbf{k}$, where $\{\mathbf{i}, \mathbf{j}, \mathbf{k}\}$ is a Cartesian basis. In orthogonal curvilinear coordinates,

$$\mathbf{V} = v_1\mathbf{e}_1 + v_2\mathbf{e}_2 + w\mathbf{N}, \tag{A.14}$$

where w is the component v_3 along the normal $\mathbf{e}_3 = \mathbf{N}$ to surface (S).

---

[4] In some cases, the surface motion is such that (A.6) do not depend explicitly on time. The points of (S) will then obey the relations $x = x(x_1, x_2, x_3)$, $y = y(x_1, x_2, x_3)$, $z = z(x_1, x_2, x_3)$, $x_3 = \phi(t)$. This is the case, for example, when (S) is always a sphere centered on O that only varies in its radius: $x_3 = R(t)$. However, this simplification is not valid for the general case.

[5] Projection operators can be used to obtain normal and parallel components of vectors and tensors [94]. We define $\mathbf{1}_{\perp} = \mathbf{e}_3 \otimes \mathbf{e}_3$ and $\mathbf{1}_{//} = \mathbf{1} - \mathbf{1}_{\perp}$. A vector $\mathbf{V}$ then has two components $\mathbf{V}_{\perp} = \mathbf{1}_{\perp} \cdot \mathbf{V} = V_{\perp}\mathbf{e}_3$ and $\mathbf{V}_{//} = \mathbf{1}_{//} \cdot \mathbf{V}$, and we have: $\mathbf{V} = \mathbf{V}_{\perp} + \mathbf{V}_{//}$.

The differential $d\mathbf{V}$ utilizes the vectors $d\mathbf{e}_j$. We have [10]:

$$\begin{cases} d\mathbf{e}_1 = d\omega_2\mathbf{N} - d\omega_3\mathbf{e}_2, \\ d\mathbf{e}_2 = d\omega_3\mathbf{e}_1 - d\omega_1\mathbf{N}, \\ d\mathbf{N} = d\omega_1\mathbf{e}_2 - d\omega_2\mathbf{e}_1, \end{cases} \tag{A.15}$$

where the differential forms $d\omega_j$ are given by

$$\begin{cases} d\omega_1 = \frac{h_{2,3}}{h_3}dx_2 - \frac{h_{3,2}}{h_2}dx_3, \\ d\omega_2 = \frac{h_{3,1}}{h_1}dx_3 - \frac{h_{1,3}}{h_3}dx_1, \\ d\omega_3 = \frac{h_{1,2}}{h_2}dx_1 - \frac{h_{2,1}}{h_1}dx_2. \end{cases} \tag{A.16}$$

Thus,

$$d\mathbf{V} = (\boldsymbol{\nabla} \otimes \mathbf{V}) \cdot d\mathbf{x}. \tag{A.17}$$

In curvilinear coordinates,

$$d\mathbf{x} = h_1\, dx_1\, \mathbf{e}_1 + h_2\, dx_2\, \mathbf{e}_2 + h_3\, dx_3\, \mathbf{N}. \tag{A.18}$$

These results lead us, after some calculations, to the expression for the tensor gradient $\boldsymbol{\nabla} \otimes \mathbf{V}$ in orthogonal curvilinear coordinates, which is provided by the following matrix:

$$\begin{Vmatrix} \frac{1}{h_1}\left(\frac{\partial v_1}{\partial x_1} + \frac{v_2}{h_2}\frac{\partial h_1}{dx_2} + \frac{w}{h_3}\frac{\partial h_1}{\partial x_3}\right) & \frac{1}{h_2}\left(\frac{\partial v_1}{\partial x_2} - \frac{v_2}{h_1}\frac{\partial h_2}{\partial x_1}\right) & \frac{1}{h_3}\left(\frac{\partial v_1}{\partial x_3} - \frac{w}{h_1}\frac{\partial h_3}{\partial x_1}\right) \\ \frac{1}{h_1}\left(\frac{\partial v_2}{\partial x_1} - \frac{v_1}{h_2}\frac{\partial h_1}{\partial x_2}\right) & \frac{1}{h_2}\left(\frac{\partial v_2}{\partial x_2} + \frac{w}{h_3}\frac{\partial h_2}{dx_3} + \frac{v_1}{h_1}\frac{\partial h_2}{\partial x_1}\right) & \frac{1}{h_3}\left(\frac{\partial v_2}{\partial x_3} - \frac{w}{h_2}\frac{\partial h_3}{\partial x_2}\right) \\ \frac{1}{h_1}\left(\frac{\partial w}{\partial x_1} - \frac{v_1}{h_3}\frac{\partial h_1}{\partial x_3}\right) & \frac{1}{h_2}\left(\frac{\partial w}{\partial x_2} - \frac{v_2}{h_3}\frac{\partial h_2}{\partial x_3}\right) & \frac{1}{h_3}\left(\frac{\partial w}{\partial x_3} + \frac{v_1}{h_1}\frac{\partial h_3}{dx_1} + \frac{v_2}{h_2}\frac{\partial h_3}{\partial x_2}\right) \end{Vmatrix}. \tag{A.19}$$

Moreover, we define the parallel gradient of $\mathbf{V}$ (parallel to (S); i.e., when the point P varies along a curve, x_3=const.), $\boldsymbol{\nabla}_{//} \otimes \mathbf{V}$, as follows:

$$\begin{Vmatrix} \frac{1}{h_1}\left(\frac{\partial v_1}{\partial x_1} + \frac{v_2}{h_2}\frac{\partial h_1}{dx_2} + \frac{w}{h_3}\frac{\partial h_1}{\partial x_3}\right) & \frac{1}{h_2}\left(\frac{\partial v_1}{\partial x_2} - \frac{v_2}{h_1}\frac{\partial h_2}{\partial x_1}\right) & 0 \\ \frac{1}{h_1}\left(\frac{\partial v_2}{\partial x_1} - \frac{v_1}{h_2}\frac{\partial h_1}{\partial x_2}\right) & \frac{1}{h_2}\left(\frac{\partial v_2}{\partial x_2} + \frac{w}{h_3}\frac{\partial h_2}{dx_3} + \frac{v_1}{h_1}\frac{\partial h_2}{\partial x_1}\right) & 0 \\ \frac{1}{h_1}\left(\frac{\partial w}{\partial x_1} - \frac{v_1}{h_3}\frac{\partial h_1}{\partial x_3}\right) & \frac{1}{h_2}\left(\frac{\partial w}{\partial x_2} - \frac{v_2}{h_3}\frac{\partial h_2}{\partial x_3}\right) & 0 \end{Vmatrix}. \tag{A.20}$$

The divergence of $\mathbf{V}$ is the trace of the tensor gradient.[6] This is expressed as follows:

$$\begin{aligned} \boldsymbol{\nabla} \cdot \mathbf{V} = {} & \frac{1}{h_1}\left(\frac{\partial v_1}{\partial x_1} + \frac{v_2}{h_2}\frac{\partial h_1}{dx_2} + \frac{w}{h_3}\frac{\partial h_1}{\partial x_3}\right) + \frac{1}{h_2}\left(\frac{\partial v_2}{\partial x_2} + \frac{w}{h_3}\frac{\partial h_2}{dx_3} + \frac{v_1}{h_1}\frac{\partial h_2}{\partial x_1}\right) \\ & + \frac{1}{h_3}\left(\frac{\partial w}{\partial x_3} + \frac{v_1}{h_1}\frac{\partial h_3}{dx_1} + \frac{v_2}{h_2}\frac{\partial h_3}{\partial x_2}\right), \end{aligned} \tag{A.21}$$

[6]We can also use the differential form $\omega = V_x dy \wedge dz + V_y dz \wedge dx + V_z dx \wedge dy$, written in orthogonal curvilinear coordinates, to provide our demonstration.

and the parallel divergence of $\mathbf{V}$ becomes

$$\nabla_{//}\cdot\mathbf{V} = \frac{1}{h_1}(\frac{\partial v_1}{\partial x_1} + \frac{v_2}{h_2}\frac{\partial h_1}{dx_2} + \frac{w}{h_3}\frac{\partial h_1}{\partial x_3}) + \frac{1}{h_2}(\frac{\partial v_2}{\partial x_2} + \frac{w}{h_3}\frac{\partial h_2}{dx_3} + \frac{v_1}{h_1}\frac{\partial h_2}{\partial x_1}). \quad (A.22)$$

Curvature

The curves (C_1), (C_2) and (C_3) are generally nonplanar curves. The curvature of (C_1), for example, is provided by $\partial\mathbf{e}_1/\partial x_1$:

$$\frac{\partial\mathbf{e}_1}{\partial X_1} = \frac{1}{h_1}\frac{\partial\mathbf{e}_1}{\partial x_1} = \frac{1}{h_1}(\frac{\partial\omega_2}{\partial x_1}\mathbf{N} - \frac{\partial\omega_3}{\partial x_1}\mathbf{e}_2); \quad (A.23)$$

in other words,

$$\frac{\partial\mathbf{e}_1}{\partial X_1} = -\frac{1}{h_1}(\frac{h_{1,3}}{h_3}\mathbf{N} + \frac{h_{1,2}}{h_2}\mathbf{e}_2). \quad (A.24)$$

To define the mean normal curvature of a surface (S) at a point P, it is necessary to consider two orthogonal planes that contain the vector $\mathbf{e}_3 = \mathbf{N}$. We then add the curvatures obtained in each plane. For a coordinate surface (S) that is normal to $\mathbf{e}_3$, we obtain

$$\frac{1}{R} = \frac{1}{R_1} + \frac{1}{R_2} = \frac{h_{1,3}}{h_1h_3} + \frac{h_{2,3}}{h_2h_3}. \quad (A.25)$$

This quantity is also equal to $\nabla\cdot\mathbf{N}$ according to (A.21). However, $\nabla\cdot\mathbf{N}$ is invariant at the considered point P and does not depend on the curvilinear coordinates used to describe (S) [5]. It follows that (A.25) remains valid in the curvilinear system defined by (A.6), and that we always have

$$\frac{1}{R} = \nabla\cdot\mathbf{N} = \frac{h_{1,3}}{h_1h_3} + \frac{h_{2,3}}{h_2h_3}. \quad (A.26)$$

Note that, according to (A.22), if $v_1 = v_2 = 0$ and $w = 1$ then

$$\nabla\cdot\mathbf{N} = \nabla_{//}\cdot\mathbf{N}. \quad (A.27)$$

Divergence of a Second-Order Tensor

We limit ourselves here to a "parallel" tensor; i.e., a tensor such that

$$\mathbf{P}_{//} = P_{11}\mathbf{e}_1\otimes\mathbf{e}_1 + P_{12}\mathbf{e}_1\otimes\mathbf{e}_2 + P_{21}\mathbf{e}_2\otimes\mathbf{e}_1 + P_{22}\mathbf{e}_2\otimes\mathbf{e}_2.$$

We then find, according to (A.15), that

$$
\begin{aligned}
d\mathbf{P}_{//} &= d\omega_1(-P_{12}\mathbf{e}_1\otimes\mathbf{N} - P_{21}\mathbf{N}\otimes\mathbf{e}_1\\
&-P_{22}\mathbf{N}\otimes\mathbf{e}_2 - P_{22}\mathbf{e}_2\otimes\mathbf{N}) + d\omega_2(P_{11}\mathbf{N}\otimes\mathbf{e}_1 + P_{11}\mathbf{N}\otimes\mathbf{e}_1\\
&+P_{12}\mathbf{N}\otimes\mathbf{e}_2 + P_{21}\mathbf{e}_2\otimes\mathbf{N}) + d\omega_3(-P_{11}\mathbf{e}_2\otimes\mathbf{e}_1 - P_{11}\mathbf{e}_1\otimes\mathbf{e}_2\\
&-P_{12}\mathbf{e}_2\otimes\mathbf{e}_2 + P_{12}\mathbf{e}_1\otimes\mathbf{e}_1 + P_{21}\mathbf{e}_1\otimes\mathbf{e}_1 - P_{21}\mathbf{e}_2\otimes\mathbf{e}_2\\
&+P_{22}\mathbf{e}_1\otimes\mathbf{e}_2 + P_{22}\mathbf{e}_2\otimes\mathbf{e}_2) + \textstyle\sum_{i,j} dP_{ij}\,\mathbf{e}_i\otimes\mathbf{e}_j.
\end{aligned}
\tag{A.28}
$$

By expressing $d\omega_j$ with the aid of (A.16), we thus obtain the components of $\boldsymbol{\nabla}\cdot\mathbf{P}_{//}$:

$$
\begin{aligned}
&\boldsymbol{\nabla}\cdot\mathbf{P}_{//} =\\
&\left\{\tfrac{1}{h_1h_2h_3}[(P_{11}h_2h_3)_{,1} + (P_{12}h_3h_1)_{,2}] + \tfrac{1}{h_1h_2}[P_{2,1}h_{1,2} - P_{22}h_{2,1}]\right\}\mathbf{e}_1\\
&+\left\{\tfrac{1}{h_1h_2h_3}[(P_{21}h_2h_3)_{,1} + (P_{22}h_3h_1)_{,2}] + \tfrac{1}{h_1h_2}[P_{12}h_{2,1} - P_{11}h_{1,2}]\right\}\mathbf{e}_2\\
&-(\tfrac{1}{h_3h_1}P_{11}h_{1,3} + \tfrac{1}{h_2h_3}P_{22}h_{2,3})\mathbf{N}.
\end{aligned}
\tag{A.29}
$$

The vector $\boldsymbol{\nabla}_{//}\cdot\mathbf{P}_{//}$ is obtained from

$$
\boldsymbol{\nabla}_{//}\cdot\mathbf{P}_{//} = \frac{\partial\mathbf{P}_{//}}{\partial X_1}\cdot\mathbf{e}_1 + \frac{\partial\mathbf{P}_{//}}{\partial X_2}\cdot\mathbf{e}_2,
\tag{A.30}
$$

which leads to

$$
\begin{aligned}
&\boldsymbol{\nabla}_{//}\cdot\mathbf{P}_{//} =\\
&\left[\tfrac{P_{11,1}}{h_1} + \tfrac{P_{12,2}}{h_2} + \tfrac{h_{1,2}}{h_1h_2}(P_{12}+P_{21}) + \tfrac{h_{2,1}}{h_2h_1}(P_{11}-P_{22})\right]\mathbf{e}_1\\
&\left[\tfrac{P_{21,1}}{h_1} + \tfrac{P_{22,2}}{h_2} + \tfrac{h_{1,2}}{h_1h_2}(P_{22}-P_{11}) + \tfrac{h_{2,1}}{h_2h_1}(P_{12}+P_{21})\right]\mathbf{e}_2\\
&-\tfrac{1}{h_3}(\tfrac{P_{11}h_{1,3}}{h_1} + \tfrac{P_{22}h_{2,3}}{h_2})\mathbf{N}.
\end{aligned}
\tag{A.31}
$$

In the case of a "cylindrical" tensor

$$
\mathbf{P}_{//} = P\,(\mathbf{e}_1\otimes\mathbf{e}_1 + \mathbf{e}_2\otimes\mathbf{e}_2)
\tag{A.32}
$$

or

$$
\mathbf{P}_{//} = P\,(\mathbf{1} - \mathbf{N}\otimes\mathbf{N}),
\tag{A.33}
$$

we obtain

$$
\boldsymbol{\nabla}_{//}\cdot\mathbf{P}_{//} = \frac{P_{,1}}{h_1}\mathbf{e}_1 + \frac{P_{,2}}{h_2}\mathbf{e}_2 - \frac{P}{h_3}(\frac{h_{1,3}}{h_1} + \frac{h_{2,3}}{h_2})\mathbf{N}.
\tag{A.34}
$$

In other words,

$$
\boldsymbol{\nabla}_{//}\cdot\mathbf{P}_{//} = \boldsymbol{\nabla}_{//}\cdot\mathbf{P} - P\,\boldsymbol{\nabla}\cdot\mathbf{N}\,\mathbf{N}.
\tag{A.35}
$$

Calculation of $\nabla_{//} \cdot (\mathbf{V} \cdot \mathbf{P}_{//})$

We have

$$\begin{aligned}\nabla_{//} \cdot (\mathbf{V} \cdot \mathbf{P}_{//}) = \tfrac{1}{h_1} \left[(v_1 P_{11} + v_2 P_{21})_{,1} + \tfrac{h_{1,2}}{h_2}(v_1 P_{12} + v_2 P_{22})\right] \\ + \tfrac{1}{h_2} \left[(v_1 P_{12} + v_2 P_{22})_{,2} + \tfrac{h_{2,1}}{h_1}(v_1 P_{11} + v_2 P_{21})\right].\end{aligned} \tag{A.30}$$

Moreover,

$$\begin{aligned}\mathbf{P}_{//}{:}(\nabla_{//} \otimes \mathbf{V}) = \tfrac{P_{11}}{h_1}(v_{1,1} + v_2 \tfrac{h_{1,2}}{h_2} + w \tfrac{h_{1,3}}{h_3}) \\ + \tfrac{P_{21}}{h_2}(v_{1,2} + v_2 \tfrac{h_{2,1}}{h_1}) + \tfrac{P_{12}}{h_1}(v_{2,1} - v_1 \tfrac{h_{1,2}}{h_2}) \\ + \tfrac{P_{22}}{h_2}(v_{2,2} + w \tfrac{h_{2,3}}{h_3} + v_1 \tfrac{h_{2,1}}{h_1}),\end{aligned} \tag{A.37}$$

and

$$\begin{aligned}\mathbf{V} \cdot (\nabla_{//} \cdot P_{//}) = [\tfrac{P_{11,1}}{h_1} + \tfrac{P_{12,2}}{h_2} + \tfrac{h1,2}{h_1 h_2}(P_{12} + P_{21}) \\ + \tfrac{h_{2,1}}{h_1 h_2}(P_{11} - P_{12})]v_1 + [\tfrac{P_{21,1}}{h_1} + \tfrac{P_{22,2}}{h_2} + \tfrac{h1,2}{h_1 h_2}(P_{22} - P_{11}) \\ + \tfrac{h_{2,1}}{h_1 h_2}(P_{12} + P_{21})]v_2 - \tfrac{1}{h_3}(P_{11} \tfrac{h_{1,3}}{h_1} + P_2 \tfrac{h_{2,3}}{h_2}).\end{aligned} \tag{A.38}$$

It is easy to prove that

$$\nabla_{//} \cdot (\mathbf{V} \cdot \mathbf{P}_{//}) = \mathbf{P}_{//}{:}(\nabla_{//} \otimes \mathbf{V}) + \mathbf{V} \cdot (\nabla_{//} \cdot P_{//}). \tag{A.39}$$

Laplacian of a Scalar

The expression for the Laplacian in orthogonal curvilinear coordinates is as follows:

$$\begin{aligned}\nabla^2 a = \Delta a = \tfrac{1}{h_1 h_2 h_3}[(\tfrac{h_2 h_3 a_{,1}}{h_1})_{,1} + (\tfrac{h_3 h_1 a_{,2}}{h_2})_{,2}] \\ + \tfrac{(h_1 h_2)_{,3}}{h_1 h_2 h_3} \tfrac{a_{,3}}{h_3} + \tfrac{1}{h_3}(\tfrac{a_{,3}}{h_3})_{,3}\end{aligned} \tag{A.40}$$

or

$$\nabla^2 a = \nabla \cdot (\nabla_{//} a) + \nabla \cdot \mathbf{N} \frac{\partial a}{\partial N} + \frac{\partial^2 a}{\partial N^2}. \tag{A.41}$$

If the normal gradients $\partial/\partial N$ are very large compared to the tangential gradients, and $\nabla \cdot \mathbf{N}$ is on the order of 1 at the scale considered,

$$\nabla^2 a \cong \nabla \cdot \mathbf{N} \frac{\partial a}{\partial N} + \frac{\partial^2 a}{\partial N^2}. \tag{A.42}$$

A.4 Kinematics of Interfaces and Interfacial Layers

Strain Rate of a Surface

Suppose that point P on (S) is given a velocity

$$\mathbf{V} = \mathbf{v}_{//} + w\mathbf{N}, \tag{A.43}$$

where $\mathbf{v}_{//}$ is present in the plane tangent to (S) at point P:

$$\mathbf{v}_{//} \cdot \mathbf{N} = 0. \tag{A.44}$$

If we follow the motion of velocity $\mathbf{V}$, the material derivative of the volume element $d\mathcal{V}$ is (see Sect. 4.2)

$$\frac{d_{\mathbf{V}}(d\mathcal{V})}{dt} = \boldsymbol{\nabla} \cdot \mathbf{V}\, d\mathcal{V}, \tag{A.45}$$

where $\boldsymbol{\nabla} \cdot \mathbf{V}$ is given by expression (A.21).

Let us now set, in curvilinear coordinates,

$$dN = h_3 dx_3 = dX_3,\ d\Sigma = h_1 h_2 dx_1 dx_2 = dX_1 dX_2. \tag{A.46}$$

Thus,

$$d\mathcal{V} = d\Sigma dN,\ \frac{d_{\mathbf{V}}(d\mathcal{V})}{dt} = \frac{d_{\mathbf{V}}(d\Sigma)}{dt} dN + \frac{d_{\mathbf{V}}(dN)}{dt} d\Sigma. \tag{A.47}$$

Consider the vector element $d\mathbf{P} = \mathbf{N}\ dN$:

$$\frac{d_{\mathbf{V}}(\mathbf{N}dN)}{dt} = (\boldsymbol{\nabla} \otimes \mathbf{V}) \cdot \mathbf{N}dN = \frac{1}{h_3} \left\| \begin{matrix} v_{1,3} - w\frac{h_{3,1}}{h_1} \\ v_{2,3} - w\frac{h_{3,2}}{h_2} \\ w_{,3} + v_1\frac{h_{3,1}}{h_1} + v_2\frac{h_{3,2}}{h_2} \end{matrix} \right\|. \tag{A.48}$$

Thus,

$$\mathbf{N} \cdot \frac{d_{\mathbf{V}}(\mathbf{N}dN)}{dt} = \frac{d_{\mathbf{V}}(dN)}{dt} = \frac{1}{h_3}(w_{,3} + v_1\frac{h_{3,1}}{h_1} + v_2\frac{h_{3,2}}{h_2}) = \boldsymbol{\nabla}_{\perp} \cdot \mathbf{V} dN. \tag{A.49}$$

Consequently,

$$\frac{d_{\mathbf{V}}(d\Sigma)}{dt} = \frac{1}{h_1 h_2 h_3} h_3[(v_1 h_2)_{,1} + v_2 h_1)_{,2}] + w(h_1 h_2)_{,3} d\Sigma. \tag{A.50}$$

We recognize that the right hand side contains the expression for the parallel derivative of $\mathbf{V}$ (i.e., $\boldsymbol{\nabla}_{//} \cdot \mathbf{V}$, as given by Eq. A.22).

Thus, the surface strain rate (or the "flame stretch") is

$$\frac{d_{\mathbf{V}}(d\Sigma)}{dt} = \boldsymbol{\nabla}_{//} \cdot \mathbf{V}\, d\Sigma. \tag{A.51}$$

The following formulae are easily validated:

$$\boldsymbol{\nabla}_{//} \cdot \mathbf{V} = \boldsymbol{\nabla}_{//} \cdot \mathbf{v}_{//} + \boldsymbol{\nabla}_{//} \cdot (w\mathbf{N}), \tag{A.52}$$

$$\boldsymbol{\nabla}_{//} \cdot (w\mathbf{N}) = w\boldsymbol{\nabla}_{//} \cdot \mathbf{N} = w\boldsymbol{\nabla} \cdot \mathbf{N}. \tag{A.53}$$

Material Derivative of a Local Quantity

If $\mathbf{v}$ is the material velocity of a point M and $a(\mathbf{M}, t)$ is a quantity that depends on position and time, and we follow the motion velocity $\mathbf{v}$, we get

$$\frac{da}{dt} = \frac{\partial a}{\partial t} + \frac{v_1}{h_1} a_{,1} + \frac{v_2}{h_2} a_{,2} + \frac{v_3}{h_3} a_{,3}. \tag{A.54}$$

Speed of a Surface

Surface (S) obeys (A.1). In parametric form, it is defined by system (A.6) with $x_3 = c$, which gives

$$\begin{cases} x = x(x_1,\, x_2,\, c,\, t) \\ y = y(x_1,\, x_2,\, c,\, t) \\ z = z(x_1,\, x_2,\, c,\, t) \end{cases} \tag{A.55}$$

or

$$\mathbf{P} = \mathbf{P}(x_1,\, x_2,\, c,\, t). \tag{A.56}$$

During the motion, the quantities x_1 and x_2 vary, so

$$d\mathbf{P} = h_1 \mathbf{e}_1 dx_1 + h_2 \mathbf{e}_2 dx_2 + \frac{\partial \mathbf{P}}{\partial t}. \tag{A.57}$$

We therefore obtain

$$w = \mathbf{N} \cdot \frac{d\mathbf{P}}{dt} = \mathbf{N} \cdot \frac{\partial \mathbf{P}}{\partial t} \tag{A.58}$$

for the normal velocity that characterizes the surface motion.

Moreover, the tangential velocity components of point P are equal to

$$\begin{cases} v_1 = h_1 \frac{dx_1}{dt} + \mathbf{e}_1 \cdot \frac{\partial \mathbf{P}}{\partial t}, \\ v_2 = h_2 \frac{dx_2}{dt} + \mathbf{e}_2 \cdot \frac{\partial \mathbf{P}}{\partial t}. \end{cases} \tag{A.59}$$

We thus find that

$$\mathbf{V} = v_1 \mathbf{e}_1 + v_2 \mathbf{e}_2 + w\mathbf{N} \tag{A.60}$$

represents the velocity of point P attached to the deforming surface.

Motion of an Interfacial Layer

The interfacial layer is delimited by two surfaces (S_-) and (S_+) that correspond to $x_3 = c_-$ and $x_3 = c_+$ when the medium is described in orthogonal curvilinear coordinates.

We can distinguish several cases that will give rise to assumptions corresponding to actual situations.

If the surfaces (S) are parallel between (S_-) and (S_+), the normal to a surface (S_0) at any point P is also normal to all of the surfaces (S) of the interfacial layer. Thus,

$$\frac{\partial \mathbf{e}_3}{\partial x_3} = 0 \tag{A.61}$$

or

$$-\frac{h_{3,2}}{h_2}\mathbf{e}_2 - \frac{h_{3,1}}{h_1}\mathbf{e}_1 = 0; \tag{A.62}$$

i.e., h_3 depends only on x_3. This leads to

$$\begin{cases} \partial \mathbf{e}_2/\partial x_3 = -(h_{3,2}/h_2)\mathbf{N} = \mathbf{0}, \\ \partial \mathbf{e}_1/\partial x_3 = -(h_{3,1}/h_1)\mathbf{N} = \mathbf{0}. \end{cases} \tag{A.63}$$

The basis $\{\mathbf{e}_1, \mathbf{e}_2, \mathbf{e}_3\}$ is thus invariant for given values of x_1, x_2 and t.

Therefore, if $N - N_-$ indicates the distance from (S_-),

$$\begin{cases} N - N_- = \int_{c_-}^{c} h_3(x_3, t)\, dx_3, \\ N_+ - N_- = \int_{c_-}^{c_+} h_3(x_3, t)\, dx_3. \end{cases} \tag{A.64}$$

The last result shows that the interfacial zone has a uniform thickness.

If, moreover, *the velocity component w does not depend on x_3*, the thickness of the interface is constant. Thus, $\partial \mathbf{e}_3/\partial x_3 = 0$ and $\partial w/\partial x_3 = 0$ lead to

$$\boldsymbol{\nabla}_\perp \cdot \mathbf{V} = 0 \tag{A.65}$$

according to (A.49). This relation signifies a constant and uniform thickness.

The thickness of the interfacial layer corresponds to the distance along which certain physical parameters vary in a significant way. If we divide the lengths by a hydrodynamic scale L, we get (without changing the notation)

$$N_+ - N_- = \epsilon \ll 1, \tag{A.66}$$

where the normal coordinate N is dimensionless this time.

Let us assume that the dimensionless radius of curvature R of the surface is $O(1)$. Without assuming parallel surfaces, but supposing that $h_1 h_2$ is $O(1)$, we can set

$$dN = \epsilon\, dn, \tag{A.67}$$

where the thickness $n_+ - n_-$ is $O(1)$ this time.

The variations of $h_1 h_2$ are then $O(\epsilon)$ in the interfacial zone, so

$$d(h_1 h_2) = \frac{h_1 h_2}{R} h_3 \, dx_3 = \frac{h_1 h_2}{R} \, dN; \tag{A.68}$$

i.e.,

$$d(h_1 h_2) = \epsilon \frac{h_1 h_2}{R} \, dn. \tag{A.69}$$

Thus, at the location $(c_- + dx_3)$ we have

$$h_1 h_2 \cong (h_1 h_2)_- (1 + \frac{\epsilon}{R} \, dn), \tag{A.70}$$

$$d\Sigma \cong (d\Sigma)_- (1 + \frac{\epsilon}{R} \, dn). \tag{A.71}$$

Equation A.71 gives the order of magnitude of the variation in $(d\Sigma)$ across an interfacial layer.

A.5 Supercritical Fluids

A.5.1 Thermodynamic Properties of Supercritical Fluids

Thermostatics of a One-Component Fluid Near the Critical Point

At the liquid–vapor critical point C (Fig. 2.3), the first and second partial derivatives of pressure with respect to volume at constant temperature vanish:

$$(\partial p / \partial \overline{V})_T = (\partial^2 p / \partial \overline{V}^2)_T = 0,$$

and some physical quantities diverge according to (A.72).

The coordinates of the critical points for some chemical species [93] are listed in Table A.1.

Body	**Critical temperature** T_C **(K)**	**Critical pressure** p_C **(MPa)**	**Critical density (kg m^{-3})**
Water	647.1	22.06	322.2
Carbon dioxide	304.14	7.378	467.8
Oxygen	154.58	5.043	436.2
Nitrogen	126.24	3.398	313.9
Ammonia	405.4	11.1	235
Helium	3.316	0.114	41.45

Table A.1. Critical point coordinates for some species [93]

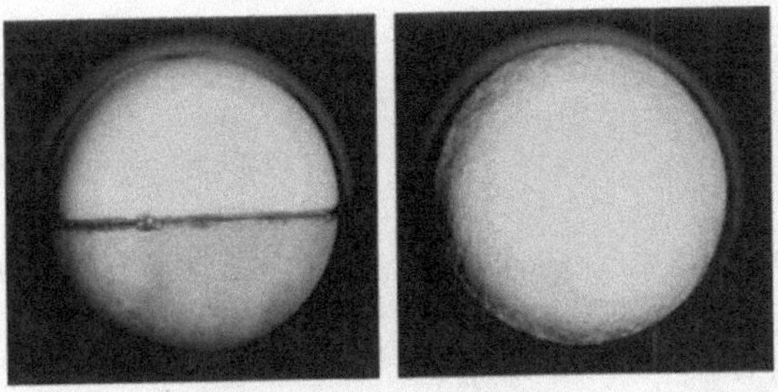

(a) (b)

Fig. A.3. From the subcritical (**a**) to the supercritical (**b**) state, using the example of CO_2 (reprinted with the permission of C.M. Rayner, A.A. Clifford and K.D. Bartle, Department of Chemistry, University of Leeds, UK)

If sufficient pressure is applied to a one-component fluid, the usual meniscus that separates the liquid phase from the gaseous phase vanishes, resulting in one continuous "supercritical" fluid phase (SCF) (Fig. A.3).

Supercritical fluids, which are very compressible and dilatable in the vicinity of the critical point but are also very dense, are sensitive to gravity, meaning that experimentation in microgravity is often necessary to understand and to model their behavior [35, 32, 187, 17]. Moreover, at the critical point, some physical properties diverge (such as the isothermal compressibility $K_T = (\partial \ln \rho / \partial p)_T$ and the specific heat c_p at constant pressure).

Classical laws (for example that of van der Waals) often use mean field theory [237]. The term "mean field" is applied to such models because they all assume that any given molecule in a system participates in the average molecular interaction for the whole system; i.e., the location of the molecule with respect to its neighbors is not accounted for. It is common to ignore the spatial distribution of molecules in an analysis of molecular interactions in the field of statistical physics. Such an approach leads to analytical expressions that are easy to use but not very accurate in the vicinity of the critical point.

Thus, the conditions $(\partial p/\partial \overline{V})_T = (\partial^2 p/\partial \overline{V}^2)_T = 0$ at the critical point lead to the relations $\overline{V}_C = 3b$, $T_C = 8\,a/27\,Rb$, $p_C = a/27\,b^2$, which yield $a = 9/8\,RT_C\overline{V}_C$, $b = \overline{V}_C/3$ and $p_C = 3\,RT_C/8\,\overline{V}_C$. This last relation is *not obeyed in reality.* Thus, for a van der Waals fluid, the isothermal compressibility $K_T = (\partial \ln \rho/\partial p)_T$ becomes $K_T = -4\frac{(\overline{V}_C)^2}{9\,RT_C}(\frac{T-T_C}{T_C})^{-1}$; a law that is also *not obeyed in reality.*

On the other hand, "renormalization group theory" makes it possible to deduce correct values close to the critical point [255]. If we denote a physical quantity by φ, we can then consider evolution according to a given thermodynamic path (for example at constant density that is equal to the critical density ρ_C). The physical quantity is then only a function of the relative temperature T/T_C. Formulae of the form

$$\frac{\varphi}{\varphi_r} \propto (\frac{T}{T_C} - 1)^s \tag{A.72}$$

are obtained, where s is a universal constant coefficient that is valid for a given type of critical transformation, and φ_r is a reference value for φ.

If s is positive, the quantity considered tends towards zero when the temperature tends towards its critical value. This is the case for the surface tension, which vanishes at the critical point ($s = 1.5$ is found using the mean field theory [237] and experiments give $s \approx 1.25$).

If s has a negative value there is a divergence at the critical point. This is the case for the heat capacity at constant pressure, which is also a divergent quantity.

Some examples of the results obtained for several quantities using the two theories (renormalization group and mean field) are given below [255]:

- Isothermal compressibility at critical density $K_T = \Gamma|\frac{T-T_C}{T_C}|^{-\gamma}$. Mean field: $\gamma = 1$. Renormalization group: $\gamma = 1.239 \pm 0.002$.
- Specific heat capacity at constant volume and critical density $c_v = A|\frac{T-T_C}{T_C}|^{-\alpha}$. Mean field: $\alpha = 0$. Renormalization group: $\alpha = 0.110 \pm 0.003$.
- Pressure evolution along the critical isotherm $p - p_C \propto |\frac{\rho-\rho_C}{\rho_C}|^{\delta}$. Mean field: $\delta = 3$. Renormalization group: $\delta = 4.8 \pm 0.02$.

Pseudo-critical State

We can even define a pseudo-critical state for a fluid mixture. This pseudo-critical state results from the assumption that the mixture behaves like a pure substance in the vicinity of a certain temperature T_C and pressure p_C; i.e., such that $(\partial p/\partial \overline{V})_{T_C} = 0$, $\partial^2 p/\partial \overline{V}^2)_{T_C} = 0$. The coordinates of this pseudo-critical point are obtained from empirical rules. We can for example assume the following rule:

$$\begin{cases} T_{Cm} = \sum_j X_j T_{Cj}, \\ p_{Cm} = R(\sum_j X_j Z_{Cj}) T_{Cm} / \sum_j X_j \overline{V}_{Cj}, \\ \omega_m = \sum_j X_j \omega_j. \end{cases} \tag{A.73}$$

Here, ω_j is the acentric factor,[7] and $Z_{Cj} = p_{Cj}\overline{V}_{Cj}/RT_{Cj}$ for each species. Other rules are possible, such as $T_{Cm} = \sum_{i,j} X_i X_j T_{Cij}$ with $T_{Cii} = T_{Ci}$, $T_{Ci \neq j} = k^*_{i,j}$ or $T_{Cij} = \sqrt{T_{Ci}T_{Cj}}(1 - k_{i,j})$. Specific $k_{i,j}$ values are provided for many pairs of species for each law.

[7] ω occurs in (2.50), the expression for the coefficients a, for constituents that obey the Peng–Robinson law of state; here, ω_m corresponds to the acentric factor of the mixture near the pseudo-critical state.

Transfer Coefficients at the Critical Point

The Theoretical Approach

Two different theoretical approaches are widely used to comprehend the singular behavior of the transfer coefficients at the critical point [164].

The dynamic renormalization group theory, first formulated by Halpering et al. [113], is an extension of the ideas of the static renormalization group theory of Wilson and Fisher [88, 292] to dynamic critical phenomena. Its strength lies in its predictions of the asymptotic critical behavior of transport coefficients and its elucidation of phenomenological concepts such as dynamic scaling and dynamic universality [119].

Mode coupling theory, on the other hand, originated from the idea of Fixman [89] that the critical anomalies of transport coefficients can be understood as being due to nonlinear coupling between the hydrodynamic modes of the system. One advantage of mode coupling theory [134] is that it can readily be applied outside the near-critical region. Starting with the work of Olchowy and Gelbart [193], mode coupling calculations and the relative decoupled mode theory of Ferrell and Perl [87, 200] have been employed to provide expressions for transport coefficients that are valid outside the asymptotic critical region [254].

Thermal Diffusivity

Thermal conductivity tends to infinity at the critical point, but the thermal diffusivity $\kappa = \lambda/\rho c_p$ vanishes as it approaches the critical point, so heat transfer by conduction becomes almost impossible. Another heat transfer mode termed the "piston effect" (see Sect. A.5.2) also occurs near the critical point.

Evolution at Constant Pressure

In order to study droplet evaporation and the combustion of supercritical fluids in rocket engines (i.e., cryogenic engines), where the pressure is nearly constant (Sect. 12.4.5), we need formulae that are valid for isobaric evolution.

In the usual formulae, the fluid density that is generally considered is the critical density ρ_C, so the physical quantities φ that are studied are only functions of the relative temperature T/T_C. This yields formulae of the form:

$$K_T = \Gamma|\frac{T-T_C}{T_C}|^{-\gamma},\ C_v = A|\frac{T-T_C}{T_C}|^{-\alpha},\ \xi = \xi_0|\frac{T-T_C}{T_C}|^{-\nu},$$

where ξ is the correlation length. We also obtain the relation $\rho_L - \rho_G = 2B|\frac{T-T_C}{T_C}|^{-\beta}$ along the liquid–vapor coexistence curve.

Near the critical point, the pressure is related to the density by $p - p_C \propto (|\rho - \rho_C|/\rho_C)^\delta$ (see Sect. A.5.1). For isobaric evolution at $p = p_C$, we can write [4]

$$\begin{cases} \rho - \rho_c \cong -\rho_0[(T - T_C)/T_C]^{1/\delta}, \ c_p = c_0|[(T - T_C)/T_C]|^{-\gamma/\beta\delta}, \\ \lambda \cong \rho_0 c_0 \kappa_0|[(T - T_C)/T_C]|^{-(\gamma-\nu)/\beta\delta}. \end{cases} \tag{A.74}$$

The following expression for the thermal conductivity, which is valid at a constant pressure and contains three terms that each vary in importance depending on the distance from the critical point, was provided by Arias-Zugasti et al. [4]:

$$\lambda = \lambda^{gas}(T) + \lambda^E(\rho) + \lambda^C(T). \tag{A.75}$$

Here, $\lambda^{gas}(T)$ is the limit at low density, $\lambda^E(\rho)$ is the excess thermal conductivity, and $\lambda^C(T)$ is the critical divergence. $\lambda^{gas}(T) = \lambda_\infty T/T_\infty$ (a linear dependence on temperature), or $\lambda^{gas}(T) = \lambda_\infty\sqrt{T/T_\infty}$ (as given by the kinetic theory of gases). $\lambda^E(\rho) = \lambda_1\rho/\rho_C$ (a linear dependence on density), $\lambda^C(T) = \lambda_0|\frac{T-T_C}{T_C}|^{1/3}$ (average field theory).

A.5.2 Supercritical Fluid Flows

The Piston Effect

We will now provide a qualitative description of the piston effect (P.E.) [28]. Thermal diffusivity tends towards zero at the liquid–vapor critical point (Sect. A.5.1), so we would expect that heat is transported very slowly around the critical point. In fact, as shown by experiments performed in space, this is not the case [190]. (Microgravity experiments are needed to investigate phenomena in critical fluids because the significant dilatability of these fluids results in large variations in density, which generates intense convective motions, hindering precise observations.) In 1990, three independent teams [26, 191, 298] provided the explanation for this phenomenon. Onuki showed that the classical heat transfer equation

$$\frac{\partial T}{\partial t} - \kappa\nabla^2 T = 0$$

is not relevant for a fluid maintained in a fixed volume. A change of entropy at any point in the fluid bulk causes a change of pressure, which generates an adiabatic change in temperature over the whole volume. This phenomenon had already been described in the literature, but as it is negligible for fluids with usual values of compressibility, it was unfortunately overlooked when studying fluids close to their critical points.

In an incompressible fluid, variations in internal energy are only due to temperature variations arising from the heat flow imposed on the fluid, whereas the adiabatic variations in temperature and density due to temporal variations in pressure must also be considered for a compressible fluid [26].

These adiabatic effects predominate in a fluid close to its critical point due to the divergence of its compressibility. A new mechanism for heat transfer within supercritical fluids was highlighted: the piston effect, which acts when a cell filled with critical fluid is subjected to local heating. The French team [298] primarily applied singular perturbation methods to the Navier–Stokes equations for a van der Waals fluid.

When a wall is heated, the P.E. mechanism proceeds as follows (see also Fig. A.4):

1. A thin layer of the fluid is heated by conduction from a heat source (Fig. A.4a)
2. The fluid subjected to this heating dilates strongly and compresses the remainder of the fluid in an isentropic way (Fig. A.4b)
3. The fluid then undergoes a homogeneous temperature increase (Fig. A.4c).

The dilated fluid in the boundary layer acts like a piston, since a system of compression waves is created that propagates throughout the whole fluid, and the first response of the system to a thermal disturbance in a thin layer of the fluid is mechanical. The passage of the continuously produced sound field throughout the fluid causes the fluid volume to slowly move, and it compresses the whole volume.

Thus, through a purely dynamic mechanism, a fluid that diffuses heat very poorly can nevertheless transport thermal energy at very high speeds. This

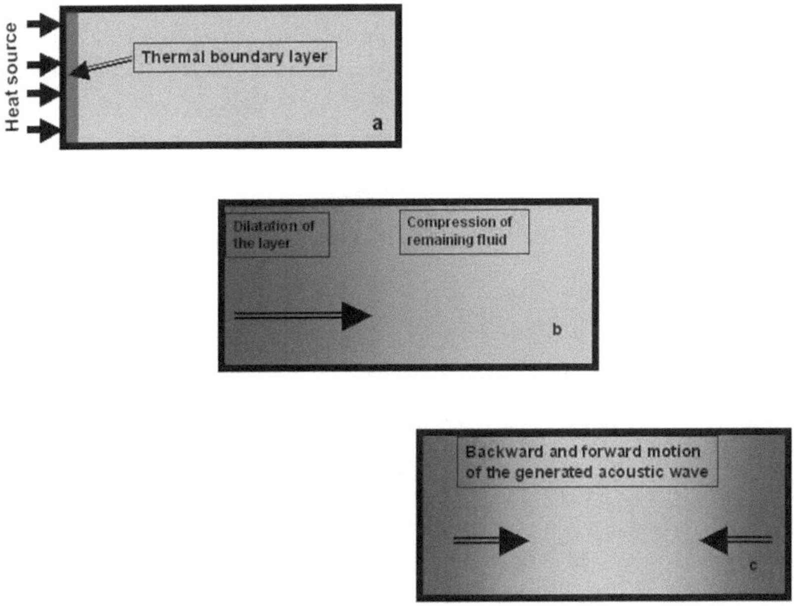

Fig. A.4. The piston effect is due to a three-step thermomechanical couple

means that temperature relaxation speeds up in a perfectly isobaric cell as the critical point is approached. This phenomenon, which contrasts with the deceleration of the diffusion process near the critical point, is named critical acceleration (or "critical speeding up").

When a temperature gradient is resorbed in a cell, density gradients are generated, and these are also resorbed, but over a longer timescale (on the order of the diffusion time). Heat transport thus takes place on P.E. timescale, but more time is required for the fluid to become homogeneous again and to reach equilibrium through diffusion.

The results of the numerical simulations presented in [299] make it possible to rule out the false assumption that the P.E. disappears in the presence of gravity. This study shows that the P.E. is not inhibited by natural convection; in contrast, it remains the fastest temperature relaxation mechanism—faster than natural convection.

Expansion of a Droplet of Critical Fluid

Distinct liquid and vapor phases do not exist in pure substances at supercriti cal pressures: the whole field is monophasic. The density relaxation of a dense cluster in a isobaric atmosphere at a higher temperature is nevertheless similar to the process of vaporizing a subcritical liquid droplet, provided that the thermal diffusivity of the liquid pocket is lower than that of the surrounding atmosphere.

A zone with high density gradients and diffusion times isolates the dense cluster from the surrounding atmosphere [212, 244]. The dense cluster is thus called a supercritical "droplet" and the zone with a high density gradient is like a thick "interface."[8] For a critical cluster (i.e., a cluster at critical density and critical temperature) introduced into a hotter atmosphere at the critical pressure, the previously mentioned diffusivity condition is valid for all atmospheric temperatures due to the critical divergence of the diffusivity in the critical pocket.

At pressures that are clearly supercritical, this state of diffusivity is more restrictive, and requires (assuming an ideal fluid) that $T_\infty \gg T_i$, where T_∞ is the temperature of the atmosphere and T_i is the initial temperature of the dense cluster.

Since the "interface" is thick, two limits must be selected to define the reference boundary of the supercritical "droplet." We first choose an isodensity

[8] At the critical point there is no surface tension. Thus, in principle, the "droplet" does not remain spherical. Indeed, we know that the cluster can maintain a certain coherence, just as a cold liquid drop does when present in the same liquid at a higher temperature [137]. This phenomenon can be explained by the existence of an effective surface tension resulting from a zone with a high temperature gradient. This was noted for a pure heated supercritical fluid in a numerical study of interfacial layer instability [300]. The same phenomenon can also be observed for miscible fluids [203].

ρ_m located at the end of the zone with a high density gradient in the fluid pocket. Note that this surface located at the end of the zone with a high density gradient is the sphere ($r = r_m$) where the unit mass flow rate is maximal. The second limit is the reference isochore ρ_s for the beginning of the zone with a strong density gradient at $r = r_s$.

Between these two limits (i.e., in the transition zone or interfacial layer), we consider the reduced radial coordinate $X = (r - r_s)/(r_m - r_s)$, with distances then being divided by the interfacial layer thickness. Quasi-steady evolution is then assumed for this new reduced coordinate. Outside this transition zone, the parameters are constant for $r < r_s$ (interior zone), and the evolution is quasi-steady in natural coordinates for $r > r_m$ (external zone).

These considerations permit an analytical solution. Without changing the notation, we now consider dimensionless time and length. The reference time is the lifetime of the drop, and the reference length is the initial droplet diameter (i.e., at time $t = 0$). If d_m is the diameter of the domain limited by the isodensity $r = r_m$ and d is the diameter of the domain limited by the isodensity ρ_s, it can be shown [211, 212] that the diameter d_m obeys a d^2 law:

$$d_m^2 = 1 - \tau.$$

The diameter d obeys the approximate relation

$$d^\beta - \beta d = (1 - \beta)\sqrt{1 - \tau}. \tag{A.76}$$

The parameter β is approximately 1 when the pressure is slightly sub- or supercritical: the d^2 law is not obeyed for such pressures, as shown by the (A.76). When the pressure definitely becomes subcritical, β becomes large and we again obtain the classical d^2 law.

Since dividing the diameter d by the initial diameter of the droplet gives a dimensionless variable, $d(\tau) \leq 1$ and, for high values of β, the preceding equation becomes $d^2 = 1 - \tau$ to the first order. Thus, the d^2 law is again observed at pressures that are clearly subcritical.

Numerical solution is based on a finite volume method for a geometry with spherical symmetry. We used the algorithm SIMPLER to solve the equations, which are filtered to eliminate acoustic waves. The analytical results are in very good agreement with the numerical ones.

A.6 More on Transfer Coefficient Determination

A.6.1 Collision Integrals and Cross-sections

Let us now consider two molecules j and k and their trajectories in the collision plane (Fig. A.5). The molecular velocities are $\mathbf{C}_j$, $\mathbf{C}_k$ before collision and $\mathbf{C'}_j$, $\mathbf{C'}_k$ after collision [10, 46, 118].

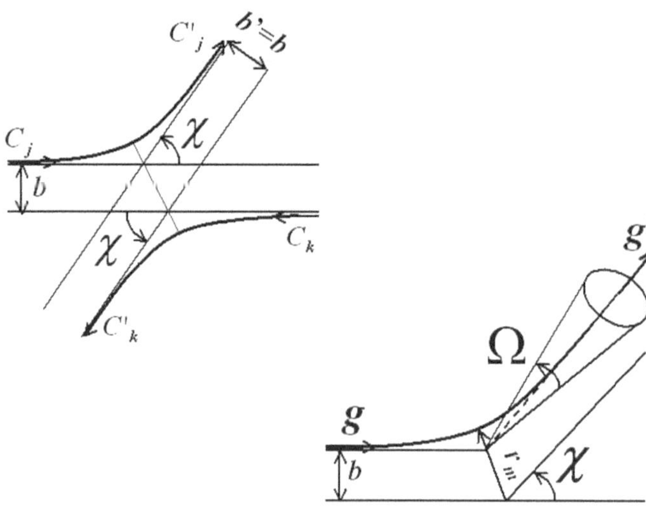

Fig. A.5. Collision between two molecules j and k: trajectories of the centers of the moving molecules

This two-body problem can be solved as a one-body problem with the reduced mass $M_{jk} = M_j M_k/(M_j + M_k)$ and the relative velocity $\mathbf{g} = \mathbf{C}_j - \mathbf{C}_k$ (where g is the modulus). Let us denote the distance between the two parallel trajectories as b, the deviation angle as χ, the intermolecular force as $F = -d\varphi(r)/dr$, the distance between the molecules as r, and the potential function as $\varphi(r)$. The deviation χ is

$$\chi = \pi - 2b \int_{r_m}^{\infty} \frac{1}{r^2}(1 - \frac{\varphi(r)}{M_{jk}g^2/2} - \frac{b^2}{r^2})^{-1/2} dr. \tag{A.77}$$

Distance r_m, which corresponds to the minimum distance between the two centers of the molecules, is the solution to the following equation:[9]

$$\varphi(r_m) = M_{jk} g^2 (1 - b^2/r_m^2)/2. \tag{A.78}$$

The post-collisional relative velocity $\mathbf{g'}$ occurs in the solid angle $d\Omega$ (Fig. A.5), and the collisional cross-sections $\sigma(g, \chi)$ are defined such that the number of collisions per unit volume and unit time is $N_{C_j} N_{C_k} g \sigma(g, \chi) d\Omega$, where

[9] Example of potential: for a potential $\varphi = ar^{-\alpha}$, the intermolecular force is $F = \alpha a r^{-(\alpha+1)}$, so $\chi = \pi - 2\int_0^{b/r_m} [1 - (b/r)^2 - (2/\alpha)(b/\beta r)^\alpha]^{-1/2} d(b/r)$, where $\beta = b(M_{jk}g^2/\alpha a)^{1/\alpha}$ and r_m is the solution to the equation $1 - (b/r_m)^2 - (2/\alpha)(b/\beta r_m)^\alpha = 0$.

$N_{C_j} = Nf(C_j)d\mathcal{V}_{C_j}$, $N_{C_k} = Nf(C_k)d\mathcal{V}_{C_k}$. The function $f(C)$ is the Boltzmann function, and $d\mathcal{V}_C = dC_1\, dC_2\, dC_3$ is the volume element of the velocity space.

In other words, if we consider an initial flux of molecules Φ_j, the number of molecules with deviations that fall within the solid angle $d\Omega$ per unit time is proportional to Φ_j and $d\Omega$:

$$\frac{d\mathcal{N}_j}{dt} = \Phi_j \sigma(g, \chi) d\Omega.$$

The collision integral for the Boltzmann equation is thus

$$\left[\frac{\partial(Nf)}{\partial t}\right]_{coll.} = \int_{-\infty}^{+\infty} \int_0^{4\pi} N^2[f(C_j')f(C_k') - f(C_j)f(C_k)] g\sigma d\Omega d\mathcal{V}_{C_k}. \quad \text{(A.79)}$$

The total collisional cross-section is defined by[10]

$$\sigma_T = \int_0^{4\pi} \sigma d\Omega = 2\pi \int_0^{\pi} \sigma \sin\chi d\chi.$$

The momentum jump during a collision in the direction of relative motion is proportional to $1 - \cos\chi = 2\sin^2(\chi/2)$, and the corresponding collisional cross-section σ_M is

$$\sigma_M = \int_0^{4\pi} (1 - \cos\chi)\sigma d\Omega = 2\pi \int_0^{4\pi} \sigma(1 - \cos\chi)\sin\chi d\chi.$$

The cross-section σ_M is important for diffusion phenomena.[11] The definition of viscosity uses the cross-section σ_μ, defined by:

$$\sigma_\mu = \int_0^{4\pi} \sin^2\chi \sigma d\Omega = 2\pi \int_0^{4\pi} \sigma \sin^3\chi d\chi.$$

If we introduce the cross-sections

$$\sigma_{j,\,k}^{(l)}(g) = 2\pi \int_0^{\infty} (1 - \cos^l\chi) b\, db, \quad \text{(A.80)}$$

the collision integrals can be written in the form

$$\Omega_{j,k}^{(l,s)} = \sqrt{kT/2\pi M_{jk}} \int_0^{\infty} e^{-\gamma^2} \gamma^{2s+3} \sigma_{j,\,k}^{(l)}(g)\, d\gamma, \quad \text{(A.81)}$$

where $\gamma^2 = 1/2\, M_{jk}\, g^2/kT$. $s = l = 1$ for diffusion, and $s = l = 2$ for viscosity. The collision integrals depend on the $\varphi(r)$ potential chosen; if the Lennard–Jones potential $\varphi(r) = 4\epsilon[(d/r)^{12} - (d/r)^6]$ is used, the collision integral is

[10]For elastic spheres, $\sigma_{el.sph.} = d^2/4$, $\sigma_{T_{el.sph.}} = \pi d^2$.

[11]In the case of elastic spheres, $\sigma_{M_{el.sph.}} = \pi d^2$ and $\sigma_{\mu_{el.sph.}} = 2/3\, \pi d^2$.

a function of the reduced temperature $T^* = kT/\epsilon$. These collision integrals have been tabulated as functions of T^* [118].

These collision cross-sections and integrals are often compared to those obtained with the elastic sphere assumption, and the ratios obtained have also been tabulated [118]. Let us set $G^{(l)} = 1 - (1/2)\,(1 + (-1)^l)/(1 + l)$. The cross-section ratio is then

$$\sigma^{(l)*} = \frac{\sigma^{(l)}}{\sigma^{(l)}_{el.sph.}} = \frac{\sigma^{(l)}}{\pi\, d^2\, G^{(l)}}, \tag{A.82}$$

and the collision integral ratio is

$$\Omega^{(l,s)*} = \frac{\Omega^{(l,s)}}{\Omega^{(l,s)}_{el.sph.}} = \frac{\Omega^{(l,s)}\,\sqrt{2\,\pi\, M_{jk}/k\,T}}{\pi\, d^2\, G(l)(s+1)!/2}. \tag{A.83}$$

A.6.2 Transfer Coefficients for Pure Gases

Diffusion

The diffusion of chemical species intervenes in system (3.46) and the equations (3.47), which correspond to the simplified theory of gases. The diffusion coefficient introduced is more a self-diffusion coefficient than a multicomponent diffusion coefficient (see Sect. A.6.3). The corresponding value obtained in a more precise estimation is

$$D = \frac{3}{16}\,\frac{k\,T}{N\,M\,\Omega^{1,1}}. \tag{A.84}$$

Viscosity

An approximate expression for the shear viscosity coefficient was given in system (3.31): $\mu = M\,\bar{C}/2\,\sqrt{2}\,\pi\, d^2$ (assuming that the radius of the sphere of influence of a molecule is equal to the diameter d of the molecule). The diameter of the sphere of influence d_{eff} is given by the relation

$$d^2_{eff} = d^2\,[1 - \frac{4}{m}\overline{(\frac{1}{g^2})} \int_d^\infty F(r)\,dr],$$

where g is the relative velocity of the molecules, $\overline{(1/g^2)}$ is the mean value of $1/g^2$, and $F(r)$ is the intermolecular force at r. As $\overline{(1/g^2)}$ is proportional to $1/T$, we have

$$d^2_{eff} = d^2\,(1 + S/T)$$

such that the relation that gives the viscosity coefficient divided by a reference value at a reference temperature is

$$\mu/\mu_{ref} = \sqrt{T/T_{ref}}\,(1 + S/T_{ref})/(1 + S/T).$$

This is known as the Sutherland relation [270].

If we write $C_1 = \mu_{ref}(T_{ref}+S)/(T_{ref})^{3/2}$, the following values are obtained for air: $\mu_{ref} = 1.716 \times 10^{-5}$ kg m^{-1} s^{-1}, $T_{ref} = 273.15$ K, $S = 110.4$ K, $C_1 = 1.458 \times 10^{-6}$ kg m^{-1} s^{-1} K$^{-1/2}$.

A more precise kinetic theory using collision integrals [118, 290] leads to

$$\mu = \frac{5}{8}\frac{k\,T}{\Omega^{(2,2)}} = \frac{5}{16}\frac{\sqrt{\pi\,M\,k\,T}}{\pi\,d^2\,\Omega^{(2,2)*}}, \tag{A.85}$$

or, with μ in g cm^{-1} s^{-1}, $\mathcal{M}$ in g/mol, d in Angströms, and $T^* = kT/\epsilon$,

$$\mu = 2.669 \times 10^{-5}\frac{\sqrt{\mathcal{M}\,T}}{d^2\,\Omega^{(2,2)*}\,(T^*)}.$$

Bulk Viscosity

For perfect monatomic gases, the bulk viscosity coefficient η is zero. It is nonzero for polyatomic gases in the case of relaxing flows. Assuming that there are i modes of internal motion that lead to relaxation, η is defined by the relationship

$$\eta = \sum_i \eta_i = (N\,\mathcal{N}_0\,k^2 T/C_v^2)\sum_i C_{vi}\,\tau_i,$$

where C_v is the specific heat at constant volume, and η_i, C_{vi}, τ_i are the volume viscosity coefficient, specific heat and relaxation time corresponding to mode i, respectively. The thermal energy of rotation per unit volume, defined by $\rho\,e\,C_{v\,rot}/C_v$, is an important influence on η; Lighthill has shown that $\eta = (\gamma - 1)^2\rho\,e\,(C_{v\,rot}/C_v)\tau_{rot}$. The bulk viscosity coefficient η intervenes, for example, in studies of sound propagation.

Thermal Conduction

The heat conduction coefficient is

$$\lambda = \frac{25}{32}c_v\frac{\sqrt{\pi\,M\,k\,T}}{\pi\,d^2\,\Omega^{(2,2)*}}, \tag{A.86}$$

or, with λ in cal cm^{-1} s^{-1} K^{-1}, d in Angströms, and $T^* = kT/\epsilon$,

$$\lambda = 1.989 \times 10^{-4}\frac{\sqrt{T/\mathcal{M}}}{d^2\,\Omega^{(2,2)*}\,(T^*)} = \frac{15}{4}\frac{R}{\mathcal{M}}\mu.$$

A.6.3 Transfer Coefficients for Gas Mixtures

Mass Diffusion in a Mixture

We define the reduced mass of the interacting species j, k as

$$M_{jk} = M_j\, M_k/(M_j + M_k).$$

By analogy with the definition of the coefficient of binary diffusion $D_{12} = X_1\, X_2\, p/M_{12}\, \nu_{12}$, a coefficient of species diffusion in a mixture of more than two species can be defined as $D_{jk} = X_j\, X_k\, p/M_{jk}\, \nu_{jk}$, where ν_{jk} is the total number of collisions per unit volume and time for molecules i and j. We then have $\nu_{12} = N_1\, N_2\, \sigma_{12}\, C_{12} = X_1 X_2 N^2\, \sigma_{12}\, C_{12}$ if there are two species (binary diffusion). Assuming the Maxwell distribution $C_{12} = \sqrt{8kT/\pi M_{12}}$, the diffusion coefficient can be written

$$D_{12} = X_1\, X_2\, p/M_{12}\, \nu_{12} = \sqrt{2\pi k^3 T^3/M_{12}}/4p\, \sigma_{12},$$

and with a more accurate theory we find that

$$D_{12} = \frac{3}{16}\, \frac{k\, T}{N\, M_{12}\, \Omega_{12}^{1,1}} = \frac{3}{16}\, \frac{\sqrt{2\pi k^3 T^3/M_{12}}}{p\, \pi d_{12}^2 \Omega_{12}^{(11)*}}, \tag{A.87}$$

or, if the molar masses $\mathcal{M}_j$ are in g/mol, p is in atm, T is in K, and d_{ij} is in Angströms, we obtain

$$D_{12} = 2.628 \times 10^{-3} \frac{\sqrt{T^3(\mathcal{M}_1 + \mathcal{M}_2)/2\mathcal{M}_1\, \mathcal{M}_2}}{p\, d_{12}^2\, \Omega_{12}^{(11)*}\, T_{12^*}}$$

(in $\text{cm}^2\ \text{s}^{-1}$), where $d_{12} = (d_1 + d_2)/2$ and $T_{12^*} = kT/\sqrt{\epsilon_1 \epsilon_2}$. Some ϵ and d values are given in Table A.2.

Species	d **(Angströms)**	ϵ/k **(K)**
N_2	3.681	91.5
O_2	3.433	113.0
H_2	2.968	33.3
CO_2	3.996	190.0
CO	3.590	110.0
NO	3.470	119.0
OH	3.110	93.8
O	3.068	102.2
H	2.497	99.8
Ar	3.418	124.0

Table A.2. Some ϵ/k and d values

First approximations of d and ϵ can be obtained from the critical data

$$\epsilon/k \cong 0.77\,T_{cr}, \;\; d \cong 0.833\,\mathcal{V}_{cr}^{1/3},$$

where T_{cr} and $\mathcal{V}_{cr}$ are the critical temperature and molar volume, respectively. The self-diffusion coefficient is deduced by making $i = j$, and (A.84) can also be written as [120]

$$D_j = 2.628 \times 10^{-3} \frac{\sqrt{T^3/\mathcal{M}_j}}{p\,d_j^2\,\Omega^{(11)*}\,T_j^*}.$$

Viscosity of a Mixture

The coefficient of viscosity of a mixture in an environment with many species can be deduced from the following relation:

$$\mu = \sum_{j-1}^{N} \frac{X_j^2}{X_j^2/\mu_j + 1.385 \sum_{k=1,\,k\neq j}^{N} X_j X_k kT/pM_j D_{jk}}.$$

For a binary mixture, the following relation was proposed:

$$1/\sqrt{\mu} = X_1/\sqrt{\mu_1} + X_2/\sqrt{\mu_2}.$$

However, for most applications that do not require extreme precision, we can use the relation

$$1/\mu = \sum_{j=1}^{N} Y_j/\mu_j.$$

Thermal Flux

Based on the phenomenological relations of Onsager (3.30), the calculation for $\mathbf{q}$ contains a sum of three terms:

1. Heat conduction in the gas, which is given by flux $-\lambda\boldsymbol{\nabla}T$, where λ is the thermal conductivity. The thermal conductivity can be readily expressed in terms of viscosity coefficients. This is one reason why it is computationally efficient to start by evaluating the latter.
2. The energy supplied by the molecular diffusion of species j, which is equal to the product of the diffusion flux $\mathbf{J}_{Dj} = \rho_j\,\mathbf{V}_j$ and the enthalpy per unit mass of species j. This term for all species becomes $\sum_{j=1}^{N} \rho_j\,\mathbf{V}_j\,h_j$.
3. A term corresponding to the Dufour effect, which is a coupling term resulting from concentration, pressure and temperature gradients. This term is $RT\sum_{j=1}^{N}\sum_{k=1}^{N}(X_k D_{T,j}/M_j D_{jk})(\mathbf{V}_j - \mathbf{V}_k)$.

Thermal Conduction of a Mixture

For each species j we have the relationship $\lambda_j = (15\,R/4\mathcal{M}_j)\,\mu_j$.

The determination of the coefficient of thermal conduction for a mixture is extremely complex, and different relations have been proposed. The simplest calculation that is an adequate approximation is based on the extension of Eucken's relation to mixtures. A preliminary calculation that determines the viscosity of the mixture μ, the specific heat at constant pressure $c_p = (\partial h/\partial T)_p$, and the molecular weight $\mathcal{M} = \sum_{j=1}^{N} Y_j \mathcal{M}_j$ leads to the relationship

$$\lambda = \mu(c_p + \frac{5}{4}\frac{R}{\mathcal{M}}).$$

Thermal Diffusion

The coefficients of thermal diffusion D_{Tj} can be obtained via a complex calculation from the exact kinetic theory [118], and they depend on the pressure, temperature and concentrations. They can be written in the nondimensional form of a thermal diffusion ratio $k_{Tj} = D_{Tj}/\rho D_{jk}$. The first approximation for the thermal diffusion ratio of a binary gas mixture is given by the following expression:

$$k_{T1} = (6C_{12}^* - 5)(X_1 X_2/6[\lambda_{12}]_1)(X_1 S^{(1)} - X_2 S^{(2)})(X_\lambda + Y_\lambda), \qquad \text{(A.88)}$$

where

$$S^{(1)} = (\mathcal{M}_1 + \mathcal{M}_2)[\lambda_{12}]_1/2\mathcal{M}_2[\lambda_1]_1 - (15/4A_{12}^*)(\mathcal{M}_2 - \mathcal{M}_1)/2\mathcal{M}_1 - 1,$$

$$S^{(2)} = (\mathcal{M}_1 + \mathcal{M}_2)[\lambda_{12}]_1/2\mathcal{M}_1[\lambda_1]_2 - (15/4A_{12}^*)(\mathcal{M}_1 - \mathcal{M}_2)/2\mathcal{M}_2 - 1.$$

Here, X_1 and X_2 are the mole fractions of species 1 and 2; $\mathcal{M}_1$ and $\mathcal{M}_2$ are the molar weights of species 1 and 2; $[\lambda_1]_1$ and $[\lambda_1]_2$ are their thermal conductivities; $[\lambda_{12}]_1 = (25/32)(3k/2)\sqrt{\pi kT/2M_{12}}/\pi d_{12}\Omega_{12}^{(22)*}T_{12}^*$; $A_{12}^* = \Omega_{12}^{22*}/\Omega_{12}^{11*}$; $C_{12}^* = \Omega_{12}^{12*}/\Omega_{12}^{11*}$; and X_λ and Y_λ are defined in [118] (p. 535).

k_{T1} is generally on the order of 10^{-1}, so thermal diffusion is ignored in many applications involving mixtures of gases.

References

1. Abramzon, B., Sirignano, W.A.: Droplet vaporization model for spray combustion calculations, Int. J. Heat Mass Transfer, **32**, 9, 1605–1618 (1989)
2. André, J.-C., Barrère, M.: *Turbulence fluide* (cours d'option), Ecole Polytechnique, Paris, 1980
3. Antkowiak, A., Bremond, N., Le Dizès, S., Villermaux, E.: Short-term dynamics of a density interface following an impact, J. Fluid Mech., **577**, 241–250 (2007)
4. Arias-Zugasti, M., García-Ybarra, P., Castillo, J.L.: Droplet vaporization at critical conditions: long-time convective-diffusive profiles along the critical isobar, Phys. Rev. E, **60**, 3, 2930–2941 (1999)
5. Aris, R.: *Introduction to the analysis of chemical reactors*, Prentice-Hall, Englewood Cliffs, 1965
6. Aris, R., Amundson, N.R.: An analysis of chemical reactor stability and control, Parts I–III, Chem. Eng. Sci. **7**, 121–155 (1958)
7. Barrère, M.: *Rocket propulsion*, Van Nostrand, New York, 1960
8. Barrère, M.: Modèles de combustion turbulente, Rev. Gén. Therm., **148**, 295–308 (1974)
9. Barrère, M., Prud'homme, R.: *Aérothermochimie des écoulements homogènes*, Gauthier-Villars, Paris, 1970
10. Barrère, M., Prud'homme, R.: *Equations fondamentales de l'aérothermochimie*, Masson, Paris, 1973
11. Bechtold, J.K., Matalon, M.: Hydrodynamic and diffusion effects on the stability of spherically expanding flames, Combust. Flame, **67**, 77–90 (1987)
12. Bénard, H.: Les tourbillons cellulaires dans une nappe liquide, Rev. Gén. Sci. Pures Appl. **11**, 1261–1271, 1309–1328 (1900)
13. Benton, E.R.: On the flow due to a rotating disk, J. Fluid Mech., **24**, 4, 781–800 (1966)
14. Berthelot, J.-M.: *Mécanique des solides rigides*, 2ème édn., Lavoisier, Paris, 2006
15. Berthelot, J.-M.: *Mechanics of rigid bodies*, CompoMechAsia website, www.compomechasia.com (2008)
16. Bertrand, G., Prud'homme, R.: Possibilities of surface reaction coupling with transport phenomena, Int. J. Quant. Chem., **12**, Suppl. 2, 159–168 (1977)
17. Beysens, D.: Near-critical fluids under micro gravity: highlights and perspectives for Europe, J. Phys. IV France, **11**, Pr6-7-22 (2001)
18. Bilous, O., Amundson, N.R.: Chemical reactor stability and sensitivity, AIChE J., **1**, 513 (1955)

19. Bockhorn, H. (ed): A short introduction. In: *Soot formation in combustion—mechanisms and models*. Springer, Berlin, 1994, pp 3–8
20. Bond, M., Struchtrup, H.: Mean evaporation and condensation coefficients based on energy dependent condensation probability, Phys. Rev. E **70**, 061605 (2004)
21. Borghi, R.: Mise au point sur la structure des flammes turbulentes, J. Chim. Phys., **81**, 6, 361–370 (1984)
22. Borghi, R., Champion, M.: *Modélisation et théorie des flammes*, Ed. Technip, Paris, 2000
23. Borghi, R., Destriau, M.: *La combustion et les flammes*, Ed. Technip, Paris, 1995
24. Borghi, R., Lacas, F.: Modeling of liquid-propellant spray combustion in rocket engine combustor. In: Proc. 2nd Int. Symp. on Liquid Rocket Propulsion, ONERA-Chatillon, France, 19–21 June 1995
25. Bose, A., Palmer, H.J.: Interfacial stability of binary mixtures evaporating at reduced pressure, J. Fluid Mech., **126**, 491–506 (1983)
26. Boukari, H., Shaumeyer, J.N, Briggs, M.E., Gammon, R.W.: Critical speeding up in pure fluids. Phys. Rev. A, **41**, 2260 (1990)
27. Boussinesq, J.: *Théorie analytique de la chaleur*, vol. 2, Gauthier-Villars, Paris, 1903, p. 172
28. Boutrouft, K.: *Instabilités thermoconvectives de type Rayleigh Taylor dans les fluides supercritiques* (doctorat). ENSAM, Angers, 2006
29. Bouttes, J.: *Mécanique des fluides*, Ellipses, Paris, 1998
30. Bozier, O., Veyssière, B.: Influence of suspension generation on dust explosion parameters, Comb. Sci. Technol., **178**, 1927–1955 (2006)
31. Bray, K.N.C.: Atomic recombination in a hypersonic wind-tunnel nozzle, J. Fluid Mech., **6**, 1, July (1959)
32. Bright, F.V., McNally, M.E. (eds): *Supercritical fluid technology*, American Chemical Society, Washington, 1992
33. Bruhat, G.: *Thermodynamique*, Masson, Paris, 1962
34. Brun, E.A., Martinot-Lagarde, A., Mathieu, J.: *Mécanique des fluides*, Tome III, Dunod, Paris, 1970
35. Bruno, T.J., Ely, J.F. (ed): *Supercritical fluid technology: reviews in modern theory and applications*, CRC, Boca Raton (1991)
36. Burgers, J.M.: A mathematical model illustrating the theory of turbulence, Adv. Appl. Mech., **1**, 171–199 (1948)
37. Burke, S.P., Schumann, T.E.W.: Diffusion flames. In: First Symposium on Combustion, Swampscott, MA, 10–14 Sept. 1928
38. Callen, H.B.: *Thermodynamics*, Wiley, New York, 1966
39. Candel, S., Poinsot, Th.: Flame stretch and the balance equation for the flame area, Combust. Sci. Tech., **70**, 1–15 (1990)
40. Candel, S., Veynante, D., Lacas, F., Maistret, E., Darabiha, N.: Coherent flame model: applications and recent extensions. In: Larrouturou, B. (ed): *Advances in combustion modeling*, World Scientific, Singapore, 1990
41. Carlson, D.J., Hoglund, R.F.: Particle drag and heat transfer in rocket nozzles, AIAA J. **2**, 11, 1980–1984 (1964)
42. Carrière, P.: *Thermodynamique et dynamique des gaz*, Cours ENSTA (1979)
43. Chandrasekhar, S.: *Hydrodynamic and hydromagnetic stability*, Clarendon, Oxford, 1961

44. Chao, K.C., Robinson, R.L. (eds): *Equations of state in engineering and research*, American Chemical Society, Washington, 1979
45. Chapman, D.L.: On the rate of explosion in gases, Phil. Mag., **47**, 90–104 (1899)
46. Chapman, S., Cowling, T.W.: *The mathematical theory of nonuniform gases*, Cambridge University Press, Cambridge, 1959
47. Chauveau, C., Gökalp, I., Morin, C.: Vaporisation de gouttes en conditions supercritiques: analyse préliminaire des résultats obtenus lors des expériences en fusées sondes Texus 38. In: Prud'homme, R., Langevin, D., Faivre, G. (eds): *Sciences de la matière et microgravité*, J. Phys. IV France, **11**, Pr6-315-324 (2001)
48. Chiang, C.H., Raju, M.S., Sirignano, W.A.: Numerical analysis of convecting, vaporizing fuel droplet with variable properties, Int. J. Heat Mass Transfer, **35**, 5, 1307–1324 (1992)
49. Chin, J.S., Lefebvre A.H.: The role of the heat-up period in fuel drop evaporation, Int. J. Turbo Jet Engines, **2**, 315–325 (1985)
50. Chin, L.P., Switzer, G., Tankin, R.S., Jackson, T., Stutrud, J.: Bi-modal size distribution predicted by maximum entropy are compared with experiments in sprays. Comb. Sc. and Tech. 109, 35-52 (1995).
51. Clavin, P.: Dynamics behavior of premixed flame fronts in laminar and turbulent flows, Prog. Energy Combust. Sci., **11**, 1–59 (1985)
52. Clavin, P., Guyon, E.: La flamme, La Recherche, 94, Nov. (1978)
53. Clavin, P., Joulin, G.: Premixed flames in large and high intensity turbulent flow, J. Phys. Lett., **44**, 1 (1983)
54. Clavin, P., Williams, F.A.: Effects of molecular diffusion and of thermal expansion on the structure and dynamics of premixed flames in turbulent flows of large scale at low density, J. Fluid Mech., **116**, 251–282 (1982)
55. Clift, R., Grace, J.R., Weber, M.E.: *Bubbles, drops and particles*, Academic, New York, 1978, pp. 306–314
56. Cousin, J.: *Prédiction des distributions granulométriques par le formalisme d'entropie maximum. Application à plusieurs injecteurs mécaniques* (thèse), Université de Rouen, Mont-Saint-Aignan, 1996
57. Corsin, S., Lumley, J.: On the equation of motion for a particle in a turbulent fluid, Appl. Sci. Res. A, **6**, 114–116 (1956)
58. Cousin, J., Dumouchel, C.: Effect of viscosity on the linear instability of a flat liquid sheet, Atomiz. Sprays, **6**, 563–576 (1996)
59. Cousin, J., Yoon, S.J., Dumouchel, C.: Coupling of classical linear theory and maximum entropy formalism for prediction of drop size distribution in sprays: application to pressure-swirl atomizers. Atomiz. Sprays, **6**, 601–622 (1996)
60. Crespo, A., Liñan, A.: Unsteady effects in droplet evaporation and combustion, Combust. Sci. Technol., **11**, 9–18 (1975)
61. Damköhler, G.: Der Eifluss der Turbulenz auf die Flammengeschwindigkeit in Gasgemischen, Z. Elektrochem. **46**, 601–652 (1940)
62. Darrozès, J.S., François, C.: *Mécanique des fluides incompressibles*, Springer, Berlin, 1982
63. Darrozès, J.S., Monavon, A.: *Analyse phénoménologique des écoulements* (cours), Université Pierre et Marie Curie, Paris, 1996
64. Deardorff, J.W.: A numerical study of three-dimensional turbulent channel flow at large Reynolds number, J. Fluid Mech., **41**, 453–480 (1970)

65. Defay, R., Prigogine, I.: *Tension superficielle et adsorption*, Ed. Desoer, Liège, 1966
66. Delplanque, J.-P., Sirignano, W.A.: Boundary-layer stripping effects on droplet transcritical convective vaporization, Atomiz. Sprays, **4**, 325–349 (1994)
67. Delplanque, J.-P., Sirignano, W.A.: Transcritical liquid oxygen droplet vaporization: effect on rocket combustion instability, J. Propul. Power, **12**, 2, 349–357 (1996)
68. Dietrich, D.L., Struk, P.M., Kitano, K., Ikegami, M.: Combustion of interacting droplet arrays in a micro-gravity environment. In: Proc. 5th Int. Micro-gravity Combustion Workshop, Cleveland, OH, 18–20 May 1999, pp. 281–284
69. Dodemand, E., Prud'homme, R., Kuentzmann, P.: Influence of unsteady forces acting on a particle in a suspension. Application to the sound propagation. Int. J. Multiphase Flow, **21**, 1, 27–51 (1995)
70. Dombrowski, N., Hooper, P.C.: The effect of ambient density on drop formation in sprays, Chem. Eng. Sci. **17**, 291–305 (1962)
71. de Donder, Th.: *Leçons de thermodynamique et de chimie physique*, Gauthier-Villars, Paris, 1920
72. Dubois, I., Habiballah, M., Lecourt, R.: Numerical analysis of liquid rocket engine combustion and stability (Paper AIAA-95-0607). In: 33rd AIAA Aerospace Sciences Meeting and Exhibit, Reno, NV, 9–12 Jan. 1995
73. Ducros, F., Comte, P., Lesieur, M.: Large-eddy simulation of transition to turbulence in a boundary layer developing spatially over a flat plate. J. Fluid Mech., **326**, 1–36 (1996)
74. Dumouchel, C.: Experimental analysis of a liquid atomization. Process at low Weber number. In: Proc. Spray 05—Heat and Mass Transfer in Spray Systems, Antalya, Turkey, 5–10 June 2005
75. Dupoirieux, F.: Calcul numérique d'écoulements turbulents réactifs et comparaison avec des résultats expérimentaux, Rech. Aérospatiale, **6**, 443–453 (1986)
76. Durox, D., Prud'homme, R.: Polyhedral flames at low pressure, Combust. Flame, **70**, 243–249 (1987)
77. Durox, D., Baillot, F., Scouflaire, P., Prud'homme, R.: Some effects of gravity on the behaviour of premixed flames, Combust. Flame, **82**, 66–74 (1990)
78. Einstein, A.: *Schallgeschwindigkeit in tellweise dissoziertengasen*, Sitzungsbericht Preussische Akademie der Wissenschaft, Berlin, 1920
79. El Ganaoui, M.: *Modélisation numérique de la convection thermique instationnaire en présence d'un front de solidification déformable* (thèse), Université d'Aix-Marseille, Marseille, 1997
80. Emmons, H.W.: The film combustion of liquid fuel, Z. Angew. Math. Mech., **36**, 60 (1956)
81. Faeth, G.M.: Evaporation and combustion of sprays, Prog. Energy Combust. Sci., **9**, 1–76 (1983)
82. Fauve, S., Libchaber, A.: Rayleigh–Bénard experiments. In: Bruter, C.P. (ed): *Bifurcation theory, mechanics and physics*, Kluwer, Dordrecht, 2001
83. Favre, A.J.: Equations statistiques des gaz turbulents, C.R. Ac. Sci. **246**, 2576, 2723, 2839, 3216 (1958)
84. Favre, A.J.: Statistical equations for fluctuations of temperature, entropy, concentration and vorticity in the compressible turbulent flow of a gas (Instituto Naz. Alta Matematica, Symposia Matematica **9**), Academic, New York, pp. 371–390; and C.R. Ac. Sci. Sér. A, **273**, 1087, 1289, 1971 (1971)

85. Fer, F.: *Thermodynamique macroscopique*, Gordon and Breach, Paris, 1970
86. Fernandez-Pello, A.C.: Flame spread modeling, Combust. Sci. Tech., **39**, 119–134 (1984)
87. Ferrell, R.A.: Decoupled-mode dynamical scaling theory of the binary-liquid phase transition, Phys. Rev. Lett., **24**, 1169–1172 (1970)
88. Fisher, M.E.: The renormalization group in the theory of critical behaviour, Rev. Mod. Phys., **46**, 597-616 (1974)
89. Fixman, J.: Viscosity of critical mixtures, J. Chem. Phys., **36**, 2, 310 (1962)
90. Fortier, A.: *Mécanique des suspensions*, Masson, Paris, 1967
91. Fortier, A.: *Mécanique des fluides et transferts de chaleur et de masse par convection*, Masson, Paris, 1975
92. Gañán-Calvo, A.M., Lasheras, J.C.: The dynamics and mixing of small spherical particles in a plane, free shear layer. Phys. Fluids A **3**, 1207–1217 (1991)
93. Garrabos, Y., Leneindre, B., Subra, P., Cansel, F., Pommier, C.: Fluides critiques et gravité, fluides supercritiques et matériaux. In: Fluides, Matériaux et Microgravité, Ann. Chim. Fr., **17**, 55–90 (1992)
94. Gatignol, R., Prud'homme, R.: *Mechanical and thermodynamical modeling of fluid interfaces* (Series on Advances in Mathematics for Applied Sciences, vol. 58), World Scientific, Singapore, 2001
95. Gatignol, R., Seppecher, P.: Modelisation of fluid-fluid interfaces with material properties, J. Méc. Théor. Appl., Num. Spéc., 225–247 (1986)
96. Germain, P.: *Mécanique des milieux continus*, Masson, Paris, 1962
97. Germain, P.: *Mécanique des milieux continus*, Tome 1, Masson, Paris, 1973
98. Germain, P.: La méthode des puissances virtuelles en mécanique des milieux continus, J. Mécanique, **12**, 235–274 (1973)
99. Germain, P.: *Mécanique*, Ellipses, Paris, 1987
100. Germano, M., Piomelli, U., Moin, P., Cabot, W.H.: A dynamic subgrid-scale eddy viscosity model. Phys. Fluids A **3**, 7, 1760–1765 (1991)
101. Germano, M., Maffio, M., Sello, S., Mariotti, G.: On the extension of the dynamic modelling procedure to turbulent reacting flows. In: *Direct and large eddy simulation II*, Kluwer, Dordrecht, 1997, pp. 291–300
102. Ghez, R.: A generalized Gibbsian surface, Surf. Sci., **4**, 125–140 (1960)
103. Ghosal, S., Moin, P.: The basic equations for the large-eddy simulation of turbulent flows in complex geometry. J. Comput. Phys., **118**, 1, 24–37 (1995)
104. Ghosh Moulic, S., Yao, L.S.: Taylor–Couette instability of travelling waves with a continuous spectrum, J. Fluid Mech., **324**, 181–198 (1996)
105. Glasstone, S., Laidler, K.J., Eyring, H.: *The theory of rate processes*, McGraw-Hill, New York, 1941
106. Godsave, G.A.E.: Studies of the combustion of drops in a fuel spray. The burning of single drops of fuel. In: Combustion Institute (ed): *Fourth Symposium on Combustion*, Williams and Wilkins Co., Baltimore, MD (1953), pp. 818–830
107. Goldstein, S.: The steady flow of viscous fluid past a fixed spherical obstacle at small Reynolds number, Proc. Roy. Soc. A, **123**, 225–235 (1929)
108. de Groot, S.R., Mazur, P.: *Nonequilibrium thermodynamics*, North Holland, Amsterdam, 1969
109. Gugenheim, G.: *Thermodynamique*, Dunod, Paris, 1965
110. Guyon, E., Hulin, J.-P., Petit, L.: *Hydrodynamique Physique* (Savoirs Actuels), InterEditions/Editions du CNRS, Paris, 1991
111. Haase, R.: *Thermodynamics of irreversible processes*, Addison-Wesley, Reading, 1963

112. Haldenwang, P., Nicoli, C., Daou, J.: High pressure vaporization of LOX droplet crossing the critical conditions, Int. J. Heat Mass Transfer **39**, 3453–3464 (1996)
113. Halpering, B.I., Hohenberg, P.C., Ma, S.: Calculation of dynamic critical properties. Using Wilson's expansion methods, Phys. Rev. Lett., **29**, 1548–1551 (1972)
114. Hauke, G., Dopazo, C., Lozano, A., Barreras, F, Hernández, A.H.: Linear stability analysis of a viscous liquid sheet in a high-speed viscous gas. Flow Turbul. Combust. **67**, 235-265 (2001)
115. Heidmann, M.F.: Frequency response of a vaporization process to distorded acoustic disturbances, NASA Tech. Note D-6806 (1972)
116. Heidmann, M.F., Wieber, P.R.: Analysis of frequency response characteristics of propellant vaporisation, NASA Tech. Note D-3749 (1966)
117. Hinze, J.O.: *Turbulence*, 2nd edn., McGraw-Hill, New York, 1975, pp. 460–471
118. Hirschfelder, J.O., Curtiss, Ch.F., Bird, R.B.: *Molecular theory of gases and liquids*, Wiley, New York, 1954
119. Hohenberg, P.C., Halpering, B.I.: Theory of dynamic critical phenomena, Rev. Mod. Phys., **49**, 3, 435–479 (1977)
120. Holsen, N.J., Strunck, M.R.: Binary diffusion coefficients in non polar gases, IEC Fundam., **3**, 2, 143–147 (1964)
121. Jackson, D., Launder, B.: Osborne Reynolds and the publication of his papers on turbulent flow, Annu. Rev. Fluid Mech., **39**, 19–35 (2007)
122. Jamet, D., Petitjeans, Ph.: A physical justification of a phase-field model for interfaces separating miscible fluids. In: Int. Workshop on Miscible Interfaces, ESPCI, Paris, 2–5 July 2001
123. Jamet, D., Lebaigue, O., Coutris, N., Delhaye, J.-M.: A numerical description of a liquid–vapor interface based on the second gradient theory, Fluid Mech. Res., **22**, 1, 1–14 (1995)
124. Jaumotte, A.L.: *Chocs et ondes de choc*, Masson, Paris, 1971
125. Jaynes, E.T.: Information theory and statistical mechanics. I and II, Phys. Rew., 106 and 108, 171-190 (1957).
126. Jeong, J., Hussain, F.: On the identification of a vortex, J. Fluid Mech., **285**, 69–94 (1995)
127. Johns Jr., L.E., Beckmann, R.B.: Mechanism of dispersed-phase mass transfer in viscous, single-drop extraction systems, AIChE J., **12**, 1, 10–16 (1966)
128. Jomaas, G., Law, C.K., Bechtold, J.K.: On transition to cellularity in expanding spherical flames, J. Fluid Mech., **583**, 1–26 (2007)
129. Joseph, D.D.: *Stability of fluid motion I*, Springer, Berlin, 1976
130. Jou, D., Casas-Vásquez, J., Lebon, G.: *Extended irreversible thermodynamics*, Springer, Berlin, 2001
131. Jouguet, J.C.E.: Sur la propagation des réactions chimiques dans les gaz, J. Math. Pures Appl. Ser. 6, **1**, 347–425 (1905); continued in **2**, 5–85 (1906)
132. Karagozian, A.R., Marble, F.E.: Study of a diffusion flame in a stretched vortex, Comb. Sci. Technol. **45**, 65–84 (1986)
133. von Kármán, Th.: Über laminare und turbulente Reibung, ZAMM **1**, 233–252 (1921)
134. Kawasaki, K.: Mode coupling and critical dynamics. In: Domb, C., Green, M.S. (eds): *Phase transition and critical phenomena*, vol. 5A, Academic, New York, 1976, pp. 165–403

135. Klimov, A.M.: Laminar flame in a turbulent flow, Zh. Prikl. Mekh. Tekh. Fiz., **3**, 49 (1963)
136. Koiter, W.T.: *The theory of thin elastic shells*, North-Holland, Amsterdam, 1960
137. Kojima, M., Hinch, E.J., Acrivos, A.: The formation and expansion of a toroidal drop moving in a viscous fluid. Phys. Fluids, **27** (1), 19–32 (1984)
138. Kolev, N.I.: *Multiphase flow dynamics 1 and 2*, Springer, Berlin, 2002
139. Kraichnan, R.H.: Eddy viscosity in two and three dimensions, J. Atmos. Sci., **33**, 1521–1536 (1976)
140. Kuentzmann, P.: *Aérothermochimie des suspensions* (Mémoire des Sciences Physiques, Fasc. 72), Gauthiers-Villars, Paris, 1973
141. Kumagai, S., Sakai, T., Okajima, S.: Combustion of free fuel droplets in a freely falling chamber. In: 13th Symp. on Combustion, Salt Lake City, UT, 23–29 Aug. 1970
142. Kuo, K.K.-Y.: *Principles of combustion*, 2nd edn., Wiley, New York (2005)
143. Lafon, P.: *Modélisation et simulation numérique de l'évaporation et de la combustion de gouttes haute pression* (Ph.D. thesis), ONERA-University Paris VI, Paris, 1994
144. Lafon, P., Prud'homme, R.: Modèles de combustion d'une goutte avec condensation des produits brûlés, Rech. Aérosp., **1**, 67–82 (1994)
145. Lamb, H.: *Hydrodynamics*, Cambridge University Press, Cambridge, 1945
146. Landau, L.D., Levich, V.G.: Dragging of a liquid by a moving plate, Acta Physicochim. URSS, **17**, 42–54 (1942)
147. Landau, L., Lifschitz, E.: *Mécanique des fluides*, Ed. MIR, Moscow, 1971
148. Lasheras, J.C., Tio, K.K.: Dynamics of a small spherical particle in steady two-dimensional vortex flows, Appl. Mech. Rev. **47**, S61–S69 (1994)
149. Laverdan, A.M., Candel, S.: Computation of diffusion and premixed flames rolled up in vortex structure, J. Prop. Power, **5**, 2, 134–143 (1989)
150. Law, C.K.: *Combustion physics*, Cambridge University Press, Cambridge, 2006
151. Law, C.K., Sirignano, W.A.: Unsteady droplet combustion with droplet heating—II: conduction limit, Combust. Flame, **28**, 175–186 (1977)
152. Law, C.K., Jomaas, G., Bechtold, J.K.: Cellular instabilities of expanding hydrogen/propane spherical flames at elevated pressure: theory and experiment, Proc. Combust. Inst., **30**, 159–167 (2005)
153. Lee, S.Y., Tankin, R.S.: Study of liquid spray (water) in a non-condensable environment (air). Int. J. Heat Mass Transfer, 27, n 3, 51-361 (1984).
154. Lefebvre, A.H.: *Atomization and sprays*, Hemisphere, Manchester, 1989
155. Legros, G., Fuentes, A., Baillargeat, J., Joulain, P., Vantelon, J.P., Torero, J.L.: Three-dimensional recomposition of the absorption field inside a non-buoyant sooting diffusion flame, Opt. Lett., **30**, 3311–3313 (2005)
156. Legros, G., Joulain, P., Valenton, J.P., Bertheau, D., Fuentes, A., Torero, J.L.: Soot volume fraction measurements in a three-dimensional laminar diffusion flame esthablished in microgravity, Combust. Sci. Tech., **178**, 813–835 (2006)
157. Legros, G., Fuentes, A., Rouvreau, S., Joulain, P., Porterie, B., Torero, J.L. : Transport mechanisms controlling soot production inside a non-buoyant laminar diffusion flame, Proc. Combust. Inst., **32**, 2461–2470 (2009)
158. Lesieur, M.: *Turbulence in fluids*, Kluwer, Dordrecht, 1997
159. Levich, V.G.: *Physicochemical hydrodynamics*, Prentice-Hall, Englewood Cliffs, 1962

160. Li, X., Tankin, R.S.: Derivation of droplet size distribution in sprays by using information theory. Comb. Sc. and Techno. 60, 345-357 (1988).
161. Libby, P.A., Chen, K.: Laminar boundary layer with uniform injection, Phys. Fluids, **8**, 4, 568–574 (1965)
162. Libby, P.A., Williams, F.A.: *Turbulent reacting flows*, Springer, New York, 1980
163. Ludford, G.S.S., Scheurer, S.: Problèmes mathématiques en combustion. In: *Ecoles CEA/INRIA/EDF*, Ed. INRIA, Paris, 1982
164. Luettmer-Strathmann, L., Sengers, J.V., Olchowi, G.A.: Non-asymptotic critical behavior of the transport properties of fluids, J. Chem. Phys., **103**, 17, 7482–7501 (1995)
165. Manneville, P.: Rayleigh–Bénard convection: thirty years of experimental, theoretical, and modeling work. In: Mutabazi, I., Wesfreid, J.E., Guyon, E. (eds): *Dynamics of spatio-temporal cellular structures—Henri Bénard Centenary Review* (Springer Tracts in Modern Physics, vol. 207), Springer, Berlin, 2006, pp. 41–65
166. Marble, F.E.: Growth of a diffusion flame in the field of a vortex. In: Casci, C. (ed): *Recent advances in aerospace sciences*, Plenum, New York, 1985, pp. 395–413
167. Marble, F.E.: Mixing, diffusion and chemical reaction of liquids in a vortex field. In: Moreau, M., Turg, P. (eds): *Chemical reactivity of liquids: fundamental aspects*, Plenum, New York, 1988, pp. 591–596
168. Marble, F.E., Broadwell, J.E.: *The coherent flame model for turbulent chemical reactions* (Project Squid Tech. Rep. TRW-9-PU), Project Squid Headquarters, Purdue University, West Lafayette, 1977
169. Marcu, B., Meiburg, E., Newton, P.K.: Dynamics of heavy particles in a Burger vortex. Phys. Fluids, **7**, 400–410 (1995)
170. Marek, R., Straub, J.: Analysis of the evaporation coefficient and the condensation coefficient of water, Int. J. Heat Mass Transfer, **44**, 39–53 (2001)
171. Markstein, G.H.: *Nonsteady flame propagation*, Pergamon, New York, 1964
172. Marmotant, P.H., Villermaux, E.: On spray formation, J. Fluid Mech., **498**, 73–111 (2004)
173. Matalon, M., Matkowsky, B.J.: Flames as dynamic discontinuities, J. Fluid Mech., **124**, 239–259 (1982)
174. Maugin, G.A., Drouot, R., Sidoroff, F. (eds): *Continuum thermomechanics* (Paul Germain's anniversary volume; Solid Mechanics and its Applications, vol. 76), Kluwer, Dordrecht, 2000
175. Maxey, M.R., Riley, J.J.: Equation of motion for a small rigid sphere in a uniform flow, Phys. Fluids, **26**, 4, 883–889 (1983)
176. Métais, O., Lesieur, M.: Spectral large-eddy simulation of isotropic and stably stratified turbulence, J. Fluid Mech., **239**, 157–194 (1992)
177. Moin, P., Squires, K., Cabot, W., Lee, S.: A dynamic subgrid-scale model for compressible turbulence and scalar transport, Phys. Fluids A, **3**, 11, 2746–2757 (1991)
178. Naja, G.: Definition of an experiment on combustion of two droplets in microgravity (ST-8511). In: 36th IAF Congr., Stockholm, Sweden, 7–12 Oct. 1985
179. Napolitano, L.G.: Unsteady wave propagation in a multi-reacting gas mixture. In: 15th Int. Congr. Astronautics, Warsaw, Poland, 11 Sept. 1964
180. Napolitano, L.G.: Transonic approximations for reacting mixtures, Israel J. Technol., **4**, 1, 159–171 (1966)

181. Napolitano, L.G.: *Thermodynamique des systèmes composites en équilibre ou hors d'équilibre* (Mémoires de Sciences Physiques, vol. 71), Gauthier-Villars, Paris, 1971
182. Nicodin I., Gatignol R.: Unsteady half-space evaporation and condensation problems on the basis of the discrete kinetic theory, Phys. Fluids, **18**, 127101–127111 (2006)
183. Nicoli, C., Haldenwang, P., Denet, B.: Ignition in a reactive mixing layer. In: Zappoli, B., Prud'homme, R. (eds): *Journées d'Oléron du GDR PR2M*, J. Chem. Phys., **96**, 1016–1021 (1999)
184. Nicoli, C., Haldenwang, P., Suard, S.: Analysis of pulsating spray flames propagating in lean two-phase mixtures with unity Lewis number, Combust. Flame, **143**, 299–312 (2005)
185. Nicoli, C., Haldenwang, P., Suard, S.: Effect of substituting fuel spray for fuel gas on flame stability in lean premixtures, Combust. Flame, **149**, 295–313 (2007)
186. Nield, D.A.: Surface tension and buoyancy effects in cellular convection, J. Fluid Mech., **19**, 341–352 (1964)
187. Nieto de Castro, C.A. In: Bruno, T.J., Ely, J.F. (eds): *Supercritical fluid technology*, CRC, Boca Raton, 1991, p. 335
188. Nigmatulin, R.I.: *Dynamics of multiphase media*, Hemisphere, Manchester, 1991
189. Niordson, F.I.: *Shell theory*, North-Holland, Amsterdam, 1985
190. Nitsche, K., Straub, J.: The critical "hump" of C_v under microgravity (result from the D-1 Spacelab experiment "Warmekapazitat"). In: Proc. 6th Eur. Symp. on Material Science Under Microgravity Conditions (ESA SP-256), Bordeaux, France, 2–5 Dec. 1986
191. Onuki, A., Hong, H., Ferrell, R.A.: Fast adiabatic equilibration in a single-component fluid near the liquid–vapor critical point, Phys. Rev. A, **41**, 2256 (1990)
192. Orlandi, P.: *Fluid flow phenomena*, Kluwer, Dordrecht, 2001
193. Oxtoby, D.W., Gelbart, W.M.: Shear viscosity and order parameter dynamics of fluids near the critical point, J. Chem. Phys., **61**, 7, 2957–2963 (1974)
194. Palmer, H.J.: The hydrodynamic stability of rapidly evaporating liquids at reduced pressure, J. Fluid Mech., **75**, 487–511 (1976)
195. Papamoschou, D.: Structure of the compressible turbulent shear layer. In: 27th Aerospace Science Meeting, Reno, NV, 9–19 Jan. 1989
196. Pareja, G., Reilly, M.J.: Dynamic effects of recycle elements in tubular reactor systems, IEC Fundam., **8**, 442–448 (1969)
197. Pearson, J.R.A.: On convection cells induced by surface tension, J. Fluid Mech., **4**, 489–500 (1958)
198. Pelce, P., Clavin, P.: Influence of hydrodynamics and diffusion upon the stability limits of laminar flames, J. Fluid Mech., **124**, 219–237 (1982)
199. Penner, S.S.: *Chemistry problems in jet propulsion*, Pergamon, New York, 1957
200. Perl, R., Ferrell, R.A.: Decoupled-mode theory of critical viscosity and diffusion in the binary-liquid phase transition, Phys. Rev. A, **6**, 2358–2369 (1972)
201. Peters, N.: The premixed turbulent flame in the limit of a large activation energy, J. Non-Equil. Thermody., **7**, 25–38 (1982)
202. Peters, N.: *Turbulent combustion* (Cambridge Monographs on Mechanics), Cambridge University Press, Cambridge, 2000

203. Petitjeans, P.: Une tension de surface pour les fluides miscibles, C.R. Acad. Sci. Paris II B, **322**, 673–679 (1996)
204. Pitzer, K.S., Curl, R.F.: The volumetric and thermodynamic properties of fluids, III: empirical equation for the second virial coefficient, J. Am. Chem. Soc., **79**, 2369–2370 (1957)
205. Plapp, M., Karma, A.: Eutectic colony formation: a phase field model, arXiv:cond-mat/0112194 (2001)
206. Poinsot, Th., Veynante, D.: *Theoretical and numerical combustion*, 2nd edn., Edwards, Flourtown, 2005
207. Poinsot, T., Echekki, T., Mungal, M.G.: A study of the laminar flame tip and implications for premixed turbulent combustion, Combust. Sci. Technol., **81**, 45–55 (1992)
208. Pomeau, Y.: Moving contact line, J. Phys. IV France, **11**, Pr6, 199–212 (2001)
209. Pope, D.N., Howard, D., Lu, K., Gogos, G.: Combustion of moving droplets and suspended droplets: transient numerical results, J. Thermophys. Heat Transfer, **19**, 2 (2005)
210. Prausnitz, J., Rüdiger, M., Lichtenthaler, N., Gomes de Azevedo, E.: *Molecular thermodynamics of fluid-phase equilibria*, 3rd edn., Prentice-Hall, Upper Saddle River, 1999
211. Préau, S.: *Etude théorique et numérique de l' "évaporation" d'une "goutte" de densité critique le long d'une isobare légèrement supercritique* (Ph.D. thesis), University Paris VI, Paris, 2003
212. Préau, S., Prud'homme, R., Ouazzani, J., Zappoli, B.: Supercritical density relaxation as a new approach of droplet vaporization, Phys. Fluids, **16**, 4075–4087 (2004)
213. Prigogine, I.: *Introduction à la thermodynamique des processus irréversibles*, Dunod, Paris, 1968
214. Prigogine, I., Defay, R.: *Chemical thermodynamics*, Longman, London, 1954
215. Proudman, I., Pearson, J.R.A.: Expansions at small Reynolds numbers for the flow past a sphere and a circular cylinder, J. Fluid Mech., **2**, 237–262 (1957)
216. Prud'homme, R.: Sur l'équation de dispersion des mélanges gazeux multiréactifs, C.R. Acad. Sci., **266**, A, 300–301 (1968)
217. Prud'homme, R.: *Ecoulements relaxés dans les tuyères* (doctorat d'Etat), ONERA–Faculté des Sciences de Paris, Paris, 1969
218. Prud'homme, R.: Analysis of transonic flow with chemical reactions by the small perturbation method. The one and two-dimensional problem, Astronaut. Acta, **15**, 575–586 (1970)
219. Prud'homme, R.: *Fluides hétérogènes et réactifs: écoulements et transferts* (Lecture Notes in Physics, vol. 304), Springer, Berlin, 1988
220. Prud'homme, R.: *La méthode du maximum d'entropie appliquée à l'atomisation des jets liquides* (RT 112/6128 DEFA/N), ONERA, Paris, 1998
221. Prud'homme, R.: Evaporation et combustion de gouttes dans les moteurs (Traité de Mécanique BM-2-521), Editions Techniques de l'Ingénieur, Paris, 2009
222. Prud'homme, R., Lequoy, C.: *Vitesses spécifiques des réactions chimiques utilisables en aérothermochimie* (ONERA Note Tech. 147), ONERA, Paris, 1969
223. Prud'homme, R., Baillot, F., Durox, D.: Modélisation interfaciale des flammes minces, J. Méc. Théor. Appl., Num. Spéc., 193–233 (1986)

224. Prud'homme, R., Gottesdiener, L., Dodemand, E., Kuentzmann, P.: Asymptotic analysis of vortex flows in suspensions, Int. J. Multiphase Flow, **24**, 3, 453–478 (1998)
225. Prud'homme, R., Habiballah, M., Nicole, A., Mauriot, Y.: Instabilités liées au phénomène d'évaporation: Réponse dynamique d'une goutte un champ acoustique. In: 17ème Congr. Français de Mécanique, Troyes, France, 29 Aug.–2 Sept. 2005
226. Prud'homme, R., Habiballah, M., Matuszewski, L., Mauriot, Y., Nicole, A.: Theoretical analysis of dynamic response of a vaporizing droplet to acoustic oscillations, J. Propul. Power, **26**, 1, 74-83 (2010)
227. Rajagopal, K.R., Tao, L.: *Mechanics of mixtures* (Series on Advances in Mathematics for Applied Sciences, vol. 35), World Scientific, Singapore, 1995
228. Rangel, R.H., Sirignano, W.A.: Combustion of parallel fuel droplet streams, Combust. Flame, **75**, 241–254 (1989)
229. Ravet, F., Baudouin, C., Schultz, J.L.: Modélisation numérique des écoulements réactifs dans les foyers de turboréacteurs, Rev. Gén. Thermique, **36**, 1, 5–16 (1997)
230. Rayleigh, J.: On the convective curents in a horizontal layer of fluid when the higher temperature is on the underside, Phil. Mag., **32**, 529–546 (1916)
231. Rayleigh, J.: *The theory of sound*, Macmillan, London, 1945
232. Raynal, L.: *Instabilité et entraînement à l'interface d'une couche de mélange liquide gaz* (thèse). Université Joseph Fourier (Grenoble 1), Grenoble, 1997
233. Reid, R.C., Prausnitz, J.M., Poling, B.E.: *The properties of gases and liquids*, 4th edn., McGraw-Hill, Boston, 1987
234. Renard, P.-H., Thévenin, D., Rolon, J.C., Candel, S.: Dynamics of flame-vortex interaction, Progr. Energy Combust. Sci., **26**, 3, 225–282 (2000)
235. Réveillon, J., Vervisch, L.: Response of the dynamic LES model to heat release induced effects, Phys. Fluids, **8**, 8, 2248–2250 (1996)
236. Reynolds, O.: An experimental investigation of the circumstances which determine whether the motion of water shall be direct and sinuous, and the law of resistance in parallel channels, Phil. Trans. Roy. Soc., 51–105 (1883)
237. Rocard, Y.: *Thermodynamique*, Masson et Cie, Paris, 1967
238. Roux, S., Lartigue, G., Poinsot, T., Meier, U., Bérat, C.: Studies of mean and unsteady flow in a swirl combustor using experiments, acoustic analysis, and large eddy simulations, Combust. Flame, **141**, 40–54 (2005)
239. Ruckenstein, E.: Scaling in laminar and turbulent heat and mass transfer. In: *Handbook for heat and mass transfer operations*, Gulf, Houston, 1984
240. Rutland, C.J., Ferziger, J.H., El Tahry, S.H.: Full numerical simulation and modelling of turbulent premixed flames. In: Proc. 23nd Symp. on Combustion, Orléans, France, 22–27 July 1990, pp. 621–627
241. Sagaut, P.: *Large Eddy simulation for imcompressible flows* (Scientific Computation), Springer, Berlin, 2001
242. Sagaut, P., Deck, S., Terracol, M.: *Multiscale and multiresolution approaches in turbulence*, Imperial College Press, London, 2006
243. Said, R., Borghi, R.: A simulation with a "cellular automaton" for turbulent combustion modelling. In: Proc. 22nd Symp. on Combustion, Seattle, WA, 14–19 Aug. 1988, pp. 569–577
244. Sanchez-Tarifa, C., Crespo, A., Fraga, E.A.: Theoretical model for the combustion of droplets in super-critical conditions and gas pockets, Astron. Acta, **17**, 685 (1972)

245. Sanfeld, A., Sefiane, K., Benielli, D., Steinchen, A.: Does capillarity influence chemical reactions in drops and bubbles? A thermodynamic approach, Adv. Colloid Interf. Sci., **86**, 153–193 (2000)
246. Sankagiri, N., Ruff, G.A.: Extension of spray nozzle correlations to the prediction of drop size distribution using principles of maximum entropy. 31st Aerospace Science and Exhibit, 11-14 jan., Reno NV (1993).
247. Schannon, C.E.: A mathematical theory of communication. Bell System Tech. J., 27, 379-423 / 623-659 (1948).
248. Scherrer, D.: Combustion d'une goutte en milieu réactif avec décomposition exothermique préalable du combustible. In: INRIA (ed) *Numerical simulation of combustion phenomena* (Lecture Notes in Physics), Springer, Berlin, 1985
249. Schlichting, H., Gersten, K.: *Boundary layer theory*, 8th edn., Springer, Berlin, 2000
250. Scriven, L.E., Sternling, C.V.: On cellular convection driven by surface-tension gradients: effects of mean surface tension and surface viscosity, Chem. Eng. Mech., **19**, 321–340 (1964)
251. Sedov, L.I.: *Similarity and dimensional methods in mechanics*, Infosearch, London, 1959
252. Selle, L., Lartigue, G., Poinsot, T., Koch, R., Schildmacher, K.-U., Krebs, W., Prade, B., Kaufmann, P., Veynante, D.: Compressible large eddy simulation in complex geometry on unstructured meshes, Combust. Flame, **137**, 489–505 (2004)
253. Sellens, R. W., Brzustowski, T.A.: A prediction of the drop size distribution in a spray from first principles. Atomization and spray technol., 1, 89-102 (1985)
254. Sengers, J.V.: Transport properties of fluids near critical points, Int. J. Thermophys., **6**, 3, 203–232 (1985)
255. Sengers, J.V., Levelt Sengers, J.M.H.: Critical phenomena in classical fluids. In: Croxton, C. (ed): *Progress in liquid physics*, Wiley, Chichester, 1978
256. Seppecher, P.: *Etude d'une modélisation des zones capillaires fluides: interfaces et lignes de contact* (thèse de doctorat), Université Pierre et Marie Curie, Paris, 1987
257. Seppecher, P.: Equilibrium of Cahn and Hilliard fluid on a wall: influence of the wetting properties of the fluid upon the stability of a thin liquid film, Eur. J. Mech. B/Fluids, **12**, 69–84 (1993)
258. Shivashinsky, G.I.: Structure of Bunsen flames, J. Chem. Phys., **62**, 2 (1975)
259. Slattery, J.C.: General balance equations for a plane interface, IEC Fundam., **6**, 1 (1967)
260. Slattery, J.C.: *Interfacial transport phenomena*, Springer, Berlin, 1990
261. Smagorinsky, J.: General circulation experiments with the primitive equations, Mon. Weth. Rev., **91**, 3, 99–164 (1963)
262. Smolderen, J.: Structure des chocs et théorie cinétique des gaz. In: Jaumotte, L. (ed): *Chocs et ondes de choc, tome 1: aspects fondamentaux*, Masson et Cie, Paris, 1971, pp. 153–191
263. Soo, S.L.: *Multiphase fluid dynamics*, Science Press/Gower Technical, Aldershot, 1990
264. Spalding, D.B.: Studies of the combustion of drops in a fuel spray. The burning of single drops of fuel. In: Combustion Institute (ed): *Fourth Symposium on Combustion*, Williams and Wilkins, Baltimore, 1953, pp. 847–864

265. Spalding, D.B.: *Turbulence models and their experimental verification: turbulence effects on combustion* (HTS 73/27), Imperial College of Science and Technology, London, 1973
266. Sparrow, E.M., Quack, H., Boerner, C.J.: Local nonsimilarity boundary-layer solutions, AIAA J., **8**, 11 (1970)
267. Stephenson, J.: A study of fluid argon via the constant volume specific heat, the isothermal compressibility, and the spead of sound and their extremum properties along isotherms, Can. J. Phys., **53**, 1367–1384 (1975)
268. Suard, S., Haldenwang, P., Nicoli, C.: Different spreading regimes of spray-flames, C.R. Mecanique, **332**, 387–396 (2004)
269. Sullivan, P.P., Horst, T.W., Lenschow, D.H., Moeng, C.H., Weil, J.C.: Structure of subfilter-scale fluxes in the atmospheric surface layer with application to large eddy simulation modeling, J. Fluid Mech., **482**, 101–139 (2003)
270. Sutherland, W.: The viscosity of gases and molecular force, Phil. Mag., S5, **36**, 507–531 (1893)
271. Tchen, C.M.: *Mean value and correlation problems connected with the motion of small particles suspended in a turbulent fluid* (Ph.D. thesis), Univ. Delft, Delft, 1947
272. Timoshenko, S.: *Theory of plates and shells*, 2nd edn., McGraw-Hill, New York, 1959
273. Tio, K.K., Liñan, A., Lasheras, J.C., Gañán-Calvo, A.M.: On the dynamics of buoyant and heavy particles in a periodic Stuart vortex flow, J. Fluid Mech., **254**, 671–699 (1993)
274. Tio, K.K., Gañán-Calvo, A.M., Lasheras, J.C.: The dynamics of small, heavy, rigid spherical particles in a periodic Stuart vortex flow, Phys. Fluids A, **5**, 1679–1692 (1993)
275. Torero, J.L., Vietoris, T., Legros, G., Joulain, P.: Estimation of a total mass transfer number from the stand-off distance of a spreading flame, Comb. Sci. Technol., **174**, 11–12, 187–203 (2002)
276. Trambouze, P., Euzen, J.-P.: Chemical reactors: from design to operation, 2nd edn., Technip, New York (2004)
277. Trouvé, A., Poinsot, Th.: The evolution equation for the flame surface density in turbulent premixed combustion, J. Fluid Mech., **278**, 1–31 (1994)
278. Truesdel, C.: *Rational thermodynamics* (Modern Applied Mathematics series), McGraw-Hill, New York, 1969
279. Turpin, G.: *Simulation numérique de statoréacteurs* (thèse), Univ. Paris VI, Paris, 2001
280. Van der Geld, C.W.M., Vermeer, H.: Prediction of drop size distributions in sprays in using the maximum entropy formalism: the effect of satellite formation. Int. J. Multiphase Flow, 20, 2, 363-381 (1994).
281. Van Dyke, M.: *An album of fluid motions*, Parabolic, Stanford, 1982
282. Veynante, D., Trouvé, A., Bray, K.N.C., Mantel, T.: Gradient and counter-gradient scalar transport in turbulent premixed flames, J. Fluid Mech., **332**, 263–293 (1997)
283. Villermaux, J.: *Génie de la réaction chimique* (Technique et Documentation), Lavoisier, Paris, 1982
284. Virepinte, J.F., Adam, O., Lavergne, G., Bicos, Y.: Droplet spacing on drag measurement and burning rate for isothermal and reacting conditions, J. Propul. Power, **15**, 1, 97–102 (1999)

285. Wang, H.Y., Joulain, K., El-Rabii, H.: Modélisation numérique et étude expérimentale sur la propagation d'une flamme de diffusion sous gravité réduite à travers du transfert radiatif. In: Colloque du GDR Micropesanteur Fondamentale et appliquée (CD CNES No. 038490-64), Fréjus, France, 26–28 Nov 2007
286. Wegener, P.P., Chu, B.-T., Klikoff, W.A.: Weak waves in relaxing flows, J. Fluid Mech., **23**, 787–800 (1965)
287. Wesfreid, E.: Henri Bénard. In: Mutabazi, I., Wesfreid, J.E., Guyon, E. (eds): *Dynamics of spatio-temporal cellular structures—Henri Bénard Centenary Review* (Springer Tracts in Modern Physics, vol. 207), Springer, Berlin, 2006, pp. 41–65
288. Williams, F.A.: An approach to turbulent flame theory, J. Fluid Mech., **40**, 2, 401–421 (1970)
289. Williams, F.A.: Recent advances in theoretical descriptions of turbulent diffusion flames. In: Murthy, S.N.B. (ed): *Turbulent mixing in nonreactive and reactive flows*, Plenum, New York, 1975, pp. 189–208
290. Williams, F.A.: *Combustion theory*, 2nd edn., Benjamin Cummings, Menlo Park, 1985
291. Williams, F.A.: Turbulent combustion, In: Buckmaster, J. (ed) *The mathematics of combustion*, SIAM, Philadelphia, 1985, pp. 97–131
292. Wilson, K.G., Fisher, M.E.: Critical exponents in 3.99 dimensions, Phys. Rev. Lett., **28**, 240–243 (1972)
293. Yang, V., Lin, N.N., Shuen, J.S.: Vaporization of liquid oxygen (LOX) droplets at supercritical conditions. In: AIAA 30th Aerospace Sciences Meeting (AIAA-92-0103), Reno, NV, 6–9 Jan. 1992
294. Yang, V., Lafon, P., Hsiao, G.C., Habiballah, M., Zhuang, F.-C.: Liquid-propellant droplet vaporization and combustion. In: Proc. 2nd Int. Symp. on Liquid Rocket Propulsion, Chatillon, France, 19–21 June 1995
295. Young, J.B.: The fundamental equations of gas-droplet multiphase flow, Int. J. Multiphase Flow, **21**, 2 (1995)
296. Ytrehus, T., Østmo, S.: Kinetic theory approach to interphase processes, Int. J. Multiphase Flow, **22**, 1, 133–155 (1996)
297. Zaleski S.: Efficient numerical methods for atomization studies, JSME centennial meeting, Tokyo 17-19 june 1997, Matsumoto and Prosperetti editors.
298. Zappoli, B., Bailly, D., Garrabos, Y., Le Neindre, B., Guenoun, P., Beysens, D.: Anomalous heat transport by the piston effect in supercritical fluids under zero gravity. Phys. Rev. A, **41**, 4, 2264–2267 (1990)
299. Zappoli, B., Amiroudine, S., Carlès, P., Ouazzani, J.: Thermo acoustic and buoyancy-driven transport in a square side-heated cavity filled with a near-critical fluid, J. Fluid Mech., **316**, 53–72 (1996)
300. Zappoli, B., Amiroudine, S., Gauthier, S.: Instabilité gravitationnelle dans un fluide supercritique pur, C.R. Acad. Sci. Paris Ser. IIb, **325**, 1–6 (1997)
301. Zel'dovich, Y.B., Barenblatt, G.: Theory of flame propagation, Combust. Flame, **3**, 61–74 (1959)
302. Zel'dovich, Y.B., Kompaneets, A.S.: *Theory of detonation*, Academic, New York, 1960

Index